THINK LOCALLY,
ACT GLOBALLY

THINK LOCALLY, ACT GLOBALLY

*Polish farmers in the global era
of sustainability and resilience*

EDITED BY
Krzysztof Gorlach and Zbigniew Drąg

IN COLLABORATION WITH
Anna Jastrzębiec-Witowska and David Ritter

JAGIELLONIAN UNIVERSITY PRESS

This publication is financed through the scientific project "Think Locally, Act Globally: Polish farmers in the global era of sustainability and resilience," National Science Centre, Poland, MAESTRO7 Program; grant # 2015/18/A/HS6/00114.

Reviewers
Prof. dr hab. Andrzej Kaleta
Prof. dr hab. Paweł Starosta

Cover design
Marta Jaszczuk

Cover photos
Joanna Gorlach, Tomasz Ćwiertnia

Translation
Anna Jastrzębiec-Witowska

Proofreading
David Ritter

ISBN 978-83-233-4949-5
ISBN 978-83-233-7195-3 (e-book)

www.wuj.pl

Jagiellonian University Press
Editorial Offices: Michałowskiego 9/2, 31-126 Kraków
Phone: +48 12 663 23 80, Fax: +48 12 663 23 83
Distribution: Phone: +48 12 631 01 97, Fax: +48 12 631 01 98
Cell Phone: +48 506 006 674, e-mail: sprzedaz@wuj.pl
Bank: PEKAO SA, IBAN PL 80 1240 4722 1111 0000 4856 3325

Table of Contents

Part Three. Diversification of Farmers' Strategies

Part Four. Some Independent Studies

Think Locally, Act Globally: Polish farmers in the global era of sustainability and resilience, ed. by Krzysztof Gorlach and Zbigniew Drąg in collaboration with Anna Jastrzębiec-Witowska and David Ritter
Jagiellonian University Press, Kraków 2021, pp. 9–21
ISBN 978-83-233-4949-5
DOI: http://dxdoi.org/10.4467/K7195.199/20.20.12723

Family Farming: A foreword

Jan Douwe van der Ploeg

At the level of everyday language, the notion of the family farm is fairly straightforward and therefore widely shared. It is, first, a farm owned (or at least controlled) by the family. Second, it is worked mainly (and often exclusively) by members of the family. And third, the main decisions regarding the farm are taken within the family. The farm may be small or large, the family may differ also very much in size and composition, but whatever the variation, it is time and again the organic unity of family and farm that is decisive. The advantage of this folk definition is that it is easy applicable and mostly uncontested (although borderline cases may provoke considerable debate). The indicated features also figure in the definition used by FAO (2014): "A family farm is an agricultural holding which is managed and operated by a household and where farm labor is largely supplied by that household" (2014). At the same time, though, this definition sometimes turns out to be problematic, partly due to its simplistic nature. It takes just two or three characteristics out of a much wider set of features that together constitute family farming.[1]

---

[1] As argued at the European Conference on Family Farming: "The concept of family farming encompasses sociological, economic and cultural elements. From a sociological perspective, family farming is associated with family values such as solidarity, continuity and commitment; in economic terms, it is identified with specific entrepreneurial skills, business ownership and management, resilience and individual achievement. Family farming is also often more than a professional occupation because it reflects a lifestyle based on beliefs and traditions about living and work" (Council of the European Union, 2013, p. 34).

Some conceptual clarification

In current debates the notion of *family farming* often associates with those of *peasant farms* and *small holdings*. Although there always has been, in practice, considerable overlap, it is important to clearly distinguish these notions conceptually. This applies especially since the widening distances between the three contribute to much of the current problems in agriculture, the countryside and food provisioning—and this applies as much to Europe as it is true for other parts of the world. Whilst once being a more or less organic unity (deviations at the fringes apart), family farming, peasant agriculture and small holdings are now increasingly constituted into separated entities and this turns "family farming" increasingly into a contradictory, confusing and Janus-faced phenomenon. Agricultural policies play an important role in this and this is again the case in both Europe and elsewhere.

Strictly speaking, the notion of family farming refers to *legal* relations. Who *owns* (and/or controls) the land and other main resources? Who provides the labor power? Who takes the decisions? In contrast to this, the notion of peasant agriculture focuses on what is *done* with the main resources: How are land, labor, animals, crops, water, machines, buildings, knowledge, networks, etc., *combined* into a specific style of farming and *put into operation*? That is: how is farming as productive practice concretely organized? And how is this way of farming *developed* further? Central here are the socio-material resources and they constitute together a specific process of production. There are many different ways to farm—peasant agriculture is one of them (Ploeg, 2018). With the notion of smallholdings and smallholders the attention shifts to a higher level of aggregation: to the *distribution* (and concentration) of the main resources at the level of the agricultural sector as a whole. It is about who is having *what* and especially *how much of it?* Some farms might dispose of many resources, others will be poorly equipped. There might be many reasons for this—but whatever the particular reason, the latter are identified as small holdings which are, consequently, operated by smallholders.

Dissimilarities

When it comes to family farming, the differences within Europe are not a few—they are many. They embrace different historical trajectories, different locations in the space defined by the scale and intensity of farming, different shares in the total work force, different structures relations between farming and markets, just as the food markets themselves are differently structured. Then there are considerable differences in technology, just as the systems for Research and Development (that generate new agricultural technologies) differ considerably. The many-sided and richly checkered relations between the state and the family farming sector equally are highly distinctive and mutually different. Etc., etc. And for each and every of these differences it applies that they are not small.

Let us first look at the *numbers*: In the 28 countries that together compose the European Union (at least so far) there are nearly 12 million family farms (Davidova & Thomson, 2013; Eurostat, 2017 refers instead to 10.8 million family farms in 2013).[2] In some European countries the number of family farms is relatively small; in others, like Poland and Romania, the numbers are elevated. As part of the Economically Active Population, the number of family farmers oscillates around an average of 10% whilst in countries as the Netherlands it is as low as 1%. When attention is focused on the *size* of the average family farm the impression of disproportionality is seemingly unavoidable. The average size amounts to 16.1 hectare of land, which hides huge differences across the continent. The Czech Republic recorded an average of 133 hectares, the second highest average size was noted in the UK: 94 hectares. Six Member States reported averages sizes below 10 hectares.

The respective *histories* equally differ. At the periphery of the big Empires, the "free farmer" has been a reality for thousands of years. After the French revolution, the "free family farm" (i.e., liberated from feudal ties) covered more parts of Europe (although there remained to be important exceptions). Throughout the centuries the agrarian populations have been struggling for what the great rural historian Slicher van Bath (1978) has called "farmers' freedom." This is freedom *from* and simultaneously freedom *to*. It is freedom from oppressive relations, asphyxiating ties, overexploitation and impoverishment. It is also freedom to—to farm in a way that corresponds with the possibilities, limitations, experiences and aspirations of the farming family. Through the many-sided struggles for such farmers' freedom, the family farm became the main land-labor institution in Europe. Nowadays, 97% of all farms in Europe (EU28) are family farms.

---

[2] Such differences are inherent to agricultural statistics. They merge from a different operationalization of the family farm concept and, above all, from different "lower ceilings" (implying the inclusion or exclusion of very small family farms).

Commonalities

Although the differences between family farms in Europe are overwhelming and the historical trajectories out of which they emerged differ strikingly, there are important commonalities as well. The first commonality I will discuss resides in the social significance of the family farm. Those who are part of it will defend their farm at all cost. The impressive history of Poland during and after the Communist period is just one expression of this (Gorlach & Mooney, 2004; Halamska, 2004). Equally, the many laws that protect, throughout Europe, the family farm testify of past struggles that aimed for security and good prospects. And then there are many non-agricultural actors who long for a small farm and who are doing their utmost best to construct and develop it (see, e.g., Milone, 2015; Morel, 2016). In the South of Poland but also in e.g., Tras-os-Montes (in the North of Portugal), migrant workers returned to their villages and engaged in farming after many years of urban labor, bringing with them savings, new knowledge and new networks that strongly contributed to the dynamics of the rural economy. This resembles, in a way, the phenomenon of pluriactivity. In e.g., the Netherlands, in 80% of the family farms the man or the wife (or the two of them) is having a paid job outside the farm. Without the additional income thus generated, it would be fairly impossible to continue the farm (Ventura, 2013). Pluriactivity thus is one out of a wide range of mechanisms for defending the family farm—just as migrant labor is for many peasants in the Global South a mechanism to defend both their farm and their family.

In short: the family farm is the outcome of long and sometimes short but explosive processes of emancipation. It equally is the object of everyday life struggles. Having a family farm has been an urge that moved both large and small parts of history. This centrality of the family farm is due, I think, to the many-sided attractiveness that family farming shows in the eyes of its beholders. One might distinguish up to ten (potential) *qualities* of family farming (see also figure 1). Individually, but especially when taken together, these constitute the attractiveness of the family farm. Not all of these features are always present: this depends on time and place. The family farm may unfold completely (then all qualities are strongly present). It may, however, as well suffer from processes of erosion (then only a few features remain).

(1) **Autonomy:** a first defining feature of the family farm is that the farming family retains control over the main resources used in the farm. This control is often (but not necessarily) rooted in property rights. The resource base includes the land, together with the animals, the crops, the genetic material, the house, buildings, machinery, labor force and, in a more general sense, the know-how that specifies how

these resources need to be utilized and combined. Access to networks and markets, as well as co-ownership of co-operatives, are also important resources. Many of these resources have been constructed and/or acquired through long processes often covering different generations. Mostly, the resources are the result of hard work, dedication and the hope for a better future (often spanning several generations). Family farmers use these resources, not to make a profit, but to make a living; to acquire an income that provides them with a decent life and, if possible, allows them to make investments that will further develop the farm.

The ownership and control of these resources provides the farming family with an important degree of (relative) autonomy. It allows them to face difficult times as well as to construct, with their own work, attractive prospects for the future. In this respect, their control over the resource base indeed is a quality, the importance of which is reflected in the value of being independent (of "being one's own boss").

At the same time the level of autonomy that farmers have over their resource base is far from self-evident. In reality it is being eroded in many places. High

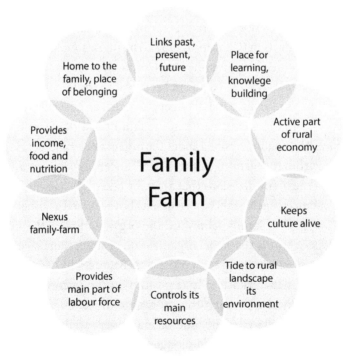

Figure 1. The multiple features of the family farm
Source: derived from Ploeg, 2013; see also Chatel, 2014.

degrees of indebtedness imply that the real control shifts to banks and/or process-ing industries. Contract farming often has similar consequences. Strict regulatory regimes imposed by state apparatuses and/or processing industries also reduce farm-ers' control over their resource base. Through external prescription and sanctioning of the production process, real control shifts to external institutions.

(2) **Self-employment:** A second defining feature relates to the farming family providing the main part of the labor force. This makes the farm into a place of self-employment and of progress for the family. It is through their dedication, pas-sion and hard work that the farm is further developed, and the family's livelihood is improved. The farm is there to meet the many needs of the family, whilst the family provides the possibilities, the means and also the limitations for the farm. This fea-ture often emerges as a quality. For the family members working on the family farm work often involves an attractive set of highly diversified activities that take place in the fields, the stables, at the kitchen table, on the markets, etc. It is working with living nature, often in the open air, and it is possible to avoid a rigid organization of the working days. The work is also attractive because it builds on an organic unity of mental and manual labor. Available surveys all show that those who work on family farms appreciate these aspects of their work. However, all is not unbridled attractiveness and satisfaction. In practice family farming always implies a balance of drudgery and satisfaction (Chayanov, 1966). The important fact, however, is that—at least in the ideal situation—it is the involved actors themselves who determine, often in a coordinated way, these balances. Equally important, farmers are ingenious at finding novel ways to reduce the drudgery as much as possible.

(3) **Co-evolution:** The nexus between the family and the farm implies a third defining feature. The major decisions about the organization and development of the farm are taken within the farming family itself. The family's interests and prospects are at the core of the many decisions that are to be taken. Decision-making involves equilibrating balances, such as those discussed above, as well as others, such as that between the "supply" of family labor and the organization of the farm. Another important balance regards the social organization of time: short term actions need to be well-coordinated with long term prospects. Balances like these tie family and farm together and make each family farm into a unique constellation. The two co-evolve simultaneously. Through this co-evolution each defines and molds the other. But again: the unity of family and farm can very well become disarrayed or disrupted. This might occur when prolonged days of hard labor and heavy work, combined with low levels of remuneration create a seemingly hopeless, endless, and inescapable routine. At this point family farming turns into (voluntary) enslavement. As with the other features, co-evolution can have a bright, as well as a dark, side.

It also means that the parents are working for their children. They want to give the next generation a solid starting point whether within, or outside, agriculture. Although in Europe farm succession is increasingly an option—and becoming a farmer or not a choice instead of an obligation or right away one's destiny (White, 2011)—those working in (and for) the farm do their utmost best to make the farm as attractive as possible so as to stimulate and facilitate succession. Since the farm is the outcome of the work and dedication of this, and previous, generations, there is often pride. There can also be anger if others try to damage or even destroy the jointly constructed farm. "Keeping the [family] name on the land" has been important throughout agrarian history (see Arensberg & Kimball, 1940) and helps to explain at least a part of farmers' resilience to outside pressures (Falkowski, 2018).

(7) **A place for learning:** The family farm is the place where experience is accumulated, where learning takes place and knowledge is passed on, often in subtle but strong ways, to the next generation. This regards an amazing cultural and intellectual wealth—of knowing how to produce intensively and in sustainable ways in often difficult ecological circumstances. The family farm is also a node in wider networks in which new insights, practices, seeds, etc., circulate. Thus, the farm becomes a place that produces agricultural know-how which is combined with innovativeness (Osti, 1991) and novelty production (Wiskerke & Ploeg, 2004). In a classical contribution to the literature, Giampietro and Pimentel (1993) refer to the "dual nature of agriculture": it provides society's needs and deals with the natural ecosystem and must ensure compatibility between the two (see also Toledo, 1990). Agriculture depends on, uses and transforms the ecosystem and the associated natural resources. It is grounded on these phenomena. Due to the highly heterogeneous nature of ecosystems and constant changes in them over time (both short and long term), the interactions between family farmers and their fields, animals, crops and the climate, require ongoing cycles of observation, interpretation, intervention and evaluation. In other words, *human agency* is central in family farming and with it comes the strategic importance of skills, practical knowledge (or: *art de la localité* as Mendras, 1976 beautifully phrases it) and ongoing learning. The family farm and, at a higher level of aggregation, the rural village, are decisive mechanisms. It goes without saying, however, that the centrality of human agency can also get lost. The capacity to learn is particularly important on smaller family farms. "The Bolton Committee (1971) found that small farms were innovative and responsive to change. In the farm context, Priebe (1969) suggested that the strength of family farms lay in their ability to react quickly to changing conditions" (Gasson et al., 1988, p. 16).

(8) **A carrier of culture:** the family farm is not just an economic enterprise that focuses mainly, or only, on incomes, but a place where continuity and culture are

equally important. The farming family is part of a wider rural community, and sometimes part of networks that extend into cities. As such, the family farm is a place where culture is generated, kept alive and transmitted to others and to future generations. Many farms are places of cultural heritage. As the Regional Conference (FAO, 2014, pp. 5, 23) observed: "Family farmers . . . preserve traditional cultures. The existence of family farms, particularly small-scale ones, is a significant part of national cultural heritage, customs, dress, music, cuisine and habitats."

(9) **A cornerstone of the rural economy**[3]: the family and the farm are also part of the wider rural economy, they are tied to the locality, carrying the cultural codes of the local community. Thus, family farms can strengthen the local rural economy through what they buy and spend their money on and by engaging in other activities (see Haggblade, Hazell, & Dorosh, 2007 for a general discussion). The strategic role of family farming for the regional rural economy is one of the major components of the "European Model of Agriculture," endorsed by the European Council in 1997. It aims at: "A farming sector that serves rural communities, reflecting their rich tradition and diversity, and whose role is not only to produce food but also to guarantee the survival of the countryside as a place to live and work, and as an environment in itself" (European Commission, 2004). We can find specific examples of this: the *mezzadri* in Italy who transferred their networking capacities into the towns and gave rise to a blossoming SME sector (Bagnasco, 1988) and part-time farmers in Norway who, having a strong fall-back position, were largely responsible for the introduction of strong labor unions and democratic relations in Norwegian society (Brox, 2006). A more recent example is the "social buffer" function exerted by the family farming sector in Russia and other Eastern European countries following the demise of the Soviet system (Petrick & Weingarten, 2004).

(10) **Building on nature and a part of the landscape.** Whomever has seen and enjoyed the impressive man-made landscapes in different parts of Europe (as the *bocages* in Normandie, the *dehesa* in Spain, the *alms* in Austria and Italy, and the *polders* in the Netherlands) has to acknowledge that the family farm is a much-needed ingredient of wider rural landscapes. The family farmer may work with, rather than against nature, using ecological processes and balances instead of disrupting them, thus preserving the beauty and integrity of landscapes. When family farmers do this, they also contribute to conserving biodiversity and to fighting global warming. Farmers' work implies an ongoing interaction with living nature—a feature that is highly

---

[3] The notion of the "cornerstone of the rural economy" is derived from the Commission on Small Farms of the Congress of the USA (see HLPE, 2013, p. 28).

valued by the actors themselves. This quality is being recognized and supported by the Second Pillar of the Common Agricultural Policy of the European Union.

The features described above constitute the family farm as a major land-labor institution. From an economical point of view the search for autonomy (1), self-employment (2) and the farm being a place for learning (7) systematically translate into *a lowering of transaction costs* and *an increase in technical efficiency*. This latter aspect implies that the function of production is moved upwards.[4] When it comes to *food security* the combination of co-evolution (3), the creation of value added (4) and the flow linking past, present and future (6) translate into *continuity, robustness and resilient food production*. And being a cornerstone of the rural economy (9), a carrier of culture (8), a co-shaper of landscapes (10) and domus (5) strongly support the *quality of life in rural areas* just as they *strengthen the regional rural economy*.

If, however, these features are weakened (that is, if family farming is eroded or even disappears), production and productivity will go down; food security will come under considerable threat; food provisioning become more expensive (partly due to increased transaction costs); and the strength of the regional rural economy and the quality of rural society will decrease—possibly irreversibly.

Both upward and downward trends are possible. Sometimes the upward ones gain weight and momentum, in other times the downward movements might start to dominate the scene. Often, though, both upward and downward tendencies are co-existing, thus contributing to the turmoil that today seems to characterize family farming both in Europe as elsewhere in the world.

At the beginning of this foreword reference was made to family farming as being, at the level of everyday language, a fairly straightforward *notion*. This is true. Nonetheless, *in practice* family farming is a highly ambiguous reality. It is a promise, an emancipatory notion (that sometimes even is made true). But often considerable parts of this promise are denied in practice. This implies that the daily reality of family farming often is perceived, by the involved actors themselves, as a daily struggle: a struggle to further improve the quality of resources, to enlarge the income gained, to build opportunities for the next generation, etc., etc. Therefore, the family farm as emancipatory notion is not just subjectivity. In as far as it inspires the daily, mostly invisible struggles (and sometimes the highly visible rebellions) of millions of people, it is a *material force*. A force that brought Europe both prosperity and problems.

---

[4] This phenomenon is referred to in the economic literature as "disembodied technical change" (Salter, 1966).

Currently, Europe is reconsidering the societal value of small family farms (Ciolos, 2013; Council of the European Union, 2013; European Parliament, 2014).[5] This occurs after decades of modernization policies (which moved growing parts of family farming away from peasant agriculture and smallholdings) and, more recently, the creation of mega-farms within and around the European Union (Visser, Mamonova, & Spoor, 2012; Ploeg, Franco, & Borras, 2015)—processes that come (as today's situation shows) with a high price and considerable risks for the future. To mention just two aspects: It is increasingly questioned whether this type of *industrialized* agriculture (that formally still operates under the guise of family farming) can support the much-required food sovereignty. It is also questioned whether this type of agriculture (that has moved out human labor and brought in a systematic dependency on fossil energy) can ever help to meet the goals set in the fight against global heating. Fortunately, there still is a relatively large, *real* family farming sector in Europe. It has been neglected and denied for decades by policy makers, but now it enters again public debate as one of the promises for the future.

REFERENCES

Arensberg, C.M. & Kimball, S.T. (1940). *Family and Community in Ireland*. Harvard University Press.

Arkleton Trust (1985). *Part-Time Farming in the Rural Development of Industrialized Countries*. Langholm: Dumfriesshire.

Bagnasco, A. (1988). *La Costruzione Sociale del Mercato, studi sullo sviluppo di piccole imprese in Italia* [The social construction of the market: Studies on the development of small businesses in Italy]. Il Mulino.

Bolton Committee (1971). *Report of the Committee of Enquiry on Small Firms*. HMSO, Cmnd. 4811, London.

Brox, O. (2006). *The Political Economy of Rural Development: Modernisation without centralisation?* J. Bryden & R. Storey (eds.). Eburon.

Ciolos, D. (2013). *L'agriculture familiale : Pour une agriculture plus durable et plus compétitive en Europe et dans le monde* [Family farming: For a more sustainable and more competitive agriculture in Europe and in the world], speech given to the conference "Family Farming: A dialogue towards more sustainable and resilient farming in Europe and the world." Brussels (Speech 13/998).

Chatel, B. (2014). Collective challenges. *Spore*, Special Issue (Aug.): *Family Farming: The beginning of a renaissance*, CTA, pp. 4–7.

Chayanov, A.V. (1966). *The Theory of Peasant Economy*, D. Thorner, B. Kerblay, & R.E.F. Smith (eds.). Manchester University Press.

---

[5] A similar reconsideration is taking place at the international level. See, e.g., HLPE, 2013.

Council of the European Union (2013). *Family Farming Prospects in the Context of Globalization*, working document prepared by the Lithuanian Presidency for the informal meeting of the Ministers of Agriculture, AGRI 516, 12786/13, Brussels.

Davidova, S. & Thomson K. (2013). *Family Farming. A Europe and Central Asia Perspective: Background Report for Regional Dialogue on Family Farming: Working towards a strategic approach to promote food security and nutrition* [draft], FAO, Brussels, 11–12 December.

European Commission (2004). *What Is the European Model of Agriculture?* http://ec.europa.eu/agriculture/faq/index_en.htm (access: October 2, 2006).

European Parliament (2014). *Small Agricultural Holdings. European Parliament Resolution of 4 February 2014 on the Future of Small Agricultural Holdings*, (2013/2096 (INI)), P7_TA-PROV(2014)0066. European Parliament, Brussels.

Eurostat (2017). *Farm Structure Statistics*, http://ec.europa.eu/eurostat/statistics-explained/index.php/Farm_structure_statistics#The_size_of_agricultural_holdings (access: June 5, 2018).

Falkowski, J. (2018). Together we stand, divided we fall? Smallholder access to political power and their place in Poland's agricultural system. *Journal of Agrarian Change* 18(4), pp. 893–903.

FAO [Food and Agriculture Organization of the United Nations] (2014). Regional Conference for Europe (ERC/14/5) FAO.

Gasson, R., Crow, G., Errington, A., Hutson, J., Marsden, T., & Winter, D.M. (1988). The farm as a family business: A review. *Journal of Agricultural Economics* 39(1), pp. 1–41.

Giampietro, M. & Pimentel, D. (1993). *The Tightening Conflict: Population, energy use, and the ecology of agriculture*, http://www.rpg.org/forum_series/tightening_conflict.htm (access: October 23, 2017).

Gorlach, K. & Mooney, P. (2004). *European Union Expansion: The impacts of integration on social relations and social movements in rural Poland*. Cornell University Mellon Sawyer Seminar, Ithaca, NY.

Haggblade, S., Hazell P., & Dorosh, P. (2007). Sectoral growth linkages between agriculture and the rural nonfarm economy. In: S. Haggblade, P. Hazell, & T. Reardon (eds.), *Transforming the Rural Nonfarm Economy*, pp. 141–182. John Hopkins University Press.

Halamska, M. (2004). A different end of the peasant. *Polish Sociological Review* 3(147), pp. 245–268.

HLPE [High Level Panel of Experts] (2013). *Investing in Smallholder Agriculture for Food Security*. Report by the High-Level Panel of Experts on Food Security and Nutrition. Report 6, CFS/FAO.

Mendras, H. (1976). *Sociétés paysannes : Eléments pour une théorie de la paysannerie* [Peasant societies: Elements for a theory of the peasantry]. Colin.

Milone, P. (2015). *I giovani e l'agricoltura tra innovazione e contadinità, Progetto: Giovani agricoltori e innovazioni per la sostenibilità* [Young people and agriculture between innovation and farmers, Project: Young farmers and innovations for sustainability], finanziato dal MiPAAF, http://www.ismea.it/flex/cm/pages/ServeBLOB.php/L/IT/IDPagina/9630 (access: October 23, 2017).

Morel, K. (2016). *Viabilité des microfermes maraicheres biologiques. Une étude inductive combinant méthodes qualitatives et modélisation* [Viability of organic market gardening micro-farms. An inductive study combining qualitative methods and modeling], These de Doctorat [doctoral thesis], L'Université Paris Saclay.

Osti, G. (1991). *Gli innovatori della periferia, la figura sociale dell'innovatore nell'agricoltura di montagna* [The innovators of the periphery, the social figure of the innovator in mountain agriculture]. Reverdito Edizioni.

Petrick, M. & Weingarten, P. (2004). *The role of agriculture in Central and Eastern European rural development: Engine of change or social buffer?* IAMO (Studies on the Agricultural and Food Sector in Central and Eastern Europe, Vol. 25).

Ploeg, J.D. van der (2013). Ten qualities of family farming. *Farming Matters* 29(4), pp. 8–11.

Ploeg, J.D. van der (2018). *The New Peasantries: Rural development in times of globalization (2nd edition).* Routledge.

Ploeg, J.D. van der, Franco, J.C., & Borras, S.M. (2015). Land concentration and land grabbing in Europe: A preliminary analysis. *Canadian Journal of Development* 36(2), pp. 147–162.

Priebe, H. (1969). The modern family farm and its problems: With particular reference to the Federal German Republic. In: U. Papi & C. Nunn (eds.), *Economic Problems of Agriculture in Industrial Societies,* pp. 251–282. Macmillan.

Rooij, S.J.G de, Ventura, F., Milone, P., & Ploeg, J.D. van der (2014). Sustaining food production through multifunctionality: The dynamics of large farms in Italy. *Sociologia Ruralis* 54(3), pp. 303–320.

Salter, W.E.G. (1966). *Productivity and Technical Change.* Cambridge University Press.

Slicher van Bath, B.H. (1978). Over boerenvrijheid (inaugurele rede Groningen, 1948). In: B.H. Slicher van Bath & A.C. van Oss (eds.), *Geschiedenis van Maatschappij en Cultuur,* pp. 71–92. Basisboeken Ambo.

Toledo, V.M. (1990). The ecological rationality of peasant production. In: M. Altieri & S. Hecht (eds.), *Agroecology and Small Farm Development,* pp. 53–60. CRC Press.

Ventura, F. (2013). *Part-time in agricoltura: Caratteristiche ed importanza del fenomeno per lo sviluppo delle aree rurali Italiane* [Part-time in agriculture: characteristics and importance of the phenomenon for the development of Italian rural areas]. ISMEA.

Visser, O., Mamonova, N., & Spoor, M. (2012). Oligarchs, megafarms and land reserves: Understanding land grabbing in Russia. *The Journal of Peasant Studies* 39(3–4), pp. 899–931.

White, B. (2011). *Who Will own the Countryside? Dispossession, rural youth, and the future of farming.* International Institute of Social Studies.

Wiskerke, J.S.C. & Ploeg, J.D. van der (2004). *Seeds of Transition: Essays on novelty production, niches, and regimes in agriculture.* Royal van Gorcum.

Think Locally, Act Globally: Polish farmers in the global era of sustainability and resilience, ed. by Krzysztof Gorlach and Zbigniew Drąg in collaboration with Anna Jastrzębiec-Witowska and David Ritter
Jagiellonian University Press, Kraków 2021, pp. 23–26
ISBN 978-83-233-4949-5
DOI: http://dxdoi.org/10.4467/K7195.199/20.20.12724

An Introductory Letter from the First Editor: Where the horses, cows, and even cats had their own names

Krzysztof Gorlach https://orcid.org/0000-0003-1578-7400/

The following monograph should be seen as an attempt to present changes affecting the category of family farm owners in Poland over the last 70 years, since the end of World War II. These changes brought significant social transformations, including the dismantling of the landowner class (who had large farms in their possession), moving the state border westward and changing the multiethnic Polish society into one close to ethnic homogeneity. The main goal of this reflection is to recount ways in which family farms coped with various unfavorable forces and factors in order to remain in operation. One could say that the entire study can be viewed as a manifestation of the well-known phrase that served as the title of the James C. Scott (1990) book: *Domination and the Arts of Resistance*. It should not be surprising that the following work gives a great deal of attention to the functioning of family (peasant) farms during Communism. From 1944 to 1989, the primary goal of family farms facing the not-so-friendly, and sometimes even openly hostile, policies of the state was to simply maintain their existence. This was described in detail in my early article published in *Sociologia Ruralis* (see Gorlach, 1989). After 1989, the year which brought political breakthrough not just in Poland but also in other countries of Central and Eastern Europe, the situation did not appear to be more comfortable. Family farms were forced to fight for survival in the reality of the market economy. Their situation was presented in my other works that were based on empirical research (see Gorlach, 1995, 2001, 2009).

The monograph presented here refers to these analyses stemming from another edition of sociological research, completed within the framework of the MAESTRO project financed by the National Science Center of Poland. The main goal of the project was to depict the functioning of family farms as the traditional sector of agriculture in Poland in the contemporary context of globalization processes. The farms were examined in terms of the principles of sustainable development as well as flexibility and resilience in reaction to various crises.

The monograph is divided into four essential parts. The first part is devoted to the theoretical issues and methodological groundwork for the entire publication. To use a painting metaphor, this part could be compared to a primer paint, on which particular features of the picture are placed. The second part of the book aims to capture the changes that took place from 1994 (when the first edition of the research described in the subject literature was conducted) to 2017, when the research of the current project was carried out. This time frame covered almost a quarter of a century (23 years to be precise), which was an adequate period to encompass the changes and metamorphoses that mostly happened as a result of two things: the regime transformation which began in 1990, and Poland's accession to the European Union on May 1, 2004. Returning to the painting analogy, it could be said that this part provides the main outline, form, and structure of the whole work. The third part deals with the crucial issues of regional variations, mostly in regard to life strategies and strategies of operating farms. In this part, the overarching question is how diverse Polish agriculture is and whether it can be discussed as a homogenous phenomenon or various types of reality that cannot really be compared. For our painting analogy, it could be said that this is the most crucial layer and the actual substance of the painted picture. Finally, there is a fourth part which places the focus on select themes, such as rural lifestyles, food safety and security, farmers' utilization of new computer and IT resources, and the potential for socio-political mobilization. This part could be seen as a special flourish of vivid colors that enrich the painting.

The presented monograph is not just the work of one person, in the sense that my previous publications were (see Gorlach, 1989, 1995, 2001, 2009). This is an authors' monograph as it has several co-authors and, in some cases, authors of the book chapters for which they are accountable. The number of authors in this case is quite substantial, and I would like to name them. First of all, I must mention my colleague Zbigniew Drąg, PhD, who was in charge of the logistics for the entire project. He was a co-editor of the following monograph and the author of statistical analyses, as well as co-author of many chapters. I should also highlight the considerable role of two other collaborators greatly involved in the editing process of this book, namely Anna Jastrzębiec-Witowska and David Ritter. They committed to the task of translating this volume from Polish to English, as well as the proofreading job done bravely by David. Finally, I should present the team of people working and/or cooperating with the Department of Social Structure at the Institute of Sociology of Jagiellonian University, as well as those who were employed specifically for this project. They are: Piotr Nowak, Marta Klekotko, Adam Mielczarek, Adam Dąbrowski, Martyna Wierzba-Kubat, Maria Kotkiewicz, Daria Łucka and Grzegorz Foryś. Furthermore, the work of two foreign collaborators should be properly noted. Jan Douwe van der Ploeg, professor emeritus of rural sociology at the Agricultural University at Wageningen in the Netherlands, and Patrick H. Mooney, sociology professor at the University of Kentucky at Lexington, KY (USA), two distinguished academics, played the role of external evaluators of the project's theoretical content and therefore were asked to prepare the foreword and the afterword.

Finally, I would like to add a very personal note. I have decided to dedicate this book to my paternal grandparents, Aniela (grandma) and Jan (grandpa) Gorlach, who were farmers that worked a small, 3-hectare farm. As a child in the 1960s, I used to spend summer vacations on their farm, where they taught me the principles of farmwork, how to cultivate good relations with neighbors, as well as how to take responsibility for what happens on the farm and in the local rural community. They also provided a great example of familial attachment to farm animals, wherein horses and cows each had their own names, as well as the household cats (The farm also had pigs, chickens, and pigeons, but they went unnamed.) Today, in the spring of 2021, such a farm can probably no longer be found in Poland but, for me, its image remains an important point of reference for all of the academic reflections in this book. And in my memory my grandparents' farm will remain forever.

REFERENCES

Gorlach, K. (1989). On repressive tolerance: State and peasant farm in Poland. *Sociologia Ruralis* 29(1), pp. 23–33.

Gorlach, K. (1995). *Chłopi, rolnicy, przedsiębiorcy: „Kłopotliwa klasa" w Polsce postkomunistycznej* [Peasants, farmers, entrepreneurs: "The awkward class" in postcommunist Poland]. Wydawnictwo Uniwersytetu Jagiellońskiego.

Gorlach, K. (2001). *Świat na progu domu: Rodzinne gospodarstwa rolne w Polsce w obliczu globalizacji* [The world at the doorstep: Polish family farms in the face of globalization]. Wydawnictwo Uniwersytetu Jagiellońskiego.

Gorlach, K. (2009). *W poszukiwaniu równowagi: Polskie rodzinne gospodarstwa rolne w Unii Europejskiej* [Searching for the Balance: Polish Family Farms in European Union]. Wydawnictwo Uniwersytetu Jagiellońskiego.

Scott, J.C. (1990). *Domination and the Arts of Resistance: Hidden transcripts*. Yale University Press.

PART ONE

THEORETICAL
AND METHODOLOGICAL
CONSIDERATIONS

Think Locally, Act Globally: Polish farmers in the global era of sustainability and resilience, ed. by Krzysztof Gorlach and Zbigniew Drąg in collaboration with Anna Jastrzębiec-Witowska and David Ritter
Jagiellonian University Press, Kraków 2021, pp. 29–30
ISBN 978-83-233-4949-5
DOI: http://dxdoi.org/10.4467/K7195.199/20.20.12725

Some Introductory Remarks by the First Editor to Part One

Krzysztof Gorlach   https://orcid.org/0000-0003-1578-7400/

The first part of this publication contains two, rather extensive, chapters. The first provides a justification for the perspective of the entire research project, which is based on the inversion of the popular slogan originally coined as "think globally, act locally." Our concept is: "Think locally, act globally." The inversion aims to transpose the entire tradition of conducting analyses on farmers' place in the contemporary world within the social sciences. In our view, the activities of individual farmers, small groups of farmers, and families stem from local experience and family tradition or a tradition of local community. For that reason, this chapter covers a variety of issues and examines the related statements that are quite commonly used by sociologists.

To start, this chapter confronts the thesis that the world becomes "flat" as a result of globalization. Quite to the contrary, we, the authors Krzysztof Gorlach, Marta Klekotko, Anna Jastrzębiec-Witowska, Grzegorz Foryś, Daria Łucka, and Piotr Nowak, assert that globalization is conducive to highlighting various "non-global" matters, such as local, regional, or even "national" differences. As a result, globalization processes are treated as sequences of intense local-global relations. To conceptualize these relations, in Chapter 1 there are numerous references to neo-endogenous and sustainable development concepts. Specific aspects of these concepts are stressed in the context of changes occurring in rural areas, and particularly in agriculture. The changes affecting agriculture are viewed through the lens of sustainable development, which leads directly to certain aspects of the functioning of family farms and farmers' decision-making.

Chapter 2, which finalizes this part of the publication, is devoted to the research methodology, presenting two important matters. Firstly, the reader's attention is directed to the multitude of methods applied in this publication. There are both quantitative and qualitative methods, creating a certain mix, with the goal of providing a multifaceted work on the studied reality of family farms in Poland. Secondly, such a methodological blend leads to the type of examination that has been called sustainable analysis. It aims to achieve a multidimensional picture of the analyzed reality, which can be maximally close to reality.

When introducing this part of the book, two other matters should be mentioned. Chapter 1 provides the theoretical perspective for the entire publication and is a result of the work of several authors. Among them are people who comprised the research team, as well as others not directly connected with this initiative. Chapter 2 only has two authors. As the head of the entire project, I am also the second author. The first author is my colleague Zbigniew Drąg, who is responsible for the methodological side of the publication.

Think Locally, Act Globally: Polish farmers in the global era of sustainability and resilience, ed. by Krzysztof Gorlach and Zbigniew Drąg in collaboration with Anna Jastrzębiec-Witowska and David Ritter
Jagiellonian University Press, Kraków 2021, pp. 31–93
ISBN 978-83-233-4949-5
DOI: http://dxdoi.org/10.4467/K7195.199/20.20.12726

Chapter One: Thinking and Acting in the Age of Local–Global Relations: Think locally and act globally*

Krzysztof Gorlach https://orcid.org/0000-0003-1578-7400/

Marta Klekotko https://orcid.org/0000-0001-5945-3061/

Anna Jastrzębiec-Witowska

Grzegorz Foryś https://orcid.org/0000-0002-9411-2681/

Daria Łucka https://orcid.org/0000-0001-6861-5122/

Piotr Nowak https://orcid.org/0000-0001-7991-5534/

1.1. Introductory Remarks

The expression "think globally, act locally" has been widely considered a kind of symbolic statement encapsulating a behavioral strategy in the contemporary globalized world. It means that each individual, as well as social collective, should take into consideration some broader and larger spectrum of factors, ideas, and values reflecting much more than the immediate issues of their community or place in order

---

*   This chapter is an expanded version of the text entitled "Think locally and act globally: Understanding human development in the era of globalization" written by Marta Klekotko, Anna Jastrzębiec--Witowska, Krzysztof Gorlach, and Piotr Nowak, published in 2018 (*Eastern European Countryside* 24, pp. 111–141).

to enact particular strategies for activities performed in their close environment. The basic idea underlying such reasoning is the assumption that every community and every place in the contemporary social space presents a specific permutation and/or particular implementation of more general events, processes, structures, institutions, and solutions relating to social problems.

This perspective draws significantly from the concept of globalization which might be understood, by at least some contemporary authors, as a process of uniformization, or even standardization, of human society, despite some observable social—and especially cultural—differences. The global economy has been noted as a leading force in such an approach.

The phrase "think globally and act locally" has been used frequently in reference to environmental issues. Certain efforts resulting, for example, in the global lowering of carbon dioxide emissions should not necessarily be grounded in some local activities and solutions, but such local initiatives, undertaken in many places and communities, might have an aggregate effect on the global (trans-local) scale. Such reasoning has overemphasized that the cumulative effect of some local activities might result in some global (trans-local) changes. But this problem might be considered in an entirely opposite way. The global-local problem also requires, for example, the analysis of behavior patterns performed by global corporations within some local milieus. As Amey (2013) stresses:

> The local action plans of these companies are nothing short of selling their "global" products in "packaging" that may appeal to the preferences of local consumers or fulfill their needs which have been shaped by local context and local traditions (Klekotko et al., 2018, p. 113).

From this perspective, the "global" does not seem to be a kind of "sum of locals." Quite to the contrary, the local seems to be a consequence of the global impact in a particular context. It is not an exaggeration to say that the global dominates the local, either with a cumulative effect or by its particular implications for local matters. In this context, the local seems to be completely erased or, at least, reduced to a simple mutation and implementation of the global. Therefore, such a social world might be perceived as a complete top-down project regulated by global actors having an essential impact on the activities of local actors. As we stressed before: "Limiting the importance of local systems and the actions of local actors to mere tools for implementation of a global idea casts individuals as mere dutiful performers and deprives them of any meaningful impact on their surrounding social reality" (Klekotko et al., 2018, p. 113).

This is, however, only one part of the entire, complex, and multi-dimensional problem of global–local relations. The other part has been focused on the idea of human development as well as on the discourse associated with it. There are, in our opinion, at least several motives in this discourse. The first one relates to issues of the environment and the message which has been formulated to articulate that people must improve their impact on the environment in their local community and/or in their place. Such activity has to be an effect of the so-called thinking about the many threats to the natural environment on a global scale. As distilled in the aforementioned article by Marta Klekotko and others from Jagiellonian University in Krakow, ". . . it is more effective for an individual to reduce their own energy consumption than wait for a global action" (Klekotko et al., 2018, p. 112). Therefore, this particular connection between the local and the global seems to have a kind of twofold relationship. First, it is an intellectual activity bringing together issues on the global scale that have a strong tendency to manifest themselves on the local. Second, it is the assumption that solving some problems on the local level brings some cumulative effects on the global.

However, the problems of global-local relations might also be considered in the context of business. As was stressed by Amey (2013): "Big companies such as McDonalds and Honda are successful on a global scale, while their products were tailored to the requirements of individual countries." Such reasoning has been based on the idea of sharing knowledge, values, and ideas concerning some global issues and, at the same time, being part of a network of institutions and practices rooted in some local spaces and places, but also belonging to some extra-local spaces and a set of practices (Klekotko et al., 2018, p. 113). Moreover, in this analysis, Amey (2013) considers in a more detailed way some strategies taken by large transnational corporations in how they approach local consumers. Some "global" products have been tailored or "packaged" in such a way (using the best tactics of public relations and marketing) as to better appeal to the desires and preferences of local consumers. This, however, does not mean that real needs will be met. To quote Klekotko and others: ". . . alluding to environmental protection and ecological values indicates that local communities are expected to forego the fulfillment of their own needs without any reassurances that global action will be taken on their behalf" (Klekotko et al., 2018, p. 113). One might observe rising opposition towards the expectation of such asymmetrical concessions.

1.2. Is the Contemporary World "Flat"?

As Amey (2013) stresses again in his influential article, such thinking might, in fact, be a manifestation of the idea to "think globally and act globally." In such a perspective there is no place for local matters, preferences, and desires. The world seems to be completely "flat" (see also: Friedman, 2005), as indicated by various authors engaged in this subject. This seems to be a recurring message presented by authors analyzing various aspects of the globalization processes.

The first aspect (of globalization processes) might be a purely political statement, presenting Europe as a cosmopolitan entity (compare to Beck & Grande, 2004). The authors highlight that, despite various specific regional, local, and national characteristics, as well as those related to the core of the political system, Europe possesses a certain set of characteristics that assert its cosmopolitan character, furthering its perception as a unified entity.

The second aspect is the economic one, and it is even more crucial than the first. According to the adherents of this approach to globalization, the "flattening" or "uniforming" of the world has its source in the logic of economic processes and, most of all, in the dominance of transnational corporations over the economy. But such statements are not without some ambiguity, as there are at least three perspectives on the fundamental thesis of this approach (compare Waters, 1995, p. 4 and more). The first perspective stipulates that globalization started with the beginning of commodity exchange, fostered by the great geographic discoveries of the 15th and 16th centuries which, in addition to economic exchange, also brought some confrontation of various cultural values. Another perspective points to the birth of capitalism as the beginning of globalization (see Wallerstein, 1974, 1979). This approach stresses capitalism as the very essence of globalization, which causes the expansion of economic relations beyond all divisions, be they geographical, political, or cultural. There is also a third perspective, which highlights the uniqueness of globalization in regard to changes experienced by society in previous time periods as part of the processes of modernization, development of capitalism, or the formation of the world system. In this view, globalization stems from the postmodern state of societal development, expressed not only in the new phase of postmodern culture but also in the transition from an industrial society towards a service-oriented society.

Similar observations are made by Philip McMichael (2000, pp. 150–187), who names several crucial characteristics of the globalization project. The main one can be described as the dominance of regulations stemming from market forces over those regulations which are the result of state policies and activities. If the dominance of states is discussed it is only limited to those with the largest economic potential, namely the states of the so-called West. Furthermore, the economic policy conducted by international institutions (such as the International Monetary Fund, World Bank, or World Trade Organization) is the consequence of dictates of the aforementioned countries. At the same time, the international economic market is dominated by transnational corporations whose assets and resources are often greater than the assets of small and medium-sized states. Therefore, nation-states seem to be more integral components of the wider, international political order rather than their own self-governing entities (Hirst & Thompson, 1995, pp. 408–442). Some authors, such as Brecher and Costello (1994, p. 5), go even further by stating that the phenomenon of the nation-state has been to some extent absorbed by the frame of inexplicable globalization processes. In this context, one can speak of a new world economy based on four crucial pillars, namely: the global cultural bazaar, the global shopping mall, the global workplace, and the global financial network (Barnet & Cavanagh, 1994, pp. 15–21). These processes result in new social divisions. As indicated by Tarrow (1998, p. 176 and others), a new dominant class emerges, encompassing managerial elites of international institutions as well as elites at the top of transnational corporations and elites of the nation-states connected with them, especially states dominating over others due to their economic and military potential. On the opposite end of the spectrum there is a contemporary dominated class, consisting of relatively well qualified employees, who are generally hired on temporary contracts (a precarious class), as well as unqualified and poorly qualified workers (including the workers in countries recognized as developed, as well as the members of the underclass and members of marginalized societies functioning somewhere on the peripheries of the global economic system.) One of the reactions to such a situation is social unrest caused by more or less severe socioeconomic austerity measures introduced under the pressure of global institutions such as the World Bank or the International Monetary Fund (see Walton & Seddon, 1994).

It should be said that the debated "flat" character of the globalized world overlooks various irregularities which make it rather "non-flat." Various authors address this matter. For example, David Korten (1995, pp. 221 and more) explains that the largest transnational corporations comprise at least half of the world's hundred largest economies. The total value of the sold production of the ten largest transnational corporations exceeded the gross domestic product of the hundred smallest countries. Furthermore, the 500 largest industrial corporations made 25% of the world's production, employing only 0.5 per mil of the world human population. The resources of the largest commercial banks and various financial institutions comprised

60% of the global value of production capital. It should be noted, however, that these data were collected in the late 1980s and the early 1990s. What is the situation today?

The situation is likely to worsen if increasing inequalities are considered to be a negative process. This phenomenon has at least two separate dimensions. The first one is the evolution of global wealth rankings. As one economist notes, "Between 1987 and 2013, the number of USD $ billionaires rose according to Forbes from 140 to 1,400 and their total wealth rose from 300 to 5,400 billion dollars" (Piketty, 2014, p. 433). Moreover, the same economist also stresses that "Between 1987 and 2013, the share of the top 1/20 million fractile rose from 0.3 percent to 0.9 percent of world wealth and the share of the top of 1/100 million fractile rose from 0.1 percent to 0.4 percent" (Piketty, 2014, p. 436). Another component is a change from rankings of billionaires to "global wealth reports." As it has also been put:

> Concretely, the wealthiest 0.1 percent of people on the planet, some 4.5 million out of an adult population of 4.5 billion, apparently possess fortunes on the order of 10 million euros on average, or nearly 200 times average global wealth of 60 000 euros per adult, amounting in aggregate to nearly 20 percent of total global wealth. The wealthiest 1 percent—45 million people of out of 4.5 billion—has about 3 million euros apiece on average (broadly speaking this group consists of those individuals whose personal fortunes exceed 1 million euros). This is about 50 times the size of the average global fortune, or 50 percent of total global wealth in aggregate (Piketty, 2014, p. 438).

Other publications confirm the above statements, addressing the tendency for economic and social inequalities to increase. For example, such observations can be found in the World Inequality Report (Alvarado et al., 2018; DOA: July 24, 2019), with a general introductory remark that reads as follows: "Since 1980, income inequality has increased rapidly in North America and Asia, grown moderately in Europe, and stabilized at an extremely high level in the Middle East, sub-Saharan Africa, and Brazil" (Alvarado et al., 2018, p. 40, DOA: July 24, 2019).

Assertion of this general tendency is supported by several other statements formulated on the basis of the data collected in various regions of today's world. Their importance lies in how they present the variety of ways in which globalization can be reflected. In other words, they provide evidence that the contemporary world, even as it is undergoing the processes of globalization, does not become "flat." To quote the first statement: "The poorest half of the global population has seen its income grow significantly thanks to high growth in Asia. But the top 0.1% has captured as much growth as the bottom half of the world adult population since 1980" (Chancel, in: Alvarado et al., 2018, p. 40, DOA: July 24, 2019). Another statement stresses other aspects of the studied processes: "Income growth has been sluggish or even nil for individuals between the bottom 50% and top 1%. This includes North American and European lower- and middle-income groups" (Chancel, in: Alvarado et al., 2018, p. 40, DOA: July 24, 2019). The subsequent statement addresses the irregular trajectory of the processes of growing economic inequalities:

> The rise of global inequality has not been steady. While the global top 1% income share increased from 16% in 1980 to 22% in 2000, it declined slightly thereafter to 20%. The trend break after 2000 is due to a reduction in between-country average income inequality, as within-country inequality has continued to increase (Chancel, in: Alvarado et al., 2018, p. 40, DOA: July 24, 2019).

Consequently, the summarizing statement of the text could be the following:

> Global income growth dynamics are driven by strong forces of convergence between countries and divergence within countries. Standard economic trade models fail to explain these dynamics properly—in particular, the rise of inequality at the very top and within emerging countries. **Global dynamics are shaped by a variety of national institutional and political contexts** [underlined by the authors: K.G., M.K., A.J.W., G.F., D.Ł, P.N.] (Chancel, in: Alvarado et al., 2018, p. 40, DOA: July 24, 2019).

In such a system, what is defined as "local" can only appear as a certain application of global rules and principles in a defined, fragmented context. Such an approach, despite being "officially" supported by the slogan "act locally," might, in fact, be a kind of threat to the local, including various local actors, as well as local systems. It may act quite contrary to the empowerment of locals and result in their declining agency. Therefore, as was stated in the previously quoted work by Klekotko and others: "Undoubtedly, change and development are not possible without the involvement of local systems and arrangements" (Klekotko et al., 2018, p. 114). Many authors claim that local agents (both individuals as well as collectives) have to play an important part in the whole mechanism and cannot be reduced exclusively to the role of followers or even to those who only imitate the strategies and solutions prepared by the global (extra-local) agents. For example, Hickey and Mohan (2004) have developed a concept of "participation" that cannot be limited to actions, activities, solutions, and strategies prepared and implemented by someone else. The local actors have to be part of the network that is in charge of developing the activities and actions mentioned above. This means that local actors should be empowered, as has been stressed by many other authors (see, e.g., Chambers 1997, 2005; Pieterse, 2010; Klekotko, 2012).

Therefore, we would like to stress that the statement "think globally, act locally" has been based simply on the concealed idea of the domination of the global over the local. Therefore, there is an urgent need to reexamine the complex relations between the global and the local in order to most accurately and precisely formulate the opposite strategy as encapsulated in the advisory statement of "Think locally and act globally" which serves as the leading message of this volume. This also seems to be a new mechanism of the current complex processes of human development. The first step in this strategy, as we indicated earlier, might be ". . . the new way of viewing the local in a society exposed to the globalization processes" (Klekotko et al., 2018, p. 115).

1.3. On Global-Local relations[1]

Reverse thinking in regard to global-local relations promotes a reversal of the entire way of thinking underlying the development of the contemporary world. Thoughts and perceptions of reality are directed towards the subjective vision of functioning and the meaning of social change (see, e.g., Sztompka, 1991). They are concentrated on the actions related to the implementation of values or fulfilling the needs of individuals and communities as sources of social dynamics. Actions and activities that make certain adjustments to greater phenomena or social processes as their main goals are of secondary importance. What seems more crucial here pertains to the characteristics of what is local, such as natural features and traits or certain local traditions. Global or even extra-local characteristics stemming from general processes of socio-economic, political, or cultural transformations are considered less crucial. Following this thought, we would argue that, in starting to theorize about global-local relations, one has to consider the sense, the meaning, and the concept of place in a way similar to how the more traditional idea of community (especially local community) has been conceptualized before. As has been mentioned in one of the previous articles:

---

[1] For this reflection on global-local relationships, the authors are indebted to our colleague and friend Marta Klekotko, Ph.D., from the Institute of Sociology of Jagiellonian University, the first author of the work "Think locally and act globally: Understanding human development in the era of globalization," published in 2018 (*Eastern European Countryside* 24, pp. 111–141), so frequently cited in this chapter.

The traditional outlook on local communities usually deals with certain ensembles of people as subjects settled within the framework of particular fragments of space. Here originates the idea of connecting such communities with concrete fragments of the space, namely with "places" (Klekotko et al., 2018, p. 115).

This means that some particular places become the spatial foundations of particular neighborhoods, collectivities or communities. They form more or less clear boundaries concurrently dividing the spatial and the social world into two significant parts, specifically the friendly "inside" and the sometimes more or less alien "outside." Moreover, such a division usually results in another one, namely between "us" and "them." Such tendencies might be mitigated, to some extent, by modernization processes in society, leading to a rise in spatial and social mobility. Furthermore, they can be weakened even more by the processes of globalization and post-modernization (see, e.g., Pakulski, 2009) that result in a situation in which we claim, following Wiborg (2004), that

> . . . particular fragments of geographic space are subjected to far more intense and thorough social construction than what could have been observed in the era of traditional societies or societies affected only by modernization. This characteristic of the concept of place is the most significant and decisive (Klekotko et al., 2018, p. 115).

However, analyses of "the places" seem to be a much more complex procedure. As we see in the proposal developed by Perkins (2006), places might be located in the commodification perspective, forming portions of space in which food production might be observed, as well as the creation of other commodities, such as natural landscape and cultural heritage, which might be visible as well. Such processes have also been related to the discourse among various actors who sometimes support quite contradictory ideas concerning the development of a particular place. Moreover, various processes which link meaning and identities with particular places might be observed, as well (Woods, 2006). Such a perspective opens the issue of "social construction of place" processes that have been rooted in some established classic sociological considerations developed by Berger and Luckmann (1971) in their idea of the "social construction of reality." A similar perspective might also be found in the works of Polish authors (see, e.g., Lubaś, 2010), known as the concept of the "social creation of places." This approach has contained several, more specific, assumptions. First of all, people (both individuals and collectives) shape some parts of space using their ideas and activities, as well as through their organized social relations. Moreover, people create place as a fragment of space where global, national, and local features and forces meet and play together. Therefore, place might be considered as a multi-scale piece of social space. Finally, such places seem to be contemporary mutations of what has been identified before as local communities. Following Giddens (1992), one might stress that ". . . the concept of place not so much refutes the concept of the local but gives it a more complex meaning which emphasizes the quite specific role of supra-local factors" (Klekotko et al., 2018, p. 117).

What are these complex meanings? One of them has been linked to the network society, as defined by Mark Shucksmith (2010). In such an approach, a place might be viewed as a social construction of constant shaping and redesign activities performed by actors observed in a particular context (Klekotko & Gorlach, 2011). However, some different approaches might also be used here (Vanclay, Higgins, & Blackshaw, 2008). Places seem not to be connected to a particular fragment of space in a physical sense, but rather to some meanings related to some locations. Therefore, places seem to be rather a set of meanings and not the direct contemporary equivalents of what has been formerly known as local communities. Moreover, emotions might be perceived as particular characteristics in the processes of the social construction of places. As we presented before: "Emotions and feelings are natural ingredients of human experience" (Klekotko et al., 2018, p. 118). For that reason, anger, frustration, friendship, love, fascination, joy, hatred, hostility, or resignation are reflected in sociological analyses. These emotions can be seen in various social interactions and situations. "Therefore, the processes of space socialization, construction of place generate emotions" (Klekotko et al., 2018, p. 118). Emotions have been connected to the issues of memories. Moreover, they might give some support to special meanings as well as the processes of understanding, perceiving, and interpreting those various special meanings. For example, one might note in these contexts ecological problems, environmental threats, animal rights, and other important issues (Smith et al., 2009).

Emotions seem to be very important in the issue of "place" due to two of its different basic characteristics. The first one relates to the natural aspects of the place, namely its natural landscape, space qualities, animals, etc. However, the second one might be related to some social and cultural phenomena, namely, family tradition, neighborhoods, local institutions, culture, etc. (Klekotko & Gorlach, 2011, pp. 30–31). It is also worth noting that not only positive emotions might be visible in this context.

In the area of agricultural and rural studies, food issues seem to be another important element contextualizing the ideas of "the place." This is especially visible in the rural sociological literature founded on the sentiments surrounding the anti-industrial type of agricultural production. From such a perspective, food production seems to be especially connected to particular places and serves as a kind of banner for them. It becomes both an activity and a product which seem to be in tune with local traditions, with particular places and their peculiar meanings (see, e.g., Fonte & Papadopoulos, 2010).

Therefore, we might theorize further concerning some connections between the place and locality on the one hand, and their relations with social transformations of space under the current regime of globalization on the other. As has been already stressed by Klekotko and other authors:

> Place becomes a context for a society affected by the processes of globalization with a special link to "locality," understood as a set of certain physical characteristics, related to a certain fragment of space and a supra-local realm. Here, we mean references to values, meanings, and actions located within the context of locality (Klekotko et al., 2018, p. 119).

However, what should be stressed here immediately is that there is another very powerful concept present in the relevant contemporary social science discourse, namely the concept of "glocalization."

It should be noted that "glocalization" is conceptualized in slightly different ways by various authors. Following the consideration presented earlier, we want to use this term in the sense of "glocal community" according to a perspective observed in the sociological literature. Here, two positions are worth mentioning: Robertson (1992), as well as Bauman (1997), who introduced it to Polish sociological literature. In Poland, Marta Klekotko, in her influential book (Klekotko, 2012), has concentrated on some similarities and common assumptions of these two approaches disregarding differences between them and treating some local phenomena and processes as being under the influence of global phenomena and processes in the contemporary world.

However, it should be stressed as well that some other authors have suggested different approaches to the issue. Authors such as (again) Robertson (1995), Khonder (2004) or Giddens (1992) address certain mutual or reciprocal relations between global and local processes. From such a perspective, the local is not only a kind of passive recipient of global impact but, at the same time, a kind of agent that has itself been able to have an impact on the global phenomena and processes. Moreover, one might say that the local seems to make an influential and important contribution shaping global issues. Then, from this perspective, Robertson argues that one might observe a mutual movement of local values, symbols, and meanings to the global level as well as the opposite, i.e., a movement of global values, symbols, and meaning to the local level. In such a context, globalization might even mean a certain renaissance of locality.

Moreover, glocalization might also contain quite different meanings than those mentioned above. As we wrote:

> It is not so much an analytical concept meant to describe and explain various phenomena and social processes, but more of an applied project. In this perspective, locality refers to processes of the dynamic symmetry between the resources located in specific communities and societies and those on the supranational (global) level (Klekotko et al., 2018, p. 120).

As has been explained by Nigro (2003, p. 5), "glocalization" might be treated as a particular type of strategy linking the benefits of global dimensions, especially in the areas of technology, information, and economy, with local realities. What is more important here is a peculiar type of mechanism which is based on a bottom-up system for the governance of globalization. Such a mechanism should result in greater equality in the distribution of resources, as well as a social and cultural rebirth of poor and disadvantaged localities (i.e., communities and societies).

Again, we might recall here Marta Klekotko (2012), who, in her already mentioned book, brings our attention to three particular types of changes occurring in local communities undergoing the processes of globalization. These types of changes, which Klekotko calls scenarios, are identified as: cosmopolitan, fundamentalist, and constructionist. The first scenario is a result of the elimination of local specificities that had resulted from the processes of locality dissolving into or blending with the global society. In turn, the second one, cosmopolitan, has been based on the emphasis of various defensive strategies undertaken by the local community as a whole or in part and focused on strong protection of the local social fabric from the impact of global forces. Moreover, the third scenario, constructionist, might be perceived as a kind of middle-ground strategy focused on the adaptation of locality to a new globalized context through various techniques focused on reconstruction of the local social fabric. More detailed analysis of the various aspects of globalization scenarios for local communities can be found in another publication (Klekotko et al., 2018, pp. 120–123).

However, what appears to be an important result of such an extensive discussion might be located in the area of "new local community." Such a concept contains several characteristics. The first of these is "openness" which means that local communities ". . . must be able to use the opportunities brought by globalization to their advantage" (Klekotko et al., 2018, p. 122). Here enters the issue of mobilization. In such a perspective, local communities might be perceived, to some extent, as social movements, i.e., some ephemeral phenomena focused on achieving particular goals. As was highlighted before: "What matters in local communities of the new type is not tight friendship between neighbors, but functional proximity related to planned and anticipated achievement of previously defined goals" (Klekotko & Gorlach, 2011, p. 42).

Based on such analyses, we might say that such contemporary local communities "... can be characterized by a new type of social bond" (Klekotko et al., 2018, p. 112). They are more similar to voluntary associations than to traditional types of localities mentioned in sociological literature. As mentioned by Klekotko and the other authors, "Local communities of the new type appear mostly within the public realm. This also influences the character of existing bonds. They are not meant to impose control over all aspects of human life but are more focused on civic engagement observed in the public realm" (Klekotko et al., 2018, p. 122). The same source addresses the civil aspect of local community in the following way:

> ... the idea of civil society is a foundation for the model of local community of the new type. The public realm where this community exists provides the area for civil activities, understood in this case as a combination of engaged people and institutions who take responsibility for the functioning of the community, not just in the present moment but also with having future generations in mind. It can be stated that the local community of the new type appreciates the significance of civic participation in actions taken for its development (Klekotko & Gorlach, 2011, p. 43).

Therefore, one might observe the peculiar process of reintegration of local communities within the processes of modernization and globalization. They might be perceived as effects of civic mobilization based on their social and cultural resources and, therefore, able to frame citizens' participation and people's abilities to cooperate, leading to collective activities resulting in observable changes in the contemporary globalized society.

Now it seems warranted to go back to the initial idea and title of this volume, namely "think locally, act globally" and revisit "... our main idea of inverting the popular slogan of thinking globally and acting locally into the combination of local thought and global actions" (Klekotko et al., 2018, p. 125). In this argument, we want to mention the particular role of what has been called in the literature "new local communities" or "local communities of a new type." Their key role seems to be based on their abilities to develop and

> ... support local ideas and initiatives with more global action strategies happening on the national and supranational levels. As we have discussed earlier, the empowerment

of citizens through authentic and full participation is a key to good understanding of relations between the global and the local (Klekotko et al., 2018, p. 125).

Following some arguments developed by Hickey and Mohan (2004) in their notable book about participation, transformation, and development, we suggest the analysis of four scenarios already presented in the article by Klekotko and other authors (Klekotko et al., 2018, pp. 125–130). In this chapter we are focused only on some of the most important characteristics of the four basic types of human development scenarios based on the interaction between the local and the global. They are as follows: think globally and act locally, think locally and act locally, think globally and act globally, and, finally, think locally and act globally.

Let us start with the first scenario, i.e., "think globally and act locally." "In this concept, the participation of local actors is usually moved to the final part of the change process, namely, to the stage of implementation" (Klekotko et al., 2018, p. 125). It means that local individuals and/or communities are not involved in the construction process of developmental strategy, as well as some extended discourses concerning it. Quite to the contrary, it might be stressed that such strategies, including their scopes, dimensions, and goals, have already been created by some global forces. The only role of the local has been perceived as the implementation and/or execution of such strategies constructed by extra-local actors. Therefore, such a perspective has to be framed almost completely as a top-down approach to human development and processes of social change. In turn, the second scenario under the name "Think locally and act locally" ". . . is an example of a twofold activity that originates in local needs (problems and challenges, use of local resources) and is conducted in the local arena via social participation" (Klekotko et al., 2018, p. 126). At first glance, such a strategy might be perceived as a tendency towards autarchy and/or the development of the mechanism for de-linking from external actors and forces. However, in the contemporary globalized world, such a situation seems to be extremely rare. In fact, the contemporary model of a new local community or, as we mentioned earlier, a local community of a new type with a strong mechanism for adaptation to the global context, has been extremely important. As Marta Klekotko (2012) stresses in another work, such communities have been based on self-reliance, developmental ideas with a strong tendency toward the so-called filtering of external values and resources through local needs and goals. Therefore, such a type of human development might also be perceived as a kind of top-down approach, but with a more active role played by the selection processes of externalities coming from the top. In this model, local strategies cannot be treated as simple imitations of the global ones, but rather as their own reconstructions in particular localities. The third scenario, identified as "think globally and act globally" ". . . could well be exemplified by farmers who, aware of the principles of the global game (i.e., Common Agricultural Policy), take the challenge and use available resources, mobilize them in order to be players on the supra-local level—and that may include the global arena" (Klekotko et al., 2018, p. 127). In such a perspective the participation, i.e., being a member of a supra-local network, matters. Local culture and local community are hardly present in this situation. Quite to the contrary, pure economic rationality, a corporate approach, and strong instrumental treatment of everything that seems to be local becomes increasingly visible. Global market logic has been a dominant force shaping all activities that might be observed. Local actors and local cultures seem to lose their empowerment and are no longer visible in the processes of human development. Resulting from this, the world does appear to become "flat."

The last perspective, which is favored by the authors of this chapter, is grounded in the logic that has been a direct challenge to the three others described briefly above. As indicated earlier in this chapter, the problems of resource imbalance and limited participation of local communities are truly global in that they can be observed all over the world with their implied threat to the empowerment of local systems. Therefore, the authors of this chapter propose a re-formulation of this famous slogan [think globally and act locally: K.G., M.K., A.J.W., G.F., D.Ł., P.N.], and

> . . . consequently, a change in thinking about development with local communities being at the heart of this process. Although the impact of local communities on global decisions and processes is quite limited, it does not mean that their abilities for 'global' action are limited to merely the actions and deeds of the most industrious individuals, while the overall community is fated to passive participation in global programs (Klekotko et al., 2018, pp. 127–128).

In this perspective, the role of the local idea and/or concept as the starting point for the particular program has to be mentioned. However, what should also be stressed in a contemporary globalized society is that such concepts and ideas might be developed and implemented only with the use of some external (extra-local, global) resources and cooperation with some external (extra-local, global) actors.

> Contrary to the "think globally—act locally" approach . . . the goals, needs, and developmental problems in our modified approach are not defined by outsider parties. The actions taken are not limited to the act of joining the EU program or "hitching a ride" . . . What particularly comes to mind is securing project funds from national

ministries and other entities, as well as the European Union, for locally conducted projects that deal with cultural identities and local traditions (Klekotko et al., 2018, p. 128).

Some examples discussed in the publication quoted often here (Klekotko et al., 2018) show two important issues connected to such a development strategy. The first one has been based on the use of external resources via application initiatives. The second issue, however, has been based on particular local abilities said to follow some ideas from the areas of social movement theories, such as political opportunity structure. We might rename this issue—for the particular goals of this project— simply as "local opportunity structures." It means that local communities and/or simple local actors have to have some initial resources at their disposal as a starting point to promote local assets, local activities and, finally, some local actors at the extra-local level of social organization.

To be honest, the idea of "think locally and act globally" does not seem entirely new in the social science discourse concerning human development issues. Some elements had already been conceptualized in the international literature (see, e.g., Sztompka, 1993; McMichael, 2010; Pieterse, 2010) but they were not framed under this idea contradicting the well-known slogan, which was reviewed earlier by the authors of the present chapter. The matters analyzed in the above-cited publications refer to the context of human agency, the ideas of alternative development, as well as the role of culture in recent human activities confronting the tendencies and issues of globalization in various parts of the world. Therefore, the present chapter aims to highlight at least two main theoretical perspectives that might form the background for debating the thesis on "think locally and act globally." These two perspectives can be identified as the ideas of sustainable development and a particular take on neo-endogenous development. Their presence in contemporary sociological literature is quite significant.

The two concepts mentioned earlier differ primarily in terms of their complexity levels. Therefore, it makes sense to start the reflection from the less complex concept, that of the proposal formulated by Christopher Ray (1999, 2006). Its "simple" character is an attempt to combine relations occurring on the local as well as on the global level within the framework of one development mechanism. However, the concept of sustainable development is somewhat more complex, and the development mechanism included within it requires reference to various issues that will be discussed later in this and the following chapter.

1.4. Neo-endogenous and Sustainable Development: A preliminary presentation

1.4.1. The Idea of Neo-endogenous Development

In the view of the authors of the present chapter, the concept of sustainable development seems to be quite weak if one would like to analyze merely its mechanism. They are more inclined to hold the opinion that the sustainable development perspective is a more normative type of thinking, i.e., to be focused on some values and/or goals and/or human desires. Therefore, when looking at mechanisms of development, one must turn to other types of sociological reasoning. One of them can be found in the perspective of so-called neo-endogenous development. This particular approach might be treated as a kind of solution resulting from the divergent conclusions of endogenous as well as exogenous concepts of human development. There are two fundamental assumptions in such an approach. As has been underscored, following the important works of Christopher Ray (1999, 2006, pp. 278–291), "On the one hand, there is an endogenous element according to which the development processes must be based on local initiative and be related to an approach known from academic literature and social practice as *a bottom-up approach*" (Klekotko et al., 2018, p. 135). However, there is another factor (assumption) that must be considered. Again, as was mentioned in the publication quoted above,

> The "neo" component . . . indicates that extra-local factors must play an important role in the development process . . . The neo-endogenous approach to development has two primary characteristics. First, the activities that occur in rural areas in pursuit of economic development are reoriented to maximize the retention of benefits within the local territory. This happens by valorizing and exploiting local resources . . . Second, the activities that lead to development are . . . contextualized by focusing on the needs, capacities, and life perspectives of local people (Klekotko et al., 2018, p. 135).

In such a perspective, two types of forces have been related together, namely: local resources, as well as the local rhetoric surrounding them, stressing the issues of self-perception and, even more, self-reflexivity and some development mechanisms initiated by some external forces. Such external forces might be perceived as stimulants for certain processes occurring within the local and extra-local network, as well as some controlling and regulating mirrors concerning all of the dynamics mentioned above.

From his perspective of neo-endogenous development, Ray (1999, 2006) presents some analytical frames that might be used in the regulation of such a very complex set of local and extra-local activities. He defines these frames as social economy, economic coordination, and multi-level governance. All of them are tightly connected to each other. However, it should be stressed that each of the elements can be considered in a more detailed way. The social economy issue has to be treated as a view of economic processes in the particular social, as well as cultural, context. Therefore, economic actors have to be contextualized and perceived as units following a substantial, rather than a formal (according to Weberian terms), type of rationality. As presented by Klekotko and other authors: ". . . all initiatives and plans related to economic changes or economic development must in all cases recognize the local context, the needs of residents, and also be able to predict possible effects of the activities taken" (Klekotko et al., 2018, pp. 135–136). Moreover, such a perspective has been supported by a special type of economic coordination. This particular approach has been focused on preventing all of the negative aspects of the neoliberal model of a market economy, namely the lack of solidarity, the degradation of local natural environments and cultures, as well as the lack of market exchanges in favor of local communities. Such a statement leads to a peculiar concept of human development that reaches beyond the one-dimensional as well as single-layer perspective on social changes and their mechanisms. Quite to the contrary, human development in such a context has been perceived as a multi-dimensional, as well as multi-layered, set of processes which result in a particular perspective that emphasizes the interests of local communities.

> Therefore, this territorial, and not a sectorial, aspect becomes a crucial component of the concept of neo-endogenous development. Multi-level governance means considering various actors, both local and extra-local, functioning within the network, where debates and decisions on making and implementing policies take place (Klekotko et al., 2018, p. 136).

The territorial component of human development mentioned above has been carefully examined under Ray's approach. Its complex character has been composed of three distinctive elements, namely, intra-territorial, territorial, and inter-territorial issues. It should be stressed that the political and administrative institutional framework seems to be especially important only in the second, namely, the territorial, element. As was highlighted earlier,

> The territorial perspective in the politico-administrative context indicates the need to analyze all local development initiatives as a chain of political and administrative

dependencies with the crucial role of the administration of nation-states and larger supranational structures (i.e., European Union) (Klekotko et al., 2018, p. 136).

That means a look into governance actors and relations that might be involved in particular developmental projects. However, two other elements—the intra- and inter-territorial—must be recognized as important accompanying frames. In the first one, types of activity as well as the social capital observed in the given territory (community, village, town, etc.) must be taken into consideration. In the second, the specific relations between particular territories have to be analyzed. In such a perspective, one should be focused on the flow of goods, consumers, values, and cultural content, as well as various matters that enable and facilitate the awareness and activities of participation in larger economic, social, and cultural units, such as regions, nation-states, unions (for example, the European Union), and global societies. Therefore, it might be said that, under the current regime of globalization, all individual and local types of thinking and activities must be connected to the larger relationships and networks forming the background of the contemporary society.

1.4.2. The Idea of Sustainable Development

The above reflection has been focused on Christopher Ray's concept, which could be considered a simple concept of human development and a starting point for further reflection, as well as a foundation for presenting a more complex conceptualization of sustainable development. It should be stressed that the concept of sustainable development might be treated as a certain explication of the ideas already presented by Ray, connected to issues broadly defined as the subject of human development, and not reduced to a simple concept of development as economic growth. It should be noted here that, in the present work, the concept of human development is prepossessed exactly as sustainable development and treated as an expanded version of the concept of neo-endogenous development[2] described above.

Sustainable development grows out from some form of confronting the human development concept as reduced to the mere idea of economic growth (Elliott, 2013). Such a perspective was reflected upon to its fullest by Walt Rostow (1960)

---

[2] It should be noted that before the emergence of the sustainable development concept and related matters, Polish academic literature was published advocating a development model consistent with ecological principles and enhancing social relations. In a sense, they were precursors to contemporary sustainable development issues (see Wierzbicki, 1973, 1986).

in the idea of growth stages. However, after twenty years, such a one-dimensional approach, focused solely on GDP increase, was strongly questioned. The challenging of this way of thinking first took form in the 1984 United Nations appointment of a team of 22 individuals from developed countries given the task to elaborate a long-term strategy of economic development that would also prevent degradation of the natural environment. The team was spearheaded by the Norwegian politician Gro Harlem Brundtland. The report by this group of experts was presented three years later, and it was the first time the term "sustainable development" was used. It was described as "development that meets the needs of the present without compromising the ability of future generations to meet their own needs" (WCED, 1987, p. 43).

It is worth highlighting that the idea of sustainable development discussed here came into being as a result of the discourse of world developmental perspectives and as a certain effect of discussion that mostly involved politicians, and not as a direct outcome of university seminars and academic research. Therefore, the meaningful steps in the development of this concept were closely connected with political events, such as "The Earth Summit" organized in 1992 in Rio de Janeiro, and "The World Summit on Sustainable Development" organized a decade later in 2002 in Johannesburg. The first event resulted in publication of the document entitled "Agenda 21," in which, in consideration of the approaching 21st century, attention was given to such factors as linking economic development with social changes; the necessity to maintain, protect, and appropriately manage natural resources; and the need to include underprivileged groups, like women, youth, indigenous people, etc., in the decision-making related to development processes. It also advocated connecting the implementation of development programs with education and the participation of stakeholders, and not on solutions imposed from above, outside of the context of local or regional communities. Even before the meeting in Johannesburg, the United Nations announced the 2000 manifesto, entitled Millennium Development Goals. It included eight crucial goals for the international community (Elliott, 2013, pp. 12–14):

1) eradication of extreme poverty and hunger,
2) achievement of universal primary education,
3) promotion of gender equality and empowerment of women,
4) reduction of child mortality,
5) improvement in maternal health,
6) combating HIV/AIDS, malaria and other diseases,
7) the active assurance of environmental sustainability, as well as
8) advancement of a global partnership for development.

The subsequent UN Summit was named "RIO +20" and took place in 2012. It served a role to summarize previous meetings, reflections, and disputes, leading to the all-encompassing term of sustainable development, understood as economic changes that are based in economic growth, but which also address the preservation of natural ecosystems and the eradication of poverty. The mechanism for such a development model is linked to the concept of multi-governance, wherein the dialogue between business, nation-state governments and non-governmental organizations takes place on both the global and local scales.

Ambitious goals, connected with the turn of the new millennium, were not only included in the official documents or declarations formulated during international summits and high-level conferences, but also in academic publications. Two of them should be given some attention here (Cheru & Bradford, 2005; Sachs, 2005). In the first, the authors focus on drafting certain principles of economic governance on the global scale. This requires some coordination of economic policies conducted by nation-states, as well as international institutions. The author of the second book enumerates necessary endeavors that could serve as tools for the eradication of poverty on the global scale. He calls for strengthening the role and improving the reputation of such international organizations as the United Nations, the International Monetary Fund, and the World Bank. What is also important in Sachs's view is the use of science and global democratization to include and give voice to countries where the issues of poverty and food shortages are significant social problems.

It is worth considering that around 2000 more academic publications devoted to the issues of sustainable development were published. The most important ones link the issues of development with globalization and the expectations formed by representatives of the 21st century (compare to Lee, Holland, & McNeill, 2000). Various authors stress the importance of the global character of natural environment issues. Minimizing the destruction of the natural environment appears to be an important goal of various subjects oriented toward the realization of the sustainable development idea. In this context, the problem of uneven development between the countries of the Global North and the Global South is not to be ignored. The Global North countries have already achieved such a level of development that the majority of people's needs are being met there. This was possible due to intensive economic development of the previous decades without significant attention being given to environmental and social aspects of economic processes. Therefore, the concept of global sustainable development should consider the different future prospects for affluent and non-affluent societies. It is also quite crucial to work up incrementally to such solutions to environmental causes and making sure that they would not lose in confrontation with the fulfilment of more direct and more ostensibly urgent needs of individuals and social communities. The reflection on sustainable development oftentimes goes even further, addressing the matters of traditional forms of ownership and freedom of expression and existence of various cultural identities, which should not be lost in the process of economic development within a market economy. A particular problem with measuring the level of advancement of sustainable development pertains mostly to highly developed countries. There are no suitable indicators of sustainability characteristics in the development processes. As Alan Holland (2000, p. 6) stresses: "This is a serious problem indeed, since the use of indicators is crucial to the effectiveness of the prevailing paradigm, and their absence provides governments with every excuse for backsliding." This matter will be further discussed later in this chapter.

Another issue worth further exploration is the problem of equity. It is noted in this context that the equity of free trade relations often leads to an imbalance of economic relations between the countries of the Global North and Global South, and an inequitable distribution of power and resources. There are also questions as to whether the economy—or to put it slightly differently, the perspective of economic rationality—is an appropriate context for consideration of the problems of environmental protection. For the purposes of this publication, this is a matter of fundamental importance, as imposing any logic other than the logic of economic rationality in regards to natural, environmental resources might be, in the view of the authors of this very work, a strategy leading to nowhere. It is only when the needs related to environmental protection are incorporated into the logic of economic activities that action strategies, oriented towards economic growth (albeit not as fast and spectacular as in the past), allow for achievement of environmental goals as well as, indirectly, social goals.

However, in the opinion of the authors of the present work, the need to consider economic rationality should not function inside the idea of sustainable development without the debate context, dialogue, confrontation of multiple viewpoints, and various kinds of knowledge. Eco-feminism may be a great example here, as the female perspective and experiences get included in the debate on goals and development methods. Similarly, various types of local and/or indigenous knowledge bring a solid contribution of local experience to developmental programs (Holland, 2000, p. 7). It is also necessary to include the environmental and social aspects of the sustainable development concept, including human population development and strengthening the role and meaning of local communities and, especially, community empowerment.

In the context of the above reflection one more issue deserves attention, and it has not yet been adequately articulated.[3] It should be remembered here that various characteristics of sustainable development often indicate the participation of many actors in mapping out developmental goals. As was already mentioned, the role of the representatives of local communities, women, and other underprivileged groups is particularly desired here as they should participate in the processes of formulating developmental goals or in the implementation of solutions leading to these goals. Therefore, this concept can be described as dialogical. Dialogical aspects are crucial to the intellectual approach known as communitarianism, which is often viewed as some kind of philosophy and/or political ideology. Such an approach puts an emphasis on the role of the local community in the society's functioning and social changes.

The renowned representative of the communitarian approach is Amitai Etzioni (2004, p. 23) who writes: "Community is to be considered as a good only when its social order is balanced with carefully laid protections of autonomy." The distinguished feature of the concept described here is indeed its dialogical character. Communitarians seek balance in all aspects of social life, trying to overcome the traditional way of thinking in the typical categories: I vs. we, individual vs. community, rights vs. duties and obligations, freedom vs. social order, individual rights and freedoms vs. social responsibility, etc. They point out the limitations of other approaches, such as liberalism or conservatism, which put emphasis on, respectively, the individual understood as absolute personal freedom or community and social order limiting the individual's freedom of conduct. The communitarian preference is given to a more middle of the road concept with an individual being strongly rooted in the community, taking from its heritage and having the abilities (and opportunities) to distance himself or herself from it or even question it (compare to Etzioni, 2018). It should be emphasized that, due to these characteristics, the concept of communitarianism corresponds well with the essential ideas and principles of sustainable development, which—until very recently—has not been adequately accentuated in the literature of social sciences.

It is worth noting that the concept of sustainable development has a multi-level character. It is not just some proposal within the theory of social change but also a certain strategy for implementing changes affecting certain fragments of social reality. Some authors (see, e.g., O'Riordan & Voisey, 1998, p. 279 and next) indicate

---

[3] The authors are grateful for these remarks made by our colleague Daria Łucka, Ph.D., from the Institute of Sociology of Jagiellonian University, whose research is devoted to matters of communitarianism.

that even the language of this concept seems to be obscure and contradictory. This means that even the theoretical layer of the sustainable development concept might provoke some doubts due to its lack of clarity in terminology. Various doubts also appear when its developmental strategies are formulated. The first kind of uncertainties might be connected with the construction of the budget and certain indicators meant to measure the progress of developmental processes. The uncertainties of the second kind relate to the various types of input in initiating and implementing the developmental processes. This applies not only to various academic disciplines that are practically applied in formulating developmental goals and finding ways to achieve them but also about ethical issues or concrete administrative solutions (i.e., addressing the ecological issues with tax policy) as well. The ecological issues, per se, comprise a crucial but not unambiguously perceived component of the sustainable development concept. The issues of tax policy are also closely related, as it is important whether to tax income (it being a dominant trend today) or to put emphasis on taxing consumption. The tax policy should influence the regulation of business practices in such a way as to foster eco-efficiency and clean technology, neither of which is unequivocally accepted today. The quoted authors are skeptical of further economic development according to the principles of sustainable development unless there is some significant mental breakthrough in entrepreneurial circles. As they write:

> Frankly, we do not feel a great optimism for significant development along the pathways for business and sustainability until the boardrooms contain genuine and knowledgeable advocates who can communicate in the language of "what is in it for me?" until the consumer openly demands product stewardship at the point of purchase, we are a long, long way from that. Yet, that is probably the best one can do in the controversial realm of sustainable consumption (O'Riordan & Voisey, 1998, p. 281).

Aside from these issues, and perhaps thanks to their existence, other authors try to construct an approach to sustainable development that concentrates on more explicit terms or specific quantitative indicators. These indicators must be, as is required by methodological tradition in social sciences,

> . . . specific (must clearly relate to outcomes); measurable (implies that there must be a quantitative indicator); usable (practical); sensitive (must readily change as circumstances change); available (it must be relatively straightforward to collect the necessary data for measurement); and cost-effective (it should not be a very expensive task to access the necessary data) (Bell & Morse, 2003, p. 31).

In the construction of indicators, the choice of the dimensions of social reality to which they apply to is obviously quite crucial. The authors mentioned above identify a list of such dimensions of tangible reality: "resources and waste, pollution, biodiversity, local needs, basic needs, satisfying work, health, access, living without fear, empowerment, culture and aesthetics" (Bell & Morse, 2003, pp. 37–38). To each of these dimensions, at least several specific indicators can be applied, and their integration becomes an essential problem. At this stage, researchers and experts who prepare the research material in the form of increasingly condensed data for policy-makers and managers set up the indicators related to particular directions of development. The measure of community or society moving in a certain direction which can be described as sustainable development is the improvement (i.e., approaching ideal values) of indicators, presenting a state of particular dimensions, characterizing such a community or society. Therefore, in this perspective, sustainable development is treated as a variable which may take different values. Consequently, sustainability may be described by the following value terms: sustainable, potentially sustainable, intermediate, potentially unsustainable, and unsustainable, which are the effects of an integrated influence of indicators related to human well-being, as well as ecosystem well-being (see Bell & Morse, 2003, p. 43).

It should be remembered that the presentation of some general processes through quantitative measurements might lead to simplification through superficial or only partial presentation of the described issues. As Stephen Morse stresses in his work (Morse, 2004, p. 183):

> We tend to see development indicators and indices as devices for "good" but they do reflect narrow visions of reality and there is a strong, and perhaps dangerously hidden, power axis at the heart of their creation and presentation to potential users. Making the index methodology transparent, while relatively easy to achieve, may not necessarily help to address this power differential. The problem is that it is all too easy to create development indices, and we can become blind to their limitations as they pander to a desire for quick fixes to complex problems. The world is complex and the human race faces many challenges. Of course, we all want to "do something" to help, but moving a country up the HDI [Human Development Index: K.G., M.K., A.J.W., G.F., D.Ł., P.N.] table, while being good politics, may not necessarily benefit most of the people who live there.

The authors of the present work maintain the position that sustainable development should be analyzed not necessarily by referring to the society, per se, but through concrete aspects of social life and social activities, among which agriculture is extremely important. Before discussing sustainability in agriculture, a recap of the matters discussed so far in this chapter is in order.

The idea of sustainable development has originally been associated with the natural environment and its use in such a way that does not lead to its devastation and the exhaustion of natural resources. As has been stressed in the history of the sustainable development idea, one might observe a continual broadening of this approach. Starting with the stress on the natural environment issue relating to the idea of ecological modernization, it was later supplemented with socioeconomic development, not merely reduced to issues of economic growth and, finally, it was accompanied by the concept of human development not leading to economic and social marginalization and exclusion. As has been emphasized in the publication by Klekotko and other authors, the literature on the subject of sustainable development presents three perspectives.

> Firstly, the natural environment is one of the three pillars of the development concept, the other two being economic and social factors. Secondly, the idea of sustainable development can be found within the array of issues concerning economic development, social development, and environmental protection. A focus on sustainability within the processes of change and transformation means that economic growth will be achieved with no significant destruction to the social tissue and no destruction of natural resources. Finally, there is the third approach to development, which considers environmental protection—understood as conservation of existing natural resources—as a determinant of sustainability. Economic and social changes are acceptable as long as they do not exceed a certain level of depletion of natural resources which are being utilized (Klekotko et al., 2018, pp. 132–133).

From such a perspective, the general idea of sustainable development seems to be a rather complex one. Therefore, in more recent literature, authors have been focused on more detailed and empirical indicators of sustainability. One such approach that seems to be quite convincing to the authors of this chapter has been formed as a kind of New Decalogue [Ten Commandments]. In our opinion, it has been formulated most clearly by Cavanagh and Mander (2004, pp. 79–102) in their well-known book. The so-called Decalogue contains the following principles: 1) the maintenance of participatory democracy (new democracy), 2) subsidiarity, 3) ecological sustainability, 4) common heritage, 5) preservation of diversity, 6) human rights, 7) job market protections and employment, 8) food security and safety, 9) equity and 10) the precautionary principle.

Participatory democracy has been focused, in this context, on the processes of cooperation among various actors. In turn, "subsidiarity" has been conceptualized here as a non-exploitive approach to local resources by external actors. Moreover, ecological stability has to be understood as a kind of "wise" consumption of resources that enables their regeneration. Therefore, such ". . . consumption should not go beyond abilities and capacities for renewal" (Klekotko et al., 2018, p. 133). It should also be remembered that preservation of common heritage cannot be reduced only to cultural issues, i.e., preservation of local culture, traditions, etc. The idea also contains knowledge as well as natural resources with a special emphasis on the issue of diversities in the cultural and biological senses. Human rights form another part of the discussed Decalogue. Quite contrary to the traditional (Western) perspective, they are not limited to political and civil rights. In the idea of sustainable development, they have been treated in a broader perspective, including also cultural, economic, and social rights, including access to safe food and water, among other matters. And last but not least, one must again stress the idea of equity:

> . . . contesting the effects of globalization leading to income disparities on a global scale and in particular societies. The disparities are not just between "rich" and "poor" but also between men and women; corporate mega-farms and relatively small family farms; ethnic groups that play a dominant role in given societies and immigrants, etc. (Klekotko et al., 2018, p. 134).

The precautionary principle has been linked to this issue as well, demanding that early evaluations of potential innovations are conducted in order to eliminate potential negative effects to the greatest extent possible. It is particularly important to avoid social marginalization and exclusion that could be the byproducts of even the most well-intentioned innovations.

Having a somewhat casual draft of the concept of sustainable development, it should be applied to the subject of changes taking place in agriculture and rural areas. There are at least two reasons to continue the reflection on sustainability focusing on agriculture. First, it should be emphasized that the concept of sustainable development is a general approach, which should lead to more detailed consideration of particular sets of problems and phenomena in various areas of social life. Second, the main interest of the analyses presented in this monograph are topics related to Polish rural areas and agriculture. Thus, the somewhat natural choice was made to continue the reflection on sustainable development in the area of social reality that pertains to agriculture and rural areas.

1.5. Sustainable Development in the Context of Rural and Agricultural Changes

The concept of sustainability in agriculture entails a balanced interplay of three components: economic, social, and environmental. Thus, sustainable agriculture is expected to be profitable for the farmer, beneficial to the local community, and lacking any threat of irreparable damages to the environment. Sustainable farming should generate decent income for farmers, provide the local community with high quality agricultural products, mostly foods, as well as maintain soil conservation and judicious use of water. This and the following chapters will also attempt to demonstrate the ways in which Polish family farmers fit this sustainability profile.

As Polish and European agriculture follow Common Agricultural Policy (CAP), the place of sustainability within this policy framework needs to be noted, as it is frequently used in documents pertaining to CAP. The Directorate-General for Agriculture and Rural Development of the European Commission, the body in charge of CAP on the EU level, has sustainable development in its mission statement. One may come to the conclusion that the concept of sustainability within Common Agriculture Policy is much appreciated at the moment. This was not always so. Established at the beginning of the 1960s, the Common Agricultural Policy was meant to provide food security for member states of the European Community. In post-war Europe, food and agricultural products were not taken for granted and familiarity with hunger was not unusual. Initially, European agriculture under CAP focused on high productivity and farm modernization, which are the essence of the agro-industrial model of agricultural development (Allen, 1993; Marsden, 2003; Buttel, 2006). Following this model, the European Community was able to successfully advance from being a food-insecure place to an agricultural superpower. This did not come without a price: the transformation of family farms into factory farms located in rural areas led to social and economic marginalization. As farms grew larger and more modernized, jobs became scarcer and the quality of life for rural residents decreased. The limits of the agro-industrial model of development became apparent and the quest for alternative models of rural development began.

Here, the work of Terry Marsden is worth considering, as his ideas of rural sustainable development are closely linked to resistance of the agro-industrial development concept. Marsden identifies and analyzes the pertinent development issues referring to two axes, namely the post-productivist dynamic and the rural sustainable dynamic (Marsden, 2003, pp. 91–255). From the perspective of the authors of the present work, the concept of the post-productivist dynamic could be seen as some kind of introduction to a certain level of general development leading to the level of sustainable development. It is crucial to understand here that rural space is not just an area of production but also a space of consumption. Another important aspect of this way of thinking is the realization that rural space becomes a diversified area. It is asserted that only part of the countryside becomes an enterprise for food production with the use of industrial production methods. The emergence of agri-environmental measures might be quite important in this context, as well as the presence of internal and external networks and social actors representing various, oftentimes contradictory, interests related to the processes of change in rural areas. In this sense, this particular fragment of Marsden's concept is, to some extent, comparable to Ray's previously discussed proposal of neo-endogenous development. However, the neo-endogenous development as presented by Ray was mostly based on the network of local (internal) actors, as well as actors whose activities went beyond local, thus making them external actors. Marsden goes further and adds the elements of territorial differentiation and the problems of agro-ecological production processes to this developmental perspective.

However, this is not the entirety of the concept elaborated by Terry Marsden. It should be remembered that this author also introduced the concept of agrarian rural development, primarily understood as ecological modernization (see Marsden, 2003, pp. 159–160) with its processes directly leading to sustainable society. As emphasized by Marsden (2003, p. 175),

> . . . parameters of agro-ecology begin to re-position agriculture as a key driver for achieving sustainable societies more generally. Rather than a further segregation of agro-food from rural society, agroecology proposes to reinforce interlinkages and foster the multifunctionality of farming systems.

This thesis asserts the connection between agriculture and rural areas in a very particular way. It occurs on the local level in the context of a particular rural community. To quote Marsden again (2003, p. 211),

> In particular, we can assess how struggles around new definitions of quality can empower local producers, such that they can develop sustained levels of value from production of food products. The development of new supply chain relationships is a key mode of rural development for achieving this; but it is one which suggests a heterogeneous and diverse set of rural development outcomes rather than a clearly generalizable model of spatial development.

His overall reflection indicates that the attempts to implement the sustainable development model should always be placed in a particular, local context. Therefore, sustainable development, in one way or another, should be antithetical to globalization. Social transformations are occurring—from this perspective—not in the area of a "flat world" but mostly in specific local contexts.

Sustainability has become a crucial issue in the development strategies of various sectors of the economy and social life. These strategies only allow transformation of the existing status quo if it does not destroy economic, social, and environmental resources of the areas where developments are being planned. The three facets of sustainability are well explained in the work of Robert Goodland (1995), who also described three types of sustainable development: social, economic, and ecological (environmental). For Goodland, social sustainable development was based on engagement of stakeholders in the strategy of transformation. In other words, people who live in the area where development is taking place need to have a say in its direction and participate in its implementation. Economic sustainable development means that capital is being preserved and consumption does not exceed the profits. The economic strategy (for sustainable development) here will not be expansive and the risk of losing initial capital must be low. In the third type of sustainable development, ecological, natural resources are being preserved as playing an important role in ensuring quality of life in certain areas. These three models of development are ostensibly intertwined—it seems that it would be difficult to find social sustainable development without sound economic sense and without a responsible approach to the environment. Therefore, for the purposes of this chapter, these three models can be lumped together into one three-dimensional model where the economy, social interactions, and environment co-exist within an optimal configuration.

Once this configuration is achieved, the system needs to aim to preserve it. Gorlach and Adamski (2007) see maintenance of the system as one of the three conditions that facilitate sustainable development when economic, social, and environmental aspects are already present. Maintenance of the system implies its flexibility. Sustainable development is achieved more easily in areas where various economic activities are being conducted and various forms of social engagement—such as associations, circles of interests, and NGOs—exist within the community. It allows for safe transformation of the entire system even if individual elements may not be in the best shape. For example, a loss of agricultural income for a local community can be compensated for by agro-tourism and sales of local artwork. Another example can be found in the case of natural disasters. Communities marked by high levels of social engagement are able to mobilize to organize relief efforts. Emphasis on maintenance of the system goes against all concepts of development that are one-dimensional and focused on one economic activity or one type of service. Sustainable rural development is not possible without farmers and taking agriculture out of rural areas and converting them into recreational parks is not sustainable. In Europe, but especially in Poland, sustainable rural development can only occur in areas where farmers have a strong presence in the countryside and in the community. Perhaps the requisites of sustainability are something Polish farmers follow intuitively without fully realizing it consciously, as they learned them years ago. It is possible that there is a significant overlap between sustainable farming practices and following traditional agricultural knowledge. Permanence, self-maintenance, and integration of internal and external resources are not new to Polish farmers who have mastered these skills in various circumstances that Polish agriculture has confronted throughout its history. The desire to have a farm, maintain it, and leave it to the next generation has described Polish farmers for centuries, even during Poland's partition years, throughout the socialist economy, and especially the last twenty years of the previous regime's upheaval and transformation. If such an historical context might be treated as a kind of "advantage of backwardness" the research might also be extended to this particular issue.

The "backwardness" mentioned above is often associated with "peasantness." It is the peasants who often serve as an example of farm managers operating in a not fully rational manner. They do not always follow the principle of formal rationality, which is based on the tally of incurred costs and expected benefits of each activity. It turns out, however, that peasantness can be an important factor characterizing the idea of sustainability, which places the problem of rational behaviors and actions in the context of not only direct profits, but also wider consequences related to preservation of the natural environmental resources, as well as safeguarding the social tissue from any dramatic increase in economic inequalities, as well as the accompanying social and political inequalities. This idea was mostly advanced in several works—important and influential for contemporary rural sociology—by the Dutch author Jan Douwe van der Ploeg.

The first of the van der Ploeg (2003) publications that are of interest to the authors of this chapter has a rather surprising title: "Past, present, and future of the Dutch peasantry." For someone who spent their professional life studying the process whereby Poland's peasants transitioned into farmers, his description of Dutch farmers, who represent the most modernized and commercialized type of agriculture in the contemporary world, as "peasantry" is somewhat eyebrow-raising. It is worth exploring why, and on what basis, van der Ploeg uses this term. One crucial explanatory element of his use of this term lies in van der Ploeg's interest in a rather specific approach to changes occurring in Dutch agriculture over the time. In his view, they come from certain decisions made by farmers (Ploeg, 2003, pp. 3–43). Therefore, the modernization process is not some homogenous scheme of transition from a traditional peasant society to one which is modernized and dominated by farmer-entrepreneurs. As the author emphasizes,

> Dutch farmers are, I will argue in this book, not so much the entrepreneurs they should be according to the agricultural expert system, but **peasants**: [underlined by: K.G., M.K., A.J.-W., G.F., D.Ł., P.N.] producers who, for the sake of their own survival, actively withdraw the processes of farm management and farm development from the logic of markets which seem to ignore their survival (Ploeg, 2003, p. 41).

It can be inferred here that the concept of farmer-entrepreneur is treated by this author as a "virtual farmer," meaning a certain model used mostly by economists describing the changes of agricultural modernization. To the contrary, van der Ploeg's proposal is a sociological concept which is not so concerned with understanding the processes of farmers adjusting to the model of modernized agriculture as created by experts, but with their motivations, decision mechanisms, and the consequences of these decisions within changing rural communities, creating social networks that undergo transformations and form the context for activities undertaken by particular farmers. These activities may lead to varied results, as they are the manifestations of agency. This agency is understood by the author as:

> . . . the capability to recognize, to utilize, to bridge, or to re- conceptualize discontinuities as essential demarcations. Insight into the interaction between intended plans, the interaction between presupposed positions, actions, reactions, outcomes, benefits, costs, and their allocation is decisive in this (Ploeg, 2003, p. 29).

Therefore, there is not only one globalized type of modernized agriculture as there never was only one type of traditional agriculture. There are numerous regional or local types of modern agriculture that often refer to regional or local traditions.

The 2008 publication by this author focuses on the characteristics of peasant agriculture in a broader perspective which goes beyond national categories (Ploeg, 2008). The reflection, although extensively devoted to matters of Dutch agriculture, also includes matters concerning farming in Latin America and some European countries as well (e.g., Italy). A new analytical perspective appears in this context and van der Ploeg names it "repeasantization." He treats the concept of repeasantization in the following manner:

... in essence, a modern expression of the *fight for autonomy and survival in a context of deprivation and dependency*. The peasant condition is definitely not static. It represents a flow through time, with upward as well as downward movements. Just as corporate farming is continuously evolving (expanding and simultaneously changing in a qualitative sense—that is, through a further industrialization of the processes of production and labor), as peasant farming is also changing. And one of the many changes is *repeasantization* (Ploeg, 2008, p. 7).

It is worth remembering that similar ideas laid the groundwork for this reflection, relating to the changing situation of Polish farmers, who were very attached to the peasant tradition and resistant to modernization. Importantly, modernization in the not-so-distant past came in the form of communist collectivization (Gorlach & Starosta, 2001).

With all the previous arguments made, the concept of repeasantization according to van der Ploeg deserves a closer look. In the broader sense, this concept pointed to some

... set of dialectical relations between the environment in which peasants have to operate and their actively constructed responses aimed at creating degrees of autonomy ... in order to deal with then patterns of dependency, deprivation and marginalization entailed in the environment (Ploeg, 2008, p. 261).

Farmers (new peasants) in the contemporary world have to confront neoliberal globalization which, following the footsteps of Noam Chomsky,[4] he calls the Empire. The Empire is ubiquitous around the world and its manifestations can be observed in various concrete forms in particular regions, societies, or even in local communities. The author provides various examples of the Empire's significance and states the following:

Empire is expressed as Parmalat, as the usurpation of water in Bajo Piura and as the tightening of the squeeze exerted on European agriculture. It is present in the Dutch Manure Law and in the associated "global cow." But Empire is also expressed in many forms not discussed in this book, such as the reduction of the world grain reserves, genetic manipulation, and in the curricula and research pursued in many agricultural schools and universities. Empire is, in short, multicentred (Ploeg, 2008, p. 262).

---

[4]   The authors refer here to the Spanish-language work by Chomsky (2005).

It can be inferred that the Empire simply stands for global capitalism. In other words, this globalized capitalism, which created rules mostly in economic life, also affects social life and the cultural realm, not to mention the rules of political life. Farmers who operate family farms must be able to navigate these rules. Often, they try to adjust to them and sometimes they make attempts to change them. There are also other action strategies that farmers follow in anticipation of some positive results. In other words, farmers struggle with these rules, fight with them and try to tame them, if and when possible. In this context, van der Ploeg defines contemporary farmers as peasants as they try to survive in a contemporary globalized society, functioning under the dictation of global capitalism's principles. Their efforts are quite similar to the ones their more traditional counterparts made in the past.

Its addition to the presentation of the repeasantization process, van der Ploeg refers to some traditional theoretical concepts, such as the theory of the peasant economy as formulated by Russian economist Alexander Chayanov (Ploeg, 2013). The peasant farm—in van der Ploeg's interpretation—is a farm unit oriented toward maintaining its balances in several crucial dimensions. These dimensions are represented by specific dual characteristics such as no wages and no capital balance, the labor-consumer balance, the balance of utility and drudgery, the balance between people and living nature, the balance of production and reproduction, the balance of internal and external relations, the balance of autonomy and dependence, and the balance of scale and integrity, all of which lead to the emergence of various farming styles. What seems important here is that the core of the functioning of a peasant farm, which can simply mean a family farm, are various decisions made by family farm members, all relating to their personal fate and destinies, as well as the destinies of the farm, which is closely tied to the household or sometimes even blended with it. It can be summed up as follows: "The peasant farm is the complex and dynamic outcome of the strategic deliberations and considerations of the farming family" (Ploeg, 2013, p. 69). Furthermore:

> The art of farming, that is, the deliberate and strategically grounded construction of a farm and the many elements that constitute it, does not separate the farm from its politico-economic environment. Part of the art of carefully equilibrating many of the balances involves taking into account the parameters, opportunities and threats coming from this environment. These threats, opportunities and parameters are not translated in a straightforward linear way into the farm. They are, instead, **always mediated by the farmer** [underlined by: K.G., M.K., A.J.W., G.F., D.Ł., P.N.] who considers different ups and downs. They are part of a balance that is equilibrated in a singular way by the farming family. Hence, general environmental tendencies will very often translate into

differentiated effects. The art of farming is intrinsically interwoven with the reproduction of heterogeneity. The more so since the resulting heterogeneity becomes part and parcel of the deliberations: it provokes debates (which practices perform better?) and might induce changes (when ruptures occur the most resilient practices might inspire others and thus become a beacon for more extensive transitions) (Ploeg, 2013, p. 70).

Therefore, from this perspective, peasant agriculture seems to be a factor that fosters sustainable development. The Dutch author quoted above identifies six characteristics of this type of agriculture which, in his view, are essential within this context. Farms functioning according to the model of a peasant economy contribute to economic growth, but they do so with relatively limited resources. The subsequent characteristics, on one hand, stress the importance of labor and capital, implying labor intensive agriculture yet, on the other hand, emphasize a certain combination of physical and mental qualities of the farmer. The next two characteristics allude to the future of farms. Whether the peasant model will be followed depends, on one hand, on the effects of the employed labor and, on the other, on market influences (Ploeg, 2013, pp. 71–73). It should be added that other authors, such as Strange (1988), Friedmann (1981), and Mooney (1988), pay attention to these aspects of farming that are characteristic of family farms of the peasant type.

The message that can be inferred from van der Ploeg's reflection is that family farms (including those in highly developed countries) functioning on the basis of peasant farm rules and principles make a certain front of allies aiming and acting towards the implementation of the sustainable development idea. The farms do not conduct grand-scale operations and intensive production which do not take into consideration limits of the natural environment and mostly focus on maximizing profits in the economic sense without paying any attention to the social consequences of this process. In this approach, the decision-making processes of the farm user, who is rooted in certain family relations and conditions, is quite important. This fosters the need of such farms to diversify their portfolio, to a greater or lesser extent, in order to fulfill the criteria of sustainable development and follow its ideas. Whether these criteria are fulfilled is—as understood in van der Ploeg's research—a matter of empirical examination in each case. And such examination was among the goals of the research project presented in this very publication.

Besides "peasantization" the discourse described above included other concepts connected with the ideas of sustainable agriculture. It is worth emphasizing that the reflection on the non-industrial, or post-productivist, model of agriculture also entails the concept of multifunctionality. The analyses regarding this matter were made on the basis of a farm survey conducted in Italy (795 cases) which led to the conclusion that multifunctionality is a strategy for large and small farms alike. For both types of farms, multifunctionality can provide a farm support in terms of supplementing the income that comes exclusively from agricultural production. As the authors noticed, "There are ... many indications that multifunctional activities provide a 'life-jacket' that helps to strengthen (through additional incomes, savings, more positive expectations, etc.) the agricultural part of the multifunctional enterprise" (Rooij et al., 2014, p. 316).

Yet multifunctionality is not the only term that, in the agricultural context, gets associated with the idea of sustainable agriculture. Local food chains and organic production comprise other important matters. As indicated by the authors who analyzed this aspect of agricultural production:

> Indeed, these platforms [organic and local production: K.G., M.K., A.J.S., G.F., D.Ł., P.N.] were created to compete with (multi)national food service providers and those who chose to participate are more or less engaged in this David-versus-Goliath struggle of "small," "local" organic networks against the "big unscrupulous companies" that rely on global, depersonalized food markets (Rogers & Fraszczak, 2014, p. 337).

Another issue that relates to the concept of sustainable development is tacit knowledge. This line of thinking has a rather long tradition and the 2009 Bruckmeimer and Tovey publication describes this matter quite well. The book analyzes three types of knowledge and links them to the concept of sustainable development in a variety of ways. The described types of knowledge are the following: "expert knowledge," "managerial knowledge," as well as "local"/"lay"/"tacit" knowledge. The message of the editors of the above volume presents the last type of knowledge as that which is closest to the sustainable development concept. What is crucial here is the generalized local experience that this type of knowledge brings to light. If a development program is to consider using local resources, it must also combine local knowledge with expert and managerial types of knowledge. The editors of the discussed volume stress the following:

> Projects for rural sustainable development are likely to be successful, then, when they develop informal network mechanisms for territorial integration of actors and knowledge, and, moreover, when they dare to address the difficulties of bringing together and combining expert and lay knowledges (Bruckmeier & Tovey, 2009, p. 279).

This corresponds well with the general mission statement of this publication's core project: think locally, act globally. These global (supra-local) actions mean using both expert and managerial knowledge which, in one way or another, must be linked to local knowledge and local ideas. As argued by Curry and Kirwan (2014: p. 356): "If sustainable agriculture is to embrace the social, cultural and environmental fully, tacit knowledge is likely to have a clear role to play in shaping courses of action."

Here, it might be worthwhile to go back to the concept of repeasantization described earlier. According to some authors, this theoretical perspective resonates well with other perspectives focusing on more diverse and more ecological activities and operations conducted by farmers. It applies even to agriculture in the United States (Nelson & Stock, 2018) which, in the popular imagination, is a prime example of industrial agriculture, built upon the idea of economies of scale. However, ecological tendencies and diversification of farm economic activities to supplement agricultural profits are currently marking their presence in the American agricultural sector. As Nelson and Stock put it:

> Specifically, we argue that an examination of entrepreneurial farmers, not peasant farmers, may be more useful in understanding how sustainable practices may emerge in seemingly hostile settings . . . Here we focus on farmers caught in between this "alternative" sphere and the conventional, industrialized model in Kansas, who nevertheless identify some of the practices associated with van der Ploeg's peasant principles as important to them and thereby contribute to repeasantization in the USA . . . (Nelson & Stock, 2018, p. 84).

The same authors add, ". . . our conclusion argues for the usefulness of the repeasantization thesis to understanding the negotiation and transformation of practices in highly industrialized and neoliberalized spaces" (Nelson & Stock, 2018, p. 84). Repeasantization is therefore understood here as resistance towards neoliberal tendencies and enrichment of farmers' activities with issues related to protection of the natural and social environment. Repeasantization is, to some extent, treated as the preservation of a certain autonomy of farms, as well as the diversification of production and, oftentimes, emergence of the process of co-production. Autonomy in this context means maintaining responsibility and agency in regard to farms, which are the qualities prone to erosion due to the dominant forces of global and corporate capitalism. Co-production, in essence, is a gentler treatment of natural resources, which—contrary to the neoliberal approach to production matters—are not infinite and cannot be treated just like any other commodities. Furthermore, diversification entails farmers' engagement in various economic activities, including non-agricultural ones which, in return, could benefit the overall conditions of the farms, thanks to farmers' additional incomes. Importantly, such reflection presents repeasantization as not some return to the past or an attempt to thwart the ongoing social and economic changes. As clarified by the authors, "The opposite of capitalist agriculture is not romantic agrarianism, but a pragmatic, peasant-inspired concern for the resources available and an appreciation of what might be possible and emergent" (Nelson & Stock, 2018, p. 98). Here, the intergenerational transfer of farms is very important. As indicated by Chiswell (2018, p. 122):

> Intergenerational farm transfer is of international significance and its prominence alone warrants continued research effort. Furthermore . . . prospective successors are the farmers of tomorrow; given the numerous demands on the agricultural industry, it is imperative that we understand how the patterns of transfer in place now could influence the sustainability and resilience of family farming systems, as well as the ability of the industry to respond to the challenges of the future.

Sustainability is closely linked to several other matters that have impact on the functioning of agriculture, and climate challenges are among them (Campbell, Singh, & Sharp, 2016; Meyer, 2016), as well as the transformations of local communities that stem from them. Similar observations can be made about the new characteristics of the contemporary agrarian question, including the functioning of family farms (Brunori & Bartolini, 2016), fisheries (Symes & Phillipson, 2016), institutional capacity and social dynamics (Schaft, 2016), literacy (Donehower & Green, 2016), manufacturing (Rosenfeld & Wojan, 2016), and the contemporary form of rural-urban relations (Noguera & Freshwater, 2016). The role of innovation is particularly meaningful in this context. Innovations fostering sustainability become the voice of resistance against the dominant market economy based on neoliberal ideology. To quote the British author Damian Maye, "Within agriculture the pressure to innovate is in response to the challenge to increase food production sustainably" (Maye, 2018, p. 332).

All of the above issues obviously apply to Polish agriculture, as well. According to Polish authors, the idea of sustainable agriculture plays a tremendous role in agriculture, as the system of agricultural production (including technology of production) determines the quality of agricultural products, as well as their nutritional value and safety. Agriculture has an impact on the natural environment, especially on soil conditions and its physical qualities, but also on water and air quality. For that reason, the evaluation of agricultural economy cannot just be narrowed to "economic categories" (Wrzaszcz, 2012, p. 211). Here, it is worth noting that, regarding the concept of food safety, which is part of the sustainable development approach to agriculture, a lot of attention is being given to biological and chemical contamination (Gulbicka, 2012). Another matter, quite important to implementation of the idea of sustainable development, is connected to the economic development of farms which prioritize the values of environmental protection (compare to Bołtro-miuk, 2011).

1.6. Broader Perspective of Sustainable Development

However, we are also aware that the idea of sustainability has been strongly criticized in the new body of literature focused on resiliency. We agree with this new approach that stresses the static character of sustainability. However, we are also aware of some shortcomings of this new approach since it has been focused mainly on the ecological dimension of sustainability. We agree with Almas and Campbell, who warn that ". . . ecological sustainability cannot be reduced to a static set of systemic conditions called 'sustainability'" (Almas & Campbell, 2012, p. 7; see also: Gunderson, 2000; Gunderson & Holling, 2002). The authors suggest that sustainability emerges from the resilience processes in socio-ecological systems. Clearly, adaptation and change form the basis of these mechanisms that help socio-ecological systems to survive over time (Folke, 2006). Considering the issue of resilience in a more detailed way, Almas and Campbell theorize about the importance of the ability of both natural and socio-ecological systems ". . . to withstand shocks to their function and integrity" (Almas & Campbell, 2012, p. 7; see also: Walker & Salt, 2006; Darnhofer, Faithweather, & Moller, 2010). In this perspective, the concept of resilience leads the ideas concerning society and agriculture back to the frame of systems, functions, integrity, etc., that formed the dominant part of sociological discourse in the earlier period of the second half of the 20th century, under the structural-functionalist framework. In the opinion of the authors of this chapter, the resilience framework seems to be a kind of return to this rather old-fashioned and strongly criticized aspect of social theory. Moreover, it mainly stresses the link between society and nature (the eco-sociological approach) that definitely misses the most important characteristics of a sustainability approach based on the economy, society, and nature triangle. Therefore, in our view, the skepticism about this systemic approach seems quite justified. The authors of this chapter are much more in favor of the agency approach (bringing individuals back into the idea) that helps to see society with its functioning and changes "from below," i.e., through particular agents (in this case, farm operators) and their farming styles. Sustainability has been perceived in that sense not as a "static set of systemic conditions" but, quite to the contrary, as kinds of mindsets constructed by farmers which then become integral parts of their farming styles, i.e., their strategies for running farms.

What can be expected from a farmer whose way of farming we perceive as sustainable? We will argue for a "sustainability and resilience mindset," which such farmers may have, but may not necessarily recognize as such. Nevertheless, his or her way of doing things indicates a striving for sustainability. This may not be a satisfying definition of "sustainability mindset," but rural sociologists have already conceded that the whole concept of sustainability is rather vague. O'Riordan and Voisey (1998) argue that sustainability is an ideal that cannot be fully achieved or even clearly envisioned, although people are obliged to pursue it as both societal quest and personal exploration. If they don't attain it, this is because the personal exploration is not followed by political behavior. We want to explore whether "sustainability and resilience mindset" on the level of particular farms might be converted into a more economically viable, socially just, and environmentally sound farm system in Poland. Sustainable development is a type of development that meets the needs of the present without compromising the ability of future generations to meet their own needs. In the agricultural sector "future generation" takes a very concrete form. In many European countries, family farmers leave their farms to their children when they retire, although getting an appropriately reciprocal commitment from their sons and daughters is becoming increasingly difficult. As it is a daunting task, farmers only reluctantly allow farm sales outside of their families. This is indicative of family farmers thinking long-term about their agricultural holdings as they want to pass them on to the next generation in the best possible condition. This is still very much the case with Polish family farms. We hope to find a significant link between expected transfer of the farm to children and sustainable farming. While asking explicitly about "sustainability mindset" may not produce adequately detailed answers (as subjects may be intimidated by such academic concepts), we will ask indirect questions about certain aspects of sustainability. The researcher will ask such questions only after gaining the trust of the farmer. The questions should be incorporated into a genial conversation about agriculture, avoiding sociological theories.

It can be argued that some mechanisms of human development presented above, as well as neo-endogenous development, might be perceived as contradicting the well-known, ubiquitously employed maxim to "think globally and act locally." The traditional perception of this idea has been based on the assumption that one has to fix in localities some effects of global changes. The risk of such a perspective is that, in fact, one must consider some kinds of local versions of global solutions. Therefore, localities might be perceived only as suppressed levels of social order simply responding to the effects of global issues. However, in this book we want to present a perspective that fully contradicts such reasoning. This particular type of reasoning has been based exactly on the Decalogue of sustainable development mentioned earlier in this chapter, as well as neo-endogenous development. As stated by Klekotko and other authors,

> The idea of neo-endogenous development appreciates the use of local resources and co-operation with extra-local actors, while the idea of sustainable development emphasizes preservation of local distinctiveness during the process of broad participation from various actors adhering to the principles of subsidiarity and precaution in assessing possible negative phenomena. Both of them prove the usefulness of the thesis about the need to supplement thinking locally with activities oriented towards what is extra-local and even global. Only this combination of thought and action is a key to success. It is, in our view, the right time to introduce the principle of "think locally, act globally" to the discourse on the principles and processes of development in contemporary globalized society (Klekotko et al., 2018, p. 138).

Such a perspective has been framed as an idea of analyzing farmers' strategies that are perceived in this book as attempts to act globally (or at least extra-locally) using some local resources located on farms and in their neighborhoods. Stephen Paul Haigh stresses: "Originality is a rare thing, and to presume too much originality . . . would be to speak out of turn" (Haigh, 2013, p. 197). Following this statement, we would like to shed new light onto patterns of farms functioning in contemporary Poland. The new light mentioned here is two-fold. It is focused on the idea of acting globally in the framework of local thinking and, at the same time, focusing on farmers and their role in the current development of rural areas. The title of the following proposal is purposely tricky. At first glance, it may suggest a misquotation of the popular slogan used by environmentalists to present the essence and logic of how the globalized world ought to function. As already mentioned in this chapter, Julian Amey (2013) states the following:

> The phrase "Think Global, Act Local" was first used in the context of environmental challenges—in order to improve our impact on the environment; it is more effective for an individual to reduce their own energy consumption than wait for global action. What about the context of big business? Big companies such as McDonalds and Honda are successful on a global scale, while their products were tailored to the requirements of individual countries.

Further in his article, the author suggests that the phrase "think globally and act locally" needs to be supplemented with "share your knowledge and be part of a network" for an adequate consideration of the larger picture. The slogan—to put it simply—means that a real contribution to the life of a particular place and community defined as local can be achieved by taking up issues that connect with global matters. As we stressed earlier in this chapter, these issues are manifestations of more general problems, characteristic of various societies all over the world. Let us take a closer look at the example of large transnational corporations entering local markets, as mentioned by Amey. Local action plans of these companies are nothing short of selling their "global" products in "packaging" that may appeal to preferences of local consumers or fulfill their needs as shaped by local conditions and traditions. The other example alluding to environmental protection indicates that local communities are expected to forego the fulfilment of their needs without any reassurances that global action will be taken on their behalf.

In the following chapter, we would like to reflect thoroughly on the relationship between local and global in reference to the statement "think globally, act locally." In our opinion this statement contains the hidden assumption of the domination of the global over the local. What is local presents itself as an implementation of a certain global project, only in some ways tailored to the local needs. Here, critics would say that it is done in a very questionable way, if done at all. We agree with this criticism. However, in our analysis we would like to go a little further and propose a reversed way of thinking through a play on words: "think locally, act globally." Justification of this message requires a closer look at the intricate relations between the local and the global and the realization that a complex development mechanism constitutes the core of the globalized world. In order to adequately present this mechanism, it is necessary to capture the new way of viewing the local in a society exposed to the globalization processes (Ball, 2004, p. 36). However, the globalization of farm practices brings some risks, and has an impact on some economic, as well as psycho-physical, conditions of farmers and their families (Beck, 2002; Bauman, 2006, 2007a, 2007b; Beck, Giddens, & Lash, 2009; Giddens, 2009).

1.7. Bringing Farmers Back In

Let us start with the following statement: "If a serious effort is made to construct theories that will even begin to explain social phenomena, it turns out that their general propositions are not about the equilibrium of societies but about the behavior of men" (Homans, 1964, p. 818). We will try to incorporate this message of one of the greatest sociologists into our project as such: sociologists should focus on particular individuals, studying their activities in order to describe and explain what is going on within societies. Just as Homans tried to bring men back into sociological analysis nearly fifty years ago, challenging some dominant tendencies at that time, we will attempt to do the same, that is, by bringing farmers back into the sociological analysis—quite contrary to dominant tendencies in contemporary rural studies. We have found that this particular social group, namely farmers, have been marginalized for the last twenty years. They have become a rather rare and minor object of interest among rural sociologists. What might seem to be a rather controversial statement can be supported even by a brief perusal of publications that dominate the field of rural sociology focused on advanced countries of Western Europe and North America. The discipline has been linked to the idea of de-agrarization of rural areas resulting from the intensity of agricultural production and emergence of the industrial model of food production. Processes of production concentration, the enlargement of farms, as well as the declining number of farms and farmers have been observed as well. In such a context, farmers have become a minority among rural inhabitants, and therefore are perceived and conceptualized as less significant and less interesting by rural sociologists.

The tendency to marginalize farmers within rural sociology is especially evident when browsing the "Handbook of Rural Studies" composed by a group of almost exclusively British and American authors (Cloke, Marsden, & Mooney, 2006). The 35-chapter volume has set the standard for much of the panorama of approaches to rural social matters and they are not necessarily sociological approaches. Among the identified approaches are: the sociological analysis of rural knowledge, some types of conceptualization of the rural, problems of rural resource governance, investigations of rural space, research concerning rural society, as well as rural economies, rural policies, and planning. The authors have also tried to point out different theoretical coordinates, which are, in fact, key perspectives on particular research problems. Among these coordinates one can find the issues of cultural representations, nature, sustainability, new economies, power, new consumerism, identity, as well as exclusion. According to some authors (Bell, 2006; DuPuis, 2006; Short, 2006), cultural representations form the most important perspective in rural sociological research. Based on the idea of the cultural turn (Ray & Sayer, 1999), they stress the traditional duality of nature and culture, while on the other hand, claim cultural representation of nature is an effect of social construction based on certain "desires." Such desires have been based on the romantic nostalgia for a kind of passing ideal nature (DuPuis, 2006, p. 125). However, it is important that the desires mentioned above form a tendency to create a "false" or, at least, one-sided image of rural reality and/or rural community that has some strong connection to the ideas constructed and shared by the middle class. It has been based on the idea of rural areas as a space of consumption (Marsden 1999, 2003) and not production, an area of leisure rather than hard work and the symbol of a desirable place to live. Nevertheless, the precise content of such an "idyll" by itself might be the issue of debates and controversies, especially because its cultural dimension—relating to harmony, safety, stability, internal power, freshness, and renewal—has been changed in the processes of modernization. It was modernization, including industrialization and urbanization that had an impact on the image of the rural idyll, contrasting it against urban modernity. Summing up, the issue of cultural representation of rurality (i.e., the rural) has been perceived in literature today as a complex discourse, composed of various issues and different social actors but, at the same time, marginalizing farmers and their ideas about the rural. Other authors (Castree & Braun, 2006; Jones, 2006; Murdoch, 2006) claim that nature, and not culture, has become an important object of rural studies.

This new approach has been based on the assumption that the traditional ontological separation between the world of human actors and the world of non-humans should be rejected. Most recent constructionist theories (Castree & Braun, 2006) have been focused on the conditions and contexts in which all the organisms that inhabit the rural world exist. The paradox of this perspective, however, seems to lie in constructivism's disintegration, which might be divided into three parts, namely: material (nature as a physical domain), discursive (ideas, representations and images of nature), as well as material-semiotic (various schemes concerning materiality and discourse). In such a way, one can speak about the particular de-naturalization of nature which becomes a social construct, not a "real" social reality, but as an element of socio-natural relations among various human and non-human actors. This is especially highlighted in the Actor-Network-Theory (Latour, 1993) that demystifies the privileged role of human actors. However, this theory has also been criticized as putting too much stress on networks and too little attention on terrains and places; as well as issues of power, inequality, and domination (Jones, 2006, pp. 187–188). Moreover, the network (relational) approach to rurality might be observed as well in a slightly different perspective than that which it has been based, not on the idea of relations between humans and non-humans, but rather on the idea of a heterogenic network of constellations that replace stable socio-economic structures in a contemporary society. In this particular perspective, the network has been conceptualized not as categories of relations but rather as a system of flows of resources (capital, money, commodities, labor, information, and images) between points located on various levels, namely: global, national, regional, and local. This perspective stresses the main effect of this system of flows concerning various trajectories of development of particular rural communities, regions, and areas (Murdoch, 2006, p. 172).

In such a context, a growing diversification of rural areas also results from different interests presented by various actors trying to dominate particular rural areas and communities in the case of the so-called contested countryside (Marsden, 2003). However, what is important from the perspective of our research plan is that the growing diversification of rural areas requires a redefining of the roles of farmers. In turn, such a re-definition requires more research focused on farmers and their perceptions of this changing situation. Therefore, we concentrate our research on farm operators embedded in a particular national milieu (in Poland) and only starting from that level can we later take into consideration some other problems focused on links between farmers and external actors and processes via particular issues, such as information technology (IT), innovations, and food security. The more detailed information on the rationale for this project should now be presented.

For many researchers, studying the economy in contemporary rural studies does not mean studying farmers and agriculture. In fact, farmers seem to be a minor or trivial object of studies. This area of interest has been broadened or enlarged into at least two major fields: commodification processes (Ritzer, 1996; Perkins, 2006; Pakulski, 2009) and mechanisms of development (Fukuyama, 1995; Marini & Mooney, 2006; Ray, 2006). In other words, this area of study might be called "rural as commodity" or "rural as cultural representation" and "rural as network." Therefore, rural economies are now based not only on agriculture but also on values of local resources and amenities. Such commodification leads to the development of different rural economies. Some authors (Marini & Mooney, 2006, pp. 96–99) enumerate three types: the rent-seeking economy, the dependent economy, and the entrepreneurial one. The authors refer to many types of economic activities, but agriculture is not perceived as being a significant one. In contrast, we argue that agriculture as an economic activity, as well as farmers as agents performing such activity, should be taken into account when considering the emergence and function of the rural economy types mentioned. We might hypothesize about dependent agriculture and dependent farmers, as well as entrepreneurial agriculture and entrepreneurial farmers, along with treating agriculture and farmers as a part of the rent-seeking system. The most important goal of this project is to recognize various farming styles that might lead to construction of various types of farmers and agricultural activity. Moreover, considering rural economies as various systems operating in different contexts and under the pressure of different factors, one should delineate various mechanisms of development. Some authors stress the role of mechanisms that have to work together in order to maintain the process of economic and social development. As has been stressed above, such a mechanism has been called a neo-endogenous one (Ray, 2006, pp. 278–291). Neo-endogenous development contains both traditional sub-mechanisms, perceived as endogenous, as well as exogenous ones. Our initial project frame, namely the idea of "think locally, act globally," views farms and farm operators from the perspective of neo-endogenous development. The nature of neo-endogenous development might be explained briefly as follows. Any process of change starts with the recognition of particular local amenities, advantages, etc. Based on that recognition, actors involved have to possess external resources (material, symbolic, social) through extra-local networks in order to converse initially recognized amenities into goods that allow changing the situation of the local. Therefore, the neo-endogenous development might be treated as the basic mechanism of a "think locally, act globally" perspective on social transformation. Such a perspective is deeply connected with the issue of globalization. Non-local actors and non-local forces taking part in the process of neo-endogenous development can be perceived as actors of globalization. In rural sociological research, authors (Lyson, 2006) focus on issues of power, domination, and regulation that might be treated as types of globalization impacts on rural communities. Such impacts seem to be predominantly negative since they produce negative effects at the economic and political levels, as well as the social life of rural communities, including effects of a neoliberal economy, economic efficiency, as well as productivity, economic growth, and orientation towards mass production and consumption. Other authors (Goodman & Watts, 1997; Goodwin, 2006) broaden this perspective by focusing on regulatory processes and state actions. Such a conceptualization is supported by analyses of the nation-state and its policies aimed at rural and agricultural interest groups (including farmers but also transnational corporations) that also use the nation-state as a platform to put forward their particular interests (Sheingate, 2001). Importantly, pressure from inter- and transnational forces might result in the establishment of regulatory regimes that affect regional and local levels (Bonnano, 2006, p. 326). Thus, two issues should be taken into consideration: the specific role of farmers that have been put under international (global) pressure, and their ability to form new constellations of networks and interests in order to defend themselves against globalization forces.

Concerning mechanisms of rural development, we are also interested in the transfer of IT (information technology) into farms and, more generally, the agricultural and food system. In the literature, two roles of IT in agriculture have been considered (World Bank, 2006): as a tool of productivity and to assist in the decision-making processes. Such studies have also been conducted in Poland (Cupiał, 2006, 2008). However, these analyses are focused only on technical aspects of software and its application in farm operations. Quite to the contrary, we are interested in social aspects of the issue, that is, the attitudes of farmers towards computerization of their businesses, as well as relations between farm operators, computers, software, and software dealers as inspired by the Actor-Network-Theory (Latour, 2010; Noe & Alroe, 2012). Usage of IT is an important indicator of extra-local inferences occurring. Rural areas might also be studied through at least three key concepts in contemporary sociological literature: consumption, identity, and the problem of exclusion. It should be stressed that in each of these perspectives, farmers have not been recognized as basic actors nor have farmers' activities been a main area of consideration (Urry, 1995; Salamon, 2006). Farmers are perceived as those in conflict with new groups in rural areas. Such actors, especially those who come to rural areas to consume rural resources or practice this consumption beyond rural areas (Miele, 2006; Fonte & Papadopoulos, 2010), seem to be the object of primary interest among rural sociologists. But the issue of consumption is also connected to problems such as food security and food safety frameworks. In order to stress the role of farmers, we plan to analyze their role in the food security system. In the contemporary literature on this subject, the role of farmers is perceived as a marginal one, which is rather surprising. Farmers are treated merely as producers that should supply consumers with high quality and healthy food.

Similar observation about farmers' marginalization can be observed when food safety and its three primary dimensions (biological, chemical, and physical) are being discussed (Lawley, Curtis, & Davis, 2008; Kołożyn-Krajewska & Sikora, 2010; Knechtges, 2011; Mikuła, 2012; Pałasiński & Juszczak, 2014). Farmers ought to identify risks in processes of food production, they should apply rules of quality management, and they should apply for certification of their products. The issue of food safety/security can be perceived in a broader manner, as a legal system consisting of various actors (Leśkiewicz, 2012) and as a peculiar type of multi-actor discourse, concerning starvation and malnutrition, the creation of local and global food chains, and the various risks connected with systems of production and distribution of food (Mooney & Hunt, 2009). Again, there has been no place in literature for consideration of farmers' role. We intend to overcome these analytical and theoretical shortcomings. We plan to study farmers' knowledge and opinions concerning food security and food safety issues. Moreover, we will prepare innovative analyses of legal and technological aspects of food security/safety, but with a focus on farmers' roles in them.

Similar reservations should be formed in the case of identity and exclusion studies. In the former, the issues of "gender" as well as "otherness" seem to form a core of recent rural studies (Little, 2002, 2006). Farmers seem to be almost completely excluded from identity studies. Therefore, our research project will tackle this particular issue, as well. The same concern can be applied to the case of exclusion (Sibley, 2006). Both ideas of identity and exclusion are connected in the notion of class position and we build on the base of theories of Weber (1978), Bourdieu (2005), and Wright (1997). We strive to describe the intricate connections between access to/exclusion from certain forms of capital and the ideal representations of positions within the rural economic field. This leads to reconstruction of the main factions within farm operators' community and their respective farming strategies.

The considerations presented above lead us to stress the innovative character of the project in its triple meaning. The first frame refers to the idea of "thinking locally, acting globally." It is a challenge to the dominant way of reasoning according to the idea of "thinking globally, acting locally," which suggests treating local phenomena as particular representations of global universalities. The second frame is based on the idea of farmers as the most important actors in processes of rural development—contrary to the dominant trend in rural studies. Their actions and attitudes should be carefully studied with use of the most recent theoretical innovations in agency, culture, and interest. In other words, we want to put farmers at the center of our research. Finally, the third frame has been based on the importance of IT innovations and food safety/security issues in farm performances. This, too, is quite counter to recent tendencies in scientific literature.

The second frame ("bringing farmers in") is the most important for further presentation of our project. As presented above, the basic goal of our project lies in an effort to counteract trends visible in current rural studies which we consider deficient. The clearest example of this unsatisfactory approach might be seen in the study by Philip Lowe (2010), wherein the author discusses major development processes, as well as relations between American and so-called "European" rural sociologies, limiting the latter to situations in Northern European countries. Quite interestingly, even in studies of agricultural globalization and sustainability, the authors seem to overlook Eastern and Central Europe. In a powerful study concerning "new peasantries" of the globalized world, one of the most outstanding contemporary rural sociologists pays attention mostly to Western Europe (as well as parts of Latin America) (Ploeg, 2008). We argue for shifting attention to "the other Europe," that means countries which have joined the European Union only recently (Gorlach, 2006). Poland has been a part of this group and we want to focus our research specifically on this particular case. The idea to limit our proposal only to this case requires an argument about Poland's peculiarity in this realm. One might recall the famous book by Theodore Shanin (1986) on the "roots of Russia's otherness." Shanin examines the peculiarity of social, economic, and political processes in Russian history that resulted not only in its backwardness compared to the development of the West, but also in the different character of its modernization process, its social structure and political institutions. Following the same pattern of sociological argument, we might claim an Eastern European otherness, especially in contrast to Western European societies and political institutions. Some sociologists claim that the major factors of Eastern European backwardness and otherness are rooted in the very deep past (Chirot, 1989, p. 3). The form by which Poland was absorbed into the world market system a few centuries ago consolidated its peripheral status. As Wallerstein (1974) noted, the marginal status of Poland in the 16th and 17th centuries might be the cause of its peculiar economic development, especially its agrarian system. The rising demand for wheat in the West forced Polish nobles to increase their exploitation of peasants. Many historians claim that Poland's partitioning at the end of the 18th century resulted in a lack of civic participation among peasants, who comprised the most populous social category at the time. The legacy of this partition period can be seen rather clearly in the contemporary government's demarcation of three macro-regions for Polish agriculture (Gorlach & Mooney, 2004). Some authors have noted this impact as an absence of civility in society, while other authors refer to it as an absence of social capital. Chloupkova, Svendsen and Svendsen (2003) argue that the vacuum in civil society was a product of Communism's effective destruction of social capital. They point to the communist regime's destruction of a once vibrant and well-rooted Polish cooperative movement as a key mechanism in the creation of contemporary low levels of social capital, public trust, and secondary associations in rural Poland. In the analysis of modernization in Eastern and Central Europe, one should take into consideration several important factors (Konrad & Szelenyi, 1979; Kochanowicz, 1992), including the role of the state in preservation of the peasant tradition in the modern era. By the end of World War II, Eastern Europe had still gone without a major modernization and transformation phase. The second half of the 20th century, however, brought to the region the modernization experienced within the communist regimes. The peculiarity of agricultural modernization is of the greatest importance for our proposal. Clearly, it was not the farmerization experienced by Western Europe in the late 19th and early 20th centuries which sped up under the productivist approach of Common Agricultural Policy and/or protectionist agricultural welfare state policies. Eastern Europe's de-peasantization took the shape of the communist collectivization policy. Peasants were transformed not by regulations and market forces, but by political and administrative pressure. They were transformed not into farmers, but into agricultural workers under either the direct control of state bureaucracy or indirect control of the collective self-government that was, in turn, also controlled by bureaucracy. This was (beside some minor differences [Swain, 1985]) a clear reflection of the Soviet pattern of collectivization undertaken in the late 1920s and early 1930s).

Poland is an exception among the Eastern European Soviet bloc countries, at least, in that particular issue. Polish peasant farms were not collectivized but the Western-type farmerization was blocked and therefore not available to them (Zieliński, 1989). That was an example of what was later called a "growth without development" policy (Kuczyński, 1981; Korboński, 1990; Lonc, 1992). The "frozen" agrarian structure was a dominant and visible characteristic of Polish agriculture (Adamski & Turski, 1990; Halamska, 1991), the practitioners of which defended their property under the threat of collectivization but, at the same time, were unable to transform themselves into Western-type farmers.

The argument presented above clearly shows the unique history of Polish peasants/farmers, not only in the context of European farming but, even more importantly, in the context of neighboring countries and their rural histories, especially in the early post-WWII period. This justifies an exclusive concentration on it, as well as the particular theoretical framework of the "family farm approach," i.e., the conceptual scheme that seems to be able to grasp various types of farming realities, namely, the peasant (post-peasant), the family farming business under the processes of abandonment and modernization, as well as the family farming business in the context of contemporary global and sustainability-oriented regulatory regimes (Mann & Dickinson, 1978; Buttel & Flinn, 1980; Friedmann, 1981; Bodenstedt, 1990; Pongratz, 1990; Lacombe, 1991; Almas, 1993; Gasson & Errington, 1993; Haan, 1993; Jollivet, 1994). Based on a discussion among authors, we chose the definition of "family farm" by Ruth Gasson and Andrew Errington. The important aspect of their approach is the need for an empirical determination of a farm's "family character." As the authors explain,

> There are six elements to our own definition of a farm family business: 1) business ownership is combined with managerial control in the hands of business principals, 2) these principals are related by kinship and marriage, 3) family members (including these business principals) provide capital to the business, 4) family members including business principals do farm work, 5) business ownership and managerial control are transferred between the generations with the passage of time, 6) the family lives on the farm (Gasson & Errington, 1993, p. 18).

Therefore, we do not use the "official" definition of family farm but, quite to the contrary, we determine the family character of a particular farm once we assess the collected data concerning every dimension of farm functioning mentioned above.

The review of literature has brought us to develop a conceptual perspective focused on the family farm while, at the same time, we are trying to grasp various peculiarities dependent on social and cultural contexts. In preparing such a conceptual toolkit we reach to major traditions of rural sociology. We consider here two particular approaches developed by van der Ploeg (1995, 2003) and Mooney (1988). Both of them make efforts to grasp the diversity of family farms in the frames of contemporary capitalism. However, both of them do so in different ways. While van der Ploeg has been focused on the so-called "farming style" idea, Mooney seems to be more interested in different types of family farmers' class locations. Despite such differences, both authors seem to share the same basic assumption. Both argue that the family farm—perceived as a peculiar category of unity between household and farm—in reality might represent various specific types of social phenomena. We argue that both theories, despite their efforts to integrate various analyses, are still one-sided. The first one focuses on human (farmer) agency, while the second one concentrates on the structural context of farm operation. An effort to overcome the one-sidedness of each perspective is required. Our proposal offers such a solution by combining both perspectives and broadening them as well. In place of van der Ploeg's "farming style" we introduce the perspective of "strategy of farming," as strategies that have been based on substantial rationality (Mooney, 1988; Ploeg, 2013). They stem from cultural "styles of farming" that contain ways of thinking and preferences about how to run a farm; relations with external institutions and processes; as well as values, beliefs, and ideas on the connection between household and farm, the definition of group interests, evaluation of rural policies, etc. We also overcome Mooney's idea of class locations and go beyond its purely structural meaning to put more emphasis on ideal factors, such as various types of class consciousness. In this particular dimension we will focus on: evaluating the defense of class interests (see also: Gorlach & Mooney, 1998; Foryś & Gorlach, 2015), farmers' self-identities, ideas about the "proper farm," social participation, etc. In contrast to both van der Ploeg and Mooney, our proposal is not based on local studies but on a random, national sample of family farm operators in Poland that will provide a foundation for further cross-national comparisons beyond this particular project.

One should also stress that the methodology for our project has been based on the idea of combining various methods and techniques, in a dual sense. We use both quantitative (multi-level statistical analyses) and qualitative (discourse analysis) methods, both classic and more contemporary approaches (see more info about this in the next chapter).

The first area contains an analysis of relations between family household and farm, especially the issue of ownership and tenancy of land, the matters of farm work performed by particular family members, neighbors' work performed under the system of rural community mutual help, hired workers, off-farm jobs taken by family members. The main goal of this part of the project is to answer the question about the "family" character of farms under consideration. Based on previous works by the principal investigator (Gorlach, 1995, 2001, 2009) where the hypothesis was put forward about the relatively weak "capitalist" character of family farms in Poland, the question then arises: Have European rural policies that made an impact on Polish farms after 2004 resulted in some changes in the character of Polish farms?

The second area under investigation contains the construction of social portraits of farm operators. We are focused on those individuals who really are perceived by family members as main decision-makers. Firstly, we are interested in their relations with the formal owners of farmlands. Moreover, we will focus on their identities, their self-perceptions of their social positions compared to other people running businesses, their decision-making processes, including sources and types of knowledge, factors taken into consideration and the general goal of farm operation, the "ideal type" of family farm as well as their attitudes towards external financial resources that might be used in farm operations. Again, based on previous studies by the principal investigator, we would offer a hypothesis about three different types of Polish farmers' identity: the yeoman (traditional peasant-type owners and operators of farms), the entrepreneur, and the marginalized person. Such identities appear to be strongly correlated with farming strategies. Then the question arises again: Did European rural policies that had an impact on Polish farms after 2004 result in some changes concerning these particular issues? Moreover, in the same area, we are also interested in gender, specifically: do male and female operators differ in respect to the issues mentioned above? The previous study (Gorlach, 2009) concluded that there is no such thing as peculiar "male" and "female" styles of farming. However, we can explore more recent developments in this area.

The third area of our interest has been focused on farmers as members of a class. In this context we are dealing especially with the issue of class diversity among farm operators. However, we conceptualized this problem differently than that in Mooney's theory, presented earlier. We use instead a Weberian (1978) type of approach to class, referring to the possession of different types of capital by farm operators which define their position within the market. We considered three types of capital: economic (the area of the farm and its level of mechanization); human/cultural (level of education, type of his/her rationality); and social capital (activity in public life, membership in associations and organizations). Then, through a Bourdieu-type analysis of combinations of types of capital we identified clusters of positions that form the three main sub-classes within the farm operators' community: the positively privileged, the middle, and the negatively privileged ones (see Bourdieu, 1977, 1984, 2005). Having identified such categories, we characterized them with respect to their dominant identities, strategies of farming, and types of class consciousness

(Wright 1989, 1997). Again, based on the previous studies by the principal investigator, we are able to put forward a hypothesis about the growing divergence between the positively and negatively privileged farmers. The question arises yet again: Did European rural policies, which have had an impact on Polish farms after 2004, result in some changes in this process?

Farmers' interests in Poland are determined by two factors. The first one, which appears to be more important, has to do with the ongoing modernization of Polish agriculture. The second factor is related to the polarization of their size structure, which leads to the disappearance of the smallest and the weakest farms or forces their owners into additional economic activity not necessarily related to agriculture. Consequently, what remains are the large and modern farms, whose owners can be described as agricultural entrepreneurs who are able to organize into strong groups and push for their own interests. Farmers in Poland are a strong social category that can effectively influence the decision-making processes, although within this category interests are diverse in two aspects. The first one is economic, and it is marked by a formation of producer groups that push for goals that are beneficial to their respective agricultural sectors. The second dimension is political, and this has to do with the existence of farmers' organizations that focus on defending the interests of agriculture as an integral part of the economy.

In the 1990s, farmers' interests were mainly economic. At the beginning of the decade, they were focused by radical economic changes in Poland, which deeply affected rural areas and agriculture in terms of income reduction, the collapse of investments in farm development, and farmers' going into debt. Farmers organized protests and they forced the government to introduce solutions for adapting agriculture to a new economic situation. In the second half of the 1990s farmers expressed their interests mostly through protest actions. Their demands were of an economic nature, but they started to voice more specific issues connected to particular agricultural producer groups (Foryś, Gorlach, 2002). In the 21st century, farmers' interests are much more diverse (Foryś, 2008; Gorlach, 2009), yet general tendencies can still be observed. Firstly, farmers' interests are predominantly economic and have a tendency to diversify and disperse, which means that particular producer groups fight for their interests independently. Secondly, farmers have changed their style of fighting for their interests. They rely less on protests, yet they engage in Europe-wide campaigns (Bush & Simi 2001). Radical actions of the past have been replaced by activities in the arena of conventional politics. Thirdly, current relations between the state and farmers' groups have taken on the form of institutionalized cooperation.

In our research we will use a combination of various methods and techniques. The first step will focus on a national random sample of farm operators, about 3000 cases. We deal here with the notion of regional diversification of family farming in Poland, which has two dimensions. The first one is connected to the characteristics of particular farm operators, while the second dimension, which is more important here, relates to the legacy of farming as a part of the socioeconomic and cultural history of the region. Therefore, we investigated around 3500 farm operators across Poland within at least 100 counties (*powiaty*). Based on such a sizeable database we were able to use HLM (hierarchical linear model) software in order to analyze the impact of regional characteristics on farm performances (see more: Gelman & Hill, 2007; Domański & Pokropek, 2011; Snijders & Bosker, 2012). This will give us an opportunity to verify, in a much more precise way than before (Bartkowski, 2003; Łuczewski, 2012), well-known hypotheses about regional diversification of farming stemming from historical legacies of the modernization processes in the 19th and early 20th centuries (i.e., the partitions period).

As we mentioned earlier, our study is a national one. There is no possibility to conduct traditional cross-national comparisons that are based on a sample of particular national cases. Such a study would need a similar project covering at least one other national case, but that would result in the significant enlargement of both the research team and the budget. There is also no possibility to broaden the study in order to supplement it with some quantitative studies focused on many other countries, since that would run against the theoretical and methodological framework of the project as an historic-contextual national case. The project should be treated as an initial step towards further national-case analyses and then comparative studies in different future frame projects. However, we employ a unique method that helps us to contextualize both the study and its results. It is a "personal theorizing evaluation" allowed by the international character of the research team. Professor Jan Douwe van der Ploeg, with his experience in studying family farms mainly in the Netherlands, as well as Professor Patrick H. Mooney, with his experience in studying family farms in the United States, mainly in the Midwest region, and various social and political issues concerning farming, food, and farmers' protests, take the roles of external theorizing evaluators at every step of our research project.

REFERENCES

Adamski, W. & Turski, R. (eds.) (1990). *Interesy i konflikt. Studia nad dynamiką struktury społecznej* [Interests and conflict: Studies on the dynamics of social structure]. Ossolineum.

Allen, P. (ed.) (1993). *Food for the Future: Conditions and contradictions of sustainability*. Wiley.

Almas, R. (1993). *Norway's Gift to Europe: Fifteen selected articles on rural persistence and change.* Centre for Rural Research.

Almas, R. & Campbell, H. (eds.) (2012). *Reframing Policy Regimes and the Future Resilience of Global Agriculture.* Emerald Books.

Alvarado, F., Chancel L., Piketty T., Saez E., & Zucman G. (2018). World Inequality Report, https://wir2018.wid.world/files/download/wir2018-summary-english.pdf (access: July 24, 2019).

Amey, J. (2013). *Think Global, Act Local.* Warwick Knowledge Base, www2.warwick.ac.uk/knowledge/business/thinklocal/ (access: May 26, 2014).

Ball, P. (2004). *Critical Mass: How one thing leads to another.* Arrow Books.

Barnet, R.J. & Cavanagh, J. (1994). *Global Dreams: Imperial corporations and the new world order.* Touchstone.

Bartkowski, J. (2003). *Tradycja i polityka: Wpływ tradycji kulturowych polskich regionów na współczesne zachowania społeczne i polityczne* [Tradition and politics: The impact of regional cultural traditions in poland on some contemporary social and political behaviors]. Żak.

Bauman, Z. (1997). *Glokalizacja, czyli komu globalizacja, a komu lokalizacja* [Glocalization: To whom globalization and to whom localization], *Studia Socjologiczne* 3(146), pp. 53–70.

Bauman, Z. (2006). *Płynna nowoczesność* [Liquid modernity]. Przekład T. Kunz. Wydawnictwo Literackie.

Bauman, Z. (2007a). *Płynne czasy: Życie w epoce niepewności* [Liquid times: Life in the era of uncertainty]. Przekład M. Żakowski. Wydawnictwo Sic!

Bauman, Z. (2007b). *Płynne życie* [Liquid life]. Przekład T. Kunz. Wydawnictwo Literackie.

Beck, U. (2002). *Społeczeństwo ryzyka. W drodze do innej nowoczesności* [Risk society: Towards a new modernity]. Scholar.

Beck, U., Giddens, A., & Lash, S. (2009). *Modernizacja refleksyjna* [Reflexive modernization]. Wydawnictwo Naukowe PWN.

Beck, U. & Grande, E. (2004). *Das kosmopolitische Europa. Gesellschaft und Politik in der zweiten Moderne.* Suhrkamp Verlag [Beck, U. & Grande, E. (2009). Europa kosmopolityczna. Społeczeństwo i polityka w drugiej nowoczesności. Przekład A. Ochocki. Scholar].

Bell, D. (2006). Variations on the rural idyll. In: P. Cloke, T. Marsden, & P.H. Mooney (eds.), *Handbook of Rural Studies*, pp. 149–160. SAGE.

Bell, S. & Morse, S. (2003). *Measuring Sustainability: Learning from Doing.* Earthscan.

Berger, P. & Luckmann, T. (1971). *The Social Construction of Reality: A treatise in the sociology of knowledge.* Penguin Books.

Bodenstedt, A. (1990). Rural culture: A new concept. *Sociologia Ruralis* 30(1), pp. 34–47.

Bołtromiuk, A. (ed.) (2011). *Uwarunkowania zrównoważonego rozwoju gmin objętych siecią Natura 2000* [The conditions of local sustainable development under the network Natura 2000]. Instytut Rozwoju Wsi i Rolnictwa PAN.

Bonnano, A. (2006). The state and rural polity. In: P. Cloke, T. Marsden, P.H. Mooney (eds.), *Handbook of Rural Studies*, pp. 317–329. SAGE.

Bourdieu, P. (1977). *Outline of a Theory of Practice.* Cambridge University Press.

Bourdieu, P. (1984). *Distinctions.* Routledge.

Bourdieu, P. (2005). *The Social Structures of Economy*. Polity Press.

Brecher, J. & Costello T. (1994). *Global Village or Global Pillage: Economic reconstruction from the bottom up*. South End Press.

Bruckmeier, K. & Tovey, H. (eds.) (2009). *Rural Sustainable Development in the Knowledge Society*. Ashgate.

Brunori, G. & Bartolini, F. (2016). The family farm: Model for the future or relic of the past? In: M. Shucksmith & D. Brown (eds.), *Routledge International Handbook of Rural Studies*, pp. 192–204. Routledge.

Bush, E. & Simi, P. (2001). European farmers and their protests. In: D. Imig & S. Tarrow (eds.), *Contentious Europeans: Protest and emerging polity*, pp. 97–121. Rowman & Littlefield.

Buttel, F.H. (2006). Sustaining the unsustainable: Agro-food systems and environment in the modern world. In: P. Cloke, T. Marsden, & P.H. Mooney (eds.), *Handbook of Rural Studies*, pp. 213–229. SAGE.

Campbell, J.T., Singh, A.S., & Sharp, J. (2016). Rural communities and responses to climate change. In: M. Shucksmith & D.L. Brown (eds.), *Routledge International Handbook of Rural Studies*, pp. 531–543. Routledge.

Castree, N. & Braun, B. (2006). Constructing rural natures. In: P. Cloke, T. Marsden, & P.H. Mooney (eds.), *Handbook of Rural Studies*, pp. 161–170. SAGE.

Cavanagh, J. & Mander, J. (eds.) (2004). *Alternatives to Economic Globalization: A better world is possible*. Berrett-Kochler.

Chambers, R. (1997). *Whose Reality Counts? Putting the first last*. ITDG Publications.

Chambers, R. (2005). *Ideas for Development*. Earthscan.

Cheru, F. & Bradford, C. Jr. (eds.) (2005). *The Millennium Development Goals: Raising resources to tackle world poverty*. ZED Books (in association with the Helsinki process on Globalisation and Democracy).

Chirot, D. (1989). *The Origins of Backwardness in Eastern Europe*. University of California Press.

Chiswell, H.M. (2018). From generation to generation: Changing dimensions of intergenerational farm transfer, *Sociologia Ruralis*, 58(1), pp. 104–125.

Chloupkova, J., Svendsen, G.L.H., & Svendsen, G.T. (2003). Building and destroying social capital: The case of cooperative movements in Denmark and Poland. *Agriculture and Human Values* 20, pp. 241–252.

Chomsky, N. (2005). *Democrazie e Impero; interviste su USA, Europa, Medio Oriente, America Latina, Roma, Italy*. Datanews Editrice.

Cloke, P., Marsden, T., & Mooney, P.H. (eds.) (2006). *Handbook of Rural Studies*. SAGE.

Cupiał, M. (2006). *System wspomagania decyzji dla gospodarstw rolniczych* [The system of assisting decisions in farms]. Polskie Towarzystwo Inżynierii Rolniczej.

Cupiał, M. (2008). Zapotrzebowanie na programy komputerowe w rolnictwie na przykładzie gospodarstw województwa małopolskiego [The need for computer programs in agriculture in the Małopolska region], *Inżynieria Rolnicza*, 9, pp. 55–60.

Curry, N. & Kirwan, J. (2014). The role of tacit knowledge in developing networks for sustainable agriculture. *Sociologia Ruralis* 54(3), pp. 341–361.

Darnhofer, I., Fairweather, J., & Moller, H. (2010). Assessing a farm's sustainability: Insights from resilience thinking. *International Journal of Agricultural Sustainability*, 8, pp. 186–198d.

Domański, H. & Pokropek, A. (2011). *Podziały terytorialne, globalizacja a nierówności społeczne: Wprowadzenie do modeli wielopoziomowych* [Territorial partitions, globalization and social inequalities: An introduction to the multilevel models]. Wydawnictwo IFiS PAN.

Donehower, K. & Green, B. (2016). Rural literacies and rural mobilities: Textual practice, relational space and social capital in a globalized world. In: M. Shucksmith & D. Brown (eds.), *Routledge International Handbook of Rural Studies*, pp. 569–579. Routledge.

DuPuis, E.M. (2006). Landscapes of desires? In: P. Cloke, T. Marsden, & P.H. Mooney (eds.), *Handbook of Rural Studies*, pp. 124–132. SAGE.

Elliott, J. (2013). *An Introduction to Sustainable Development (5th ed.).* Routledge.

Etzioni, A. (2004). *The Common Good*. Polity Press.

Etzioni, A. (2018). *Law and Society in a Populist age: Balancing individual rights and the common good*. Bristol University Press.

Flinn, W.L. & Buttel, F.H. (1980). Sociological aspects of farm size: Ideological and social consequences of scale in agriculture. *American Journal of Agricultural Economics* 62(5), pp. 946–953.

Folke, C. (2006). The economic perspective: Conservation against development versus conservation for development. *Conservation Biology* 20(3), pp. 686–688.

Fonte, M. & Papadopoulos, A.G. (2010). *Naming Food after Places: Food relocalisation and knowledge dynamics in rural development*. Ashgate.

Foryś, G. (2008). *Dynamika sporu. Protesty rolników w III Rzeczpospolitej* [The dynamics of contention. Farmers' protests in contemporary Poland]. Scholar.

Foryś, G. & Gorlach, K. (2002). The dynamics of Polish peasant protests under post-communism. *Eastern European Countryside* 8, pp. 47–65.

Foryś, G. & Gorlach, K. (2015). Defending interests: Polish farmers' protests under post-communism. In: B. Klandermans & C. van Stralen (eds.), *Movements in Times of Democratic Transition*, pp. 316–340. Temple University Press.

Friedman, T.L. (2005). *The World Is Flat: A brief history of the twenty-first century*. Farrar, Straus and Giroux.

Friedmann, H. (1981). *The Family Farm in Advanced Capitalism: Outline of a theory of simple commodity production*. Unpublished manuscript.

Fukuyama, F. (1995). *Trust: The social virtues and creation of prosperity*. The Free Press.

Gasson, R. & Errington, A. (1993). *The Farm Family Business*. CAB International.

Gelman, A. & Hill, J. (2007). *Data Analysis Using Regression and Multilevel/Hierarchical Models*. Cambridge University Press.

Giddens, A. (1992). *Modernity and Self-identity: Self and society in the late modern age*. Polity Press.

Giddens, A. (2009). *Europa w epoce globalnej* [Europe in the global age]. Wydawnictwo Naukowe PWN.

Goodland, R. (1995). The concept of environmental sustainability. *Annual Review of Ecology and Systematics* 26, pp. 1–24.

Goodman, D. & Watts, M.J. (eds.) (1997). *Globalising Food: Agrarian questions and global restructuring*. Routledge.

Goodwin, M. (2006). Regulating rurality? Rural studies and the regulation approach. In: P. Cloke, T. Marsden, & P.H. Mooney (eds.), *Handbook of Rural Studies*, pp. 304–316. SAGE.

Gorlach, K. (1995). *Chłopi, rolnicy, przedsiębiorcy – kłopotliwa klasa w Polsce postkomunistycznej* [Peasants, farmers, entrepreneurs: The awkward class in postcommunist Poland]. Wydawnictwo Uniwersytetu Jagiellońskiego.

Gorlach, K. (2001). *Świat na progu domu. Rodzinne gospodarstwa rolne w Polsce w obliczu globalizacji* [The world in my backyard. Polish family farms in the face of globalization]. Wydawnictwo Uniwersytetu Jagiellońskiego.

Gorlach, K. (2006). Between hopes and fears: Rural Eastern Europe on its way to Europe. *Eastern European Countryside* 12, pp. 5–30.

Gorlach, K. (2009). *W poszukiwaniu równowagi. Polskie rodzinne gospodarstwa rolne w Unii Europejskiej* [Searching for the balance: Polish family farms in the European Union]. Wydawnictwo Uniwersytetu Jagiellońskiego.

Gorlach, K. & Adamski, T. (2007). Neo-endogenous development and the revalidation of local knowledge. *Polish Sociological Review* 160(4), pp. 481–497.

Gorlach, K. & Mooney, P.H. (1998). Defending class interests: Polish peasants in the first years of transformation. In: J. Pickles & A. Smith (eds.), *Theorising Transition: The political economy of post-communist transformations*, pp. 262–283. Routledge.

Gorlach, K. & Mooney, P.H. (2004). *European Union Expansion: The impacts of integration on social relations and social movements in rural Poland*. Unpublished manuscript.

Gorlach, K. & Starosta, P. (2001). De-peasantization or re-peasantization? Changing rural structures in Poland after World War II. In: L. Granberg, I. Kovach, & H. Tovey (eds.), *Europe's Green Ring*, pp. 41–65. Ashgate.

Gulbicka, B. (2012). *Zanieczyszczenia biologiczne i chemiczne jako problem bezpieczeństwa żywności*. Instytut Ekonomiki Rolnictwa i Gospodarki Żywnościowej, Państwowy Instytut Badawczy.

Gunderson, L.H. (2000). Ecological resilience—in theory and application. *Annual Review of Ecology and Systematics* 31, pp. 425–439.

Gunderson, L.H. & Holling, C.S. (eds.) (2002). *Panarchy: Understanding transformations in human and natural systems*. Island Press.

Haan, H. de (1993). Images of family farming in the Netherlands. *Sociologia Ruralis* 33(2), pp. 147–166.

Haigh, S.P. (2013). *Future States: From internation to global political order*. Routledge.

Halamska, M. (1991). *Chłopi na przełomie epok* [Peasants on the verge of epochs]. IRWiR PAN.

Hickey, S. & Mohan, G. (eds.) (2004). *Participation: From tyranny to transformation? Exploring new approaches to participation in development*. Zed Books.

Hirst, P. & Thompson, G. (1995). Globalization and the future of the nation-state. *Economy and Society* 3, pp. 408–442.

Holland, A. (2000). Introduction. In: K. Lee, A. Holland, & D. McNeill (eds.), *Global Sustainable Development in the 21st Century*, pp. 1–9. Edinburgh University Press.

Homans, G.C. (1964). Bringing men back in. *American Sociological Review* 29(5), pp. 809–818.

Jollivet, M. (1994). Kapitalizm i rolnictwo [Capitalism and agriculture]. In: P. Rimbaud & Z.T. Wierzbicki (eds.), *Socjologia wsi we Francji* [Rural sociology in France], pp. 36–47. Wydawnictwo Uniwersytetu Mikołaja Kopernika.

Jones, O. (2006). Non-human rural studies. In: P. Cloke, T. Marsden, & P.H. Mooney (eds.), *Handbook of Rural Studies*, pp. 185–200. SAGE.

Khonder, H.H. (2004). Glocalisation as globalisation: Evolution of the sociological concept. *Bangladesh e-Journal of Sociology* 1(2), pp. 1–9.

Klekotko, M. (2012). *Rozwój po śląsku. Procesy kapitalizacji kultury śląskiej w społeczności górniczej* [Development in Silesia: processes of capitalization of culture in a post-mining Silesian community]. Wydawnictwo Uniwersytetu Jagiellońskiego.

Klekotko, M. & Gorlach, K. (2011). Miejsce, lokalność, globalizacja. Przyczynek do problematyki socjologii wsi (i nie tylko) w społeczeństwie ponowoczesnym [Place, locality, globalisation: An attempt (and not only) in post-modern society]. In: H. Podedworna & A. Pilichowski (eds.), *Obszary wiejskie w Polsce: różnorodności i procesy zróżnicowania* [Rural areas in Poland: Diversities and processes of diversification], pp. 25–55. Wydawnictwo IFiS PAN.

Klekotko, M., Jastrzębiec-Witowska, A., Gorlach, K., & Nowak, P. (2018). Think locally and act globally: Understanding human development in the era of globalization. *Eastern European Countryside* 24, pp. 111–141.

Knechtges, P.L. (2011). *Food Safety: Theory and practice*. Jones & Bartlett.

Kochanowicz, J. (1992). *Spór o teorie gospodarki chłopskiej: gospodarstwo chłopskie w teorii ekonomii i historii gospodarczej* [The debate on peasant economy: Peasant farms in economic theory and history of economy]. Wydawnictwo Uniwersytetu Warszawskiego.

Kołożyn-Krajewska, D. & Sikora, T. (2010). *Zarządzanie bezpieczeństwem żywności. Teoria i praktyka* [Food security management: Theory and practice]. Wydawnictwo C.H. Beck.

Konrad, G. & Szelenyi, I. (1979). *The Intellectuals on the Road to Class Power: A sociological study of the role of intelligentsia in socialism*. Harcourt Brace Jovanovich.

Korboński, A. (1990). Soldiers and peasants: Polish agriculture after martial law. In: K.E. Wadekin (ed.), *Farming in the Soviet Union and Eastern Europe*, pp. 263–278. Routledge.

Korten, D. (1995). *When Corporations Rule the World*. Kumarian.

Kuczyński, W. (1981). *Po wielkim skoku* [After the great leap]. PWE.

Lacombe, P. (1991). Farming, farms and families. In: M. Tracy (ed.), *Farmers and Politics in France*, pp. 48–64. The Arkleton Trust.

Latour, B. (1993). *We Have Never Been Modern*. Harvester/Wheatsheaf.

Latour, B. (2010). *Splatając na nowo to, co społeczne. Wprowadzenie do teorii aktora-sieci* [Reassembling the social: An introduction to actor-network theory], Universitas.

Lawley, R., Curtis, L., & Davis, J. (2008). *The Food Safety Hazard Guidebook*. Royal Society of Chemistry.

Lee, K., Holland, A., & McNeill, D. (eds.) (2000). *Global Sustainable Development in the 21ˢᵗ Century*. Edinburgh University Press.

Leśkiewicz, K. (2012). Bezpieczeństwo żywnościowe i bezpieczeństwo żywności – aspekty prawne [Food security and food safety: Legal aspects]. *Przegląd Prawa Rolnego* 10(1), pp. 179–198.

Little, J. (2002). *Gender and Rural Geography: Identity, sexuality and power in the countryside*. Pearson.

Little, J. (2006). Gender and sexuality in rural communities. In: P. Cloke, T. Marsden, & P.H. Mooney (eds.), *Handbook of Rural Studies*, pp. 365–378. SAGE.

Lonc, T. (1992). Social and economic adjustment and policies in the period of crisis and emergence from it: The experience of Poland in the 1980s. In: A. Hakkinen (ed.), *Just a Sack of Potatoes? Crisis experience in European societies, past and present*, pp. 215–226. SHS.

Lowe, P. (2010). Enacting rural sociology: Or what are the creativity claims of the engaged sciences? *Sociologia Ruralis* 50(4), pp. 311–330.

Lubaś, M. (2010). Przestrzenne skale odtwarzania miejsca. Analiza społecznej organizacji przestrzeni w zachodniej Macedonii [Spatial scales of recreation of places. An analysis of the social organization of space in Western Macedonia]. In: A. Bukowski, M. Lubaś, & J. Nowak (eds.), *Społeczne tworzenie miejsc. Globalizacja, etniczność, władza* [Social creation of places. Globalization, ethnicity, Power], pp. 41–70. Wydawnictwo Uniwersytetu Jagiellońskiego.

Lyson, T.A. (2006). Global capital and the transformation of rural communities. In: P. Cloke, T. Marsden, & P.H. Mooney (eds.), *Handbook of Rural Studies*, pp. 292–303. SAGE.

Łuczewski, M. (2012). *Odwieczny naród: Polak i katolik w Żmiącej* [The eternal nation: A Pole and a catholic in Żmiąca]. Wydawnictwo Uniwersytetu im. Mikołaja Kopernika.

Mann, S. & Dickinson, A. (1978). Obstacles to the development of a capitalist agriculture. *Journal of Peasant Studies* 5, pp. 466–481.

Marini, M.B. & Mooney, P.H. (2006). Rural economies. In: P. Cloke, T. Marsden, & P.H. Mooney (eds.), *Handbook of Rural Studies*, pp. 91–103. SAGE.

Marsden, T. (1999). Rural futures: The consumption countryside and its regulation. *Sociologia Ruralis* 39(4), pp. 501–520.

Marsden, T. (2003). *The Condition of Rural Sustainability*. Van Gorcum.

Maye, D. (2018). Examining innovation for sustainability from the bottom-up: An analysis of the permaculture community in England. *Sociologia Ruralis* 58(2), pp. 331–350.

McMichael, P. (2000). *Development and Social Change: A global perspective (2nd edition)*. Pine Forge Press.

McMichael, P. (ed.) (2010). *Contesting Development: Critical struggles for social change*. Routledge.

Meyer, M.A. (2016). Climate change, environment hazards and community sustainability. In: M. Shucksmith & D. Brown (eds.), *Routledge International Handbook of Rural Studies*, pp. 335–345. Routledge.

Miele, M. (2006). Consumption culture: The case of food. In: P. Cloke, T. Marsden, & P.H. Mooney (eds.), *Handbook of Rural Studies*, pp. 344–354. SAGE.

Mikuła, A. (2012). Bezpieczeństwo żywnościowe Polski [Food security in Poland]. *Roczniki Ekonomii Rolnictwa i Rozwoju Obszarów Wiejskich* 99(4), pp. 38–48.

Mooney, P.H. (1988). *My Own Boss? Class, rationality and the family farm*. Westview Press.

Mooney, P.H. & Hunt, S.A. (2009). Food security: The elaboration of contested claims to a consensus frame. *Rural Sociology* 74(4), pp. 469–497.

Morse, S. (2004). *Indices and Indicators in Development: An unhealthy obsession with numbers*. Earthscan.

Murdoch, J. (2006). Networking rurality: Emergent complexity in the countryside. In: P. Cloke, T. Marsden, & P.H. Mooney (eds.), *Handbook of Rural Studies*, pp. 171–184. SAGE.

Nelson, J. & Stock, P. (2018). Repeasantization in the United States. *Sociologia Ruralis* 58(1), pp. 83–103.

Nigro, S. (2003). *Glocalisation: Research study and policy recommendations*. CEFRE & Global Forum.

Noe, E. & Alroe, H.F. (2012). Observing farming systems: Insights from social systems theory. In: I. Darnhofer, D. Gibbon, & B. Dedieu (eds.), *Farming Systems Research into the 21st Century: The new dynamics*, pp. 387–403. Springer.

Noguera, J. & Freshwater, D. (2016). Rural-urban in a peri-urban context. In: M. Shucksmith & D.L. Brown (eds.), *Routledge International Handbook of Rural Studies*, pp. 511–517. Routledge.

O'Riordan, T. & Voisey, H. (eds.) (1998). *The Transition to Sustainability: The politics of agenda 21 in Europe*. Earthscan.

Pakulski, J. (2009). Postmodern social theory. In: B.S. Turner (ed.), *The New Blackwell Companion to Social Theory*, pp. 251–280. Blackwell.

Pałasiński, M. & Juszczak, L. (eds.) (2014). *Wybrane zagadnienia nauki o żywności i zarządzania jakością* [Selected problems of food science and quality management]. Wydawnictwo Uniwersytetu Rolniczego.

Perkins, H.C. (2006). Commodification: re-resourcing rural areas. In: P. Cloke, T. Marsden, & P.H. Mooney (eds.), *Handbook of Rural Studies*, pp. 243–257. SAGE.

Pieterse, J.N. (2010). *Development Theory*. SAGE.

Piketty, T. (2014). *Capital in the Twenty-First Century*. The Belknap Press of Harvard University Press.

Ploeg, J.D. van der (1995). From structural development to structural involution: The impact of new development in Dutch agriculture. In: J.D. van der Ploeg & G. van Dijk (eds.), *Beyond Modernisation: The impact of endogenous development*, pp. 109–146. Van Gorcum.

Ploeg, J.D. van der (2003). *The Virtual Farmer: Past, present and future of the Dutch peasantry*. Van Gorcum.

Ploeg, J.D. van der (2008). *The New Peasantries: Struggles for autonomy and sustainability in an era of empire and globalization*. Earthscan.

Ploeg, J.D. van der (2013). *Peasants and the Art of Farming: A Chayanovian manifesto*. Fernwood Publishing.

Pongratz, H. (1990). Cultural tradition and social change in agriculture. *Sociologia Ruralis* 30(1), pp. 5–17.

Ray, C. (1999). Towards a meta-framework of endogenous development: Repertoires, paths, democracy and rights. *Sociologia Ruralis* 4(39), pp. 521–537.

Ray, C. (2006). Neoendogenous rural development in the EU. In: P. Cloke, T. Marsden, & P.H. Mooney (eds.), *Handbook of Rural Studies*, pp. 278–291. SAGE.

Ray, L. & Sayer, R.A. (eds.) (1999). *Culture and Economy after the Cultural Turn*. SAGE.

Ritzer, G. (1996). *The McDonaldization of Society*. Pine Forge Press.

Robertson, R. (1992). *Globalisation: Social theory and global culture*. SAGE.

Robertson, R. (1995). Glocalisation: Time-space and homogeneity-heterogeneity. In: M. Featherstone, S. Lash, & R. Robertson (eds.), *Global Modernities*, pp. 25–44. SAGE.

Rogers, J. & Fraszczak, M. (2014) Like the stem connecting the cherry to the tree: The uncomfortable place of intermediaries in a local organic food chain. *Sociologia Ruralis* 54(3), pp. 321–340.

Rooij, S. de, Ventura, F., Milone, P., & Ploeg, J.D. van der (2014). Sustaining food production through multifunctionality: The dynamics of large farms in Italy. *Sociologia Ruralis* 54(3), pp. 303–320.

Rosenfeld, S.A. & Wojan, T.R. (2016). The emerging contours of rural manufacturing. In: M. Shucksmith & D. Brown (eds.), *Routledge International Handbook of Rural Studies*, pp. 120–132. Routledge.

Rostow, W. (1960). *The Stages of Economic Growth: A non-communist manifesto*. Cambridge University Press.

Sachs, J. (2005). *The End of Poverty: How we can make it happen in our lifetime*. Penguin Books.

Salamon, S. (2006). The rural household as a consumption site. In: P. Cloke, T. Marsden, & P.H. Mooney (eds.), *Handbook of Rural Studies*, pp. 330–343. SAGE.

Schaft, K.A. (2016). Social dynamics and institutional capacity: Structures, mobilities and identities beyond the periphery of the global metropolis. In: M. Shucksmith & D. Brown (eds.), *Routledge International Handbook of Rural Studies*, pp. 511–517. Routledge.

Shanin, T. (1986). *Russia as a Developing Society: The roots of otherness*. Yale University Press.

Sheingate, A. (2001). *The Rise of the Agricultural Welfare State: Institutions and interest group power in the United States, France, and Japan, Princeton and Oxford*. Princeton University Press.

Short, B. (2006). Idyllic ruralities. In: P. Cloke, T. Marsden, & P.H. Mooney (eds.), *Handbook of Rural Studies*, pp. 133–148. SAGE.

Shucksmith, M. (2010). Disintegrated rural development? Neoendogenous rural development, planning and place shaping in diffused power contexts. *Sociologia Ruralis* 50(1), pp. 1–14.

Sibley, D. (2006). Inclusions/exclusions in rural space. In: P. Cloke, T. Marsden, & P.H. Mooney (eds.), *Handbook of Rural Studies*, pp. 401–410. SAGE.

Smith, D., Davidson, J., Cameron, L., & Boni, L. (eds.) (2009). *Emotion, Place and Culture*. Ashgate.

Snijders, T. & Bosker, R. (2012). *Multilevel Analysis: An introduction to basic and advanced multilevel modeling*. SAGE.

Strange, M. (1988). *Family Farm: A new economic vision*. University of Nebraska Press.

Sztompka, P. (1991). *Society in Action: The theory of social becoming*. Cambridge University Press.

Sztompka. P. (1993). *The Sociology of Social Change*. Blackwell.

Symes, D. & Phillipson, J. (2016). Industrializing the marine commons: Adapting to change in Europe's coastal fisheries. In: M. Shucksmith & D. Brown (eds.), *Routledge International Handbook of Rural Studies*, pp. 323–334. Routledge.

Swain, N. (1985). *The Collective Farms Which Work?* Cambridge University Press.

Tarrow, S. (1998). *Power in Movement: Social movements and contentious politics*. Cambridge University Press.

Urry, J. (1995). *Consuming Places*. Routledge.

Vanclay, F., Higgins, M., & Blackshaw, A. (eds.) (2008). *Making Sense of Place: Exploring concepts and expressions of place through different senses and lenses*. National Museum of Australia.

Walker, B.H. & Salt, H. (2006). *Resilience Thinking: Sustaining ecosystems and people in a changing world*. Island Press.

Wallerstein, I. (1974). *The Modern World System*, Vol. 1: *Capitalist agriculture and the origins of the European world-economy in the sixteenth century*. Academic Press.

Wallerstein, I. (1979). *The Capitalist World Economy*. Cambridge University Press.

Walton, J. & Seddon, D. (1994). *Free Markets and Food Riots: The politics of global adjustment*. Blackwell.

Waters, M. (1995). *Globalization*. Routledge.

WCED (World Commission on Environment and Development) (1987). *Our Common Future*. Oxford University Press.

Weber, M. (1978). *Economy and Society*. University of California Press.

Wiborg, A. (2004). Place, nature and migration: Student attachment to their home places. *Sociologia Ruralis* 44(4), pp. 416–432.

Wierzbicki, Z.T. (ed.) (1973). *Aktywizacja i rozwój społeczności lokalnych* [Activation and Development of Rural Communities]. Ossolineum, Wydawnictwo PAN.

Wierzbicki, Z.T. (1986). Od ekologii i neoekologii do sozoekologii społecznej (artykuł dyskusyjny) [From ecology and neoecology to social socioecology. Discussion], *Ruch Prawniczy, Ekonomiczny i Socjologiczny*, 48(2), pp. 287–301.

Woods, M. (2006). Political articulation: The modalities of new critical politics of rural citizenship. In: P. Cloke, T. Marsden, & P.H. Mooney (eds.), *Handbook of Rural Studies*, pp. 457–471. SAGE.

World Bank (2006). *Enhancing Agricultural Innovation: How to go beyond the strengthening of research systems*. https://openknowledge.worldbank.org/handle/10986/24105 (access: November 27, 2020).

Wright, E.O. (ed.) (1989). *The Debate on Classes*. Verso.

Wright, E.O. (1997). *Class Counts: Comparative studies of class analysis*. Cambridge University Press.

Wrzaszcz, W. (2012). *Poziom zrównoważenia indywidualnych gospodarstw rolnych w Polsce (na podstawie danych FADN)* [The problem of sustainability of farms in Poland (based on FADN data)]. Instytut Ekonomiki Rolnictwa i Gospodarki Żywnościowej, Państwowy Instytut Badawczy.

Zieliński, M. (1989). Wojna chłopska w Polsce [Peasant war in Poland]. *Res Publica* 5, pp. 99–112.

Think Locally, Act Globally: Polish farmers in the global era of sustainability and resilience, ed. by Krzysztof Gorlach and Zbigniew Drąg in collaboration with Anna Jastrzębiec-Witowska and David Ritter
Jagiellonian University Press, Kraków 2021, pp. 95–147
ISBN 978-83-233-4949-5
DOI: http://dxdoi.org/10.4467/K7195.199/20.20.12727

Chapter Two: Mixed Methodologies, Sustainable Analyses

Zbigniew Drąg https://orcid.org/0000-0002-9106-7758/

Krzysztof Gorlach https://orcid.org/0000-0003-1578-7400/

2.1. Introductory Remarks

The second chapter of this monograph is devoted to methodological issues. Considering the context, readers should be reminded that sustainable farm development is at the center of this publication's academic reflection. It should also be recalled that sustainable development is not limited to the effects of economic changes but also involves environmental (mostly related to resources) and social (equality and inequality issues) dimensions. This might lead to the statement that the development process has a rather multifaceted character. Therefore, its analysis must be more complex, and take into consideration various factors, including psychological references to the actors taking part in these processes. A similar way of thinking may be applied to the methodology used in sociological studies. The use of various methods and research techniques can provide a more multidimensional picture of the studied reality. Thus, mixed methodologies can provide a better means to observe analyzed phenomena and social processes from various angles and this, in our view, might be more conducive to sustainability analyses. Such analyses are the main goal of our project and this monograph.

Chapter 1 essentially contains the draft of the theoretical concept to which the entire publication is devoted to. This chapter mostly deals with the methodological consequences of theoretical resolutions described in the previous chapter, which are important for the methods of academic research. The crucial point of reflection referred to the methodological work of Andrew Abbot[1] with its central premise of advocating the reversal of the approach to the relation between the global and local dimensions of social life, an approach widespread in academic literature. As emphasized by Abbott (2004, p. 7): "Switching questions is a powerful heuristic move." Such a move was taken when the main slogan of the entire publication was formulated by changing the order—and thus, the meaning—of the popular statement "think globally, act locally" to "think locally, act globally." This allowed for several important issues to be addressed, which became the basis for various analyses presented in this volume.

What are the consequences of such a switch? First of all, attention is given to concrete farms, grounded in certain local communities, operated by particular families. These farms are the starting point of all analyses in this monograph and the procedure for farm selection described in great detail in this chapter aimed at selecting adequate farm representation of the entire country. Careful and well-considered farm selection was the product of a conviction that local conditions are important, as is the character of the place that is referred to.

The attention of the researcher is directed to the role of global factors and actions, although it would be better to call them supralocal. The necessity to have contacts with supralocal subjects and to consider supralocal factors in operating the farm, as well as the need for farmers to have some influence on supralocal decision centers are part of the reality of farm management in contemporary society. This corresponds with the theoretical and methodological frame of the sustainable development concept, which is essential for this chapter. For the analysis of how the farms are operated, it is necessary to consider many various external factors but it is also helpful to learn how the farmers (farm operators) perceive and evaluate these factors. Among them are matters concerning the sources for farm production supply or matters related to securing sales points for what gets produced on farms. The role of information technology and the issues pertinent to food safety and food security are also important. The analysis of farm management should not steer away from various aspects related to political mobilization and the lifestyles of farmers and their families. All of these elements map out the broad, supralocal context in which concrete, locally rooted farms function.

---

[1] It should be noted that it was Professor Patrick H. Mooney, from the University of Kentucky-Lexington, USA, who recommended checking the work of Andrew Abbott as the point of reference for this publication. Professor Mooney is one of the external theorizing evaluators of this publication.

In the context of relations between the local and supralocal factors, the regional diversification of farms is also a vital matter. Special links between the types of factors that were mentioned earlier make, in effect, some interestingly diverse regional configurations ensuring that the globalized, or even just currently globalizing, world is not completely "flat." Such characteristics can be noted when adequately broad empirical material is collected, thus allowing comparisons to be made between particular research subjects. Therefore, the main method for collecting empirical data in the described research project was through personal interviews with farm operators, who were selected for a very precise population sample. The description of the sampling process and some additional comments on the sampling method can be found later in this chapter.

In contemporary academic literature there is a strong tendency to use various research methods and techniques simultaneously. The previously quoted points to five "successful methodological traditions" (Abbott, 2004, pp. 15–26). He mentions them in a rather ad hoc manner, naming them as follows: ethnography, historical narration, standard cause analysis, small N-comparisons, and formalization. Each of these research methods has its own characteristics. The first of them, "ethnography," is based on direct participation from the researcher in the reality that he studies. The second is "historical narration" with a thoroughly descriptive view of the reality. The third, "standard causal analysis" assumes the preparation of a large number of cases and application of adequate "statistical models." The fourth perspective is "small N-comparisons" oriented towards comparing particular cases in a small cluster of analyzed situations. And finally, there is the last perspective, "formalization," meaning the analyses of "mathematical relationships."

In this monograph, the team of authors used almost all of the research methods and techniques mentioned earlier. They are referred to in particular elaborations which constitute the monograph. However, this is not an author monograph in the classic sense. It is not the work product of one author but the joint effort of a team of authors, involving particular individuals preparing designated parts of the book which, when taken together, sketch a certain sociological portrait of Polish farmers during the 2017–2019 time frame. Therefore, it could be called a co-authored monograph.

The most important element of the study is a randomly selected national sample of 3500 family farms and their owners and/or farm operators, which allows for application of standard causal analysis, as well as the use of the hierarchical linear model (HLM), meaning formalization of the analyses of "mathematical relationships." The former is presented in the second part of this publication, in chapters four through nine and the latter can be seen in the third part of this publication, in chapters nine and eleven. The procedure for selecting a sample that would be the subject for analysis, in the second and the third part of the monograph, is presented later in this chapter. It is worth highlighting that Chapter 9, which closes part two, addresses the effects of more thorough qualitative analyses elaborated on the ground of ethnographic concepts, as well as historical narration, which was the basis for the non-structured interviews conducted with representatives of agricultural chambers.

The analyses based on the sample described above can be found in chapters 13–15. They serve as a background to solutions that are a result of applying other methods. What is considered here are the elements of "historical narration" (mostly seen in chapters 13 and 15), as well as "small N-comparisons" (chapters 12 and 14). It should be noted that the chapters constituting part four of this publication are the result of a separate, smaller, and independent research study undertaken by the team members. They show that the methodological dimension of the reflection presented in this co-authored monograph combines various research methods and techniques.

2.2. Research Subject and Research Problem

A focus on the phenomena and conditions determining the functioning of family farms in Poland is the core of the project "Think Locally, Act Globally: Polish farmers in the global world of sustainable development." The analyses of researchers usually focus on individual characteristics of farmers, including socio-demographic and mental characteristics, as well as physical features of the farms (their size and the equipment owned and used on them). To a lesser degree, researchers tend to analyze the impact of local conditions on farm development strategies employed by their operators. It should be stated that territorial differentiation has been recognized in most studies as important, but not to an extent which would justify investigations of the role of concrete characteristics of territorial communities (regional or local) and without stressing its meaning in that context. General statements about the meaning of the collective past, which appear relatively often in such studies, seem to be insufficient. The time has come for questions which would direct analysis not just towards the territorial differentiation of "thinking about the farm" and how farms are managed. The realization of certain development strategies and the aspiration for market success prompt questions about the characteristics of local communities which could be meaningful here.

The main research questions are the following: What determines the development strategies of family farms in Poland? Is the strategy differentiation a result of individual characteristics and preferences of farmers or is it also determined by characteristics of local communities, where the farms function? Theoretically, the first possibility, in its extreme version, could indicate a "self-creative" character of farm development strategies, elaborated exclusively by the individual intellectual and mental resources of farm operators, without considering any impact from the local and global context. The second scenario can be described as an example of the strategy of a "territorial" character, with a strong focus on local social resources. The extreme version would be a "globalized" strategy, oriented exclusively towards supralocal resources and the supralocal market, ignoring the meaning of local resources and farmers' individual characteristics.

Analyzing the relationship between development strategies and territorial differentiation requires studying farmers and their farms, as well as the local communities in which they function. Therefore, for research purposes, only farmers who actively managed and operated farms of over 1 hectare (ha) were considered for the study. It was determined that only the farmers who fulfilled such criteria could have some adequately considered developmental strategies of a long-term perspective. The local context for the study was chosen to be the county level, the second lowest level of territorial administration in Poland. The decision to establish the territorial context at the county level was made for two reasons. First, multiple offices and institutions collecting various data pertinent to the local communities in many divergent aspects of their lives are connected with the county administration. Second, the county as a unit of local government creates a foundation for the fostering of social communities.

2.3. Research Methodology

For formulating questions about the impact of individual resources and characteristics of farmers, as well as the characteristics of local communities (territorial differentiation), on the developmental strategies followed by farmers, classical research methods were applied and they included interview methods and desk research. The study on characteristics, opinions, and attitudes of farmers was conducted through field research and basic empirical material was collected from direct interviews, using questionnaires. Desk research analysis served to describe in detail the multifaceted characteristics of counties.

The study was meant to be representative of the entire population of active farms of over 1 ha in Poland. Considering land area, they were divided into 4 groups:

1 to 1.99 ha, 2 to 4.99 ha, 5 to 9.99 ha, and 10 or more hectares. Additionally, for the analysis of the impact of territorial differentiation on development strategies, a multilevel model was applied, with three steps of sampling: random selection of counties, random selection of municipalities, and a random selection of addresses of farms in municipalities of previously chosen area groups. The number of chosen counties and the ultimate size of the sample were determined by the decision to apply the multilevel model (see more on multilevel models and their application in Raudenbush & Bryk, 2002; Węziak, 2007; Twisk, 2010; Garson, 2013). As the subject literature emphasizes (see Hox, 2010, pp. 234–236; Domański & Pokropek, 2011, pp. 366–367; Hox, Moerbeek, & Schoot, 2017, pp. 213–218), a 2-level model requires at least 100 level-1 units and at least 30 level-2 units. In other words, in the analysis of the impact of country-level territorial differentiation on the development strategies preferred by farmers, application of a 2-level model would require research in at least 100 counties, with at least 30 active farms in each county. Finally, this means that the minimal sample of farms should be 3000. For that reason, the renowned Center of Sociological Research of the Institute of Philosophy and Sociology of the Polish Academy of Science was entrusted with selection of the research sample, including choosing an adequate sampling scheme. The center also conducted field research and desk research.

2.4. Selection of the Population Sample[2]

According to the previously discussed premises, the questionnaire interviews were conducted among farmers who personally managed farms, as they were thought to play a crucial role in mapping out farm development strategies. As a pre-existing sampling frame defined in such a manner is lacking in Poland, using the data of the Agricultural Census 2010 on the population of the owners of farms of over 1 ha as the point of reference seemed like an adequate solution to this problem. This was necessary as the only sampling frames available through Agricultural Census 2010 were the NUTS (Nomenclature of Territorial Units for Statistics) level 4 and level

---

[2]   The presented descriptions of the procedure of sample selection is based on the information in the report delivered by The Center of Sociological Research of the Institute of Philosophy and Sociology of the Polish Academy of Sciences: *Opis schematu próby do badania „Myśl lokalnie, działaj globalnie: Polscy rolnicy w globalnym świecie rozwoju zrównoważonego i odporności na kryzys"* (Warsaw, January 2018), written by Janusz Engel (IQS and QUANT Group) in cooperation with Wiesława Dąbała.

5, with NUTS level 4 pertinent to counties and NUTS level 5 pertinent to municipalities. Referring to the 2010 Agricultural Census was inevitable as the research was conducted in 2017 and the next scheduled agricultural census was only set to take place in 2020. Using the census listing of farm owners as a sampling frame was not contradictory to the previously assumed premise of conducting the study exclusively among actual farm operators or farm managers. The researchers followed the rule that the interview would be conducted with only those owners who had decision-making powers in managing active farms of over 1 hectare. In situations where the researcher met with a farmer who did not operate or manage a farm of more than 1 hectare, the interview was not conducted. In cases of an owner passing the farm to another person, whether for rent or in another form, the researcher was obliged to conduct the interview with the new owner/manager, if the area of farm operations was over 1 ha.

According to the 2010 Agricultural Census the number of farms in Poland in 2010 over 1 ha was 1,480,227. Farms of 1–2 ha comprised 20.3% of this number, while farms of between 2 and 5 ha made up 33.1%. The percentage of farms between 5 and 10 ha was 23.4%. Farms exceeding 10 ha accounted for 23.2% of all farms in Poland. The total number of farms within each area category varied greatly depending on the province (voivodeship). The data illustrating these differences can be seen in Table 2.4.1.

The first stage of the procedure for choosing the population sample was county sampling. The total number of counties was 380 and included 314 land counties and 66 city counties (cities with county legal status). Initially, the studied population was divided into strata and the main line of division were provinces (voivodeships) that only contained land counties. However, in the provinces where the number in the county sample was equal to or greater than four, additional strata containing one or more adjacent subregions were constructed. As a result, one province (voivodeship) could constitute one or several strata. Furthermore, the strata connected with city counties of the group of provinces (voivodeships) belonging to particular regions were also considered. Adopting the division of counties into strata on the basis of their localization in the regions, provinces, and subregions meant greater differentiation of counties in the sample, and was conducive to better recognition of differentiation in the traditions of agricultural communities within the counties.[3]

---

[3]   In 2017, during the time of the Polish farms study, the statistical division of the territory of Poland into the NUTS units (classification of nomenclature of territorial units for statistics used by Central Statistical Office in Poland and European Statistical Office—Eurostat) was the following: NUTS 1 level: Regions—6 units, NUTS 2 level: provinces (voivodeships)—16 units, NUTS 3 Level: subregions—72 units. Regions are the units grouping provinces (voivodeships), the provinces

Table 2.4.1. The number of individual farms with area exceeding 1 ha, listed by regions

Province (voivodeship)	From 1 ha to 1.99 ha	From 2 ha to 4.99 ha	From 5 ha to 9.99 ha	10 ha and more	Total
Poland	300,515	489,518	346,062	344,132	1,480,227
Lower Silesian (dolnośląskie)	12,746	17,702	13,217	16,250	59,432
Kuyavian-Pomerania (kujawsko-pomorskie)	7,913	13,171	16,387	29,290	66,847
Lubelskie (lubelskie)	32,163	66,671	52,443	34,366	187,515
Lubuskie (lubuskie)	4,673	5,975	4,079	6,621	21,593
Łódzkie (łódzkie)	19,013	42,143	40,582	27,259	129,899
Lesser Poland (małopolskie)	55,498	69,790	19,550	5,659	150,430
Mazovian (mazowieckie)	31,999	69,729	66,399	57,537	222,768
Opolskie (opolskie)	5,017	7,579	5,828	9,347	27,705
Subcarpathian (podkarpackie)	49,453	63,020	19,116	6,238	138,356
Podlaskie (podlaskie)	7,118	16,787	21,713	37,922	84,210
Pomeranian (pomorskie)	5,185	8,924	9,236	16,807	39,579
Silesian (śląskie)	21,940	23,872	10,239	6,460	59,160
Holy Cross (świętokrzyskie)	20,022	41,257	24,269	9,620	97,270
Warmian-Masurian (warmińsko--mazurskie)	5,024	7,668	7,438	22,724	44,175
Greater Poland (wielkopolskie)	18,150	28,311	30,132	45,488	121,721
West Pomeranian (zachodniopomorskie)	4,601	6,919	5,434	12,544	29,567

Source: elaborated on the basis of the BDL data: https://bdl.stat.gov.pl/BDL/dane/podgrup/tablica/ (access: June 23, 2019). Category: agriculture, forestry and hunting; group: farms; subgroup: farms according to area group of agricultural utilized land; dimensions: area groups; types of ownership; years.

(voivodeships) are groups of subregions, and subregions are groups of counties. Designated regions are the following: Central Region covering Mazovian (with 8 subregions) and Łódzkie (with 5 subregions), Southern Region encompassing Lesser Poland (with 6 subregions) and Silesian (with 8 subregions), Eastern Region encircling provinces: Lubelskie (with 4 subregions), Subcarpathian (with 4 subregions), Holy Cross (with 2 subregions) and Podlaskie (with 3 subregions), Northwestern Region encompassing the provinces: Greater Poland (with 6 subregions), West Pomerania (with 4 subregions) and Lubuskie (with 2 subregions), Southwestern Region grouping Lower Silesian (with 5 subregions) and Opolskie (with 2 subregions), and Northern Region covering the following provinces: Kuyavian-Pomeranian (with 5 subregions), Pomeranian (with 5 subregions) and Warmian-Masurian (with 3 subregions). See https://stat.gov.pl/statystyka-regionalna/jednostki-terytorialne/klasyfikacja-nuts/klasyfikacja-nuts-w-polsce/ (access: June 12, 2019).

A Hartley-Rao scheme (Hartley & Rao, 1962; also see: Rao, Hartley, & Cochran, 1962; Bansal & Singh, 1986; Rao & Wu, 1988; Kott, 2005) was applied to survey the random sample of counties from the strata. This is a collective, systematic sampling scheme, with use of additional information in the survey, with probability proportional to the sample size, without replacement. It gives the guarantee of positive sampling of counties with greater or smaller number of farmers alike. The sampling procedure used with the Hartley-Rao scheme was the following:

Step 1. Drafting of probabilities. Probability of sampling in j-county in the w-stratum:

$$p_{jw} = X_j \left(\sum_{j=1}^{N} X_{jw} \right)^{-1} = \frac{X_{jw}}{X_w} \tag{1}$$

First order probability:

$$\pi_{jw} = n_w^p p_{jw} \tag{2}$$

where:

$X_{jw} > 0$ is a number of farmers in the j-county, in the w-stratum

$$X_w = \sum_{j=1}^{N_w^p} X_{jw} \text{ —number of farmers in the stratum,}$$

$$X = \sum_{w=1}^{W^p} X_w \text{ —total number of farmers in all strata,}$$

j—number of the county,

$j = 1, 2, ..., N_w^p,$

N_w^p —number of counties in the stratum,

$$N^p = \sum_{w=1}^{W^p} N_w^p \text{ —total number of counties in Poland,}$$

w—the stratum number,

$w = 1, 2, ..., W^p$

W^p —total number of strata,

$$n_w^p \cong \frac{X_w}{\sum\limits_{w=1}^{W^p} X_w} n^p \tag{3}$$

n_w^p —number of, counties in the sample to be selected from the w-stratus,

n^p —number of counties in the sample to be selected for the entire country

Step 2. The pairs of numbers (j, π_j) were arranged separately in each stratum, and then for eachstratum cumulative series of probabilities were established:

$$\pi_{jw}^{(sk)} = \sum\limits_{r=1}^{j} \pi_{rw} \tag{4}$$

Step 3. For each stratum the draw interval was adopted as k = 1, and then random numbers U_w with uniform distribution in the (0,1) interval were generated, each for every strata. U_w is so called "random beginning" for this stratum.
The inequality needs to be checked:

$$\pi_{j-1,w}^{(sk)} < U_w + (i-1) \leq \pi_{jw}^{(sk)} \tag{5}$$

for $j = 1,2,...,N_w^p$, $i = 1,2,...,n_w^p$, $\pi_{0,w}^{(sk)} = 0$.

Counties with the j number (after random arrangement), which met the above inequality were selected for the sample (5).

Value $\dfrac{n_w^p}{X_w}$, for the w-stratum, in the pattern (4) on $\pi_{jw}^{(sk)}$ is constant and, therefore,

after the random arrangement of the counties stratum cumulative series of values

of the numbers of farmers in the counties: $X_{jw}^{sk} = \sum\limits_{r=1}^{j} X_{rw}$. Then similarly to systematic drawing:

– the length of the drawing step was designated: $\dfrac{X_w}{n_w^p} = L_w$

– random number was designated to mark "the beginning of drawing": U_w^*, where $L_w \geq U_w^* > 0$,

– the counties selected for the sample had to meet the following inequality requirement:

$$X_{j-1,w}^{(sk)} < U_w^* + (i-1)L_w \leq X_{jw}^{(sk)} \tag{6}$$

for, $j = 1, 2, ... N_w^P, i = 1, 2, ..., n_w^P, X_{0,w}^{(sk)} = 0$.

As a result of applying the above, 103 counties from all 16 provinces were sampled. The sample included 97 land counties and 6 city counties. It was assumed that for each county a sample of farmers allowing for 30 interviews would be drawn, but the sample would still be proportional to the number of individual farms in a particular county. Consequently, the actual research sample would count 3505 farms, which would correspond with the planned number of interviews to be conducted. The list of counties selected for the research sample with the planned number of interviews can be seen in Table 2.3.2. Picture 2.1. presents the territorial location of these counties.

A random drawing of municipalities from a pre-selected group of 103 counties was conducted with the use of the TERYT system as a sampling frame. Secondary sampling units (SSUs) included rural census areas and statistical areas in cities. It was determined that SSUs should contain at least four farms and for that reason rural and urban statistical areas with less than four farms were merged with adjacent areas to ensure that each SSU would contain at least four farms. All selected SSUs were grouped in strata in such a way that SSUs in particular municipalities constituted one stratum. For SSUs, a sample drawing the procedure of stratified sampling was applied to ensure the probability of choice proportional to the number of farm SSUs. They were drawn separately in each stratum according to the Hartley-Rao scheme. As a result, 863 municipalities were selected for the sample and their full list can be found in Annex No. 1.

It was determined that a drawing of farm addresses according to group areas in surveyed counties/municipalities would be conducted in each SSU that was selected for the sample, through the procedure of normal drawing without repetitions, using Central Statistical Office databases. As farms in the sample had to be larger than 1 ha in surveyed counties, four area groups were considered: 1) farms larger than 1 ha and smaller than 2 ha, 2) farms equal or larger than 2 ha but smaller than 5 ha, 3) farms sized 5 ha or more but smaller than 10 ha, and 4) farms of 10 ha and more. Detailed data can be found in Annex No. 2. The ultimate structure of the sample, taking into

Table 2.4.2. Randomly selected counties with the number of interviews with the farmers and planned number of interviews according to province

Province (voivodeship)	County	Planned number of interviews to be conducted	Planned numbers of interviews to be conducted in the province
Lower Silesian	Lubański	30	121
	Kłodzki	31	
	Polkowicki	30	
	Oleśnicki	30	
Kuyavian-Pomeranian	Rypiński	30	155
	Świecki	31	
	Wąbrzeski	30	
	Włocławski	34	
	Mogileński	30	
Lubelskie	Włodawski	30	468
	Bialski	49	
	Krasnostawski	35	
	Chełmski	38	
	Biłgorajski	49	
	Tomaszowski	41	
	Łęczyński	32	
	Lubelski	60	
	Janowski	33	
	Rycki	39	
	Puławski	30	
	City of Zamość*	30	
Lubuskie	Sulęciński	30	60
	Żarski	30	
Łódzkie	Kutnowski	32	321
	Skierniewicki	32	
	Zgierski	32	
	Radomszczański	37	
	Opoczyński	36	
	Piotrkowski	40	
	Wieluński	37	
	Sieradzki	43	
	Poddębicki	32	
Lesser Poland	Myślenicki	35	355
	Bocheński	35	
	Miechowski	33	
	Gorlicki	38	
	Limanowski	44	
	Nowotarski	45	
	Olkuski	32	
	Tarnowski	60	
	Dąbrowski	33	
Mazovian	Sierpecki	31	516
	Mławski	32	
	Pułtuski	31	
	Ostrołęcki	41	
	Ostrowski	35	
	Kozienicki	32	

	Szydłowiecki	30	
	Białobrzeski	31	
	Miński	36	
	Garwoliński	40	
	Sochaczewski	32	
	Grójecki	43	
	Sokołowski	34	
	Węgrowski	35	
	Capital City of Warsaw*	33	
Opolskie	Oleski	31	61
	Nyski	30	
Subcarpathian	Krośnieński	36	306
	Sanocki	33	
	Przemyski	34	
	Jarosławski	37	
	Kolbuszowski	33	
	Ropczycko-Sędziszowski	32	
	Dębicki	38	
	Niżański	32	
	Tarnobrzeski	31	
Podlaskie	Suwalski	31	197
	Sokólski	36	
	Bielski	34	
	Łomżyński	33	
	Kolneński	31	
	City of Białystok*	32	
Pomeranian	Tczewski	30	120
	Słupski	30	
	Kościerski	30	
	City of Gdańsk*	30	
Silesian	Cieszyński	31	157
	Gliwicki	30	
	Zawierciański	32	
	Kłobucki	34	
	City of Jastrzębie-Zdrój*	30	
Holy Cross	Konecki	31	206
	Ostrowiecki	31	
	Buski	42	
	Sandomierski	39	
	Włoszczowski	32	
	Kazimierski	31	
Warmian-Masurian	Olecki	30	90
	Bartoszycki	30	
	Iławski	30	
Greater Poland	Kępiński	30	282
	Ostrowski	33	
	Turecki	34	
	Gnieźnieński	30	
	Kolski	35	
	Wolsztyński	30	

Province (voivodeship)	County	Planned number of interviews to be conducted	Planned numbers of interviews to be conducted in the province
	Czarnkowsko--Trzcianecki	30	
	Szamotulski	30	
	City of Leszno*	30	
	Gryficki	30	
West Pomeranian	Szczecinecki	30	90
	Koszaliński	30	
Total			3505

* This symbol means city county. The remaining counties are land counties.

Source: Own work based on data in: Engel & Dąbała, 2018.

Picture 2.4.1. Territorial location of counties selected for the research sample
Source: Own work.

consideration farm size (with farms assigned to four area groups mentioned above) was the following: 19.3% were farms smaller than 2 ha (675 such farms), 33.8% were farms larger than 2 ha and smaller than 5 ha (1186 farms), 23.5% farms sized 5 to 10 ha (823 farms), and 23.4% farms larger than 10 ha (821 farms).

The detailed structure of the sample with considered farm size in each of the surveyed counties can be found in Annex No. 3.

The sample selected in such a way should be recognized as a representative sample for the total population of active farms in Poland with an area greater than 1 ha.

2.5. Realization of Sample Selection and Field Research[4]

2.5.1. Sampling Frame for Farms

Despite the best efforts of the Center of Sociological Research of the Institute of Philosophy and Sociology of the Polish Academy of Science, selecting the address sample for individual farms by Central Statistical Office in Poland proved to be impossible due to our adherence to the principles of statistical information secrecy (according to the law on public statistics from June 29, 1995 (Journal of Laws from 2016, Item 1068). For this reason, the construction of the sample took several stages. First, the Central Statistical Office provided the dataset of addresses and a sample of farms addresses was drawn in each previously selected county/municipality and then the addresses were verified by the Agency for Restructuring and Modernization of Agriculture in terms of operating or not operating the farm. If the household was indeed part of a farm, the agency provided information on utilized agricultural areas and then the farm was classified for the sample. As was indicated earlier, SSUs were to have at least four farms each, and the completed research sample was to contain at least 3505 farms. In the first stage 14,020 farm addresses were drawn from a database of the Central Statistical Office in the sampled 103 counties from the same office (3505 × 4 = 14,020) and, in the second stage, a further 15,144 addresses of farms from 68 counties were drawn. They were the counties in which the planned number of interviews in the sample constructed on the basis of the first stage drawing, and the acquired percentage of realization in these counties, was the basis for estimating the number of drawn addresses. Overall, 29,164 addresses were drawn for the sample but only 8697 of them turned out to be addresses of farms of previously determined area. The number of households selected for the sample in particular counties can be found in Annex 4.

---

[4] The presented descriptions of the procedure of sample selection and field research are based on the report mentioned in footnote number 2.

2.5.2. Level of Execution of the Sample Size

The planned number of interviews could not be conducted in two out of 103 selected counties. They were: City County of Warsaw (where none of the planned 33 interviews were conducted) and the City County of Leszno (where only 8 out of 30 planned interviews were done). Regarding the city county of Warsaw, 46 replacement interviews were conducted in neighboring counties. Overall, 3551 interviews were conducted, including an additional 101 interviews in 23 counties.

As far as farm area structure was concerned, the planned sample was not achieved in the group of farms that were over 1 ha and less than 2 ha. Overall, out of 675 planned interviews, only 327 (48.4%) were conducted and, it should be mentioned, that in as many as 11 counties not even one interview was performed. This was connected with the general tendency that the number of farms in this category is declining: while in 2010 there were 300,515 of them, by 2016 they had dropped down to 271,122. There were at least several reasons for such a situation. Among those that researchers noted during the field research were the following: abandonment of the farm for economic reasons, abandonment of farming by older people who do not have a successor and are unable to find anyone willing to operate such a small farm, leasing the entire farms or large portions of farms to other farmers and leaving only a few ares (for American readers, 1 acre = 40.47 ares) to themselves to grow plants for household needs. Detailed data on the number of interviews planned and executed on the farms according to area groups in selected counties can be found in Table 2.5.1.

Table 2.5.1. Number of planned and conducted interviews on farms, by area groups in selected counties

Province (voivodeship)	County	1 to 1.99 Planned	1 to 1.99 Conducted	2 to 4.99 Planned	2 to 4.99 Conducted	5 to 9.99 Planned	5 to 9.99 Conducted	10 and over Planned	10 and over Conducted	Total Planned	Total Conducted	Diff Planned	Diff Conducted
Lower Silesian	Lubański	5	1	14	15	5	10	6	6	30	32		+2
	Kłodzki	5	1	11	11	7	10	8	9	31	31		
	Polkowicki	6	3	10	10	6	6	8	11	30	30		
	Oleśnicki	5	2	9	11	7	7	9	10	30	30		
Kuyavian-Pomeranian	Rypiński	3	2	6	5	9	12	12	11	30	30		
	Świecki	5	4	7	8	6	6	13	13	31	31		
	Wąbrzeski	3	3	6	6	6	3	15	18	30	30		
	Włocławski	3	0	7	8	10	11	14	15	34	34		
	Mogileński	4	0	4	6	6	4	16	20	30	30		
Lubelskie	Włodawski	3	3	8	8	9	9	10	10	30	30		
	Bialski	6	0	13	9	14	15	16	25	49	49		
	Krasnostawski	5	2	12	15	11	11	7	9	35	37		+2
	Chełmski	4	5	16	9	9	13	9	11	38	38		
	Biłgorajski	8	4	20	20	16	19	5	6	49	49		
	Tomaszowski	8	3	15	18	11	14	7	6	41	41		
	Łęczyński	6	4	12	11	9	9	5	8	32	32		
	Lubelski	11	3	24	27	15	18	10	12	60	60		
	Janowski	5	5	13	17	11	10	4	2	33	34		+1
	Rycki	6	5	13	14	9	11	4	2	32	32		
	Puławski	11	7	16	19	9	11	3	2	39	39		
	City of Zamość	8	8	16	15	4	3	2	4	30	30		

| Province (voivodeship) | County | Number of interviews conducted on farms in named county, by area (in hectares) | | | | | | | | Total number of interviews | | General difference between the numbers of interviews planned and conducted | |
| | | 1 to 1.99 | | 2 to 4.99 | | 5 to 9.99 | | 10 and over | | | | | |
		Planned	Conducted	Planned	Conducted	Planned	Conducted	Planned	Conducted	Planned	Conducted	Planned	Conducted
Lubuskie	Sulęciński	6	2	8	5	6	9	10	14	30	30		
	Żarski	6	1	11	12	6	6	7	13	30	32		+2
Łódzkie	Kutnowski	4	2	6	7	9	14	13	11	32	34		+2
	Skierniewicki	5	4	10	9	11	13	6	6	32	32		
	Zgierski	4	3	9	10	11	10	8	9	32	32		
	Radomszczański	6	4	14	15	11	8	6	10	37	37		
	Opoczyński	7	2	16	19	10	12	3	3	36	36		
	Piotrkowski	5	4	13	11	14	16	8	9	40	40		
	Wieluński	7	4	14	16	11	11	5	6	37	37		
	Sieradzki	5	0	13	14	15	14	10	15	43	43		
	Poddębicki	3	0	8	11	10	10	11	11	32	32		
Lesser Poland	Myślenicki	16	5	16	27	2	3	1	0	35	35		
	Bocheński	13	2	17	25	4	7	1	1	35	35		
	Miechowski	5	1	12	8	10	12	6	12	33	33		
	Gorlicki	16	7	17	24	4	4	1	3	38	38		
	Limanowski	14	9	24	29	5	5	1	1	44	44		
	Nowotarski	14	2	22	27	7	14	2	2	45	45		
	Olkuski	11	4	15	16	5	7	1	5	32	32		
	Tarnowski	21	11	30	49	7	11	2	4	60	75		+15
	Dąbrowski	8	5	16	21	7	7	2	3	33	36		+3

Region	District												
Mazovian	Sierpecki	3	2	5	6	8	10	15	14	*31*	32		+1
	Mławski	3	4	6	5	7	9	16	14	*32*	32		
	Pułtuski	2	2	7	10	10	17	12	17	*31*	46		+15
	Ostrołęcki	2	0	7	3	13	10	19	28	*41*	41		
	Ostrowski	3	1	10	4	11	12	11	18	*35*	35		
	Kozienicki	5	2	12	9	10	11	5	10	*32*	32		
	Szydłowiecki	8	3	14	17	6	9	2	1	*30*	30		
	Białobrzeski	4	1	10	12	11	11	6	7	*31*	31		
	Miński	6	3	14	9	11	17	5	9	*36*	38		+2
	Garwoliński	6	2	15	10	14	29	5	17	*40*	58		+18
	Sochaczewski	6	1	11	13	9	10	6	8	*32*	32		
	Grójecki	5	5	14	12	16	23	8	14	*43*	54		+11
	Sokołowski	4	1	9	2	10	16	11	15	*34*	34		
	Węgrowski	9	2	15	9	7	18	4	6	*35*	35		
	Capital City of Warsaw	8	0	14	0	7	0	4	0	*33*	0	–33	
Opolskie	Oleski	5	1	10	8	8	10	8	12	*31*	31		
	Nyski	7	4	8	9	6	5	9	13	*30*	31		+1
Subcarpathian	Krośnieński	19	8	14	20	2	6	1	2	*36*	36		
	Sanocki	11	6	14	17	5	7	3	4	*33*	34		+1
	Przemyski	10	4	16	16	6	11	2	4	*34*	35		+1
	Jarosławski	10	6	17	15	7	6	3	10	*37*	37		
	Kolbuszowski	7	4	18	16	7	12	1	1	*33*	33		
	Ropczycko-Sędziszowski	10	11	17	16	4	4	1	2	*32*	33		+1
	Dębicki	12	6	19	21	6	11	1	0	*38*	38		
	Niżański	9	10	15	16	7	6	1	1	*32*	33		+1
	Tarnobrzeski	10	3	16	18	4	5	1	5	*31*	31		

| Province (voivodeship) | County | Number of interviews conducted on farms in named county, by area (in hectares) | | | | | | | | Total number of interviews | | General difference between the numbers of interviews planned and conducted | |
| | | 1 to 1.99 | | 2 to 4.99 | | 5 to 9.99 | | 10 and over | | | | | |
		Planned	Conducted	Planned	Conducted	Planned	Conducted	Planned	Conducted	Planned	Conducted	Planned	Conducted
Podlaskie	Suwalski	2	1	5	1	5	8	19	21	31	31		
	Sokólski	2	1	5	5	10	9	19	21	36	36		
	Bielski	4	2	8	10	10	14	12	8	34	34		
	Łomżyński	2	0	7	2	9	11	15	20	33	33		
	Kolneński	1	1	3	1	7	8	20	21	31	31		
	City of Białystok	7	4	14	15	7	4	4	9	32	32		
Pomeranian	Tczewski	3	3	5	4	6	7	16	17	30	31	+1	
	Słupski	5	3	7	8	6	7	12	12	30	30		
	Kościerski	3	3	6	7	7	8	14	14	30	32	+2	
	City of Gdańsk	7	7	10	5	8	10	5	8	30	30		
Silesian	Cieszyński	15	8	10	13	3	7	2	3	30	31	+1	
	Gliwicki	8	5	10	11	7	8	9	10	34	34		
	Zawierciański	7	5	11	12	8	9	6	6	32	32		
	Kłobucki	5	5	15	15	7	10	3	0	30	30		
	City of Jastrzębie-Zdrój	15	10	12	16	3	5	1	0	31	31		
Holy Cross	Konecki	8	3	14	18	7	8	2	5	31	34	+3	
	Ostrowiecki	7	4	12	11	8	10	4	6	31	31		
	Buski	8	0	18	26	12	21	4	7	42	54	+12	
	Sandomierski	9	2	19	22	9	12	2	3	39	39		
	Włoszczowski	4	3	13	8	10	9	5	12	32	32		
	Kazimierski	4	0	11	11	11	14	5	9	31	34	+3	

Region	County												
Warmian-Masurian	Olecki	4	1	*5*	6	5	5	*16*	18	*30*	30		
	Bartoszycki	3	1	*5*	7	5	6	*17*	16	*30*	30		
	Iławski	3	3	*5*	5	6	3	*16*	19	*30*	30		
Greater Poland	Kępiński	5	0	*6*	9	7	7	*12*	14	*30*	30		
	Ostrowski	6	5	*9*	7	8	9	*10*	12	*33*	33		
	Turecki	6	2	*11*	13	10	15	*7*	4	*34*	34		
	Gnieźnieński	3	0	*5*	6	5	4	*17*	20	*30*	30		
	Kolski	5	4	*9*	11	10	10	*11*	10	*35*	35		
	Wolsztyński	5	2	*7*	12	8	8	*10*	8	*30*	30		
	Czarnkowsko--Trzcianecki	4	1	*7*	9	6	3	*13*	17	*30*	30		
	Szamotulski	4	1	*5*	3	6	3	*15*	23	*30*	30		
	City of Leszno	8	3	*9*	2	5	0	*8*	3	*30*	8	−22	
West Pomeranian	Gryficki	4	1	*5*	7	5	5	*16*	17	*30*	30		
	Szczecinecki	4	2	*6*	7	5	8	*15*	13	*30*	30		
	Koszaliński	5	2	*7*	11	6	5	*12*	12	*30*	30		
Total		675	327	*1,186*	1,246	823	980	*821*	998	*3,505*	3,551	−55	+101

Elaboration: Engel & Dąbała, 2018.

2.5.3. Weights

Due to insufficient representation of farms with areas larger than 1 ha and less than 2 ha, and overrepresentation of farms in the remaining three area groups in 23 counties, the Center of Sociological Research of the Institute of Philosophy and Sociology of the Polish Academy of Sciences, which conducted the study, prepared three types of scales:

1) probabilistic weights, which have the main task of equalizing the probability of sample drawing in a particular unit of the research. In other words, these weights may serve to correct a sample which used an imperfect selection scheme, in which the probability of being selected is not equal, as seen in a simple random sample (SRS);

2) post-stratification correction weights, which are meant to provide correction of data acquired through incomplete execution of the sample and potential defects of an already selected sample;

3) combined weights, which combine different elements that were previously calculated by probabilistic weights and post-stratification weights.

The six types of probabilistic weights were:

1) Wprob_SSU is an unscaled weight describing the probability of drawing the secondary sampling unit (SSU) that comes from information in the sample selection. In cases where there was no information on probabilities, imputation was used and it meant assigning to secondary sampling units an average weight for the entire sample;

2) Wprob_p1 is an unscaled weight that describes the probability of drawing the planned sample of households in a particular county;

3) Wprob_p2 is an unscaled weight that describes the probability of drawing the sample of agricultural households in the county j and then scaled to the general number in the sample, which does not change the size of the executed sample;

4) Wprob_w1 is an unscaled weight that describes the probability of drawing the sample of agricultural households in the province (voivodeship);

5) Wprob_w2 is a weight describing the probability of drawing the sample of farms in the province, scaled to the general number in the sample of farms without hanging the size of the executed sample;

6) Wprob_hh is a weight correcting the probabilities of drawing farms according to their sizes (numbers of members). These probabilities were proportional

to the number of their members due to the use of address samples of farms in lieu of samples based on names.

There were two types of corrective post-stratification weights prepared:

1) Wstrat_size is a post-stratification weight aligning the distributions of samples executed by farm size area (grouped into four categories: 1 to1.99 ha, 2 to 4.99 ha, 5 to 9.99 ha, and 10 ha or more) to the alignment of sample distributions of farms throughout Poland. The weight is scaled to the general number in the sample of farms and, therefore, the size of the executed sample remains unchanged;

2) Wstrat_voi_size is a post-stratification weight aligning the distributions of samples executed according to farm size (area), grouped into four categories: 1 to 1.99 ha, 2 to 4.99 ha, 5 to 9.99 ha, and 10 ha or more, which are divided by provinces (voivodeships) and show the distribution of actual farm areas throughout Poland. The weight is scaled to the general number in the sample of farms and, therefore, the size of the executed sample remains unchanged.

Additionally, a combined weight was also prepared: wcomb. This weight consists of probabilistic weights (for provinces, voivodeships, counties and farm size) and post-stratification weight, combining the farm size with the division by province (voivodeship).

2.6. Desk Research

One of the main goals of the project is to analyze the territorial differentiation in functioning of family farms. The authors of the study described here were even more ambitious with their goal, going much further than just stating that the presence of such an influence exists. Using the multi-level model, they attempted to analyze which concrete characteristics of local communities (at the county level) could be the most meaningful for farmers constructing development strategies for their farms. Therefore, one of the research tasks was to collect data on counties, where the research was conducted among the managers of farms, which could become the basis for elaborating various indicators for comparative analysis for counties and operators of farms alike. Out of the dataset of initial data indicated by researchers, The Center of Sociological Research of the Institute of Philosophy and Sociology of the Polish Academy of Sciences was responsible for collecting and preparing the database on counties, primarily by using the datasets of the Bank of Local Data,[5] and constructed a dataset containing the information on county population, the number of people in towns considered to be urban areas, the number of such towns, the number of towns (both urban and rural) in the county, the physical distance from the county capital to the province capital, the structure of age and education level of the inhabitants, the number of individual farms with areas over 1 ha in the county, family employment on farms with areas over 1 ha, the number of active business entities, as well as the number of people employed outside of agriculture, hunting, forestry, and fisheries and the number of people employed outside of these four sectors and outside of industry and construction combined. Other characteristics included county GDP in current valuation; county unemployment rate; total number of households; number of households that receive help from social welfare; average result of the middle school exam in mathematics; percentage of children aged 3–4 who attend preschool; number of high schools, vocational schools, and post-high school educational facilities; number of students in high schools, vocational schools, and post-high school educational facilities; municipal and county expenses per en-rolled student of elementary and junior high schools; number of doctors employed by National Health Fund (not including dentists); numbers of clinics and health centers; number of beds in the county hospitals; number of crimes and percentage of crimes solved with the offender being captured; number of libraries, cultural centers, museums, theatres, cinemas, art galleries and other cultural institutions; number of registered cars; voter turnout in the county council election in 2014; voter turnout in parliamentary election in 2015; number of registered nongovernment organization; number of book loans in libraries; and total taxpayer income (all inhabitants of the county) according to income tax returns for 2016. The above data for each of the 103 counties selected for the research conducted among farmers in these counties are presented in the following tables: 2.6 A, 2.6 B, and 2.6 C. These data describe multiple aspects of social life and serve to elaborate indicators that measure the functioning of various areas of the lives of county communities.

---

[5]   https://bdl.stat.gov.pl/BDL.

Table 2.6.1. Essential characteristics of the surveyed counties (data for 2015)

Province/county (provinces bolded)	Population	Population in towns that operate as urban areas for having town charter/urban rights	Number of towns that operate as urban areas for having town charter/urban rights	Total number of towns and villages	Distance between the county capital and the province capital in km	Population under age 19	Population over age 65	Number of people with at least high school education (2011)	Number of individual farms with size over 1 ha (2010)	Number of people working on farms sized over 1 ha—family labor (2010)	Business entities (not including agriculture, hunting, forestry and fisheries)	Number of people who are employed (not including employment in agriculture, hunting, forestry and fisheries)	Number of people who are employed (not including the employment in agriculture, hunting, forestry, fisheries, industry and construction)
Poland	38,437,239	23,166,429	915	52,529	×	7,732,280	6,076,418	16,266,875	1,480,227	3,669,449	4,109,008	8,911,073	5,871,709
Lower Silesian	2,904,207	2,008,951	91	2,527	×	544,820	470,617	1,272,950	59,915	133,910	352,280	738,254	471,143
Kłodzki	162,465	104,395	11	186	91.3	28,393	28,447	33,885	4,693	9,730	172,20	24,865	17,457
Lubański	55,533	34,847	4	53	156.5	10,505	8,771	10,322	1,376	3,018	5,761	8,314	5,271
Oleśnicki	106,486	62,021	5	114	33.0	22,065	15,509	22,706	3,237	7,654	9,750	19,519	9,372
Polkowski	63,057	36,925	3	91	88.4	13,718	8,303	11,440	1,810	4,263	4,522	30,775	8,364
Kuyavian-Pomeranian	208,6210	1,244,129	52	3,583	×	426,394	319,704	792,496	66,761	154443	189,167	449,959	279,052
Mogileński	46,254	17,910	2	172	63.2	9,661	6,823	8,007	2,248	5,397	3,203	7,646	4,755
Rypiński	44,384	16,629	1	144	108.5	9,744	6,527	7,103	3,203	7,483	3,252	7,095	3,760
Świecki	99,764	32,129	2	260	46.6	21,637	13,681	15,617	4,177	9,482	7,278	19,181	9,522
Wąbrzeski	34,844	13,887	1	86	75.8	7,500	5,080	5,013	2,094	5,122	2,337	5,520	2,879
Włocławski	199,799	130,385	7	438	106.4	38,469	32,725	47176	7,552	16,965	17,337	40,627	23,258
Lubelskie	2,139,726	988,034	43	4,047	×	432,673	350,047	902,624	185,643	474,307	169,048	373,184	263,968
Bialski	169,921	80,189	3	416	129.1	36,552	25,289	41,668	14,699	35,759	11,999	27,394	21,153
Biłgorajski	102,647	34,232	4	188	90.1	21,372	16,452	19,567	12,524	36,218	7,258	13,231	7,845
Chełmski	143,647	68,715	2	372	69.2	27,975	22,373	31,774	10,756	24,774	9,377	20,038	15,404
Janowski	46,895	13,522	2	109	82.5	9,659	7,962	8,827	6,233	18,485	3,245	5,515	3,624
Krasnostawski	65,422	19,116	1	192	55.1	11,948	12,552	11,870	7,751	19,627	3,638	8,160	5,088
Lubelski	491,905	352,438	3	353	.0	96,271	81,714	170,326	21,037	53,391	55,147	133,806	106,583
Łęczycki	57,401	19,437	1	124	26.2	12,423	6,851	12,137	5,528	14,040	3,182	13,209	4,991

Province/county (provinces bolded)	Population	Population in towns that operate as urban areas for having town charter/urban rights	Number of towns that operate as urban areas for having town charter/urban rights	Total number of towns and villages	Distance between the county capital and the province capital in km	Population under age 19	Population over age 65	Number of people with at least high school education (2011)	Number of individual farms with size over 1 ha (2010)	Number of people working on farms sized over 1 ha—family labor (2010)	Business entities (not including agriculture, hunting, forestry and fisheries)	Number of people who are employed (not including employment in agriculture, hunting, forestry and fisheries)	Number of people who are employed (not including the employment in agriculture, hunting, forestry, fisheries, industry and construction)
Piławski	115,206	55,052	3	157	53.4	22,623	21,481	25,798	8,856	23,057	9,583	23,729	13,991
Rycki	57,388	26,608	2	108	66.4	11,792	9,333	12,689	5,542	14,728	3,777	8,315	5,472
Tomaszowski	85,705	23,931	3	238	123.6	16,874	14,450	16,197	10,016	24,788	6,301	10,404	6,896
Włodawski	39,280	13,562	1	126	86.6	7,898	6,266	8,645	3,712	8,390	2,567	4,642	3,222
City of Zamość	64,788	64,788	1	1	87.0	12,326	9,844	20,709	2,068	4,077	7,746	17,280	14,101
Lubuskie	1,018,075	661,321	42	1,297	×	206,695	150,359	408,047	21,348	46,954	107,856	217,809	130,884
Sulęciński	35,596	14,847	3	92	92.4	7,453	4,873	6,817	1,285	2,776	2,730	5,855	3,867
Żarski	98,160	59,558	4	157	45.6	19,754	14,189	17,982	2,116	4,598	9,270	19,926	9,815
Łódzkie	2,493,603	1,572,862	44	5,009	×	470,039	439,962	1,100,837	128,997	317,662	237,422	573,596	365,835
Kutnowski	99,258	57,963	3	351	60.5	17,712	17,696	21,911	5,037	11,510	7,410	20,522	11,206
Opoczyński	77,457	25,809	2	238	86.5	16,989	11,896	16,074	7,951	21,293	4,494	12,674	7,273
Piotrkowski	166,466	83,837	3	404	94.4	34,871	27,192	39,651	10,366	26,882	12,678	34,869	23,388
Poddębicki	41,679	10,739	2	329	40.0	7,977	7,317	7,441	5,685	12,632	3,082	4,932	2,986
Radomszczański	115,125	53,575	3	358	92.9	22,503	19,879	23,386	8,112	18,323	8,777	21,428	11,959
Sieradzki	119,268	51,819	4	399	68.5	24,286	19,370	26,723	11,333	29,813	8,789	20,017	13,047
Skierniewicki	86,565	48,388	1	182	78.7	17,923	13,867	22,011	6,189	16,345	7,240	15,379	9,181
Wieluński	77,290	23,169	1	213	102.9	15,947	12,599	14,389	8,113	20,800	6,406	13,678	7,420
Zgierski	165,130	116,549	5	245	13.6	31,756	28,427	41,310	5,196	12,517	15,633	32,874	18,305
Lesser Poland	3,372,618	1,634,901	61	1,951	×	715,005	515,096	1,389,739	150,497	411,106	359,934	753,532	519,941
Bocheński	105,416	32,868	2	103	49.4	24,682	14,242	20,468	7,182	19,541	8,433	17,674	9,421
Dąbrowski	59,374	16,054	2	79	103.4	12,337	8,586	8,796	6,304	17,312	3,056	6,046	4,261

Gorlicki	109,140	35,799	3	78	137.6	24,312	16,065	17,494	8,731	25,334	7,590	15,242	9,832
Limanowski	129,699	23,052	2	89	70.7	34,592	15,918	22,544	11,898	33,925	9,462	16,288	10,252
Miechowski	49,592	11,737	1	142	41.2	9,546	9,154	10,025	6,364	17,301	4,249	5,824	5,017
Myślenicki	125,020	31,353	3	71	33.0	30,238	15,393	23,930	7,397	20,353	11,973	18,330	10,538
Nowotarski	190,517	52,418	3	101	88.6	44,653	25,906	29,854	12,298	32,298	15,203	24,311	16,514
Olkuski	113,148	55,327	3	108	42.1	21,154	19,112	27,990	5,616	14,777	11,682	23,788	12,151
Tarnowski	311,244	134,730	8	180	83.5	65,760	48,088	69,630	16,570	45,997	23,029	60,317	38,615
Mazovian	5,349,114	3,438,225	86	8,532	×	1,101,826	868,178	2,613,279	225,664	530,084	757,003	1,522,274	1,188,120
Białobrzeski	33,569	8,012	2	165	73.0	7,750	4,715	6,185	4,309	9,985	2,557	3,569	2,485
Garwoliński	108,740	30,706	4	279	73.4	26,006	15,183	21,197	9,816	26,170	7,265	16,940	9,683
Grójecki	98,619	34,613	4	399	47.8	20,823	15,332	20,653	11,248	27,530	8,559	17,534	10,474
Kozienicki	61,319	17,869	1	207	93.0	12,351	9,538	12,986	5,614	13,991	4,011	10,812	5,172
Miński	151,520	69,973	5	335	41.2	34,573	21,836	36,440	7,752	17,626	13,878	22,963	14,505
Mławski	73,758	31,030	1	247	128.2	15,866	10,994	13,749	5,440	12,534	4,885	14,185	7,239
Ostrołęcki	140,949	55,939	2	358	128.5	31,756	18,668	28,474	10,944	26,997	10,982	23,569	15,112
Ostrowski	73,911	24,731	2	315	98.4	15,681	11,858	12,924	7,262	17,443	6,098	11,179	6,394
Pułtuski	51,637	19,309	1	238	61.4	11,406	7,547	11,408	4,548	11,490	3,817	6,651	4,462
Sierpecki	52,980	18,317	1	241	124.8	11,485	7,908	9,434	4,350	9,674	3,136	6,861	4,458
Sochaczewski	85,167	37,102	1	238	60.5	17,942	12,816	18,490	5,668	13,764	8,581	18,012	11,979
Sokołowski	55,152	20,915	2	250	15.1	11,087	9,719	10,784	6,879	14,926	3,713	10,072	6,206
Szydłowiecki	40,241	11,940	1	90	131.6	8,639	5,971	7,737	2,692	6,249	2,991	4,438	2,932
Węgrowski	67,005	19,622	2	240	8.9	14,426	10,786	14,317	7,498	17,564	4,515	8,392	5,388
City of Warsaw	1,744,351	1,744,351	1	1	.0	315,869	331,571	824,928	6,397	11,427	400,128	862,386	755,385
Opolski	996,011	517,177	35	1,172	×	180,281	163,838	378,610	27,771	68,993	97,696	202,139	120,840
Nyski	138,969	72,989	5	159	55.1	24,995	23,081	28,577	3,479	8,342	13,435	20,421	13,020
Oleski	65,306	23,591	4	158	47.4	11,964	10,615	10,109	4,484	12,213	5,251	10,475	5,136
Subcarpathian	2,127,657	877,671	51	1,664	×	446,611	316,791	821,722	137,827	379,802	161996	424,518	258,987
Dębicki	135,293	53,916	3	84	55.4	29,558	18,834	27,663	8,965	25,977	9137	28,260	14,801
Jarosławski	121,508	47,498	3	108	59.6	25,858	17,804	23,126	8,486	23,047	8469	21,797	13,159
Kolbuszowski	62,513	9,236	1	52	30.9	13,521	8,781	9,511	6,038	16,430	3537	8,356	5,841
Krośnieński	158,892	60,281	5	108	56.2	33,573	24,892	34,331	8,181	21616	13058	34,711	20,502
Niżański	67,178	23,637	3	78	59.2	13,419	9,457	11,261	5,220	13,641	4,036	7,329	4,300

Province/county (provinces bolded)	Population	Population in towns that operate as urban areas for having town charter/urban rights	Number of towns that operate as urban areas for having town charter/urban rights	Total number of towns and villages	Distance between the county capital and the province capital in km	Population under age 19	Population over age 65	Number of people with at least high school education (2011)	Number of individual farms with size over 1 ha (2010)	Number of people working on farms sized over 1 ha—family labor (2010)	Business entities (not including agriculture, hunting, forestry and fisheries)	Number of people who are employed (not including employment in agriculture, hunting, forestry and fisheries)	Number of people who are employed (not including the employment in agriculture, hunting, forestry, fisheries, industry and construction)
Przemyski	137,035	62,720	1	148	90.9	27,851	21,140	32,031	7,417	19,290	10,143	23,043	18,301
Ropczycko-Sędziszowski	73,756	23,245	2	47	30.7	16,966	10,232	12,032	5,304	16,093	4,916	11,671	6,132
Sanocki	95,645	43,568	2	113	75.9	19,698	14,288	19,662	6,224	16,651	6,688	20,788	10,254
Tarnobrzeski	101,527	60,660	3	37	74.8	19,360	16,325	24,431	5,735	16,113	8,197	19,975	12,035
Podlaski	1,188,800	719,890	40	3,759	× ×	235,022	191,243	484,053	83,540	199,200	96,334	214,248	147,412
Bielski	56,562	29,881	2	234	44.7	10,419	11,653	12,058	6,842	15,526	3,614	10,251	5,008
Kolneński	39,162	12,789	2	167	105.1	8,400	5,728	6,415	4,191	11,814	2,246	3,398	2,506
Łomżyński	114,176	66,561	3	299	82.1	23,786	16,175	23,620	7,664	18,949	9,076	18,678	13,576
Sokólski	69,375	29,051	4	422	40.9	13,045	12,311	13,252	7,616	19,376	3,748	8,268	5,722
Suwalskie	105,302	69,370	1	324	131.0	22,764	14,199	23,520	6,160	15,125	8,653	22,149	13,580
City of Białystok	295,981	295,981	1	1	.0	56,199	46,061	106,847	5,164	9,488	34,224	83,666	66,207
Pomerania	2,307,710	1,486,719	42	2,861	×	498,739	338,835	955,601	40,152	96,037	277,387	527,923	354,707
Kościerski	71,624	23,744	1	224	57.6	17,915	8,961	12,622	3,086	8,063	5,908	12,715	7,287
Słupski	190,636	112,140	3	343	135.6	38,269	28,660	46,963	2,812	5,915	21,383	43,016	25,079
Tczewski	115,610	75,044	3	131	37.8	25,992	15,772	22,458	1,796	4,341	10,302	27,078	12,711
City of Gdańsk	462,249	462,249	1	1	.0	83,710	85,034	172,372	1,575	2,639	73,111	158,193	124,516
Silesian	4,570,849	3,525,289	71	1,292	×	858,065	761,884	1,971,850	62,511	155,205	460,894	1,188,920	696,561
Cieszyński	177,562	80,601	5	55	81.2	37,349	27,681	34,236	4,967	13,493	19,057	37,347	22,206
Gliwicki	298,571	246,032	5	65	55.8	54,916	50,158	79,898	2,627	7,008	33,177	100,753	58,520
Kłobucki	85,256	17,570	2	182	84.9	16,786	13,809	14,936	6,954	18,312	6,890	12,947	6,025
Zawierciański	120,270	76,259	6	133	44.5	21,451	22,019	29,800	5,582	14,064	10,815	18,965	10,678

City of Jastrzębie-Zdrój	90,283	90,283	1	1	71.8	17,746	16,035	17,373	837	2,406	5,914	28,171	10,839
Holy Cross	1,257,179	561,219	32	2,493	×	239,720	215,821	518,150	95,168	247,125	108,697	232,060	150,097
Buski	73,068	17,964	2	233	48.0	13,854	13,501	13,393	10,786	27,860	5,462	10,915	8,572
Kazimierski	34,449	6,956	2	127	74.3	6,000	6,387	5,716	4,734	13,620	1,735	3,004	2,467
Konecki	82,336	25,759	2	273	48.5	15,345	14,824	15,459	4,745	10,891	6,266	13,232	7,315
Ostrowiecki	112,417	76,805	3	115	62.3	19,890	20,313	26414	4,412	10,932	10,030	18,733	11,529
Sandomierski	79,266	28,595	3	197	90.4	15,050	14,097	19,000	9,029	24,233	6,246	13,154	9,194
Włoszczowski	45,921	10,269	1	160	56.2	9,215	7,812	8,785	5,248	13,732	3,469	8,148	4,020
Warmian-Masurian	1,439,675	850,385	49	3,874	×	300,978	202,309	541,397	42,854	92,288	119,434	271,202	168,375
Bartoszycki	59,378	32,797	4	269	71.9	11,839	8,629	11,778	2,243	4,883	3,905	7,920	5,476
Iławski	92,795	53,219	5	213	72.9	20,936	12,295	16,750	3,501	7,832	6,732	20,385	8,685
Olecki	34,745	16,460	1	152	162.4	7,605	4,825	6,801	1,674	3,721	2,892	6,413	3,447
Greater Poland	3,475,323	1,906,899	111	5,450	×	742,164	508,274	1,395,358	122,081	300,646	398,511	894,926	534,204
Czarnkowsko-Trzcianecki	87,890	40,496	4	160	77.7	19,623	12,277	15,097	3,764	8,701	6,896	17,073	7,563
Gnieźnieński	145,085	90,376	5	252	55.2	31,835	20,449	31,211	3,903	9,406	15,172	26,010	15,994
Kępiński	56,427	14,469	1	117	168.7	12,399	7,829	8,843	3,080	8,137	5,677	18,785	4,910
Kolski	88,399	33,056	4	275	130.7	18,528	13,709	14,469	7,223	17,090	6,689	13,566	6,831
Ostrowski	161,435	84,660	4	209	116.5	33,580	24,208	33,425	6,441	16,985	15,564	38,400	17,951
Szamotulski	90,133	42,583	5	201	39.1	19,961	12,231	16,710	2,740	6,596	8,378	23,038	10,734
Turecki	84,366	32,526	3	214	127.6	18,361	12,316	15,754	6,803	17,035	6,004	19,614	8,170
Wolsztyński	57,012	13,359	1	103	70.8	13,349	7,587	8,851	3,047	7,960	5,433	14,349	8,070
City of Leszno	64,559	64,559	1	1	79.0	13,069	10,503	18,791	313	521	8,856	22,410	13,371
West Pomeranian	1,710,482	1,172,757	65	3,018	×	333,248	263,460	720,161	29,498	61,687	214,812	326,529	221,583
Gryficki	61,371	30,890	3	154	97.2	12,630	8,369	14,855	1,400	3,200	7,469	7,909	5,847
Koszaliński	173,814	121,739	4	299	160.6	32,675	28,287	49,081	2,503	5,279	25,141	40,305	27,074
Szczecinecki	78,578	51,296	4	241	173.6	15,944	11,930	16,833	1,591	3,360	7,714	12,590	7,913

Source: : Own work based on data collected by Centre of Sociological Research of the Institute of Philosophy and Sociology of Polish Academy of Sciences from Central Statistical Office, Local Data Bank, Provincial Tax Offices, and Education Authorities; data for provinces bolded.

Table 2.6.2. Essential characteristics of the counties (data for 2015)

Province/county	GDP in current prices in millions PLN (2014)	Number of households using the help of social welfare	Unemployment rate (%)	Average result of the junior high math exam (2016)	Percentage of children aged 3–4 years covered by preschool education	Number of high schools, vocational schools and post high school educational institutions	Number of students in high schools, vocational schools and post-high school educational institutions	Total municipal and county expenses per 1 student (grammar schools, junior high schools)	Number of doctors employed through contracts with National Health Fund (not including dentists)	Number of clinics and health centers	Number of hospital beds	Number of crimes committed	Percentage of crimes with offender being captured
Poland	1,719,704.000	1,086,936	9.7	47.89	77.26	14,416	2,301,606	8,276	88,437	20,412	186,994	799,779	64.7
Lower Silesian	145,512.000	70,871	8.5	47.00	78.43	1,036	156,377	8,530	6,540	1,350	14,841	79,395	60.3
Kłodzki	4,743.154	5,446	19.9	40.90	70.61	72	8,647	7,730	284	90	1,055	3,242	70.5
Lubański	1,905.920	1,575	14.2	40.40	69.26	21	2,466	8,890	63	20	321	1,176	66.0
Oleśnicki	4,107.551	1,941	9.8	44.70	68.58	34	5,261	7,213	144	40	248	2,471	77.7
Polkowski	4,384.124	2,031	6.3	42.10	76.68	13	1,676	10,279	39	12	0	1,532	68.5
Kuyavian-Pomeranian	76,063.000	79,978	13.2	46.00	67.65	969	130,312	7,883	4,813	818	9,846	38,832	66.2
Mogieleński	1,338.370	1,671	15.8	51.00	73.81	28	3,430	7,878	49	16	122	543	71.7
Rypiński	1,138.732	2,706	15.7	50.00	48.44	20	2,612	7,949	54	16	196	560	83.3
Świecki	3,810.404	4,099	12.3	49.00	59.21	47	4,365	7,974	137	34	307	1,511	73.7
Wąbrzeski	1,020.142	1,435	17.9	46.00	50.83	13	1,643	7,249	35	14	112	458	75.6
Włocławski	3,381.730	12,047	20.1	48.00	63.29	105	15,866	8,641	365	97	650	4,686	55.0
Lubelskie	67,074.000	64,415	11.7	49.00	74.21	912	141,721	8,644	5,319	1,186	11,307	32,981	71.0
Bialski	2,432.664	5,526	13.3	48.00	76.90	60	12,649	9,288	331	89	756	2,870	77.8
Biłgorajski	2,008.498	2,868	8.0	51.00	71.29	38	6,451	8,392	68	40	314	966	77.0
Chełmski	1,960.282	5,322	15.8	45.50	70.71	57	10,610	8,988	251	72	643	3,140	77.8
Janowski	966.492	1,506	12.2	46.00	57.68	21	2,537	8,763	70	27	301	702	86.2
Krasnostawski	1,617.544	2,323	15.1	45.00	67.97	24	2,809	8,406	87	35	290	811	76.0
Lubelski	11,527.718	12,398	9,0	52.00	81.99	231	40,879	8,807	2,738	347	4,009	9,378	58.1

Łęczycki	2,123.666	1,262	8,3	47.00	82.18	30	1,856	9,213	98	28	193	841	73.9
Piławski	4,122.453	2,987	8,9	49.00	83.68	63	9,191	8,927	214	70	604	1,976	68.4
Rycki	1,888.811	1,731	12.5	48.00	73.20	35	4,261	8,950	70	37	158	724	71.8
Tomaszowski	1,907.269	2,750	12.5	45.00	64.40	19	3,583	9,582	115	38	439	1,152	77.2
Włodawski	978.664	1,671	20.8	44.00	66.46	18	2,060	8,810	47	16	258	580	75.8
City of Zamość	2,364.861	1,995	14.0	51.00	90.08	67	15,269	8,550	384	69	754	1,101	78.8
Lubuskie	38,416.000	36,455	10.5	46.49	76.35	373	56,845	7,929	2,035	551	4,402	26,949	71.2
Sulęciński	1,155.104	2,098	11.7	47.44	71.26	9	1,148	8,373	70	15	400	1,538	85.9
Żarski	3,305.826	4,009	9.7	46.29	76.14	45	5,588	7,993	175	49	363	2,207	67.3
Łódzkie	104,951.000	77,788	10.3	49.00	78.19	968	150,568	8,236	6,676	1,562	12,985	48,321	58.8
Kutnowski	3,280.875	3,929	14.2	42.90	64.45	43	5,630	8,316	139	50	332	1,676	67.4
Opoczyński	2,448.227	2,554	10.0	50.06	58.20	38	4,598	7,946	72	25	230	1,078	68.0
Piotrkowski	3,533.507	5,840	9.6	48.90	71.74	96	16,790	9,579	361	80	590	2,673	65.4
Poddębicki	1,071.919	1,507	11.4	41.60	62.12	7	1,654	8,152	32	22	261	666	61.7
Radomszczański	4,141.350	4,470	11.2	47.20	64.52	44	7,963	8,489	175	55	451	2,062	77.9
Sieradzki	3,550.949	3,338	10.6	48.50	70.79	48	6,833	8,816	267	50	788	2,029	70.7
Skierniewicki	1,670.046	1,965	7.8	51.50	84.46	41	6,128	9,049	168	37	393	1,359	58.6
Wieluński	2,267.318	1,697	10.0	50.20	81.69	44	7,591	8,379	126	44	309	1,582	80.4
Zgierski	5,812.035	5,069	12.6	47.50	79.05	20	5,657	7,608	377	104	792	2,375	64.3
Lesser Poland	134,008.000	73,855	8.3	53.00	77.25	1,104	202,282	8,209	7,668	1,817	14,861	69,300	67.1
Bocheński	2,928.041	2,220	7.0	53.00	79.74	60	6,725	8,244	135	51	218	1,515	77.4
Dąbrowski	1,151.804	1,980	16.1	49.00	64.66	21	3,189	8,618	76	16	248	1,533	85.7
Gorlicki	2,869.010	4,118	11.3	49.00	65.73	38	5,921	8,182	175	60	466	1,089	75.0
Limanowski	3,339.385	4,238	13.5	52.00	58.95	35	7,250	8,436	124	55	459	1,247	66.9
Miechowski	1,075.153	1,146	8.8	50.00	71.31	25	2,630	8,325	97	19	315	791	70.1
Myślenicki	3,176.373	2,670	7.5	52.00	74.74	32	6,413	8,274	164	98	291	1,232	67.5
Nowotarski	4,149.882	3,248	10.7	49.00	66.30	76	11,597	8,200	310	79	712	2,213	65.3
Olkuski	4,023.627	2,219	12.3	50.00	77.65	36	6,399	9,072	171	44	501	2,038	73.1
Tarnowski	4,908.254	8,060	9.9	50.00	82.14	110	23,025	9,222	606	152	1,231	5,107	77.7
Mazovian	381,551.000	130,310	8.3	48.00	83.58	1,875	324,362	8,277	14,497	2,741	25,929	112,453	53.8
Białobrzeski	1,001.710	871	10.9	45.60	60.88	4	1,629	8,211	20	16	0	495	67.3
Garwoliński	2,828.213	2,481	13.3	49.10	71.08	40	6,997	8,389	159	49	300	1,835	82.6

Province/county	GDP in current prices in millions PLN (2014)	Number of households using the help of social welfare	Unemployment rate (%)	Average result of the junior high math exam (2016)	Percentage of children aged 3–4 years covered by preschool education	Number of high schools, vocational schools and post high school educational institutions	Number of students in high schools, vocational schools and post-high school educational institutions	Total municipal and county expenses per 1 student (grammar schools, junior high schools)	Number of doctors employed through contracts with National Health Fund (not including dentists)	Number of clinics and health centers	Number of hospital beds	Number of crimes committed	Percentage of crimes with offender being captured
Grójecki	3,227.626	1,987	4.5	48.10	82.56	44	4,183	8,352	106	38	375	1,907	51.4
Kozienicki	2,338.445	2,063	13.9	46.40	70.84	22	2,738	9,012	95	29	364	950	69.1
Miński	5,379.689	3,343	8.7	51.30	83.37	45	8,011	8,066	133	43	391	2,198	68.9
Mławski	2,359.776	2,179	12.2	44.20	56.37	44	5,715	8,796	53	40	199	997	73.0
Ostrołęcki	2,548.505	4,782	14.3	47.65	72.34	78	11,111	9,565	269	82	528	3,229	78.0
Ostrowski	2,505.091	2,320	12.5	45.70	51.02	22	4,263	8,484	85	30	226	1,103	61.7
Pułtuski	1,819.113	2,023	20.1	43.40	64.17	19	2,630	8,800	68	28	187	1,179	82.4
Sierpecki	2,171.035	2,336	20.6	46.10	50.50	22	3,145	8,619	46	24	198	981	85.3
Sochaczewski	2,700.521	2,295	8.7	45.30	72.34	31	4,906	8,587	159	36	349	1,214	63.9
Sokołowski	2,244.126	1,535	9.4	49.20	76.34	17	3,089	9,077	81	29	237	770	74.6
Szydłowiecki	903.640	1,993	30.8	42.90	56.58	10	1,230	9,357	34	15	0	550	73.6
Węgrowski	2,171.998	2,385	11.3	44.20	64.84	30	4,143	8,582	58	29	172	1,118	75.0
Capital City of Warsaw	226,356.000	28,987	3.3	61.70	99.11	532	121,261	7,629	8,778	1,169	12,044	48,982	39.3
Opolskie	36,393.000	24,684	10.1	47.00	86.54	392	53,719	8,893	1,946	538	4,604	19,927	65.8
Nyski	3,574.645	4,571	14.3	43.80	82.32	59	7,604	8,521	239	70	1,101	2,760	64.9
Oleski	1,945.300	1,332	7.4	45.80	87.90	27	3,005	9,290	83	26	199	1,216	81.0
Subcarpathian	67,350.000	69,252	13.2	50.00	71.00	732	13,5471	8,828	4,467	1,137	10,249	27,487	71.6
Dębicki	4,469.998	5,201	11.0	51.00	66.11	43	7,129	8,537	172	42	516	1,570	64.8
Jarosławski	2,940.803	3,908	15.7	46.00	64.33	52	9,223	9,049	187	53	285	1,269	80.6
Kolbuszowski	1,426.599	2,192	13.3	50.00	62.73	9	1,710	9,605	67	31	201	438	75.4

Region													
Krośnieński	2,399.186	5,119	9.9	55.50	78.26	63	11,999	9,432	349	89	664	4,315	77.6
Niżański	1,595.895	2,455	19.8	48.00	58.62	15	2,001	9,043	85	33	230	832	72.2
Przemyski	1,891.341	6,427	16.4	47.50	75.15	57	9,934	9,802	343	80	1,014	1,796	65.0
Ropczycko-Sędziszowski	2,197.881	1,863	17.0	48.00	56.17	16	3,753	8,943	71	32	81	546	78.1
Sanocki	2,726.117	3,047	9.9	50.00	64.51	33	6,332	8,530	188	50	420	1,135	72.2
Tarnobrzeskie	1,789.887	2,773	12.2	50.50	86.21	63	7,357	9,958	227	57	660	1,667	73.5
Podlaskie	38,605.000	40,995	11.8	49.00	73.94	441	74,819	8,507	3,011	760	5,933	17,470	65.4
Bielskie	1,890.870	2,180	7.6	47.00	69.16	17	2,420	8,979	87	38	247	762	72.8
Kolneński	755.323	1,467	16.8	45.00	38.00	13	1,308	8,274	23	14	139	410	79.5
Łomżyński	1,772.376	3,145	12.5	48.50	70.31	60	9,959	10,534	245	61	591	1,647	65.1
Sokólski	1,635.376	3,184	15.0	47.00	58.84	22	2,410	9,277	87	39	214	1,362	81.1
Suwalskie	1,700.317	3,575	7.8	50.00	60.60	50	8,440	9,320	231	46	494	1,875	58.3
City of Białystok	12,884.854	8,369	10.5	56.00	97.72	132	32,377	8,013	1,647	299	2,475	5,117	59.4
Pomeranian	97,833.000	65,498	8.9	48.00	71.24	847	140,657	8,417	5,142	949	9,506	49,050	59.1
Kościerski	1,932.904	2,761	10.5	46.00	67.23	25	4,418	8,014	185	22	528	1,090	65.5
Słupski	3,599.274	7,681	11.7	49.00	73.42	95	12,102	9,492	368	86	685	3,439	66.5
Tczewski	4,447.837	3,727	9.1	50.00	61.96	43	6,717	7,815	130	39	252	2,803	70.3
City of Gdańsk	27,855.455	6,949	4.0	60.00	83.87	178	33,987	9,107	2,093	214	3,101	11,203	43.0
Silesian	213,589.000	93,312	8.2	47.78	82.41	1,683	262,865	8,158	11,077	2,814	25,526	117,163	68.4
Cieszyński	7,551.937	3,391	7.4	49.49	82.86	58	8,228	8,335	358	120	1,778	5,527	80.5
Gliwicki	8,682.819	4,718	7.0	49.72	83.99	119	16,772	8,744	709	190	1,642	9,841	66.3
Kłobucki	2,720.408	1,934	11.5	47.36	79.51	13	2,045	9,458	80	42	98	1,032	80.6
Zawierciański	4,059.609	3,381	13.6	46.38	80.31	38	6,393	8,194	173	59	376	2,136	71.7
City of Jastrzębie-Zdrój	3,859.192	1,511	8.3	46.91	85.10	53	7,453	7,548	236	49	595	1,868	71.6
Holy Cross	41,304.000	47,069	12.5	48.00	72.31	530	79,017	8,645	2,916	576	6,313	24,749	76.4
Buski	2,176.721	2,462	7.1	48.00	60.58	32	3,978	8,575	181	32	538	774	69.6
Kazimierski	737.408	1,064	10.5	47.70	61.32	22	1,799	8,836	22	7	57	1,895	96.8
Konecki	2,118.545	3,731	17.2	49.20	63.67	35	5,095	8,889	208	22	612	1,062	76.5
Ostrowiecki	3,633.101	3,727	17.6	45.10	74.33	45	7,957	7,796	241	50	451	1,942	73.7
Sandomierski	2,472.898	3,392	8.9	44.70	75.09	41	4,565	8,196	184	35	464	990	73.9
Włoszczowski	1,206.286	1,388	9.0	48.90	62.68	13	2,507	8,670	71	23	271	3,255	96.8

Province/county	GDP in current prices in millions PLN (2014)	Number of households using the help of social welfare	Unemployment rate (%)	Average result of the junior high math exam (2016)	Percentage of children aged 3–4 years covered by preschool education	Number of high schools, vocational schools and post high school educational institutions	Number of students in high schools, vocational schools and post-high school educational institutions	Total municipal and county expenses per 1 student (grammar schools, junior high schools)	Number of doctors employed through contracts with National Health Fund (not including dentists)	Number of clinics and health centers	Number of hospital beds	Number of crimes committed	Percentage of crimes with offender being captured
Warmian-Masurian	46,191.000	68,973	16.2	45.00	65.17	594	86,719	7,991	2,983	831	6,668	27,120	64.9
Bartoszycki	1,771.803	4,119	24.4	43.00	55.32	22	3,061	7,985	87	34	303	900	64.2
Iławski	2,572.928	4,163	7.3	44.00	64.06	41	5,449	7,847	115	38	248	1,640	62.8
Olecki	882.758	1,770	18.5	46.00	64.18	19	2,011	7,676	43	16	124	551	79.4
Greater Poland	166,508.000	87,370	6.1	47.82	82.02	1,252	210,937	7,902	5,231	1,851	15,756	68,804	70.3
Czarnkowsko-Trzcianecki	2,590.377	2,751	8.3	44.07	79.95	37	5,009	7,752	80	58	303	967	81.6
Gnieźnieński	5,121.442	3,597	9.5	44.74	75.70	60	8,982	6,923	214	57	296	2,203	70.6
Kępiński	2,551.103	1,481	2.6	48.62	77.17	12	2,594	8,157	58	18	259	997	81.8
Kolski	2,687.541	2,501	12.8	45.04	62.18	37	4,720	8,641	60	42	224	1,402	85.9
Ostrowski	6,544.070	3,860	5.9	48.34	84.15	50	10,558	8,132	184	62	499	4,356	86.0
Szamotulski	3,754.238	1,873	5.9	44.60	77.95	34	4,213	6,717	43	42	187	1,832	85.1
Turecki	2,645.254	2,598	7.0	48.40	76.73	17	4,795	8,686	54	36	199	1,400	85.1
Wolsztyński	2,097.597	1,059	3.9	45.91	82.90	19	3,536	7,374	56	28	177	805	77.1
City of Leszno	3,654.069	2,002	5.6	52.84	97.20	44	9,636	7,461	187	46	538	1,921	84.8
West Pomeranian	6,4356.000	56,113	13.1	45.17	72.06	708	94,935	8,045	4,116	931	8,268	39,778	69.9
Gryficki	1,722.680	2,491	21.3	42.09	54.66	34	3,027	8,367	146	46	367	1,363	67.2
Koszaliński	3,618.340	4,996	14.8	44.70	78.04	76	11,742	9,445	484	83	788	4,179	66.3
Szczecinecki	2,167.693	3,610	22.2	43.03	54.14	42	5,314	7,551	91	43	220	1,204	77.8

Source: Own work based on data collected by Centre of Sociological Research of the Institute of Philosophy and Sociology of Polish Academy of Sciences from Central Statistical Office, Local Data Bank, Provincial Tax Bank, and Education Authorities data for provinces held ad

Table 2.6.3. Essential characteristics of the sampled counties (data for years 2014–2015)

Province/county	Number of libraries, cultural centers, museums, theaters, cinemas, galleries and other cultural institutions	Number of registered cars	Voter turnout for the 2014 county council election (%)	Voter turnout for the 2015 parliamentary election (%)	Number of registered nonprofit and nongovernmental organizations (2016)	Number of library book loans (2016)	Total income of all inhabitants of the county according to 2016 income tax returns, in millions of PLN	Number of households
Poland	13,816	20,723,423	47	51	149,693	110,239,610	793,301	13,567,999
Lower Silesia	1,035	1,610,385	48	49	13,250	8,309,898	64,821	1,099,505
Kłodzki	71	91,575	45	44	695	502,537	2,678	63,897
Lubański	27	32,001	48	43	205	155,274	930	21,302
Oleśnicki	29	59,587	47	46	325	277,872	2,103	34,673
Polkowski	41	31,213	55	49	292	151,946	1,717	21,412
Kuyavian-Pomeranian	713	1,103,443	48	46	7,651	5,213,346	37,993	729,226
Mogileński	13	27,284	49	42	147	132,845	742	14,925
Rypiński	22	24,707	55	38	117	140,489	638	14,411
Świecki	37	54,505	43	41	430	196,150	1,688	32,304
Wąbrzeski	18	19,586	47	38	140	65,882	511	11,714
Włocławski	57	103,095	57	38	698	539,723	3,409	71,923
Lubelskie	873	1,100,685	52	49	9,201	7,097,607	35,039	741,616
Bialski	99	95,227	54	47	624	663,235	2,610	56,609
Biłgorajski	47	49,665	53	48	391	317,203	1,253	30,798
Chełmski	62	76,937	52	38	281	361,278	2,184	52,660
Janowski	23	20,064	58	53	185	162,428	583	13,678
Krasnostawski	29	31,747	51	43	268	207,816	913	23,500
Lubelski	150	252,378	53	51	525	1,973,004	10,488	190,161
Łęczycki	28	33,436	52	47	226	206,814	1,155	19,043
Piławski	56	57,652	47	53	479	376,070	2,257	41,036
Rycki	25	33,787	55	49	207	156,483	1,020	19,041
Tomaszowski	35	42,703	52	43	349	254,550	1,061	27,720
Włodawski	22	23,800	52	43	191	110,347	565	14,016

Province/county	Number of libraries, cultural centers, museums, theaters, cinemas, galleries and other cultural institutions	Number of registered cars	Voter turnout for the 2014 county council election (%)	Voter turnout for the 2015 parliamentary election (%)	Number of registered nonprofit and nongovernmental organizations (2016)	Number of library book loans (2016)	Total income of all inhabitants of the county according to 2016 income tax returns, in millions of PLN	Number of households
City of Zamość	19	28,362	39	50	371	253,513	1,219	24,234
Lubuskie	367	577,711	47	45	4,650	2,674,133	18,994	364,789
Sulęciński	24	21,302	52	39	156	65,156	585	11,963
Żarski	36	60,774	45	40	376	254,914	1,699	35,230
Łódzkie	811	1,358,314	52	52	9,537	7,638,399	50,374	943,827
Kutnowski	39	58,142	45	45	333	301,884	1,841	37,276
Opoczyński	37	39,452	62	50	250	154,115	1,264	25,031
Piotrkowski	50	85,629	58	49	589	647,520	3,093	58,946
Poddębicki	21	24,938	56	45	193	106,163	622	14,040
Radomszczański	40	61,374	55	46	368	318,389	1,950	39,278
Sieradzki	56	70,213	52	47	497	486,253	1,901	38,930
Skierniewicki	37	46,766	57	48	351	278,503	1,848	28,935
Wieluński	47	43,047	53	47	293	320,169	1,247	24,580
Zgierski	45	87,549	48	53	540	387,658	3,425	60,720
Lesser Poland	1,383	1,725,672	51	55	13,803	9,929,854	66,847	1,080,150
Bocheński	95	50,588	55	55	362	302,429	1,834	29,305
Dąbrowski	25	29,985	49	45	186	187,076	683	17,076
Gorlicki	69	49,844	47	49	430	315,623	1,522	31,376
Limanowski	47	64,959	55	55	365	309,064	1,638	31,168
Miechowski	18	27,651	55	44	252	113,628	783	15,782
Myślenicki	36	68,486	54	55	385	294,133	2,045	31,291
Nowotarski	97	72,932	46	46	617	430,963	2,179	52,761
Olkuski	44	62,485	53	53	348	479,450	2,525	37,571
Tarnowski	140	149,968	51	51	662	953,325	5,000	93,857
Mazovian	1,491	3,170,356	54	59	26,949	14,536,679	143,038	1,943,209
Białobrzeski	12	22,616	59	47	123	58,729	497	10,224

Region							
Garwoliński	43	61,035	57	53	279	250,766	1,934	31,604
Grójecki	23	68,133	55	48	281	195,300	1,783	31,717
Kozienicki	21	36,628	58	49	186	159,814	1,199	20,232
Miński	35	77,400	50	55	406	407,900	3,380	46,658
Mławski	26	42,121	55	42	206	105,632	1,159	23,213
Ostrołęcki	55	69,822	54	43	482	208,883	2,206	43,741
Ostrowski	22	39,895	56	47	235	172,877	1,126	24,473
Pułtuski	18	31,488	62	50	190	122,891	843	16,659
Sierpecki	22	40,284	63	42	189	145,441	785	16,002
Sochaczewski	27	54,012	49	49	245	196,451	1,787	27,498
Sokołowski	24	30,553	59	53	235	98,067	903	18,462
Szydłowiecki	20	21,151	63	46	149	61,008	544	12,708
Węgrowski	33	36,295	53	49	278	129,274	1,058	21,398
Capital City of Warsaw	378	1,131,120	47	68	14,876	6,180,719	67,531	774,611
Opolskie	561	581,477	43	43	4,069	2,876,033	18,133	353,683
Nyski	79	79,853	45	43	563	391,445	2,263	48,615
Oleski	41	40,482	42	40	237	211,873	988	20,705
Subcarpathian	1,131	1,024,748	51	50	8,653	6,490,739	35,222	648,682
Dębicki	53	66,004	50	52	451	378,627	2,199	37,650
Jarosławski	68	66,185	54	49	482	223,555	1,812	35,901
Kolbuszowski	31	31,438	48	48	214	223,451	774	17,203
Krośnieński	64	74,155	51	48	691	527,058	2,813	49,314
Niżański	32	28,943	49	45	193	159,735	879	19,647
Przemyski	122	72,128	55	44	756	258,562	2,178	45,152
Ropczycko-Sędziszowski	31	34,061	57	54	234	189,741	1,111	20,008
Sanocki	62	42,679	47	45	434	235,740	1,561	30,040
Tarnobrzeski	52	47,001	52	46	445	362,727	1,831	34,301
Podlaskie	461	551,519	52	47	4,760	2,875,963	18,907	417,027
Bielskie	20	30,765	52	45	190	110,079	873	21,078
Kolneński	21	18,477	54	40	126	65,675	402	10,993

Province/county	Number of households	Total income of all inhabitants of the county according to 2016 income tax returns, in millions of PLN	Number of library book loans (2016)	Number of registered nonprofit and nongovernmental organizations (2016)	Voter turnout for the 2015 parliamentary election (%)	Voter turnout for the 2014 county council election (%)	Number of registered cars	Number of libraries, cultural centers, museums, theaters, cinemas, galleries and other cultural institutions
Łomżyński	38,023	1,644	34,0581	463	43	54	54,379	49
Sokólski	23,678	869	191,292	250	42	52	34,049	40
Suwalskie	35,378	1,744	308,481	430	41	53	51,049	42
City of Białystok	118,752	6,209	770,337	1,345	57	41	116,510	55
Pomeranian	806,158	47,131	5,384,331	9,853	52	48	1,237,891	760
Kościerski	21,107	1,083	128,224	281	49	56	40,564	30
Słupski	71,381	3,611	433,947	900	40	45	93,128	180
Tczewski	38,089	2,125	308,634	382	46	41	61,036	27
City of Gdańsk	189,174	12,474	1,197,462	340	61	40	255,176	89
Silesian	1,727,586	107,055	15,229,042	14,401	52	49	2,398,119	1,313
Cieszyński	58,553	3,768	482,005	682	53	48	97,231	62
Gliwicki	118,001	7,614	1,078,397	1,044	48	45	173,081	63
Kłobucki	25,357	1,439	182,174	239	46	52	56,371	31
Zawierciański	44,095	2,460	400,477	398	49	52	55,947	47
City of Jastrzębie-Zdrój	33,333	2,270	395,696	243	54	42	44,789	25
Holy Cross	428,504	21,145	3,302,265	4,758	47	56	629,323	461
Buski	23,574	1,032	148,516	272	44	59	38,241	48
Kazimierski	10,841	380	66,808	139	36	51	19,851	10
Konecki	28,744	1,278	191,239	257	45	57	39,802	31
Ostrowiecki	42,767	1,873	298,316	350	43	45	46,503	32
Sandomierski	25,097	1,210	193,824	374	42	55	43,166	29
Włoszczowski	14,110	718	91,503	198	46	60	23,438	19
Warmian-Masurian	515,857	24,401	4,075,911	6,545	42	48	695,656	511
Bartoszycki	21,618	891	126,351	209	39	46	30,493	29

Iławski	29	45,944	45	42	313	233,505	1,495	29,143
Olecki	15	17,432	43	37	149	82,186	510	11,561
Greater Poland	**1,159**	**2,084,773**	**49**	**50**	**14,523**	**10,496,683**	**72,439**	**1,129,008**
Czarnkowsko-Trzcianecki	42	52,861	46	44	287	274,286	1,588	27,335
Gnieźnieński	39	79,609	45	49	487	495,149	2,707	46,431
Kępiński	22	36,755	56	50	229	105,336	924	15,833
Kolski	33	52,643	51	43	296	207,686	1,397	28,986
Ostrowski	52	92,911	48	49	621	498,587	3,068	48,486
Szamotulski	35	57,777	51	47	327	265,413	1,958	26,469
Turecki	28	52,005	55	46	281	169,371	1,519	25,957
Wolsztyński	21	31,861	51	48	205	173,697	1,017	15,557
City of Leszno	17	32,116	42	53	328	419,472	1,475	23,657
West Pomeranian	**786**	**873,351**	**47**	**46**	**7,103**	**4,108,727**	**31,762**	**639,172**
Gryficki	59	32,677	53	40	197	113,428	928	20,686
Koszaliński	89	83,831	51	42	884	495,772	3,509	67,718
Szczecinecki	57	32,897	48	42	381	181,949	1,316	29,326

Source: Own work based on data collected by Centre of Sociological Research of the Institute of Philosophy and Sociology of Polish Academy of Sciences from Central Statistical Office, Local Data Bank, Provincial Tax Offices, and Education Authorities; data for provinces bolded.

REFERENCES

Abbott, A. (2004). *Methods of Discovery: Heuristics for the social sciences*. W.W. Norton & Company.

Bansal, M.L. & Singh, R. (1986). On the generalization of Rao, Hartley, and Cochran's Scheme. *Metrika* 33, pp. 307–314.

Domański, H. & Pokropek A. (2011). *Podziały terytorialne, globalizacja a nierówności społeczne. Wprowadzenie do modeli wielopoziomowych*. Wydawnictwo IFiS PAN.

Engel, J. & Dąbała, W. (2018). *Raport: Opis schematu próby do badania „Myśl lokalnie, działaj globalnie: Polscy rolnicy w globalnym świecie rozwoju zrównoważonego i odporności na kryzys."* The Centre of Sociological Research of the Institute of Philosophy and Sociology of the Polish Academy of Sciences (typescript).

Garson, G.D. (2013). *Hierarchical Linear Modeling: Guide and applications*. SAGE.

Hartley, H.O. & Rao, J.N.K. (1962). Sampling with unequal probabilities and without replacement. *The Annals of Mathematical Statistics* 33, pp. 350–374.

Hox, J.J. (2010). *Multilevel Analysis Techniques and Applications*. Routledge.

Hox, J.J., Moerbeek, M., & Schoot, R. van de (2017). *Multilevel Analysis Techniques and Applications*. Routledge.

Kott, P.S. (2005). A note on the Hartley-Rao variance estimator. *Journal of Official Statistics* 21(3), pp. 433–439.

Rao, J.N.K., Hartley, H.O., & Cochran, W.G. (1962). On a simple procedure of unequal probability sampling with replacement. *Journal of the Royal Statistical Society* (24)2, pp. 482–491.

Rao, J.N.K. & Wu, C.F.J. (1988). Resampling inference with complex survey data. *Journal of the American Statistical Association* 83, pp. 231–241.

Raudenbush, S.W. & Bryk, A.S. (2002). *Hierarchical Linear Models: Applications and data analysis methods*. SAGE.

Twisk, J.W.R. (2010). *Analiza wielopoziomowa – przykłady zastosowań. Praktyczny podręcznik biostatystyki i epidemiologii*. Przekład E. Frątczak. Szkoła Główna Handlowa w Warszawie.

Węziak, D. (2007). Wielopoziomowe modelowanie regresyjne w analizie danych. *Wiadomości Statystyczne* 9, pp. 1–12.

Internet sources

https://bdl.stat.gov.pl/BDL/dane/podgrup/tablica/. Kategoria: rolnictwo, leśnictwo i łowiectwo; Grupa: gospodarstwa rolne; Podgrupa: gospodarstwa rolne wg grup obszarowych użytków; Wymiary: grupy obszarowe; rodzaje własności; lata. (access: June 23, 2019).

https://stat.gov.pl/statystyka-regionalna/jednostki-terytorialne/klasyfikacja-nuts/klasyfikacja-nuts-w-polsce/ (access: June 12, 2019).

Annex No. 1

List of selected municipalities in the counties of the research sample, by provinces N (number of municipalities in the sample) = 863.

Province	County	Municipality
Lower Silesian	Lubański	Lubań, Świeradów-Zdrój, Leśna, Olszyna, Platerówka, Siekierczyn
	Kłodzki	Kłodzko, Kudowa-Zdrój, Nowa Ruda, Polanica-Zdrój, Bystrzyca Kłodzka, Lądek-Zdrój, Lewin Kłodzki, Międzylesie, Nowa Ruda, Radków, Stronie Śląskie, Szczytna
	Polkowicki	Chocianów, Gaworzyce, Grębocice, Polkowice, Przemków, Radwanice
	Oleśnicki	Oleśnica, Bierutów, Dobroszyce, Dziadowa Kłoda, Międzybórz, Syców, Twardogóra
Kuyavian-Pomeranian	Rypiński	Rypin, Brzuze, Rogowo, Skrwilno, Wąpielsk
	Świecki	Bukowiec, Dragacz, Drzycim, Jeżewo, Lniano, Nowe, Osie, Pruszcz, Świecie, Świekatowo, Warlubie
	Wąbrzeski	Wąbrzeźno, Dębowa, Łąka Książki Płużnica
	Włocławski	Kowal, Baruchowo, Boniewo, Brześć Kujawski, Choceń, Chodecz, Fabianki, Izbica Kujawska, Lubanie, Lubień Kujawski, Lubraniec, Włocławek
	Mogileński	Dąbrowa, Jeziora Wielkie, Mogilno, Strzelno
Lubelskie	Włodawski	Włodawa, Hanna, Hańsk, Stary Brus, Urszulin, Wola Uhruska, Wyryki
	Bialski	Międzyrzec Podlaski, Terespol, Biała Podlaska, Drelów, Janów Podlaski, Kodeń, Konstantynów, Leśna Podlaska, Łomazy, Piszczac, Rokitno, Rossosz, Sławatycze, Sosnówka, Terespol, Tuczna, Wisznice, Zalesie
	Krasnostawski	Fajsławice, Gorzków, Izbica, Krasnystaw, Kraśniczyn, Łopiennik Górny, Rudnik, Siennica Różana, Żółkiewka
	Chełmski	Rejowiec Fabryczny, Białopole, Chełm, Dorohusk, Dubienka, Kamień, Leśniowice, Rejowiec Fabryczny, Ruda-Huta, Sawin, Siedliszcze, Wierzbica, Wojsławice, Żmudź, Rejowiec
	Biłgorajski	Biłgoraj, Aleksandrów, Biszcza, Frampol, Goraj, Józefów, Księżpol, Łukowa, Obsza, Potok Górny, Tarnogród, Tereszpol, Turobin
	Tomaszowski	Tomaszów Lubelski, Bełżec, Jarczów, Krynice, Lubycza Królewska, Łaszczów, Rachanie, Susiec, Tarnawatka, Telatyn, Tyszowce, Ulhówek
	Łęczyński	Cyców, Ludwin, Łęczna, Milejów, Puchaczów, Spiczyn
	Lubelski	Bełżyce, Borzechów, Bychawa, Garbów, Głusk, Jabłonna, Jastków, Konopnica, Krzczonów, Niedrzwica Duża, Niemce, Strzyżewice, Wojciechów, Wólka, Wysokie, Zakrzew
	Janowski	Batorz, Chrzanów, Dzwola, Godziszów, Janów Lubelski, Modliborzyce, Potok Wielki
	Rycki	Dęblin, Kłoczew, Nowodwór, Ryki, Stężyca, Ułęż
	Puławski	Puławy, Baranów, Janowiec, Kazimierz Dolny, Końskowola, Kurów, Markuszów, Nałęczów, Puławy, Wąwolnica, Żyrzyn
	City of Zamość	City of Zamość
Lubuskie	Sulęciński	Krzeszyce, Lubniewice, Słońsk, Sulęcin, Torzym
	Żarski	Łęknica, Żary, Brody, Jasień, Lipinki, Łużyckie, Lubsko, Przewóz, Trzebiel, Tuplice

Province	County	Municipality
Łódzkie	Kutnowski	Kutno, Bedlno, Dąbrowice, Krośniewice, Krzyżanów, Łanięta, Nowe Ostrowy, Oporów, Strzelce, Żychlin
	Skierniewicki	Bolimów, Głuchów, Godzianów, Kowiesy, Lipce Reymontowskie, Maków, Nowy Kawęczyn, Skierniewice, Słupia
	Zgierski	Głowno, Ozorków, Aleksandrów Łódzki, Parzęczew, Stryków, Zgierz
	Radomszczański	Radomsko, Dobryszyce, Gidle, Gomunice, Kamieńsk, Kobiele Wielkie, Kodrąb, Lgota Wielka, Ładzice, Masłowice, Przedbórz, Radomsko, Wielgomłyny, Żytno
	Opoczyński	Białaczów, Drzewica, Mniszków, Opoczno, Paradyż, Poświętne, Sławno, Żarnów
	Piotrkowski	Aleksandrów, Czarnocin, Gorzkowice, Grabica, Łęki Szlacheckie, Moszczenica, Ręczno, Rozprza, Sulejów, Wola Krzysztoporska, Wolbórz
	Wieluński	Biała, Czarnożyły, Konopnica, Mokrsko, Osjaków, Ostrówek, Pątnów, Skomlin, Wieluń, Wierzchlas
	Sieradzki	Sieradz, Błaszki, Brąszewice, Brzeźnio, Burzenin, Goszczanów, Klonowa, Warta, Wróblew, Złoczew
	Poddębicki	Dalików, Pęczniew, Poddębice, Uniejów, Wartkowice, Zadzim
Lesser Poland	Myślenicki	Dobczyce, Lubień, Myślenice, Pcim, Raciechowice, Siepraw, Sułkowice, Tokarnia, Wiśniowa
	Bocheński	Bochnia, Drwinia, Lipnica Murowana, Łapanów, Nowy Wiśnicz, Rzezawa, Trzciana, Żegocina
	Miechowski	Charsznica, Gołcza, Kozłów, Książ Wielki, Miechów, Racławice, Słaboszów
	Gorlicki	Gorlice, Biecz, Bobowa, Lipinki, Łużna, Moszczenica, Ropa, Sękowa, Uście Gorlickie
	Limanowski	Limanowa, Mszana Dolna, Dobra, Jodłownik, Kamienica, Laskowa, Łukowica, Mszana Dolna, Niedźwiedź, Słopnice, Tymbark
	Nowotarski	Szczawnica, Czarny Dunajec, Czorsztyn, Jabłonka, Krościenko nad Dunajcem, Lipnica Wielka, Łapsze Niżne, Nowy Targ, Ochotnica Dolna, Raba Wyżna, Rabka-Zdrój, Spytkowice, Szaflary
	Olkuski	Bukowno, Bolesław, Klucze, Olkusz, Trzyciąż, Wolbrom
	Tarnowski	Ciężkowice, Gromnik, Lisia Góra, Pleśna, Radłów, Ryglice, Rzepiennik Strzyżewski, Skrzyszów, Tarnów, Tuchów, Wierzchosławice, Wietrzychowice, Wojnicz, Zakliczyn, Żabno, Szerzyny
	Dąbrowski	Bolesław, Dąbrowa Tarnowska, Gręboszów, Mędrzechów, Olesno, Radgoszcz, Szczucin
Mazovian	Sierpecki	Gozdowo, Mochowo, Rościszewo, Sierpc, Szczutowo, Zawidz
	Mławski	Mława, Dzierzgowo, Lipowiec Kościelny, Radzanów, Strzegowo, Stupsk, Szreńsk, Szydłowo, Wieczfnia Kościelna, Wiśniewo
	Pułtuski	Gzy, Obryte, Pokrzywnica, Pułtusk, Świercze, Winnica, Zatory
	Ostrołęcki	Baranowo, Czarnia, Czerwin, Goworowo, Kadzidło, Lelis, Łyse, Myszyniec, Olszewo-Borki, Rzekuń, Troszyn
	Ostrowski	Andrzejewo, Boguty-Pianki, Brok, Małkinia Górna, Nur, Ostrów Mazowiecka, Stary Lubotyń, Szulborze Wielkie, Wąsewo, Zaręby Kościelne
	Kozienicki	Garbatka-Letnisko, Głowaczów, Gniewoszów, Grabów nad Pilicą, Kozienice, Magnuszew, Sieciechów

	Szydłowiecki	Chlewiska, Jastrząb, Mirów, Orońsko, Szydłowiec
	Białobrzeski	Białobrzegi, Promna, Radzanów, Stara Błotnica, Stromiec, Wyśmierzyce
	Miński	Cegłów, Dębe Wielkie, Dobre, Halinów, Jakubów, Kałuszyn, Latowicz, Mińsk Mazowiecki, Mrozy, Siennica, Stanisławów, Sulejówek
	Garwoliński	Garwolin, Łaskarzew, Borowie, Górzno, Łaskarzew, Maciejowice, Miastków Kościelny, Parysów, Pilawa, Sobolew, Trojanów, Wilga, Żelechów
	Sochaczewski	Brochów, Iłów, Młodzieszyn, Nowa Sucha, Rybno, Sochaczew, Teresin
	Grójecki	Belsk Duży, Błędów, Chynów, Goszczyn, Grójec, Jasieniec, Mogielnica, Nowe Miasto nad Pilicą, Pniewy, Warka
	Sokołowski	Sokołów Podlaski, Bielany, Ceranów, Jabłonna Lacka, Kosów Lacki, Repki, Sabnie, Sterdyń
	Węgrowski	Węgrów, Grębków, Korytnica, Liw, Łochów, Miedzna, Sadowne, Stoczek, Wierzbno
	City of Warsaw	Białołęka, Ursynów, Wawer, Wilanów, Włochy
Opolskie	Oleski	Dobrodzień, Gorzów Śląski, Olesno, Praszka, Radłów, Rudniki, Zębowice
Opolskie	Nyski	Głuchołazy, Kamiennik, Korfantów, Łambinowice, Nysa, Otmuchów, Paczków, Pakosławice, Skoroszyce
Subcarpathian	Krośnieński	Chorkówka, Dukla, Iwonicz-Zdrój, Jedlicze, Korczyna, Krościenko Wyżne, Miejsce Piastowe, Rymanów, Wojaszówka, Jaśliska
Subcarpathian	Sanocki	Sanok, Besko, Bukowsko, Komańcza, Tyrawa Wołoska, Zagórz, Zarszyn
Subcarpathian	Przemyski	Bircza, Dubiecko, Fredropol, Krasiczyn, Krzywcza, Medyka, Orły, Przemyśl, Stubno, Żurawica
Subcarpathian	Jarosławski	Jarosław, Radymno, Chłopice, Laszki, Pawłosiów, Pruchnik, Radymno, Rokietnica, Roźwienica, Wiązownica
Subcarpathian	Kolbuszowski	Cmolas, Kolbuszowa, Majdan Królewski, Niwiska, Raniżów, Dzikowiec
Subcarpathian	Ropczycko- -Sędziszowski	Iwierzyce, Ostrów, Ropczyce, Sędziszów Małopolski, Wielopole Skrzyńskie
Subcarpathian	Dębicki	Dębica, Brzostek, Czarna, Jodłowa, Pilzno Żyraków
Subcarpathian	Niżański	Harasiuki, Jarocin, Jeżowe, Krzeszów, Nisko, Rudnik nad Sanem, Ulanów
Subcarpathian	Tarnobrzeski	Baranów Sandomierski, Gorzyce, Grębów, Nowa Dęba
Podlaskie	Suwalski	Bakałarzewo, Filipów, Jeleniewo, Przerośl, Raczki, Rutka-Tartak, Suwałki, Szypliszki, Wiżajny
Podlaskie	Sokólski	Dąbrowa Białostocka, Janów, Korycin, Krynki, Kuźnica, Nowy Dwór, Sidra, Sokółka, Suchowola, Szudziałowo
Podlaskie	Bielski	Bielsk Podlaski, Brańsk, Boćki, Orla, Rudka, Wyszki
Podlaskie	Łomżyński	Jedwabne, Łomża, Miastkowo, Nowogród, Piątnica, Przytuły, Śniadowo, Wizna, Zbójna
Podlaskie	Kolneński	Kolno, Grabowo, Mały Płock, Stawiski, Turośl
Podlaskie	City of Białystok	City of Białystok
Pomeranian	Tczewski	Tczew, Gniew, Morzeszczyn, Pelplin, Subkowy
Pomeranian	Słupski	Ustka, Damnica, Dębnica Kaszubska, Główczyce, Kępice, Kobylnica, Potęgowo, Słupsk, Smołdzino
Pomeranian	Kościerski	Kościerzyna, Dziemiany, Karsin, Liniewo, Lipusz, Nowa Karczma, Stara Kiszewa
Pomeranian	City of Gdańsk	City of Gdańsk

Province	County	Municipality
Silesia	Cieszyński	Cieszyn, Ustroń, Wisła, Brenna, Chybie, Dębowiec, Goleszów, Hażlach, Istebna, Skoczów, Strumień, Zebrzydowice
	Gliwicki	Knurów, Pyskowice, Gierałtowice, Pilchowice, Rudziniec, Sośnicowice, Toszek, Wielowieś
	Zawierciański	Poręba, Zawiercie, Irządze, Kroczyce, Łazy, Ogrodzieniec, Pilica, Szczekociny, Włodowice, Żarnowiec
	Kłobucki	Kłobuck, Krzepice, Lipie, Miedźno, Opatów, Panki, Popów, Przystajń, Wręczyca Wielka
	City of Jastrzębie-Zdrój	City of Jastrzębie-Zdrój
Holy Cross	Konecki	Fałków, Gowarczów, Końskie, Radoszyce, Ruda Maleniecka, Słupia (Konecka), Smyków, Stąporków
	Ostrowiecki	Ostrowiec Świętokrzyski, Bałtów, Bodzechów, Ćmielów, Kunów, Waśniów
	Buski	Busko-Zdrój, Gnojno, Nowy Korczyn, Pacanów, Solec-Zdrój, Stopnica, Tuczępy, Wiślica
	Sandomierski	Sandomierz, Dwikozy, Klimontów, Koprzywnica, Łoniów, Obrazów, Samborzec, Wilczyce, Zawichost
	Włoszczowski	Kluczewsko, Krasocin, Moskorzew, Radków, Secemin, Włoszczowa
	Kazimierski	Bejsce, Czarnocin, Kazimierza Wielka, Opatowiec, Skalbmierz
Warmian-Masurian	Olecki	Kowale Oleckie, Olecko, Świętajno, Wieliczki
	Bartoszycki	Bartoszyce, Górowo Iławeckie, Bisztynek, Sępopol
	Iławski	Iława, Lubawa, Kisielice, Susz, Zalewo
Greater Poland	Kępiński	Baranów, Bralin, Kępno, Łęka Opatowska, Perzów, Rychtal, Trzcinica
	Ostrowski	Ostrów Wielkopolski, Nowe Skalmierzyce, Odolanów, Przygodzice, Raszków, Sieroszewice, Sośnie
	Turecki	Turek, Brudzew, Dobra, Kawęczyn, Malanów, Przykona, Tuliszków, Władysławów
	Gnieźnieński	Gniezno, Czerniejewo, Kiszkowo, Kłecko, Łubowo, Mieleszyn, Niechanowo, Trzemeszno, Witkowo
	Kolski	Koło, Babiak, Chodów, Dąbie, Grzegorzew, Kłodawa, Kościelec, Olszówka, Osiek Mały, Przedecz
	Wolsztyński	Przemęt, Siedlec, Wolsztyn
	Czarnkowsko-Trzcianecki	Czarnków, Drawsko, Krzyż Wielkopolski, Lubasz, Połajewo, Trzcianka, Wieleń
	Szamotulski	Obrzycko, Duszniki, Kaźmierz, Ostroróg, Pniewy, Szamotuły, Wronki
	City of Leszno	City of Leszno
West Pomerania	Gryficki	Brojce, Gryfice, Karnice, Płoty, Rewal, Trzebiatów
	Szczecinecki	Szczecinek, Barwice, Biały Bór, Borne Sulinowo, Grzmiąca
	Koszaliński	Będzino, Biesiekierz, Bobolice, Manowo, Mielno, Polanów, Sianów, Świeszyno

Elaboration: Engel & Dąbała, 2018.

Annex No. 2

Number of individual farms with area over 1 hectare in selected counties according to four area groups. N (number of selected counties) = 103

Province (voivodeship)	County	Number of farms in the county, by area (in hectares)				Total number of farms
		1 to 1.99	2 to 4.99	5 to 9.99	10 or more	
Lower Silesian	Lubański	226	683	241	262	1,412
	Kłodzki	729	1,794	1,041	1,336	4,900
	Polkowicki	403	617	407	529	1,956
	Oleśnicki	541	962	759	1,022	3,284
Kuyavian-Pomeranian	Rypiński	315	671	1,036	1,261	3,283
	Świecki	661	966	853	1,810	4,290
	Wąbrzeski	229	386	447	1,078	2,140
	Włocławski	654	1,476	2,255	2,930	7,315
	Mogileński	265	319	453	1,254	2,291
Lubelskie	Włodawski	467	1,039	1,158	1,285	3,949
	Bialski	1,629	3,870	4,118	4,790	14,407
	Krasnostawski	1,141	2,771	2,480	1,810	8,202
	Chełmski	1,151	4,330	2,602	2,468	10,551
	Biłgorajski	2,061	5,097	4,160	1,428	12,746
	Tomaszowski	2,030	3,901	2,733	1,784	10,448
	Łęczyński	1,060	2,142	1,507	882	5,591
	Lubelski	3,265	7,037	4,393	2,881	17,576
	Janowski	928	2,437	2,138	836	6,339
	Rycki	1,089	2,293	1,639	616	5,637
	Puławski	2,402	3,604	1,922	727	8,655
	City of Zamość	119	242	64	26	451
Lubuskie	Sulęciński	249	342	292	462	1,345
	Żarski	408	772	414	516	2,110
Łódzkie	Kutnowski	573	1,053	1,417	2,025	5,068
	Skierniewicki	853	1,969	2,019	1,229	6,070
	Zgierski	658	1,510	1,704	1,211	5,083
	Radomszczański	1,442	3,091	2,475	1,419	8,427
	Opoczyński	1,585	3,590	2,275	770	8,220
	Piotrkowski	1,380	3,278	3,399	1,983	10,040
	Wieluński	1,487	3,047	2,484	1,216	8,234
	Sieradzki	1,408	3,401	3,948	2,704	11,461
	Poddębicki	507	1,536	1,951	2,024	6,018

Province (voivodeship)	County	Number of farms in the county, by area (in hectares)				Total number of farms
		1 to 1.99	2 to 4.99	5 to 9.99	10 or more	
Lesser Poland	Myślenicki	3,477	3,546	445	57	7,525
	Bocheński	2,758	3,712	783	153	7,406
	Miechowski	916	2,500	2,081	1,174	6,671
	Gorlicki	3,712	3,875	976	333	8,896
	Limanowski	3,911	6,628	1,462	120	12,121
	Nowotarski	4,018	6,024	1,832	429	12,303
	Olkuski	1,853	2,567	894	215	5,529
	Tarnowski	5,798	7,971	1,863	459	16,091
	Dąbrowski	1,541	3,056	1,383	451	6,431
Mazovian	Sierpecki	389	790	1,206	2,142	4,527
	Mławski	447	1,009	1,280	2,799	5,535
	Pułtuski	345	1,002	1,532	1,831	4,710
	Ostrołęcki	565	1,920	3,350	4,879	10,714
	Ostrowski	738	2,112	2,300	2,334	7,484
	Kozienicki	916	2,134	1,835	863	5,748
	Szydłowiecki	804	1,279	546	175	2,804
	Białobrzeski	634	1,506	1,658	901	4,699
	Miński	1,304	2,908	2,327	1,151	7,690
	Garwoliński	1,454	3,892	3,398	1,324	10,068
	Sochaczewski	1,068	2,042	1,634	1,021	5,765
	Grójecki	1,371	3,703	4,187	2,228	11,489
	Sokołowski	811	1,891	2,150	2,356	7,208
	Węgrowski	1,039	2,558	2,616	1,880	8,093
	City of Warsaw	302	499	232	144	1,177
Opolskie	Oleski	755	1,387	1,151	1,210	4,503
	Nyski	803	861	739	1,072	3,475
Subcarpathian	Krośnieński	4,134	2,958	519	263	7,874
	Sanocki	2,039	2,725	1,009	489	6,262
	Przemyski	2,152	3,360	1,127	479	7,118
	Jarosławski	2,435	3,909	1,534	717	8,595
	Kolbuszowski	1,274	3,324	1,421	172	6,191
	Ropczycko-Sędziszowski	1,724	2,791	678	174	5,367
	Dębicki	2,938	4,477	1,330	314	9,059
	Niżański	1,585	2,507	1,136	180	5,408
	Tarnobrzeski	1,429	2,267	489	90	4,275
Podlaskie	Suwalski	305	855	1,063	3,336	5,559
	Sokólski	390	1,092	2,347	4,274	8,103
	Bielski	824	1,812	2,093	2,708	7,437
	Łomżyński	468	1,449	1,905	3,285	7,107
	Kolneński	181	463	922	2,778	4,344
	City of Białystok	142	281	150	81	654

Pomeranian	Tczewski	138	286	365	942	1,731
	Słupski	441	651	532	1,100	2,724
	Kościerski	339	683	738	1,435	3,195
	City of Gdańsk	143	215	155	103	616
Silesian	Cieszyński	2,402	1,739	451	373	4,965
	Gliwicki	443	570	402	476	1,891
	Zawierciański	1,219	2,065	1,474	1,009	5,767
	Kłobucki	1,232	3,560	1,731	665	7,188
	City of Jastrzębie-Zdrój	361	295	64	34	754
Holy Cross	Konecki	1,210	2,297	1,076	362	4,945
	Ostrowiecki	955	1,569	966	563	4,053
	Buski	2,253	4,892	3,277	1,088	11,510
	Sandomierski	2,186	4,420	2,175	600	9,381
	Włoszczowski	730	2,152	1,745	756	5,383
	Kazimierski	527	1,812	1,810	727	4,876
Warmian--Masurian	Olecki	229	315	257	965	1,766
	Bartoszycki	255	337	385	1,327	2,304
	Iławski	404	592	656	1,911	3,563
Greater Poland	Kępiński	460	650	754	1,223	3,087
	Ostrowski	1,222	1,773	1,544	1,910	6,449
	Turecki	1,176	2,224	2,129	1,350	6,879
	Gnieźnieński	418	598	698	2,240	3,954
	Kolski	944	1,932	2,193	2,247	7,316
	Wolsztyński	530	744	796	989	3,059
	Czarnkowsko--Trzcianecki	565	923	795	1,590	3,873
	Szamotulski	361	514	553	1,365	2,793
	City of Leszno	49	50	32	44	175
West Pomeranian	Gryficki	174	250	236	777	1,437
	Szczecinecki	209	335	293	827	1,664
	Koszaliński	384	523	455	867	2,229
Total		**117,623**	**214,834**	**149,104**	**127,480**	**609,041**

Elaboration: Engel & Dąbała, 2018.

Annex No. 3

Number of interviews planned for individual farms with an area of over 1 ha in selected counties according to four area groups. N (number of selected counties) = 103

Province (voivodeship)	County	Number of planned interviews on farms in the county, by farm area (in ha)				Total number of interviews
		1 to 1.99	2 to 4.99	5 to 9.99	10 or more	
Lower Silesian	Lubański	5	14	5	6	30
	Kłodzki	5	11	7	8	31
	Polkowicki	6	10	6	8	30
	Oleśnicki	5	9	7	9	30
Kuyavian-Pomeranian	Rypiński	3	6	9	12	30
	Świecki	5	7	6	13	31
	Wąbrzeski	3	6	6	15	30
	Włocławski	3	7	10	14	34
	Mogileński	4	4	6	16	30
Lubelskie	Włodawski	3	8	9	10	30
	Bialski	6	13	14	16	49
	Krasnostawski	5	12	11	7	35
	Chełmski	4	16	9	9	38
	Biłgorajski	8	20	16	5	49
	Tomaszowski	8	15	11	7	41
	Łęczyński	6	12	9	5	32
	Lubelski	11	24	15	10	60
	Janowski	5	13	11	4	33
	Rycki	6	13	9	4	32
	Puławski	11	16	9	3	39
	City of Zamość	8	16	4	2	30
Lubuskie	Sulęciński	6	8	6	10	30
	Żarski	6	11	6	7	30
Łódzkie	Kutnowski	4	6	9	13	32
	Skierniewicki	5	10	11	6	32
	Zgierski	4	9	11	8	32
	Radomszczański	6	14	11	6	37
	Opoczyński	7	16	10	3	36
	Piotrkowski	5	13	14	8	40
	Wieluński	7	14	11	5	37
	Sieradzki	5	13	15	10	43
	Poddębicki	3	8	10	11	32
Lesser Poland	Myślenicki	16	16	2	1	35
	Bocheński	13	17	4	1	35
	Miechowski	5	12	10	6	33
	Gorlicki	16	17	4	1	38
	Limanowski	14	24	5	1	44
	Nowotarski	14	22	7	2	45
	Olkuski	11	15	5	1	32
	Tarnowski	21	30	7	2	60
	Dąbrowski	8	16	7	2	33

Mazovian	Sierpecki	3	5	8	15	31
	Mławski	3	6	7	16	32
	Pułtuski	2	7	10	12	31
	Ostrołęcki	2	7	13	19	41
	Ostrowski	3	10	11	11	35
	Kozienicki	5	12	10	5	32
	Szydłowiecki	8	14	6	2	30
	Białobrzeski	4	10	11	6	31
	Miński	6	14	11	5	36
	Garwoliński	6	15	14	5	40
	Sochaczewski	6	11	9	6	32
	Grójecki	5	14	16	8	43
	Sokołowski	4	9	10	11	34
	Węgrowski	9	15	7	4	35
	Capital City of Warsaw	8	14	7	4	33
Opolskie	Oleski	5	10	8	8	31
	Nyski	7	8	6	9	30
Subcarpathian	Krośnieński	19	14	2	1	36
	Sanocki	11	14	5	3	33
	Przemyski	10	16	6	2	34
	Jarosławski	10	17	7	3	37
	Kolbuszowski	7	18	7	1	33
	Ropczycko--Sędziszowski	10	17	4	1	32
	Dębicki	12	19	6	1	38
	Niżański	9	15	7	1	32
	Tarnobrzeski	10	16	4	1	31
Podlaskie	Suwalski	2	5	5	19	31
	Sokólski	2	5	10	19	36
	Bielski	4	8	10	12	34
	Łomżyński	2	7	9	15	33
	Kolneński	1	3	7	20	31
	City of Białystok	7	14	7	4	32
Pomeranian	Tczewski	3	5	6	16	30
	Słupski	5	7	6	12	30
	Kościerski	3	6	7	14	30
	City of Gdańsk	7	10	8	5	30
Silesian	Cieszyński	15	10	3	2	30
	Gliwicki	8	10	7	9	34
	Zawierciański	7	11	8	6	32
	Kłobucki	5	15	7	3	30
	City of Jastrzębie--Zdrój	15	12	3	1	31
Holy Cross	Konecki	8	14	7	2	31
	Ostrowiecki	7	12	8	4	31
	Buski	8	18	12	4	42
	Sandomierski	9	19	9	2	39
	Włoszczowski	4	13	10	5	32
	Kazimierski	4	11	11	5	31

Province (voivodeship)	County	Number of planned interviews on farms in the county, by farm area (in ha)				Total number of interviews
		1 to 1.99	2 to 4.99	5 to 9.99	10 or more	
Warmian--Masurian	Olecki	4	5	5	16	30
	Bartoszycki	3	5	5	17	30
	Iławski	3	5	6	16	30
Greater Poland	Kępiński	5	6	7	12	30
	Ostrowski	6	9	8	10	33
	Turecki	6	11	10	7	34
	Gnieźnieński	3	5	5	17	30
	Kolski	5	9	10	11	35
	Wolsztyński	5	7	8	10	30
	Czarnkowsko--Trzcianecki	4	7	6	13	30
	Szamotulski	4	5	6	15	30
	City of Leszno	8	9	5	8	30
West Pomeranian	Gryficki	4	5	5	16	30
	Szczecinecki	4	6	5	15	30
	Koszaliński	5	7	6	12	30
Total		675	1,186	823	821	3,505

Elaboration: Engel & Dąbała, 2018.

Annex No. 4

Number of households selected in particular counties

Province	County	Number of selected households (stage 1)	Number of selected households (stage 2)	Total number of selected households
Lower Silesian	Lubański	120	360	480
	Kłodzki	124	248	372
	Polkowicki	120	120	240
	Oleśnicki	120	120	240
Kuyavian--Pomerania	Rypiński	120	60	180
	Świecki	124	62	186
	Wąbrzeski	120	0	120
	Włocławski	136	68	204
	Mogileński	120	0	120
Lubelskie	Włodawski	120	60	180
	Bialski	196	0	196
	Krasnostawski	140	70	210
	Chełmski	152	152	304
	Biłgorajski	196	0	196
	Tomaszowski	164	0	164
	Łęczyński	128	64	192
	Lubelski	240	120	360
	Janowski	132	0	132
	Rycki	128	0	128
	Puławski	156	0	156
	City of Zamość	120	450	570
Lubuskie	Sulęciński	120	120	240
	Żarski	120	180	300
Łódzkie	Kutnowski	128	64	192
	Skierniewicki	128	0	128
	Zgierski	128	64	192
	Radomszczański	148	74	222
	Opoczyński	144	0	144
	Piotrkowski	160	0	160
	Wieluński	148	0	148
	Sieradzki	172	0	172
	Poddębicki	128	0	128
Lesser Poland	Myślenicki	140	140	280
	Bocheński	140	70	210
	Miechowski	132	0	132
	Gorlicki	152	0	152
	Limanowski	176	88	264
	Nowotarski	180	90	270
	Olkuski	128	64	192
	Tarnowski	240	120	360
	Dąbrowski	132	66	198

Province	County	Number of selected households (stage 1)	Number of selected households (stage 2)	Total number of selected households
Mazovian	Sierpecki	124	62	186
	Mławski	128	64	192
	Pułtuski	124	0	124
	Ostrołęcki	164	82	246
	Ostrowski	140	0	140
	Kozienicki	128	0	128
	Szydłowiecki	120	180	300
	Białobrzeski	124	0	124
	Miński	144	72	216
	Garwoliński	160	0	160
	Sochaczewski	128	64	192
	Grójecki	172	0	172
	Sokołowski	136	0	136
	Węgrowski	140	0	140
	City of Warsaw	132	1,056	1,188
Opolskie	Oleski	124	0	124
	Nyski	120	300	420
Subcarpathian	Krośnieński	144	576	720
	Sanocki	132	66	198
	Przemyski	136	68	204
	Jarosławski	148	296	444
	Kolbuszowski	132	264	396
	Ropczycko--Sędziszowski	128	512	640
	Dębicki	152	76	228
	Niżański	128	256	384
	Tarnobrzeski	124	124	248
Podlaskie	Suwalski	124	0	124
	Sokólski	144	0	144
	Bielski	136	68	204
	Łomżyński	132	0	132
	Kolneński	124	0	124
	City of Białystok	128	768	896
Pomeranian	Tczewski	120	120	240
	Słupski	120	360	480
	Kościerski	120	120	240
	City of Gdańsk	120	960	1,080
Silesian	Cieszyński	124	992	1,116
	Gliwicki	120	480	600
	Zawierciański	128	128	256
	Kłobucki	136	136	272
	City of Jastrzębie--Zdrój	120	960	1,080

Holy Cross	Konecki	124	62	186
	Ostrowiecki	124	62	186
	Buski	168	0	168
	Sandomierski	156	0	156
	Włoszczowski	128	0	128
	Kazimierski	124	0	124
Warmian--Masurian	Olecki	120	0	120
	Bartoszycki	120	60	180
	Iławski	120	120	240
Greater Poland	Kępiński	120	60	180
	Ostrowski	132	66	198
	Turecki	136	0	136
	Gnieźnieński	120	180	300
	Kolski	140	70	210
	Wolsztyński	120	60	180
	Czarnkowsko--Trzcianecki	120	300	420
	Szamotulski	120	120	240
	City of Leszno	120	960	1,080
West Pomeranian	Gryficki	120	360	480
	Szczecinecki	120	120	240
	Koszaliński	120	240	360
Total		**14,020**	**15,144**	**29,164**

Elaboration: Engel & Dąbała, 2018.

PART TWO

CHANGES IN THE POST-COMMUNIST TRANSFORMATION

Think Locally, Act Globally: Polish farmers in the global era of sustainability and resilience, ed. by Krzysztof Gorlach and Zbigniew Drąg in collaboration with Anna Jastrzębiec-Witowska and David Ritter
Jagiellonian University Press, Kraków 2021, pp. 151–152
ISBN 978-83-233-4949-5
DOI: http://dxdoi.org/10.4467/K7195.199/20.20.12728

Some Introductory Remarks by the First Editor to Part Two

Krzysztof Gorlach https://orcid.org/0000-0003-1578-7400/

The second part of the monograph is quite extensive and diverse in what it covers. It consists of seven chapters that address a variety of issues but share one common idea. All chapters emphasize the changes experienced by farm users, who are understood here as a social category, through the years 1994–2017, and describe them in some detail. Equally important are the various characteristics of family farms in Poland that were observed during the 2017 study and are analyzed in this section of the monograph.

The role of the introductory chapter of this part of the publication, which constitutes Chapter 3 of the monograph and describes and summarizes changes affecting family farms from 1994 through 2007, and also refers to my earlier research, is twofold. First, it serves as a testimony to changes occurring in Polish rural areas during the last decade of the 20th century and the first decade of the 21st century. Second, following the principles of historical sociology, the chapter presents a certain foundation and starting point for the analyses of social processes that can be observed in the studies conducted in 2017 as part of the most recent edition of the research on family farms in Poland.

The next six chapters (4 through 9) consist of analyses of research data collected during the 2017 research project, which is vital for this publication, and presents its results. Thus, Chapter 4 contains a general picture of family farms in Poland, which creates a quite useful background for all subsequent analyses, not just in this chapter but throughout the entire publication.

Chapter 5 compares two different categories of farms, namely large and small farms. This chapter addresses ideas regarding the differentiation of farms (quite common in the sociological literature) as well as the rationale and criteria for dividing the farms by size into two groups, small and large. It illustrates how small and large farms function differently within nation-state societies. Such a reflection typically leads to more observations and remarks on the concentration of agricultural production, farm stratification (i.e., the concept of the disappearing middle), or various patterns of thought and behavior, representing different farm categories.

Chapter 6 focuses on the main issue of the research study, namely the various aspects of sustainable agriculture. Of particular importance are those aspects connected with certain preferences of the surveyed farm operators which tend to influence their decision-making processes.

The two subsequent chapters (7 and 8) are devoted to women's matters. They do not make a contribution to the gender perspective, per se, but are the outcome of straightforward use of the sex/gender variable in the analysis of various aspects of farm management. On one hand, women are viewed as farm operators and/or managers but, on the other, as participants in social life.

Finally, the last chapter of this part, Chapter 9, is devoted to class analysis in regard to farm operators. It provides a historical and sociological reflection on peasants and/or farmers, if one is to use more contemporary terms, as a category in a class system of various societies. No less importantly, it addresses the meaning of class divisions and whether they still matter in the analysis of owners and the users of family farms in Poland.

To conclude these initial remarks on the second part of this work, one more thing should be addressed. All the chapters of this part are the result of the work of two main authors, who are also the lead editors of the entire volume. The exception is the first chapter in this part, which simultaneously comprises Chapter 3 for the entire publication, as it is coauthored by another person who is a member of the larger team working on the research project which is an integral part of this work.

Think Locally, Act Globally: Polish farmers in the global era of sustainability and resilience, ed. by Krzysztof Gorlach and Zbigniew Drąg in collaboration with Anna Jastrzębiec-Witowska and David Ritter
Jagiellonian University Press, Kraków 2021, pp. 153–224
ISBN 978-83-233-4949-5
DOI: http://dxdoi.org/10.4467/K7195.199/20.20.12729

Chapter Three: From Repressive Tolerance to Oppressive Freedom: The end of Polish peasantry?

Krzysztof Gorlach (iD) https://orcid.org/0000-0003-1578-7400/

Grzegorz Foryś (iD) https://orcid.org/0000-0002-9411-2681/

3.1. Introduction

The following monograph aims to present various phenomena, tendencies, and processes occurring among the group of owners (main operators) of farms in Poland that participated in the 2017 research study. It has now been over 30 years since the fundamental political and economic shift that occurred in Poland (and other countries in Central Europe) in 1989–1990. There is a sociological justification for the notion that phenomena and processes of today are well-rooted in past events and developments. What is occurring today may be their immediate continuation or, in some way, their opposition. Such a perspective is very much in tune with historically-oriented sociology, with its principles of continuity and change in social life. There is no question that the historical perspective should accompany the following reflection.

There is a question of how far back one should go when applying an historical perspective to analyze contemporary matters. In other words, what period of time should be taken into consideration in order to discuss what is happening at present? However, this is not the only thing of significant importance. The decision on the role and meaning of certain moments in time that could be treated as starting points for historical background is always a challenge, as these chosen moments may be treated as crucial in terms of shaping social phenomena and processes to be analyzed in the subsequent work. Although it would be tempting to follow in the footsteps of famous historian Norman Davies (2001), who presented one of the best synthetic histories of Poland, going against chronology, in a backward order, would be too much of an imitation. He started his scholarly reflection from contemporary times and finished by recollecting the initial stages of the formation of the Polish state. The purpose of this technique was to show how particular phenomena and processes that were characteristic of certain epochs developed on the basis of what was happening prior.

From the perspective of economic, social, and cultural transformations experienced by farmers, the most important caesura of recent times was the already mentioned collapse of the Communist regime during 1989–1990. At that time, Poland went from being a statist system, a centrally planned economy, and an oppressive political system, legitimized by both communist ideology and some elements of national ethos, to a market economic system, concurrent with the privatization of state companies and the introduction of liberal democracy. Therefore, the last 30 years can provide the historical context for the following analysis and sociological reflection.

When choosing the analytical perspective, one should be aware of two traps which need to be avoided. First of all, it does not seem helpful to engage in a more or less thorough evaluation of that period of time and the so-called regime transformation that was a consequence of the collapse of the Communist system. These evaluations seem to contain more criticism today than they did during the initial phase of changes in the 1990s (see Ost, 2005; Leder, 2014), as they focus on the social costs of the aforementioned transformation. Secondly, it is worth emphasizing that the onset of this period of accelerated changes did not happen in a vacuum but in the context of established behavioral patterns, already existing capital assets, and social relations. Therefore, it seems necessary to touch upon at least some aspects of the situation of farms in Poland that were in operation prior to the period of regime change.

3.2. On Repressive Tolerance: State and peasant farms in state socialist Poland[1]

The reflection presented in this subchapter will be applied to the time period after World War II, when Poland existed behind the Iron Curtain and under Soviet rule and domination. The country was simultaneously undergoing a rapid modernization, conducted in a forceful manner within the system of a centrally planned economy; a modernization which was sometimes called an "imposed industrialization" that neglected the necessary changes in other aspects of social life, leading as a consequence to what Ivan Szelenyi called "under-urbanization" (see Andrusz, Harloe, & Szelenyi, 2011). Most of all, it meant economic changes resulting from the focus on heavy industry and insufficient attention placed on the production of consumer goods and urban development, which led to a drain on human and economic resources in rural areas and in peasant families (see also: Szczepański, 1973).

The idea of repressive tolerance used by Herbert Marcuse (1976, p. 301) seems best suited to describe the relations between traditional peasant family farms and the state that were existing in Poland after World War II. At the time, the state was implementing the political will of the dominant party that took inspiration from the Soviet tradition and ideas of collective farms. In his concept, Marcuse points out that repressive tolerance is a configuration of relations of dominance and submission, in which a dominating actor, without using direct violence, creates the relation of submission with other actors by limiting their opportunities, means to function, and development. In our view, such a situation took place in Poland after Communist authorities backed down from the Soviet-style collectivization of agriculture in the second half of the 1950s after its unsuccessful implementation through administrative means. In effect, the peasant agricultural policy of Communist authorities was defined as "growth without development," as reflected in the works of Kuczyński (1981), well-known Polish economist, who also was one of the architects of the regime change in Poland.

There are lots of arguments to be made that Poland was not a representative country of state socialism. As presented in earlier writings:

> Poland is not typical of countries under state socialism. Western observers have repeatedly claimed that it is a country full of puzzles. While a number of characteristics of

---

[1]   This section is an abbreviated and updated version of the article previously published in the quarterly *Sociologia Ruralis* (see Gorlach, 1989).

a communist system, such as the political monopoly of the communist party and the state-directed economy, are present, Poland has kept at least two enclaves which have not lost their autonomy: the Roman Catholic Church and the peasant family farm. The latter is a unique phenomenon in view of the collectivization of farm land accomplished in neighboring and other countries. The family farm is an important battlefield in Polish society's fight for its independence from the communist system. The private farm has crippled the communist system in Poland, making it incomplete and weaker than in many other countries where the communists seized power (Gorlach, 1989, p. 23; see also: Wilkin, 1988).

As was already mentioned, forced collectivization was abandoned in the 1950s, causing the demise of already established collective farms, of which only a few were able to last until the end of the 1960s or beyond. Their durability was the result of independent decisions of peasants and had some economic justification. What seems most essential for sociological analysis here are the following processes: the shaping of peasants' consciousness, the maintenance and solidification of certain preferences and behavioral patterns in their circles and, most of all, the increase in the feeling and sense of "being on your own" and "being my own boss." It seems appropriate to refer to the earlier work: "What effects, then, did the collectivization imposed on Poland in the Stalinist era have on the consciousness of the peasantry, peasants' socio-political attitudes and more generally, their position as a social category?" (Gorlach, 1989, p. 25). Gołębiowski and Hemmerling (1982) point out that the significance of the failure of collectivization of Polish villages in peasant consciousness:

> ... has remained up till now difficult to evaluate or classify. It has strengthened ... confidence in the possibilities of private farming with its irreplaceable values, to a greater extent than did the revolutionary agrarian reform itself with its "little deeds" during the first years of the "people's rule" (Gołębiowski & Hemmerling, 1982, p. 33).

To clarify this quotation, it should be added that both quoted authors state that reinforcement of the idea of superiority of individual family farms over collective ones had a stronger influence on the resistance towards forced collectivization than the agricultural reform of 1944–1945. That reform was orchestrated by Communist authorities and its main goal was to get peasants' support for the new socio-political order and the new authorities.

Well-established peasants' attitudes towards private ownership of the land were confronted with the labile and inconsequential policy of the Communist authorities towards family farms that continued until 1989, when state socialism ended in Poland. On one hand, there was an emphasis on agricultural profits but, on the other, the ideological goal of agriculture having a "socialist character" was still seen as a vital element and a potential future prospect. This meant pursuing the orientation towards development and domination of some social, communal, or collective farms. The agricultural policy focused on providing the sufficient volume of agricultural production and, at the same time, did not abandon the long-term, future-oriented path to collectivization of agriculture. These dualist goals of agricultural policy resulted in contemporary contradictions in regard to concrete solutions, such as the ban on selling heavy agricultural machinery and equipment to peasant farms and freezing the trade of agricultural land, which prevented farmers from expanding their individual farms. At the same time, the authorities backtracked from forceful forms of collectivization. All of these efforts were taken in order to create such a situation in which collectivization (or some other way of blending individual farms together) would be the only rational solution for the owners and operators of individual farms to increase farm size. Through the collectivization path they would be able to benefit from the effects of the "economy of scale" that were already well-known in agriculture in other parts of the world. As one of the architects of the agricultural policy of the time, Augustyn Woś summarized this approach: "The point is to create a situation whereby the transition to the forms of socialist agriculture [whatever that means—K.G.] will be the only rational and fully conscious step on the part of each farmer" (Woś, 1978, p. 94).

This policy was not free from other various effects and they were especially noticeable towards the late 1960s and the beginning of the early 1970s. To put it simply, every time new possibilities for sales of agricultural products arose, or increased chances for a more permanent perspective for individual (family) farms within the framework of the state-run economy opened, farmers made special efforts to make the best use of them and displayed support for that direction of change. The deep political crisis of December 1970 can serve as an example here. The elites at the time, who had remained in power since the de-Stalinization process of 1956, were confronted by demonstrations from highly dissatisfied workers and forced to resign. The new authorities trying to gain the support of various social groups made significant efforts to win over peasants, the owners of family farms. Immediately after December of 1970, a series of important decisions changed the situation of farmers in a crucial way. One of these decisions related to the abolition of so-called "obligatory deliveries," that were really just sales of some portion of the production from farms to state agencies at set-up artificial prices that were much lower than market prices. Interestingly, before these obligatory deliveries were terminated, they had taken the form of a modified continuation of the regulations introduced by the Nazis during their occupation of Poland in order to provide food security for the German Nazi army and the areas of the Third Reich at the expense of the people living in the Polish countryside. The removal of these regulations about 25 years after the conclusion of World War II was combined with other actions, such as increasing the purchase prices of agricultural products, allowing sales of heavier machinery and equipment to family farms, and—this was particularly important—extending state health insurance to peasant populations, meaning people connected with family (individual) farms. The barriers previously imposed by the ban on selling land to peasant farmers were lowered.

Unfortunately, this was not the beginning of some radical changes favoring private agriculture. In a few years this became clear, particularly in the trade of agricultural land. From 1974 to 1976 the National Land Fund passed no more than 10% of land resources to peasant farms as compared to 1/3 of the land resources passed to private farms from 1971 to 1973. This is not all. The administrative reform of 1973, which introduced communes as administrative and territorial units in rural areas, limited the freedom of activities and development of privately owned farms. The communes were governed by chief (naczelnik) state officials, who were appointed by higher authorities. Interestingly, the prerogatives of the chief (naczelnik) regarding the functioning of farms were quite extensive. Such officials could recommend a certain (and not any other) direction of production or order the state repossession of land improperly used, either due to inadequate production or a farmers' decision to set the land aside. To sum up, this created the possibility of direct administrative intervention in the functioning of farms owned by peasant families. Maria Halamska, the famous Polish rural sociologist, describes this phenomenon in the following way:

> The village universe has been divided into two foreign and hostile worlds. One, that of farming peasants, convinced of the importance of the mission of producing bread, and at the same time underestimated and looked down upon. And the other world of bureaucratic institutions identified with the state. This line of division despite the existence of many others, has proved to be decisive (Halamska, 1981, p. 31).

The high level of politicization of the measures that the party-dominated state took towards peasant agriculture could be observed in the reaction to another political crisis that occurred in 1976. Its main component was the workers' protest in reaction to the decision by the state authorities to increase prices on numerous food products. When confronted with sharp social conflict with the workers' circles, the authorities had to retract their decision on hiking food prices and doing so was not economically justified. Having difficult relations with workers, the authorities made another attempt to improve relations with peasants. New concessions to family farming were made. First of all, the administrative barriers (introduced just a few years earlier) connected with the land transfers from the National Land Fund to peasant owners of family farms were removed. In 1976, which was a year of socio-political crisis, only 21.8 thousand ha of land were transferred to privately owned farms. Three years later, in 1979, the amount of land passed to peasant farms increased almost fivefold, to 113 thousand ha. This was not met with the full support of local authorities who, as is worth remembering, were nominated by higher authorities. Central authorities of the ruling party—who were also the state authorities—conducted a rather capricious, inconsistent, and opportunistic policy towards rural areas. Thus, they had to deal with farmers' distrust on one hand and insubordination from their underlings in local structures and organizations on the other. These local-level officials treated the decisions made by central authorities as if they were temporary, and not meant for more systematic and long-term implementation.

In the second half of 1970 (after the 1976 crisis) a state of very fragile balance was preserved in agricultural policy. On one hand, the area of agricultural land transferred from state reserves to peasant farms increased yet, on the other, the level of expenditures allocated to the modernization of agriculture in the private sector (peasant) remained unchanged. So, in 1979, peasants who operated on three fourths (75 %) of the utilized agricultural areas in Poland received only 30 percent of all financial means that were allocated for the modernization and functioning of agriculture. It is worth emphasizing that, in the same year, "... the outlay of state farms per hectare of farmland was four times as high as on private farms, but the final output per worker was only twice as high. This is unquestionable evidence of the greater economic efficiency of the peasant [i.e., private—K.G.] sector" (Gorlach, 1989, p. 30).

In the years 1980–1981 Poland experienced another sharp socio-economic crisis, which resulted in the emergence of the social movement named "Solidarność" (Solidarity). As an effect of the compromise that was reached after the negotiations with the Communist authorities, various social and professional circles were able to organize independent representation under the umbrella of trade unions. Such efforts were also made by Polish farmers. The specifics of that process are worth some attention. First of all, farmers called for a separate (different and independent from other social and professional groups) union organization from the very beginning, but the resistance from the authorities towards the formation of a registered farmers' union, independent from the state, was much stronger than it was in the case of industrial workers' unions. Secondly, initiatives taken by farmers were more diverse in terms of both ideology and territoriality. In the earliest stage, three initiatives described as

"Agricultural Solidarity," "Rural Solidarity," and "Peasant Solidarity" emerged. This could be seen as an effect of the diversified peasant economy in different regions of the country, as well as a result of various types of farmers' identity, in that sometimes they would call themselves "farmers" and at other times they would use the more traditional term "peasant" ("chłop") or "landlord" ("gospodarz"). The terms such as "agricultural producer" or "food producer" were sometimes employed as well to highlight the modern aspects of individual farm operations. The first of these organizations seemed to represent the relatively well-educated and prosperous owners of larger and modernized farms. The second organization reflected the aspirations of the owners of smaller farms who were looking for allies and collaborators among people employed by state farms and farming cooperatives, which were rather rare. Finally, the third organization included people that were focused around political and societal (national, religious) goals of the movement. In 1981, political rationality resulted in the unification of these organizations, but the secession tendencies continued to exist in the organizational configuration, which could be seen as proof of the increasingly diverse interests of farmers and the heterogeneity of farmers' circles.

Both the power and the role of the peasant movement were noticeable later after the Communist authorities introduced martial law in December 1981. As a result of such law, all organizations that were in any way independent from the state were forced to suspend their activities and then, ultimately, to terminate them. At the time, no independent peasant organization officially existed, but the state authorities formally introduced a constitutional amendment that would guarantee continuity of family ownership in agriculture.

3.3. Problems with Modernization of Peasant Farms in Poland: A short overview[2]

After the brief overview of the situation experienced by farmers and family farms during the times of state socialism, some crucial aspect of this situation should also be addressed. The struggle to maintain the rights of peasants to land ownership and keeping farms in families took place from 1948 to 1989. It was connected with the continual processes of modernization of Polish society. In other words, the modernization of Poland, which meant transformation from an agricultural and rural

[2]   This section of the monograph is an updated version of an earlier work (see Gorlach, 2001, pp. 58–69)

society to an industrial and urban one, had a different trajectory than development processes in Western Europe. In Poland, modernization was occurring within the framework of a centrally planned, statist economy and it was strongly influenced by Communist ideology. It may be worthwhile to briefly review the effects of these types of modernization that had an impact on the functioning of the agricultural economy, family farms, rural community, and rural areas. This reflection is particularly important in order to understand the changes which started to take place in Poland after the regime collapse in 1989. While pursuing the principles of historical sociology in the following work as one of the important analytical frames, it is worth noting that according to the so-called path dependency theory (Stark, 1992) some behavioral patterns established under state socialism—habits and preferences, as well as the material and institutional base already solidified by that time—played a significant role in the post-collapse changes, especially right after the time of the regime collapse.

A few remarks can be made on historical sociology. First of all, as was highlighted by Abrams (1989, p. 2), all sociological explanations have an historical character by their very nature. For that reason, historical sociology should not be treated as a specific branch of sociology and may be seen more like the essence of this academic discipline (see Delanty & Isin, 2003). Skocpol (1990, 1991, p. 374) gives more attention to another, more concrete aspect of reasoning, stating that, according to the principles of historical sociology, emphasis should be placed on the role and meaning of cause and effect chain analysis, which make up the core of social processes. In other words, the perspective of historical sociology in some way forces the researcher, who is focused on current affairs and events, to take a look back, and to some extent consider earlier characteristics of the presently occurring processes. Furthermore, it should also be mentioned that the historical sociology perspective highlights the meaning of contexts of the study's phenomena and processes. What is important here is to look at how social life is happening, rather than building some abstract models of social reality based on selectively chosen cases. This suggestion, however, should not be treated as an argument against the value of such analyses, theoretical references, or more general models and concepts of collective action or interpersonal relations (see Skocpol, 1991, p. 376; and also, Tilly, 1984; or Polish authors: Topolski, 1973; Zamorski, 2008).

Keeping the above statements in mind, it may be useful to go over the effects of the modernization of Polish agriculture and particularly the modernization of peasant family farms in Poland which took place in the second half of the 20th century (until 1989) in the already existing system of state socialism. This will allow for—according to the premises of historical sociology mentioned above—a closer look at transformations of these farms in the conditions of specific modernization process in Poland as an example of a country located in Central & Eastern Europe.

Earlier in this chapter there were several opinions of various authors on the modernization potential of peasant farms in Poland during the time of transformation. It is of course problematic that these opinions are selective and, in effect, present only some, assorted aspects of the analyzed phenomena. If considered together, they could provide a rather comprehensive and general look at the issues that are interesting and important for further reflection in the following work.

In her work published at the beginning of the 1990s, right after the regime change, one of the most prominent Polish rural sociologists, Maria Halamska named three attitudes that, in her view, could be identified among operators of peasant family farms, which should be treated as the effects of the experience of the modernization processes taking place before 1989. These three attitudes are the following: learned helplessness, collective egoism, and distrust towards the outside world. It should be added here that these characteristics were analyzed by the author in the context of Poland's then possible, yet not completely certain, accession into the European Union. Distrust towards the outside world, which Halamska lists as the last of the characteristics, is particularly important. According to the author:

> This last [the outside world—K.G.] took in farmers' mind the dimensions of the "city," "cooperative," "bank" or "state." Now, it can be coupled with a different form, one which is completely foreign to farmers: Europe with its overproduction of food, subsidized agriculture, customs restrictions, competition, and high requirements regarding quality. All of this means that farmers' [peasants'—K.G.] adaptation to a market economy is a condition of "joining Europe." It may be very dramatic and could be met with peasants' resistance. It will be the "end of peasants" in Poland and as it seems—the last peasants in Europe (Halamska, 1991a, p. 53).

This quotation leaves no doubts as to how the author perceived the role of peasants in Poland's process of accession into the European Union. This explains the role of the circles working against the process of Poland's integration with the EU. In this context, farmers were treated as carriers of conservative attitudes, who feared the changes that were already taking place. They were also experiencing anxieties about the requirements of the market economy and modern mechanisms regulating social and economic processes.

It should also be remembered, and the following work has already addressed, how Polish farmers were able to keep their farms despite the collectivization pressure of the state. This ability is often seen as proof of peasant farmers' flexibility, and in some way, their market-oriented attitudes. Here it may be helpful to refer to the work of American political scientist Stephen Cohen (Graham, Cohen, & Colton, 1992) who juxtaposed the changes after the fall of the Communist system taking place in Poland and Russia. He showed how in Poland private ownership of primary sources of agricultural land was preserved and how a specific culture of entrepreneurship came into being. Based on this opinion one can conclude in this case that peasants can be a natural ally of pro-market reforms and the strategy of integrating the Polish economy and Polish society with the structures of the European Union.

Usually, when two such divergent opinions like those of Halamska and Cohen are being compared, the question of which of them is correct or closer to the truth may emerge. Neither picture is error-free. This can be seen in both of these opinions being one-dimensional in terms of their approach to peasants, their attitudes, and values. Therefore, the main task is to take a closer look at events, processes, and phenomena, which happened within the certain historical context of modernization processes that shaped certain attitudes and action strategies and then collided with the new reality after 1989. To accomplish this, various opinions of Polish and foreign authors reflecting upon these issues will be quoted and analyzed.

Here, the process described in the Polish subject literature as a "swelling of peasant strata" should be addressed. It means reinforcement of farmers' position within the society, with simultaneous reproduction of the internal structure of this social category. This process is described by Halamska (1991a, p. 46), who, while analyzing the late period of state socialism, namely the decade before the regime change, emphasized that, "... peasants in the 1980s are a very dynamic and expansive group. The collective dynamics of this group as a whole dominate over the dynamics of processes occurring inside the group and those processes that diversify it internally" (Halamska, 1991a, p. 46). Following this thesis, the author stresses the character of the so-called incomplete modernization of peasant farms in Poland. She indicates growing supply in the technical means of production, which favors modernization of farms, but mostly in this more objective dimension. In the cultural dimension and through various aspects of peasants' awareness, the pressure related to the threatened existence of these farms due to political preferences of state socialism, causes their owners to hold on to peasant traditions despite the introduction of various elements of modern farming. Explaining this issue more in-depth, the author states:

> Farming style treated as relations between the amount of external and internal work and the effects combined with market principles and vertical and horizontal market reactions is not changing much. Despite technical advances the manner of farming changes minimally. **Even though the peasant economy is technically improving, it**

> does not change its logic just like the use of a stethoscope and surgical gloves does not change a quack into a doctor (Halamska, 1991a, p. 47).

As was already mentioned, through a half century of functioning in the system of statist and state-run economy, family farms in Poland could not escape a sense of threat. Adamski and Turski address it accurately and synthetically. As they put it:

> The phenomenon of relatively high professional activity of Polish peasants in the situation of permanently marked sense of group threat and deprivation of technical means of production (and other unfavorable circumstances) explains the following hypotheses with good probability: 1) ethos of the group threat caused in the peasant culture the reaction of group self-defense; 2) situation of threatened group interest strengthened in some peasants the traditional sense of obligation towards the land as a motherland's good; 3) cyclical, repetitive policy of social and economic discrimination of peasants forced many of them to increase their activities (Adamski & Turski, 1990, p. 148).

This is not a complete repertoire of changes and tendencies embedded in the consciousness of peasant family farm operators. Halamska presents a somewhat different approach (1991a, 1991b). Based on the results of her studies, she identifies several crucial features of peasants' consciousness that were developed as a result of their functioning within the system of state socialism. The first feature can be described as negative integration. It contains the peasants' conviction that the outside world is something foreign and that global society makes its structures available on a negative basis, only to submissive groups. Such a way of thinking may prompt not only an attitude of passive resistance (see, e.g., Szczepański, 1973), but can also—to the contrary—awaken aspirations for empowerment. Studies conducted in the 1980s (see Gorlach & Serega, 1991) revealed very strong connections between aspirations toward empowerment and being the creator of one's own fate in peasant circles. Farmers focused around owning the farm, developed the perception of peasants' own identity as operators, owners and food producers, as well as the awareness of related possibilities to articulate their group interests in the socio-economic context that could limit or even block these possibilities. Consequently, the institutional organization of the economy following state socialism continued to prevail. The second feature in this context relates to the group deficiency, widely present among peasants despite relatively beneficial changes in their social position in the 1980s. In this context, it seems appropriate to quote Halamska again: "Compared with other groups situated in the city, peasants feel less affluent, with smaller possibilities to fulfill their needs and with less significant possibilities to have political influence and achieve social respect. This is the state of multidimensional synchronization of deprivation" (Halamska, 1991a, p. 48). The same author highlights the sense of threat that accompanied peasants throughout the whole period of state socialism or—to put it mildly—uncertainty that at the time was brought about by state policy towards peasant family farms. As Halamska writes (1991a, pp. 49–50), "A sense of threat can be drawn from omnipresent distrust and uncertainty expressed in peasants' statements. . . . This uncertainty can be seen in great distrust towards agricultural policy and disorientation about its further directions." As it happens, such feelings of apprehension are so deeply rooted in the 1980s that, even after the introduction of the constitutional provision on family farms as a permanent element of the national economy, almost half of farm operators expressed feelings of uncertainty, or even felt threatened about the fate and the future of their farms (see Adamski & Turski, 1990, p. 142).

This created a situation characterized by a lack of confidence and distrust towards the political doings of state authorities, especially within the scope of agricultural or social policy. Peasant circles came up with various defense strategies, sometimes referred to as the "art of survival." One of them was a strategy of passive resistance or—in other words—"specific resistance" towards the matter at hand which alleviated for farmers the most drastic consequences of the policy conducted by state authorities. In the words of one of the most prominent Polish sociologists of the second half of the 20th century, "Peasant masses turned out to be the 'moderators' of social progress, breaking the most radical inclinations, softening the effects of radical revolutionary actions" (Szczepański, 1973, p. 213). Thanks to this defense strategy, peasants earned the appellation "veto group" (see Lane & Kolankiewicz, 1973). Another form of peasant defense was a specific form of "coping with the system." It was mostly manifested in the emergence of an elaborate configuration of complicated net links between farm owners and local representatives of the state administration and state apparatus. This meant, but was not limited to, the existence of paternalistic and clientele relations (see, e.g., Adamski, 1967), informal ties, and the prevalence of corruption among local officials representing the Communist apparatus (see, e.g., Halamska, 1991b). There were also more structural aspects, as highlighted by Wilkin (1988, p. 31), one of the most prominent Polish agricultural economists. In his words:

> . . . the primary adaptation of peasant farms to the system of socialist economy in Poland was quite effective. Very crucial qualitative differences in the characteristics of

the subsystem of the peasant economy and the dominant system of the overall socialist economy became serious obstacles in the broader development and modernization of the peasant economy.

Later in the same work the author adds,

> In several decades of existence of the system of socialist economy, slow but rather crucial changes encompassing various important features of the system were taking place. These changes helped with maintaining of peasant economy. The most important among them were the following: gradual increase in importance of market mechanisms, partial decentralization of economic decisions connected with development and resource allocation, discriminating treatment of the private sector, and increasing meaning of economic calculations. In the case of Poland these changes were to a significant extent imposed by **the very existence of the large sector of peasant economy** [emphasis—K.G.] (Wilkin, 1988, p. 32).

Based on the above consideration, it would not be appropriate to form a thesis on the homogenous character of peasantry in Poland under state socialism, whether in a structural or awareness sense. The diversification processes of peasant farms, although not occurring on a scale comparable to the phenomena occurring in Western countries and under the Western market economic system, could already be observed in the 1970s. At that time there was a small group of farms of a modern character that made use of a government program for specialized farms introduced by the state authorities. Farm diversity was reflected in emerging types of social identity presented by farmers of that time. Besides traditional terms such as "peasant" new references to the identity of "farmer" as a person performing a certain profession appeared, as well as other terms such as "producer" and "yeoman" (see Jagiełło-Łysiowa, 1967; Halamska, 1982, 1991b; Gorlach & Seręga, 1991, pp. 76–82). These last two terms referred to something other than merely specific vocational activities and also pertained to the making of certain products (such as agricultural and food products), as well as the ownership of certain means of production (land, machinery, livestock, etc.). This diversification in conceptions of social identity became quite apparent in the years 1980–1981, when the diverse "Solidarity" ("Solidarność") movement developed in farmers' circles, as was already mentioned above.

It would seem worthwhile to reflect further on the role that certain habits and behavioral patterns, consolidated at the time of state socialism, could play after the regime change of 1989 and the introduction of free market principles in the economy. It turns out that the economy during state socialism shaped such patterns of behavior, which did not fulfill positive functions in the context of the market economy. The two authors mentioned above have made a very accurate commentary on this issue:

> The policy of discrediting the validity of peasant social class significantly limited the possibility of its normal development as a social group. One of the paradoxical consequences of such a policy was imposing the state of uncertainty upon peasants while in some peasants it unintentionally created the status of artificial confidence. These latter farmers in various periods of time and for various reasons (economic, cultural, or personality-related reasons) did not feel inclined to engage in active professional activity on their farm but tried to limit it to the elementary, inertia scale. Agricultural policy of the state intervened in processes of rational competition in peasant economy and thus helped various stagnant forms of farming to survive. Additionally, their subsidized existence caused reproduction of ineffective agricultural enterprises. This strange confidence, maybe even carelessness, that the protectionism of the state policy generated in some of the most traditional and less productive farm owners and operators became a source of the crisis situation and social conflict to no lesser degree than the fears and threats of existential character that dominated the peasant social class (Adamski & Turski, 1990, p. 148).

The processes presented above led to some modernization paradoxes observed in peasant farms within the economy of state socialism. Here, the most powerful was the situation in which state policy in Poland after 1989 exacerbated the situation of uncertainty for peasant farms and simultaneously preserved the character of this sector of the economy. As observed by another renowned Polish sociologist,

> State socialism saved the Polish peasant class from the existential threat that stemmed from the development of market economy in Poland. This rescue turned out to be a historical trap but it was the only game that could be played at the time. The peasant class persisted, but it happened at the expense of developmental potential. In other words, the peasantry survived at the expense of peasant families (Mokrzycki, 1997, p. 38).

The situation that developed in the peasant (or private) sector of Polish agriculture could in the 1980s finally be seen as a direct consequence of "suppressed" modernization (see Kochanowicz, 1988), or "unfinished modernization" (see Halamska, 1991a, 1991b). The changes in the peasant economy, as explained earlier, did not, for political and ideological reasons, follow the path of modern farming (according to the model adopted by Western countries). However, the Soviet-style collectivization applied in the Soviet Union and many other countries of the Eastern Bloc was strongly resisted by peasants and could not be successfully implemented on a larger scale. Therefore, it seems appropriate to point out the most pivotal characteristics of the peasant agriculture situation at the moment of the Communist system collapse in Poland. As Kochanowicz stresses,

> Poland remains a country with an unfavorable agrarian structure. There is a situation of dualism, with only a certain portion of several million agricultural land units deserving to be perceived as farms or quasi-farms, while others do not provide significant production to either fulfill the needs of the urban population or even provide their owners and their families with adequate income to thrive or have equitable existence (Kochanowicz, 1988, p. 87).

3.4. Facing the Oppressive Freedom

After the regime change in 1989 the sense of threat experienced by farmers did not vanish, but the sources of perceived threats changed. While previous fears had to do with administrative and political pressure not so much towards collectivization but towards complete subordination of peasant economy to the system of the state-run economy, post-1989 fears had their roots in mechanisms of the market economy, which forced farm owners and operators to make decisions that would bring about productivity and profits. The market also forced farmers to actively seek market outlets for their products, as well as find ways to lower production costs, which would enhance the competitiveness of the products offered by certain farms and influence the income level (i.e., the "profits") of the farmer and his/her family. The farmer (owner, operator) gained more freedom in their decision-making. This latitude was mostly accomplished thanks to the independence that farmers had from political and administrative decisions and interventions of state apparatuses and was simultaneously limited by the nature of resources, such as the area of the land owned or operated by the farmer, level of mechanization, the state of farm buildings, their amenities, accrued financial capital, etc., and signals coming from the market. Although farm owners and operators had some formal latitude in decision-making, they still found themselves under pressure from economic factors. It was in this way that the system of "repressive tolerance" so characteristic of the era of state socialism was transformed into a system of "oppressive freedom" typical of capitalism (see also, Gorlach & Seręga, 1991).

The situation described above is experienced by both typical peasants operating agricultural land, as well as quasi-farmers and regular farmers, to use Kochanowicz's (1988) terminology. The latter, by their very nature, seem to be a suitable type of farming for a modern, market economy and thus, on principle, these farms are more sensitive to market fluctuations. Production that consumes a lot of capital and exists on a larger scale fulfills the conditions needed for potential profits, carries a larger element of risk in the event of failure. It is not a coincidence that during the 1980s farm crisis in the United States there were the most bankruptcies occurring among Polish farmers oriented towards market production and expansion of their resources (see Strange, 1988; Gorlach, 1994). Thus, farmers' demands towards the state, to protect them from the vagaries of various market mechanisms, seem justified. This is not a surprise that small farm owners and tenants in Poland, who only have a few dozen hectares at most, or even owners and operators of large farms with cattle, hogs, and poultry prefer far-reaching state interventionism (see Wielowieyska, 1997). Traditional, small-scale farms that are less market-oriented, as well as peasant farms with peasant methods of production, also call for state interventionism as they are not able to clear adequate profits and provide their owners and operators with sufficient income.

It should not be surprising that farmers were the first social group to protest "the Balcerowicz plan," which was the package of radical economic reforms stimulated by the regime change. It was meant to secure the transition from the statist, state-run economy to free market reality. Six months after the introduction of this reform package from Poland's first non-Communist government since World War II, the new authorities were dealing with the big wave of farmers' protests culminating in the two-hour nationwide roadblock organized on July 11th, 1990. According to *The Warsaw Voice* (1990) this was the first "heart attack" of that first non-Communist government in post-World War II Poland.

The protests lasted until 1993, which was when the first phase of the reforms related to the regime change concluded. They outlasted the already mentioned first non-Communist government led by former Solidarity ("Solidarność") adviser and trade union expert Tadeusz Mazowiecki. During this period, almost 1500 collective protest actions took place and only 112 of them saw initiation by, or participation from, farmers' circles. Farmers' demands were not much different from those of their counterparts in various countries. They related to profitability of agricultural production and relevant shaping of economic policy (agricultural policy in particular), which would be beneficial to farmers. Farmers called for affordable credits and loans, which could be used for expanding the farms and modernizing them, as well as suspending the bankruptcy procedure of those farms that fell into excessive debt.

The necessity of having custom tariffs prevent cheap imports which could threaten domestic food products was also brought to attention. The need to promote exports as well as to facilitate and subsidize contract production, in order to prevent unpredictable effects of the market economy, was also stressed. All of these demands were displaying a rational reaction to a rapid drop in farmers' incomes, which from 1990 to 1992 amounted to about 60% of the 1989 income level. In 1992, there was also a cut in expenditures on farm modernization, a reduction of 27.6% from the 1990 level (CBOS, 1993a). One could also point to a certain "demonstration effect," which, until the 1980s, was seen in the European Union where agriculture received special attention from particular states, which tied in well with the principles of the Common Agricultural Policy.

It should be explained that the farmers discussed here were not of this kind of social category to uncritically perceive the situation of their farms and only care about maintaining the existing status quo. Therefore, farmers could not be recognized as the most vocal social category in opposing the reforms connected with Poland's accession to the European Union and adopting free market principles. As indicated by nationwide research study (see CBOS, 1993b), Polish farmers noticed the structural disadvantages of Polish agriculture caused by its specific development path, briefly presented above. A significant number of interviewed farmers (75%) stated then that Polish agriculture was too fragmented. More than half (58%) expressed opinions that it was "backwards" and/or "not very efficient." Exactly one third (33%) said that Polish farmers were not very active or entrepreneurial. Opinions on the future of farming in Poland were also well articulated. 78% of respondents said that family farms were the future of Polish agriculture, but they also added that their size should be decidedly larger.

Even more information can be drawn from observation of farmers' behavior at the time of the research. It was evident that financial assets were mostly channeled into farm development and not so much into households, which Halamska noted (1991a). Between 20 and 30% of respondents reported a willingness to make active, market-oriented decisions. This mostly applied to young and better educated owners of larger farms. Two thirds of the interviewed farmers reported at the time of the research (early 1990s) that the Polish countryside was undergoing the process of farm diversification and fragmentation. They saw agricultural policy as unfavorable in the sense that the state's intervention in market processes was insufficient. More than half of respondents thought that such policy did not provide adequate protection to smaller and weaker farms, operated by families of multiple professions. Despite a generally unfavorable situation and the wave of protests in rural areas, 27% of respondents declared making larger investments in 1992 and as many as 35% of respondents admitted to having rather concrete investment plans for the following two years (CBOS, 1993c).

These tendencies turned out to last well into the years after the regime change as skillful agricultural policy stimulated structural changes in agriculture. For example, the Agency for Restructuring and Modernization of Agriculture offered credits and loans for establishing and equipping farms intended for individuals under the age of 40 (young "farmers," male and female). From the end of 1995 to the first quarter of 1997 the Agency signed over 33 thousand such agreements. These credits and loans offered from the Agency were used exclusively on farms of over 10 hectares, which in Polish reality meant "larger than average." It should also be remembered that the resources of such farms were then three times higher than the resources of the average farm unit in Poland. Furthermore, the indicators of their profitability were relatively high. The farms were generally not in debt, but their owners or operators were rather careful with credits and loans even though their potential was significant (W.T., 1997, p. 5). This could be related to typically peasant traditional distrust or, speaking in a more contemporary manner, one could say that at the time caution characterized the action strategies applied by owners and operators of family farms (for a more detailed analysis, see Gorlach, 2001, pp. 58–69).

3.5. "Into the Deep Water": Introductory and methodological remarks

It is time to take a better look at the functioning and maintenance of family farms after 1989. The proposed analysis is based on the results of three studies within the research project conducted from the first half of the 1990s to the first decade of the 21st century. The results of these studies were published in three monographs (see Gorlach, 1995, 2001, 2009). Their synthetic summaries are presented later in this chapter.

Here, the message in the title of this very subchapter, namely the "deep water," should be explained. It alludes to the traditional saying grounded in the Polish language, describing a new, unexpected situation, in which various subjects facing certain activities are described as being "tossed into the deep water." It also refers to the metaphor that in certain situations some people report that they do not have their "feet on the ground." Dealing with such situations often requires using all currently available, as well as new and innovative, solutions in order to prevent "drowning." An analysis of changes in family farms in Poland after the regime change in 1989, which could be seen as throwing them into the "deep water" of the market economy and limited state intervention, is a subject of the following reflection.

Methodology should be addressed firsthand. In the 1990s and in the first decade of the 21st century three series of empirical research projects, in the form of panels, were conducted. In each of the three studies the same sample of family farms was used. The first study was completed in 1994 thanks to the № 1 1628 93 03 grant financed from the Committee of Scientific Research of that time (see Gorlach, 1995). The research was conducted on a nationwide quota sample of 800 individual farms (with a minimum area of 1 hectare). Calculations of the sample were made based on data presented in Publications of the Chief Statistical Office (see CBOS, 1993a, 1993b) and the data from the list of villages in a particular municipality, while the municipalities were randomly drawn. The sample of 800 farms was divided into 54 strata. There were 9 macro-regions distinguished in the sample due to the administrative division of Poland at the time. They were the following: nation's capital, north-eastern, northern, central-western, central, central-eastern, south-eastern, southern and south-western, as well as 6 land strata (1–2 ha, 2–5 ha, 5–7 ha, 7–10 ha, 10–15 ha, 15 and more ha) (see for a more detailed information: Gorlach, 1995, pp. 1–19). As a result of these studies, financed by the Committee of the Scientific Research through the № 1 1628 92 03 grant that was acquired in the grant contest, it was possible to make an address list of individual farms. This list became the basis for two further consecutive editions of the research project conducted in 1999 and 2007. Conducting the research on the nationwide sample and making the address list of the studied farms created a very solid foundation for the panel research. It opened a possibility, and at the same time created an incentive, for starting a process for following the destinies of studied farms and their owners. In the situation of sudden and radical transformations in Polish agriculture and rural areas this provided a chance to follow these changes. In other words, it created an occasion to observe important social processes in a special kind of laboratory. After five years, another research panel was organized and its results were published in another work devoted to family farms in Poland (see Gorlach, 2001). This time the theoretical frame and terminology were expanded. References to the globalization concept were present, especially in regards to the version that was experienced by Polish society, at the time of Poland's preparations to join the European Union. Although the research theme remained unchanged, a few additional issues came to light and they were connected with the opinions expressed by interviewed farmers, mostly on the subject of protecting their interests and evaluating the policy the state authorities had on Poland's entry into the European Union. Just like in the previous edition of the research, the financial means were secured with another grant (№ 1 H02E 007 15) from the Committee of

Scientific Research and acquired through the contest. It is worth mentioning that out of 800 farm owners, who comprised the selected population sample for interviews in 1999, researchers were able to reach 687 farms but not all of them were operated by the same people as five years prior.

Eight years after the last panel it was decided that a new research study should be conducted with the address list of the 687 farms whose owners participated in the 1999 study and were interviewed. The reason for doing so was very similar to what it was in the previous study and had to do with the desire to explore the changes experienced by these farms over the period of time that had passed since the last study. It is worth mentioning that in this period of time one significant change had occurred and it was not without influence on the pace and character of changes taking place in Polish agriculture and the countryside generally. This was of course, the formal accession of Poland to the European Union which took place on May 1, 2004. From this moment forward, Polish agriculture and countryside have been involved in various programs which constitute the Common Agricultural Policy or—as it is more recently known—Common Rural Policy. The change of the name reflects the change in philosophy regarding agriculture and the countryside, not just in the countries of the European Union and not only as academic discourse. It goes well with the construction of certain policy conducted through the agenda of the European Union, and at the same time follows the government programs as well as the programs of regional authorities and various nongovernment organizations. This time it was possible to conduct interviews with the owners of 515 farms out of the 687 that were on the address list created in 1999 as a result of the research. Similar to the previous two editions, the financial support for the final study came from national funds, and this time it was the № N116 003 31/0267 grant from the "successors" of the Committee of Scientific Research, namely the Department of Scientific Research of the Ministry of Science and Higher Education that was won in 2006 in the grant contest.

3.6. The Dynamics of the "Un-family-ing" Process of the Investigated Farms

The following subchapter will start with a look at the evolution process related to two "opening" characteristics pertaining to the degree of "family" feature present in farms (see Mooney, 1988; Gasson & Errington, 1993; Gorlach, 1995). The first of these characteristics is considered exemplary of the family farm and stresses the farm ownership situation. With family farms the agricultural land that is being used is owned by the farmer. The second characteristic deals with the manner of acquiring the farm, which typically involves receiving it as an inheritance. These very issues are presented in tables 3.1 and 3.2.

Table 3.1. Farm ownership situation

Owner's category	Number of users/operators as a %		
	in 1994	in 1999	in 2007
Operator	85.6	77.1	76.7
Tenant	12.6	19.8	21.1
Renter	1.8	3.1	2.2
Total	100.0	100.0	100.0

Source: Gorlach, 2009, p. 97.

As can be inferred from Table 3.1, the evolution of particular categories is rather ambiguous. Firstly, it should be emphasized that in all three moments in time, designated by three consecutive research editions, the "exemplary" category of family ownership indicated above remains dominant. However, its predominance was gradually declining from 85.6% in 1994 to 76.7% thirteen years later. This confirms the dominance of "typical" family farms among the researched agricultural units, especially if the ownership is being analyzed. However, the pace of the transformations in that regard is also worth some attention. The scale of change is incomparably larger in the first-time interval, even though, the time frame was shorter than in the second. It is noteworthy that the first time frame only encompassed 5 years. During that period, the overall share of "typical" family farms declined by 8.5%. In the second studied time period, which lasted 8 years, the change was small and did not exceed 0.5%. Therefore, one could even venture a stabilization theory, which means that the volatility of changes in ownership relations and land transfers between farms were more obvious in the second half of the 1990s compared to later periods.

The observations confirm to some extent the changes in frequency of the occurrence of the second type of farms presented in the table, those used and operated as rental properties. The category of "owner-tenants," the percentage of which increased rapidly in the first of the analyzed periods—from 12.6% to almost 20% over eight years—experienced only a small increase of just slightly over 1%. The emergence of capitalist relations into the area of land ownership presented significant volatility in the 1990s, sharply contrasting with the later period.

It is worth noting here that the changes in the two categories described above were complementary. The extent of the presence of the family in the "model" family farm was decreasing, while the number of farms that experienced the effect of the diffusion of market relations increased. It might be said that relations typical of a market economy were unfolding in the realm of family farms but, firstly, they did not undermine the dominant position of "typical family" units and, secondly, the pace of these changes was weakening over the later period. In that context, it could be beneficial to take another look at the dynamics of the third category of farms distinguished in the table, which is not so clear-cut at first sight.

The term "owner-renters" is used to describe the group of farm owners who utilize less land than they own as they rent part of their land to other operators. In the research this category was decidedly marginal. Owner-renters comprised slightly over 3% of the studied farms. In the first research edition, the growth in this category was notable: from 1.8% in 1994 to 3.1% five years later. However, over the next eight years this percentage fell back to 2.2%. This tendency allowed the formation of a consideration related to ownership and use of farms. First of all, it should be stressed that the decline in the percentage of owners-renters in the studied population sample from 1999 to 2007 could additionally support the thesis of a certain stabilization on the property market formulated above. Secondly, it is worth highlighting that the decline might imply a growing interest in utilization and operation of agricultural land as related to the regulations of Common Agricultural Policy (CAP) being already applied in Polish rural areas. As is well known, direct subsidies are the main financial instrument of CAP available to farmers. Only farmers who actually utilize and operate the land are eligible for subsidies and not the formal owners. Therefore, this phenomenon can be seen as an indicator of the direct effects of new regulations which are connected with Poland's accession to the European Union and applied to Polish farmers. It could also serve as an argument that the effects of these regulations contributed to the stabilization of family farms, at least in the family dimension.

The second aspect of the described phenomenon is presented in Table 3.2. It deals with ways of acquiring and—to be more precise—utilizing the farm. It is worth remembering that a typical—even standard—way of taking on the role of a family farm user is through inheritance, namely, by taking over from someone from the previous generation. This process deserves a closer look. Firstly, it is noticeable that one third of the farms studied in 2007 (exactly 170) involved situations in which there was a new owner after 8 years. The extent of changes is wider than in the 1994–1999 time interval. It should be remembered that only 105 farms out of 685 studied units had changed ownership by 1999 (see Gorlach, 2001, p. 189). This was only slightly above 15% of the population sample in 1999. Considering that the second time interval is longer than the first (eight years v. five years) the extent of change in farm ownership in the second time frame can be partially explained by this factor, but not completely. The time period of 8 years is not twice as long as

5 years. It could be argued that by 1999 the pace of family farms changing ownership was more intense. Should it be treated as an indicator of increasing interest in agriculture among the representatives of the younger generation, connected on one hand with the first, volatile transformation period and, on the other, with the perspective of Poland's accession to the European Union? Such an answer seems quite possible. It should be stressed, however, that it needs to be confronted with other data compiled in the research.

Table 3.2. Ways of acquiring the farm from 1999 to 2007

How one became user of the farm	Number of farms	Percentage
Bought or rented	20	3.9
Received as inheritance	150	29.4
The same owner as in 1999	340	66.7
Total	510	100.0

Source: Gorlach, 2009, p. 99.

Here, a key point of view in the reflection on family farms must be considered. From the 170 new owners that were identified in 2007, only 20 had acquired the farm through purchase or rental. Only 12% of the respondents (out of the 170 new farm owners considered as 100%) acquired farms as the result of specific market transactions. The remainder of the new owners, meaning 88%, came into possession of their farms by 2007 in a typical or even traditional way by taking over the farm through inheritance. It could be concluded that even in the process of relatively wide generation replacement of family farm owners, the mechanisms that generally have dominant and highly meaningful position within the ownership transfer still persisted.

Considering the above observations and proposed generalizations it should be described here how the processes characterized above looked in the period of time between the two first research editions. It turned out (see Gorlach, 2001, p. 230) that the percentage of new owners, who acquired farms through market transactions (purchase, rental) was small, below 5%. It could mean that from 1999 to 2007 it was not only the extent of the ownership changes which was wider than during the 1994–1999 period. What could even be more important was the significant presence of atypical ways of acquiring the family farm which were rather unusual for the family farm model in Poland. This could be explained by increased penetration of the market mechanism in the realm of ownership relations.

Declarations of the new farm users regarding the circumstances in which they acquired their farms might throw some light on the issues described in this section.

Here, the dominant type of circumstances might be regarded as rather typical for farms functioning according to the family model. It largely involved the family tradition of being employed in agriculture and this factor was indicated by the vast majority (above ¾) of respondents, who took over their farms through inheritance. However, it should be mentioned here that the majority (here, 60%) of the new farmers who acquired farms by purchasing them or renting from others declared a strong role of family tradition in their motivations.

Table 3.3. Declarations of the new farm users on circumstances of becoming the farmer

Type of the new farm user (ways of acquiring the farm)	Type of opinion		Total
	family tradition	other	
Purchase or rental	60.0	40.0	100.0
Inheritance	76.7	23.3	100.0

Source: Gorlach, 2009, p. 100.

The matter of financing the farm from outside sources is also meaningful. In the typical family farm the financial means that are used usually come from the owner-user's own resources and the resources of the family members. Using external sources of financing, mostly involving credits/loans, is another example of the family-based agricultural economy being penetrated by market relations. The materials and data collected in three research editions could provide some interesting and valuable information in this regard.

It seems helpful to take a look at the situation of market relations in agriculture in the 1990s. The data for this period of time is presented in Table 3.4. The percentage of farms applying for credits was gradually increasing during that period of time. While in the first half of the decade only one third of respondents declared using credits, in the second half over 43% of respondents used them. It can be said that the 1990s brought about increasing financial dependency of the farms from market mechanisms. As can be recalled from earlier analyses, applying for and using the credits was more common among farms which survived the 1990s (see Gorlach, 2001, p. 184).

Table 3.4. Use of loans and credits in the 1990s

Use of loans and credits	Number of farms (in %)	
	1990–1993	1994–1998
No	462 (67.4)	386 (56.4)
Yes	223 (32.6)	299 (43.6)
Total number of farms in 1999	685 (100.0)	685 (100.0)

Source: Gorlach, 2009, p. 100.

The last statement can be supported by the data on farms that survived until 2007 (see Table 3.5). The differences between those and the farms that did not last to that year, in which the last edition of the research was conducted, are quite sharp. In the first category, almost half of the studied farms applied for and used credits in the second half of the 1990s. However, in the second category of farms, those that did not remain in operation, the percentage stayed below 30%. It turned out that the risk related to the functioning of the market economy and relying on credits was financially rewarding. Those who took risks related to applying for credits had better odds of surviving the difficult situation of the 1990s.

Table 3.5. Use of loans and credits from 1994 to 1998 (in %)

Use of credits	Farms operating in 2007	Farms not operating in 2007
Yes	250 (49.2)	49 (27.7)
No	258 (50.8)	128 (72.3)
Total	508 (100.0)	177 (100.0)

Source: Gorlach, 2009, p. 101.

Table 3.6 shows the collapse of the trend that was growing in the 1990s and related to using credits in the studied sample population. From the table it can be inferred that the situation appeared to be going back to the first half of the 1990s. In their declarations in 2007, two thirds of respondents reported not using credits from 1999 to 2006. This level is almost as high as in the first half of the 1990s. Similarly, the 2007 declarations of respondents who used the credits from 1999 to 2006 indicate the return of the early 1990s trend.

Table 3.6. Use of loans and credits from 1999 to 2006

Use of credits	Number of farms	Percentage
Yes	190	33.2
No	340	66.8
Total number of farms in the 2007 research edition	510	100.0

Source: Gorlach, 2009, p. 101.

This tendency could be connected with two factors. One of them was the growth of farmers' incomes, mostly in the early years of the 21st century. This could have led to the diminishing interest in credits, especially among those farmers who had experienced some cash influx as a result of their success in agricultural production. Another factor was even more visible and its influence was even stronger. It can be described as an "integration factor" connected with the process of Poland's accession to the European Union. Thanks to the accession, Polish farmers could benefit from the instruments of Common Agricultural Policy, which brought a significant inflow of financial means to the rural areas. In the time preceding Poland's admission as an EU member, Polish farmers became eligible for various pre-accession programs, allowing for transfer of some financial means from the EU to at least some parts of farmers' circles. All of the above factors could justify the theory that less frequent use of credits in the later period of time covered by the research might not serve as proof of weakening of the penetration of market forces into the domain of family agriculture. This is rather an effect of the changing character of this penetration, which no longer appeared to be a "clean" market mechanism like a typical credit offered by banks in a capitalist economy. The penetration took the role of a regulatory mechanism connected with the policy of the institutions responsible for the state of agriculture and its functioning in the European Union.

Another issue that needs to be recognized as interesting in the process of "de-familying" of farms traditionally connected with rural families, deals with the type of workforce used in studied farms. On a typical family farm, the farm work is being provided almost exclusively by family members. Any presence of hired work could be treated as evidence of the penetration of market relations or relations representative of the capitalist manner of production and abandonment of the model family farm situation described earlier. The subject literature stresses that the increased involvement of hired workers in the total work force of family farms confirms yielding to the logic and functioning of capitalist economic principles (see Mooney, 1988; Gasson & Errington, 1993). From the theoretical perspective of the functioning of economic institutions, the meaning of relying on this type of workforce is invaluable. Such a relation between the owner and the worker that is based on the principle of renting contracts determines the character of the institution.

A closer look at the research results presented in the tables 3.7, 3.8, and 3.9 might be helpful in answering questions on the advancement of the "de-familying" processes of a family farm in Poland during the years of transformation.

Table 3.7. Hired labor (number of workers)

Number of workers	Percentage of farms		
	1994	1999	2007
0	80.6	75.7	80.0
1	6.4	8.7	5.5
2	6.3	6.8	8.6
3 or more	6.7	8.8	5.9
Total	100.0	100.0	100.0

Source: Gorlach, 2009, p. 102.

In Table 3.7, at first, there is a tendency of increasing number of farms using hired labor in the first period of time from 1994 to 1999 but in the following time period this trend is reversed. It can be observed that from 1994 to 1999 the percentage of farms not using hired labor was declining and then from 1999 to 2007 it grew again, almost returning to the level from 13 years prior. This could serve as an argument for the entrenchment of the family type of farms during the studied period. If one would consider the effects of the processes of selection of farms in the studied population sample (from 800 in 1994 to 510 in 2007) it would be justified to risk the statement about the strengthening of farms that are typical family farms in terms of their use of labor. If a somewhat different perspective is applied, one can conclude that family labor strengthens flexibility and the ability to adjust family farms to changing conditions (see Mooney, 1988; Gorlach, 1997).

Information related to the number of workers hired on the farm could provide interesting material for analysis, as the tendencies are not unambiguous. In the first part of the studied period, the percentage of farms that hired one, two, or three workers was growing. It was clear evidence not only of the increased use of hired labor but also of its intensification in the functioning of family farms. In later years, these tendencies diversified quite visibly. The percentage of farms that hired two workers grew but the percentages of farms hiring one and three or more workers declined from 1999 to 2007. This could indicate a dual tendency in the studied population sample. On one hand, there was a stabilization of the dominant group of farms that was based exclusively on family labor and, on the other, the presence of farms that modestly used hired labor of 2 workers was also being marked. Less visible were farms that used hired labor for supplemental work and hired only one person. A similar situation could be observed with farms that presented a more capitalist character and hired three or more workers. Could this indicate the onset of some stabilization period, in which the dominant role would belong to typical family farms but also involve family farms modestly supported by hired labor?

Table 3.8. Hired labor (number of days in the year)

Number of days	Percentage of farms		
	in 1994	in 1999	in 2007
0	80.6	75.7	80.0
14 or less	9.2	15.7	7.4
15–30	6.1	5.5	6.7
More than 30	4.1	3.1	5.9
Total	100.0	100.0	100.0

Source: Gorlach, 2009, p. 103.

The above statements get additional support in the study results presented in Table 3.9, which shows that in the 1999 to 2007 time frame the number of farms that used hired labor to a minimal degree, in terms of time, decreased. This was true for the category of farms that used hired labor for a maximum of 14 days in a year. As shown in the table, there was a steep decline in that category from 1999 to 2007. Different trends were reported in the two remaining categories of farms that used hired labor. In the first part of the studied period, from 1994 to 1999, the number of farms in these two categories declined. Then, in the subsequent eight years, the percentages rose in both of these categories. This might prompt a thesis on the diversification of the studied farms. Typical family farm units comprised the dominant majority but there was also a rather stable category of family farms that were supported by hired labor, which improved their market position. They used hired labor systematically for more than 14 days a year and not just on an ad hoc basis. Such irregular use in the case of the described research meant 14 days per year, at most. These statements are also in sync with the research results presented in Table 3.7. The emergence of a group of family farms that are moderately but, at the same time, rather systematically supported by hired labor, should be noted.

Table 3.9. Types of labor used on farm

Type of labor force	Percentage of farms		
	in 1994	in 1999	in 2007
Family	53.1	51.7	65.5
Family + hired labor	11.7	14.0	11.0
Family + work exchange with other farmer/s	27.5	24.5	14.5
All types	7.7	9.8	9.0
Total	100.0	100.0	100.0

Source: Gorlach, 2009, p. 104.

Hired labor is not the only way in which needed labor resources are secured for family farms. In Poland, with its family-based agriculture, there is a certain type of labor, related to traditional, peasant agriculture, with a strong presence of bonds between relatives and neighbors which could be observed on the level of local rural communities. However, as was already described in my other publications (see Gorlach, 1995, 2001), the processes of accelerated modernization of the Polish countryside observed after 1989 caused this tradition-oriented type of labor involving relatives and neighbors to decline.

This issue is being considered in the next part of the following reflection. Table 3.9 contains the study results related to various types of labor used on farms, including the type of labor that can be described as neighborly farm assistance and its reciprocation. Juxtaposing all three types of labor (1) family labor, 2) help from neighbors, and 3) hired labor) in three different moments in time allows for formulating other statements that will be presented below.

From this perspective, the process of strengthening typical family farms in the studied community is quite visible. The percentage of such farms grew from 53.1% in 1994 to 65.5% thirteen years later. It should be emphasized that in the first part of the studied time period this percentage was rather stable and only a small decline was noted (down to 51.7% in 1999). The most obvious tendency, characteristic of a dramatic drop in the percentage of farms using the system of neighborly assistance and reciprocation, was noted in the second part of the studied period, from 1999 to 2007. This drop in reciprocal assistance seems to be contributing to a significant growth in the number of typical family farms. The relatively small change in the percentage of farms using hired help, whether in combination with family-only labor resources or with the use of reciprocated neighborly assistance, is worth attention as well. These insights allow for formulating more general observations pointing to the acceleration in the pace of modernization processes occurring in Polish agriculture after 1989, which led to farmers cutting their use of traditional forms of farm labor. There was no radical increase in the presence of the typically capitalist forms of hired labor. This shows the tendency for a certain individualization of families and individuals operating farms in the rural areas. They tend to concentrate more and more on using family labor resources, supported sometimes by hired labor and, at the same time, abandoning more traditional forms of help and neighborly cooperation.

An additional issue that provides some insight into the changes in mentality on hired labor in farmers' circles is the analysis of respondents' answers to the following questions: "Would you agree to someone from your family performing paid work for other farmers?" In 1994 (the first research edition) negative answers were decidedly dominant, with some respondents allowing such a situation only in extreme circumstances. This could be explained as an effect ". . . of a certain cultural tradition, according to which a worker hired by a farmer is always seen as somebody inferior to the farmer, with rather low social prestige. Such a person is not able to afford operation of a farm" (Gorlach, 2001, p. 228). This tradition was very characteristic for this type of peasant thinking that was concentrated around work "on one's own" and "for oneself." Comparing the study results from 1994 with those from 1999 (see Gorlach, 2001, p. 229, as well as Table 3.10) it could be inferred that the phenomenon of hired labor developed more meaning in the collective consciousness of the respondents. As was reported: "On one . . . hand, such work has been gaining wider recognition as a way to supplement income and, on the other, could be seen as having negative connotations related to situations of submissiveness and dependency on the employer" (Gorlach, 2001, p. 230). The analysis of the data collected in 2007 showed that, generally, these tendencies did not change much. While there were over 56% of respondents who did not accept such work (for various reasons) in 1994, 13 years later that percentage was higher, exceeding 60%. However, the concrete validations of particular types of declarations are the most crucial and thought-provoking (see, again, Table 3.10).

What was most striking was the percentage decline in declarations regarding the acceptance of having a family member work for other farmers for economic reasons. This can be treated as an indication of the selection processes and elimination of those farms which had engaged earlier in work for other farmers as a way to round out the household income. As far as other tendencies were concerned, the percentage of non-economic motivations in 2007 was noticeable, as it was quite close to the percentage from thirteen years prior. There was also a noticeable decline in statements indicating readiness to accept such work as an exception, in extreme situations. The growth of percentage of respondents accepting the work of their own family members for other farmers on the condition of it being "appropriate work" should not be omitted. In connection with another observation, a slight decrease in the acceptance of situations where family members provide hired labor to others, due to certain negative connotations, could be noted. One might hypothesize that paid work performed for other farmers should not be—in the respondents' opinion—just any job. The respondents expressed expectations that their family members should be well treated and well paid while doing such work. This was quite consistent with a small but not insignificant increase in the percentage of answers where such work was not acceptable, based on a "sense of dignity." This traditionally peasant way of thinking was still present in the collective consciousness of a rather small, but still visible category, among the studied subjects. Such a tendency was rather stable, present for over a decade and still existing despite the selection processes occurring among studied farms.

Before finishing this part, one more issue deserves some attention to support the thesis on the stabilization of the family character of Polish agriculture. In the studied period of time (see Table 3.11) there was a noticeable increase in the number of farm owners who declared they would not sell their land under any circumstances. Yet the percentages of these respondents who were willing to sell their farms or would do so only if necessary in dramatic circumstances also declined over the specified years. It is worth noting that the percentage drop in these last two types of declarations took place in the second part of the studied period of time, from 1999 to 2007.

Table 3.10. Evaluation of hired labor in agriculture

Category of answers	Percentage of answers		
	in 1994	in 1999	in 2007
Yes—economic justifications	24.3	29.8	18.2
Yes—non-economic justifications	5.8	2.3	5.3
Yes, but only in extreme situations	3.4	4.3	1.6
Yes, if there is "extra free labor" available on one's own farm	2.8	2.8	2.7
Yes, if such "work is appropriate"	7.5	7.2	9.8
No, due to self-esteem issues	7.6	6.8	7.1
No, due to the character of hired labor	8.4	10.8	11.6
No, enough work on their own farm	36.7	34.8	37.3
Others	3.5	1.2	6.4
Total	100.0	100.0	100.0

Source: Gorlach, 2009, p. 106.

Table 3.11. Opinions on readiness to sell farms (in %)

Type of opinion	1994	1999	2007
Not in any situation	70.4	58.9	80.0
Yes, if necessary	10.1	21.1	5.4
Yes	19.5	20.0	14.6
Total	100.0	100.0	100.0

Source: Gorlach, 2009, p. 107.

It can be suggested that the process of stabilization of the position of family farms took place during that period of time. The relativity of their value as a natural effect of agriculture entering the market economic system was already visible throughout the 1990s. At the time, the number of farmers/respondents who allowed for some possibility of selling their land grew, and the percentage of those who would not consider such a possibility declined. The reversal of this tendency in the following years is undoubtedly the result of those two processes. On one hand, this could be seen as an effect of selecting the farms, which occurred in the already mentioned 13 years. In the studied population sample of 2007, the dominant majority of respondents could report some successful achievements and had confidence about following the right path professionally and in other aspects of their lives. The effect of Poland's entering the European Union, which brought relative stabilization to the farms situation, should also be noted here. This certainly could increase the tendencies among the owners and users to reject the possibility of selling their farms.

3.7. Cultural and Economic Capital: Evolution of the investigated farms

In the following subchapter the processes of diversification of farms' market position in the 1994–2007 time frame will be discussed. Basic terms, as well as variables such as "economic capital" or "cultural capital," or "market position of the farm" that will be used were formulated in another publication (see Gorlach, 2001, pp. 231–245).[3]

The analysis should start with the already mentioned diversification process of farms according to the economic capital they have. It seems helpful to recall what was previously written on this matter from 1994 to 1999. The most important statement indicated that in 1994 the group of farms with a high level of economic capital comprised less than 7% of the studied population and five years later it was close to 12%. This tendency indicated the polarization process that eliminates farms with a moderate amount of land and a moderate (medium) amount of capital. At the same time, farms with either a high or low level of economic capital marked their presence quite well. (Gorlach, 2001, p. 238).

Table 3.12. Level of economic capital in studied farms

Level of economic capital	Number of farms in %		
	1994	1999	2007
Low	53.1	48.2	41.8
Medium	40.1	40.0	43.9
High	6.8	11.8	14.3
Total	100.0	100.0	100.0

Source: Gorlach, 2009, p. 108.

[3] Both variables, namely, "economic capital" and "cultural capital," were constructed using the data obtained during empirical research (see Gorlach, 1995, 2001, pp. 231–237). Each of them contained three variable components. In the case of economic capital they were: farm area, farm mechanization level related to plant production, and animal husbandry. In the case of cultural capital they were: level of rationality of the owner measured on a special scale, level of education, and age. The logic of constructing these two variables was the following: A higher level of economic capital ties in with a larger area of the utilized agricultural land and with possession of more sophisticated machinery and other equipment for plant and animal production. A higher level of cultural capital is usually connected with a higher level of rationality, a higher level of education, and a lower age of the surveyed farm owners. In effect, the positively privileged market position of the farm (class position of the owner) is connected with a higher level of economic and cultural capital. Further, the negatively privileged market position of the farm (class position of the owner) could be interpreted as an effect of a lower level of economic and cultural capital.

Could the conclusions formulated in 1999 still apply in 2007? The answer seems to be somewhat ambiguous. It should be noted that in 2007 farms with a low level of economic capital at the time already constituted a smaller percentage of the studied population in comparison with those with a medium level of capital (see Table 3.12). The process of restructuring could be seen in the more visible presence of farms with levels of capital designated as "medium" or "high," with their percentages growing steadily over the entire period of time from 1994 to 2007. The only category with decreasing percentages was the group of farms with a low level of economic capital.

Previous analyses led to formulating the conclusion that the level of economic capital was an important factor in determining the future of particular farms. In other words, farms equipped with a higher level of economic capital had a better chance of survival in the market economic system. Table 3.13 confirms such a constellation in the last period of time. As can be seen in the table, the percentage of farms with a high level of economic capital in 1999 was almost three times higher among the farms that stayed in operation until 2007 compared to farms that quit agricultural activities. It should be stressed that the same tendency could be seen in the category of farms with a medium level of economic capital. Only a low level of economic capital turned out to be a factor that made surviving in the following years harder when dealing with the market economy. Furthermore, it is worth highlighting that such an equation was more statically meaningful in the period of time from 1999 to 2007 ($p < 0,001$, see Table 3.13) than it was in the previous period of time ($p < 0.05$) (see Gorlach, 2001, p. 239).

Table 3.13. Level of economic capital of the farms that stayed in operation from 1999 to 2007 and the farms which ceased operations*

Level of economic capital	Percentage of farms	
	that stayed in operation	that ceased operations
Low	43.7	61.4
Medium	42.2	33.5
High	14.1	5.1
Total	100.0	100.0

* $p < 0,001$.

Source: Gorlach, 2009, p. 108.

Transformations related to the second fundamental variable, which is cultural capital of the owner-operators of the studied farms, represented some other patterns (see Table 3.14). In the first period of time, from 1994 to 1999, the percentages of those who had low or medium levels of cultural capital increased somewhat. Only in the category of high cultural capital could an opposite tendency be observed, as their percentage declined. Based on such observations a general statement could be made addressing the weakening quality of the entire populations of surveyed farms, first in 1994 and then in 1999, due to the low level of cultural capital. In 1994, 73.6% of farms could, at best, report a medium level of such capital. Five years later, the selection processes caused that percentage to rise to 76.4%.

However, in the subsequent period of time, from 1999 to 2007, these tendencies were radically reversed. First of all, a significant reduction could be seen in the category of "the weakest" farms with the lowest capital level. The percentage of such farms fell from almost 25% in 1999 to barely 10% in eight years. At the same time there was a noticeable spike in the number of farms with a medium level of capital (from 52.1% in 1999 to 60.3% in 2007). There was also a reverse tendency relating to those farms which could be considered "the best," with a high level of cultural capital. Such farms were more visible in 2007, and not just in reference to the situation from 1999 onward but, more importantly, from the starting point of the research in 1994.

Table 3.14. Level of cultural capital in studied farms

Level of capital	Number of farms (in %)		
	in 1994	in 1999	in 2007
Low	22.8	24.3	10.0
Medium	50.8	52.1	60.3
High	26.4	23.6	29.7
Total	100.0	100.0	100.0

Source: Gorlach, 2009, p. 109.

It can be stated that from 1994 to 2007 the patterns of transformation and distribution of economic and cultural capital in the studied population of family farms were characterized by a significant level of diversity. In the case of economic capital, there was an increasing trend in the categories of farms with high and medium levels of such capital. In the entire studied period of time the percentage of farms with economic capital increased. Such a tendency in the case of cultural capital appeared rather late in the second part of the studied period after 1999. This could be viewed as a certain manifestation of "cultural lagging." The initial period of time for farm selection led to survival of farms which were better equipped with economic capital but not so well situated with cultural capital. The compatibility effects of these processes could only be observed in the second period of time. This could mean that from the start (time of the research program's implementation), economic capital was a crucial factor in the positive selection of farms. Cultural capital showed its role in that regard much later.

Table 3.15. Level of cultural capital of farms that remained in operation from 1999 to 2007 and farms which ceased operations

Level of capital	Percentage of farms	
	that remained in operation	that ceased to operate
Low	23.4	27.3
Medium	51.8	53.4
High	24.8	19.3
Total	100.0	100.0

Source: Gorlach, 2009, p. 110; statistical relationship insignificant.

Seemingly, the above thesis can be supported with the conclusions stemming from the analysis of values in two tables presenting the levels of both types of capital in both categories of farms: those which stayed in business and those which ceased to operate. Values relating to economic capital can be seen in Table 3.13 described earlier. The issues connected with cultural capital are presented in Table 3.15. In the entire studied period of time, both high and medium levels of economic capital were clearly seen on farms that did not stop operating. This relationship was significant in a statistical sense.

The matters were quite different with cultural capital. The considerations were limited to the time period from 1999 to 2007, when the tendencies of the distribution of cultural capital started to resemble the processes that were observed in the case of economic capital. It turned out that the percentages of farms with low and medium cultural capital were higher in the category of farms that ceased operating during that period of time. Only the farms that stayed in operation had a higher level of cultural capital than farms which ceased to produce. This could mean that only a high level of cultural capital was a favorable factor in dealing with market mechanisms. As can be seen, a medium level of economic capital proved to be sufficient. Furthermore, unlike in the case of economic capital and processes of farm selection, it turned out that connecting the relationship between the processes of farm selection and the level of cultural capital was not significant in a statistical sense. This could be treated as an additional argument supporting the thesis that cultural capital (i.e., age, education, and type of rationality of farm owners) had a smaller impact on the future of the farm than economic capital, which was crucial for providing production means. Nevertheless, it might be justified to say that with the development of a market economy in Poland and ongoing processes of farm selection and restructuring, the impact of this factor would likely be stronger.

To support the above statements, the data presented in tables 3.16 and 3.17 should be considered. These data stem from analyses in which combined levels of both types of capital and their influence were examined. The percentage of farms with a low level of capital (economic and cultural combined) was steadily declining in the studied population. On the other hand, the percentage of farms in the opposite situation was growing. Farms with a medium level of capital were the exception that reflected the level of complication of restructuring processes, taking place in the studied population of farms. The changing impact of both types of capital determined the destinies of farms in the medium level capital category. The percentage of such farms in the first period of time (1994–1999) was declining and in the second period of time (1999–2007) experienced a quite pronounced growth.

Table 3.16. Level of combined economic and cultural capital of the studied farms

Level of combined capital	Number of farms (in %)		
	in 1994	in 1999	in 2007
Low	47.9	45.8	33.5
Medium	44.3	42.4	51.1
High	7.8	11.8	15.4
Total	100.0	100.0	100.0

Source: Gorlach, 2009, p. 111.

Table 3.17. Level of combined economic and cultural capital of farms that stayed in operation and the farms that ceased to operate in the years 1999–2007*

Level of combined capital	Percentage of farms	
	that stayed in operation	that ceased to operate
Low	42.2	56.8
Medium	44.2	36.4
High	13.6	6.8
Total	100.0	100.0

* $p < 0.01$.

Source: Gorlach, 2009, p. 111.

Furthermore, a statistically significant relationship between the variables, such as level of capital and the fate of the farm could be observed. The significance level was $p < 0.01$.

Table 3.18. Market position of the farms

Farm privilege position	Number of farms (in %)		
	in 1994	in 1999	in 2007
Negative privilege	47.9	48.1	36.2
Average privilege	34.5	30.7	35.5
Positive privilege	17.6	21.2	28.3
Total	100.0	100.0	100.0

Source: Gorlach, 2009, p. 112.

Table 3.18 presents the diversity of farms in regards to their market situation. The market position is a derivative of characteristics connected with farms being equipped in both types of capital. It is constructed as the aggregate effect of particular categories.[4] In the previous work, based on the analysis of the tendency in the time frame from 1994 to 1999, as illustrated in Table 3.18, the following generalization was formulated:

> The analysis . . . clearly points to the process of farm stratification, which could be treated as an indicator of the disintegration of the peasant class . . . generally in both extreme populations, meaning that the percentage of farms with few possibilities for adjustment to the market situation [a negatively privileged market position—K.G.] and those with extensive possibilities to adjust to it [a positively privileged market position—K.G.] increased, while the middle category of farms with medium adjustment possibilities [a market position of average position—K.G.] there was a notable decline. This tendency can be treated as a manifestation of the process of the so-called disappearing middle, which was often described in the rural sociology literature and viewed from the perspective of comprehensive farm characteristics that could not be reduced to the size of the land owned (Gorlach, 2001, p. 243).

It is worth remembering that such a generalization was formulated based on the analysis of the values presented in Table 3.18, in the second and third column. Did anything change in the subsequent period of time, from 1999 to 2007? The answer

---

[4]   Aggregation (see Gorlach, 2001, pp. 238–245) meant simple combination of certain categories of cultural and economic capital. Therefore, the farms with a high level of cultural and economic capital attained the highest market position (positively privileged) and their owners/operators highest class position. Farms with the lowest level of economic and cultural capital constituted the category with the lowest market position (a negatively privileged market position) and their owner-operators had the lowest class position. All combinations of various levels of economic and cultural capital created the category of the "averagely privileged." Additionally, combinations of levels of both types of capital, close to the high level were categorized as "positively privileged," but those close to low levels were placed in the category of "negatively privileged."

to this question could be found in the fourth column of Table 3.18. A rather evident change could be distinguished in the observed tendencies. The concept of "the disappearing middle" could not be applied any longer to the studied population sample. This is confirmed by an ostensible reduction in the percentage of farms with a negatively privileged market position. While in 1999 such farms comprised slightly over 48% of the studied population, in 2007 their percentage was much lower, falling to 36%. Secondly, farms characterized as having a market position of average privilege, which in 1999 made up less than 31% of the studied population, exceeded 35% in 2007. What could be observed here is not a "disappearing middle," but more likely an "expanding middle" and clearly this was happening at the expense of the extreme category of farms with the least potential for adjusting to the market situation.

This statement could be reinforced by a closer look at the tendency observed in the category of farms that have a positively privileged market position. Their percentage grew more noticeably in the period of time from 1999 to 2007 than in the previous time frame, from 1994 to 1999. The thesis regarding the disappearing middle mentioned earlier should therefore be reformulated. The study results from 2007 demonstrated that the processes of farm restructuring led to farm stratification. Its form was not so much of a bi-polar structure of family farms but rather—at least at that time—concentration of capital on farms that had better possibilities to adjust to the market economy. While discussing this issue it should be remembered that, due to the panel method used in the research, all processes were observed in the shrinking sample population, in which farms with a higher level of capital had better chances of survival. It should not be surprising that the farms with a market position of at least average privilege had better visibility.

In Table 3.19, there is a shift between the categories designated by market position. The analysis of the data contained in the table allows for the formulation of several more general observations, which are presented below. First of all, it could be worth inquiring as to whether a more stable situation of farms could be observed during the 1999 to 2007 time frame. Or perhaps the opposite could be true, with more intensified processes of changes in market position? In this case, the percentage of farms which remained in the same category during the studied period of time was an indicator of stability. It could easily be inferred that from 1994 to 1999 farms had a more stable situation than in subsequent years. It could be noted that in all three categories, the percentage of farms remaining in the same category was higher in the 1994 to 1999 time period than in the next, from 1999 to 2007. They were respectively: for farms with a negatively privileged market position, 74.3% and 59.1%; for farms that were averagely privileged, 49.8% and 44.2%; and for farms with a positively privileged market position, 63.3% and 55.5%.

Table 3.19. Dynamic of changes in market position of the studied farms from 1999 to 2007 (*from 1994 to 1999*) (number of farms in absolute values and in %)

Market position	Negatively privileged in 1999 *(1994)*	Averagely privileged in 1999 *(1994)*	Positively privileged in 1999 *(1994)*	Total in 1999 *(1994)*
Negatively privileged in 2007 *(1999)*	124 59.1% *243 74.3%*	42 27.0% *72 30.6%*	10 8.4% *15 12.5%*	176 36.3% *330 48.4%*
Averagely privileged in 2007 *(1999)*	61 29.0% *63 19.3%*	69 44.2% *117 49.8%*	43 35,1% *29 24.2%*	173 35.7% *209 30.6%*
Positively privileged in 2007 *(1999)*	25 11.9% *21 6.4%*	45 28.8% *46 19.6%*	66 55.5% *76 63.3%*	136 28.0% *143 21.0%*
Total in 2007 *(1999)*	210 43.3% *327 47.9%*	156 32.2% *235 34.5%*	119 24.5% *120 17.6%*	485 100.0% *682 100.0%*

Note: The number of farms that were studied involved 485 units in 2007 and 682 in 1999. Due to incomplete data it was decided that 25 and 3 farms, respectively, would be eliminated.

Source: Gorlach, 2009, p. 113.

Without looking into the categories of farms that shifted between the categories in both periods of time, the characteristic of the studied process would be one-sided. It turned out that the variables in the categories of farms with a negatively privileged market position confirmed decidedly the above thesis. It could clearly be seen that the percentages of farms with averagely and positively privileged market positions were lower in the period of time from 1994 to 1999 than from 1999 to 2007. For the averagely privileged they were 19.3% in the first period of time and 29.0% in the latter. For those positively privileged these values were 6.4% in the first period of time and 11.9% in the second period of time.

A somewhat different shifting pattern could be seen in regard to farms with an averagely privileged market position. In the first of the studied periods of time the processes of their social degradation, indicated by moving from the category of averagely privileged to the negatively privileged position, were more clearly pronounced than in the next period of time, from 1999 to 2007. The respective farm percentages in this category were 30.6% and 27%. However, the processes of social advancement viewed from the perspective of this social category presented a different logic; they were more noticeable in the second part of the analyzed period of time. From 1999 to 2007, such processes applied to almost 29% of farms as compared to less than 20% in the 1994 to 1999 period.

And what was the situation of farms with a positively privileged market position? Here, the processes of strong degradation (transition from the positively privileged position to the negatively privileged position) were more pronounced in the first studied period (experienced by 12.5% of farms) rather than in the second (8.4% of farms). The situation looked somewhat different in the case of what the research described as moderate degradation, which meant shifting from the positively privileged position to the averagely privileged position. The applicable values were 24.2% of farms in the 1994 to 1999 period and 35.1% in the 1999 to 2007 time frame.

What could be the general conclusions formulated on the basis of the above observations? Firstly, it should be stressed that the period of time from 1999 to 2007—contrary to earlier expectation—was the time of a weaker stabilization of farms' market positions. During that time, the categories of the studied population sample experienced strong restructuring processes. Furthermore, it turned out that these processes took different forms in various farm categories. The processes of the advancement of the negatively privileged farms were more noticeable in the second part of the entire period of time covered by the research. Similar statements could be made in reference to the processes of advancement of farms with an averagely privileged position. What should be emphasized here is the greater visibility of degradation of such farms in the first period of the analyzed time frame. This, in some sense, is confirmed by degradation processes occurring in the category of farms with the positively privileged market position. Strong degradation was mostly observed in the first part of the analyzed time period (from 1994 to 1999), and quite weaker in the second part of the studied period (from 1999 to 2007). Therefore, it can be said that in the second part of the analyzed period of time the stabilization of farms' market position was weaker, while the farms' advancement processes were more clearly visible.

3.8. Peasant Classes?

In this part the hypothesis regarding structural transformations of the studied family farms should be analyzed. It pertains to ways of thinking, defining a farmer's own identity and opinions about relations between employers and employees as presented by owners of these farms, which—according to the premises of the research program—could be treated as indicative of a certain class consciousness (see Gorlach, 2001, pp. 245–262). It is worth stressing that the analyses presented below mostly apply to those farms which survived the entire period of time from 1994 to 2007, during which the three editions of the research program were conducted. The number of farm units in 2007 was 510. This was how many of them remained from the original 800 which were in the observation pool in 1994 when the first edition of the research was conducted. Only these 510 farms were considered, when the data from 1994 and 1999 were presented.

This last remark is important for two reasons. Firstly, the reflection conducted in this manner might be a valuable contribution in characterizing farm destinies. Secondly, focusing on the farms that survived the entire duration of the research allows us to show the effects of the farm selection processes, as well as to illustrate various tendencies developed in farm communities related to their characteristics.

The first crucial characteristic, which should be analyzed, was the way the respondents identified their role (identity). In the analyses on class consciousness this is viewed as one of the essential levels of class consciousness (see Giddens, 1973). The perception of one's own social and professional role is the base on which various contents—aspects of consciousness—can be embedded, creating several layers, or—as Giddens put it—consecutive levels of class consciousness.

In the analyses of the study results, connected with the respondents' answers about the understanding of the situation of family farm owners, various contents were aggregated in three fundamental categories, such as: yeoman, entrepreneur, and the marginalized. The identity described as the "yeoman" type often invoked traditionally peasant aspects relating to the status of land ownership and a certain pride in possessing it. Consequently, the identity described as the "entrepreneur" type tied in with a certain modern conceptualization of ownership as the basis for various possibilities of action and creation of new values. And finally, the third type, described as the identity of the "marginalized," meant that the respondents referred to the situation of hopelessness and treated farms in their possession as a certain difficulty in their life and not as a chance for an active presence in a society, an asset that could improve their situation. In some sense, this went well with the peasant tradition and mostly with its aspects that dealt with grievances, a sense of injustice, or being doomed. Table 3.20 presents the frequency of occurrence of three types of identity in three different moments in time when consecutive research editions were conducted.

It turns out that the yeoman identity type that referred to positive aspects of the peasant tradition continued to exist in a stable way in the studied population. This type of identification of one's own social position was shared (with some minor fluctuations) by more or less one third of the respondents. The changes in the other two identity types were far more interesting. The entrepreneur identity was declared by slightly more than 40% of respondents in 1994, yet five years later by just 28%. Then, in the following eight years it became a self-definition of the social role reported by over half of farm owners who participated in the research (!). In 2007 this category was the largest. Every other (or even more than that) owner of a farm in the research considered himself to be an entrepreneur (!). This process predominantly occurred thanks to a significant decline in the percentage of respondents who identified themselves in the marginalized category. In 1994 almost every fourth respondent presented this type of identity and then in 1999 this category grew to 40% of the researched sample population. This trend was reversed by 2007 when the process of marginalization of the marginalized category took place. At the time of the third research edition the marginalized only made up 12.2% of the studied population.

Table 3.20. Way of perceiving the role of the owner in the studied farms

Way of defining the role	Number of answers (in %)		
	in 1994	in 1999	in 2007
Yeoman	33.3	30.9	34.4
Entrepreneur	40.4	28.2	53.4
Marginalized	26.3	40.9	12.2
Total	100.0	100.0	100.0

Source: Gorlach, 2009, p. 115.

The transformations of the last two categories taking place in the two studied processes in the periods of time, from 1994 to 1999 and from 1999 to 2007, were particularly important. In the case of the entrepreneur identity there was an initial drop in its frequency and then its percentage grew. In the case of the marginalized identity the situation was the reverse. First, there was a rapid increase and then a decline. Two factors could be responsible for this process. On one hand it could quite surely be the result of the farm selection process and the elimination of those farms that were not able to withstand the reality of the market economy. Owners of such farms were, understandably, most likely to represent the marginalized identity type. On the other hand, it was not without meaning that the period of time from 1994 to 1999 was the time of essential farm restructuring, to which many farmers reacted with frustration, fear, or even feelings of helplessness.

Some additional light to the following reflection can be brought through analysis of the values enclosed in Table 3.21, tying the types of identity with class position of the studied farms' owners. It was measured by market position of the farm, which was discussed earlier in this chapter. What could primarily be seen here were statistically significant relations in every case, although decidedly weaker in the 2007 study than in the two previous research editions.

Table 3.21. Ways of perceiving the role of the owner depending on the class position of the studied farmers in 1994, 1999, 2007

Way of defining one's role	Negatively privileged position (in %)			Averagely privileged position (in %)			Positively privileged position (in %)		
	1994*	1999*	2007**	1994*	1999*	2007**	1994*	1999*	2007**
Yeoman	42.4	36.3	38.3	27.9	30.4	29.5	21.1	21.8	35.8
Entrepreneur	24.7	17.0	45.7	46.4	31.7	57.2	67.4	43.5	58.4
Marginalized	32.9	46.6	16.0	25.7	37.9	13.3	11.6	34.7	5.8

* $p < 0.001$; ** $p < 0.05$.

Source: Gorlach, 2009, p. 116.

It might be worthwhile to take a closer look at the tendencies of frequency of occurrence of particular types of identity in each of the three categories of respondents, occupying different—according to the accepted scheme—class positions. The reflection on this matter should start with some remarks on the type of identity categorized as a yeoman, tied with a traditionally peasant sense of "owning the land." Among the negatively privileged, the percentage of respondents presenting this type of identity was slightly reduced during the entire timespan of the research, from 42.4% in 1994 to 38.3% in 2007. In the case of the averagely privileged, stabilization was quite visible and the applicable values in that category were, respectively, 27.9% and 29.5%. In the category of the positively privileged there was a tendency quite opposite to the one observed in the category of the negatively privileged. There was a noticeable growth in the category of farms whose owners self-identified as the yeoman type. In 1994 there were 21.1% of such respondents, and this did not change much over the next 5 years. However, in the later period of time there was significant growth in this category of self-identification, up to 35.8% in 2007. Interestingly, in this case, the 1994 percentages of owners identifying as the yeoman type diverged quite markedly from the opposing class categories, meaning the negatively and positively privileged (42.4% compared to 21.2%). Then, thirteen years later, these values became similarly common (38.3% compared to 35.8%). This could make an argument for fostering more similarities in the way of thinking, at least in that regard, in both opposite categories of the studied farmers.

What was the situation of the entrepreneur identity, which could be perceived as indicative of the modern way of thinking and characteristic of the farmer who operated the farm according to the principles of modern enterprise? In this case an interesting and rather paradoxical situation could be observed. It turned out that the percentages of respondents self-identifying as such increased significantly in the categories of the negatively and averagely privileged. In the case of this first category, the growth started from the level of 24.7% in 1994 and rose to 45.7% in 2007; while in the category of the averagely privileged it grew, respectively, from 46.4% to 57.2%. In the category of the negatively privileged this could mean noticeable influence of the ongoing processes of farm selection in the studied population sample. As was indicated earlier, the percentage of farms with a negatively privileged market position was significantly reduced in the period of time from 1999 to 2007. The owners of the farms that survived despite unfavorable conditions most likely treated them as small enterprises dealing with various problems, typical for a market economy. A similar mechanism could possibly be going on within the category of the averagely privileged.

Given the situation of farm owners with a positively privileged market position, an opposite tendency can be observed. In this category the percentage of respondents identifying themselves as entrepreneurs was lower in 2007 (58.4%) than in 1994 (67.4%). It should be noted that the lowest percentage in that category could be observed in 1999 (43.5%). Considering the last observation, it could be said that in the category of the positively privileged there was relative growth in the second part of the analyzed period. It should also be noted that in all the categories of the interviewed farm owners, the percentages representing this type of producer were at their lowest in 1999. This could bolster the argument supporting the thesis that the 1990s brought about the most traumatic situation for farmers, which caused many of them to question or even abandon the identity of the entrepreneur. This is additionally confirmed by the fact that in all farm categories the percentage of farm owners presenting the marginalized identity type was the highest in that particular year, 1999.

One essential relationship, which confirms the thesis about the links between the character of farms that have the best positions in the market economy with the "entrepreneur" identity type should be presented here. In all research editions the percentage of farm owners presenting this identity type was the highest (although in the last, 2007 edition the difference was not highly visible) in the category of farms with a positively privileged market position. The percentage decline in this category in comparison to the situation that had been observed in 1994 had to do with the increase of the percentage of farm owners representing the yeoman identity type. While this issue would certainly require additional and more in-depth studies, one can speculate that a certain portion of farm owners with a positively privileged market situation started to refer to at least some elements of peasant tradition. Perhaps this could be treated as a contribution to the thesis on representation that in recent years has been quite noticeable in the rural sociology literature (see Ploeg, 2008, 2013)?

Another issue that might deserve a closer look is a reconstruction of the class consciousness of the respondents as related to their perceived sense of distinctiveness from others. According to the proposal by Giddens (1973), this could be seen as another layer (level) of class consciousness, created on the basis of one's sense of identity. The analysis of the sense of distinctiveness was conducted from the perspective of comparative studies by juxtaposing the respondents/farm owners with the owners of other types of enterprises. It was based on the premise that, by virtue of the modernization processes, farms (even including family farms) were gradually transformed into business enterprises. The question to be asked is whether or not this was reflected in farmers' consciousness. Would the farmers have some sense of belonging in the category of enterprise owners, or at least feel a willingness to have their situation viewed in the larger context as similar to that of owners of various other types of enterprises? The answer to this question might serve as a basis for formulating statements relating to the class consciousness of farmers in modernized society. A collective memory of the social category strongly rooted—at least in the case of Polish society—in peasant past and tradition might be a significant factor working against the establishment of the entrepreneur identity in respondents. Therefore, while researching this issue it was decided that the respondents would be asked not just about their perception of similarities but also about perceived differences in relation to owners of other types of enterprises. The results of the study are presented in tables 3.22, 3.23, 3.24, and 3.25.

Table 3.22. Perception of similarities between the situation of the farm owner and the situation of the owner of other types of businesses/enterprises

Perception of similarities	Responses (in %)		
	in 1994	in 1999	in 2007
There are no similarities	55.0	62.2	40.2
There are similarities	45.0	37.8	59.8
Total	100.0	100.0	100.0

Source: Gorlach, 2009, p. 119.

Table 3.22 contains concise data on the similarities mentioned above that were inferred from the responses of the farm owners who participated in the research. The analysis of the results of three consecutive research editions provided the foundation for the thesis that over the entire duration of the research the farm owners' self-perception as being similar to other enterprise owners gradually increased. In 2007 this way of thinking was presented by 60% of the respondents, while in 1994 that percentage was lower—45%. It should be added that this process was not linear. In 1999 the percentage of farmers expressing such opinions was below 40%, lower than the analogous percentage in 1994. This could serve as another argument in support of the statement that the 1990s was a particularly traumatic decade for farmers. The responses of the interviewed farmers could also be seen as the effect of having a very specific situation and interpreting it as differentiating farmers from other categories of enterprise owners.

The data presented in Table 3.23, might shed some more light on the answers of the respondents relating to comparisons between the situation of farmers and the situation of other entrepreneurs. It should be noted here that the percentage of those who claimed that there were no differences between farmers and owners of other enterprise was the highest in 2007—14.4%. This is not a high percentage and generally it could serve to support the thesis that there is a strong sense of differentiation between farmers and other producers and there are specific characteristics of this circle. Nevertheless, this tendency can be treated as evidence of the conviction, which exists and is growing among farmers, that their situation is becoming more similar to that of owners of other enterprise types. The cessation of this trend in 1999 (decline of the percentage indicating lack of difference) was probably an effect of the already-mentioned farmers' experience of living through the particularly painful decade of the 1990s.

Table 3.23. Perception of differences between the situation of the farm owner and the situation of the owner of a different enterprise

Perception of differences	Responses (in %)		
	in 1994	in 1999	in 2007
No differences	11.3	5.5	14.4
Indications of specifics of agriculture	38.0	25.2	46.0
Feeling of inferiority	46.7	67.7	36.9
Feeling of superiority	4.0	1.6	2.7
Total	100.0	100.0	100.0

Source: Gorlach, 2009, p. 119.

The distribution of opinions related to factors causing such differences is quite interesting. The various answers of the respondents (similar to the results from previous research editions) were aggregated into three essential categories. One of these categories dealt with the specifics of agriculture as a certain sector of production or economic activity and two others had a comparative character. In these two categories there were responses indicating participants' inferiority or superiority complexes in reference to other enterprise owners. What conclusions can be formulated on the basis of the analysis of values enclosed in Table 3.23? The percentage of responses referring to the specifics of agriculture was on the increase, while the ones indicating an inferiority complex were in decline. The responses suggesting a superiority complex make up a marginal category over the entire duration of the research. It can be stated that farmers became more aware of the specifics of their own environment and at the same time the level of complexes and low self-esteem decreased. This surely was the result of farm selection processes taking place in the studied population but it also may have been a consequence of the special treatment of agriculture at the time and use of the pre-accession programs of the European Union as well as later, after Poland joined the European Union, when Polish farmers became directly affected by mechanisms of Common Agricultural Policy. This trend in responses emphasized the specific character of agriculture but also validated farmers as belonging to a certain social category.

Table 3.24. Perception of similarities between the situation of the farm owner and the situation of the owner of a different enterprise depending on the class position of the respondents in 1994, 1999, 2007

Perception of similarities	Negatively privileged position (in %)			Averagely privileged position (in %)			Positively privileged position (in %)		
	1994*	1999*	2007**	1994*	1999*	2007**	1994*	1999*	2007**
Lack of similarities	65.4	72.6	47.4	51.4	57.1	41.6	36.8	50.0	29.2
There are similarities	34.6	27.4	52.6	48.6	42.9	58.4	63.2	50.0	70.8

* p < 0.001; ** p < 0.01.

Source: Gorlach, 2009, p. 120.

The earlier thesis on the better perception of similarities rather than differences between farmers and owners of other types of enterprises appeared to be connected with the modernization process, in which farms were becoming more like family businesses or enterprises, working for profit. This was confirmed by an analysis of perceiving the similarities by farm owners occupying different market positions (see Table 3.25). It turned out that the percentages of respondents declaring similarities between the situation of farmers and the situation of owners of other types of enterprises were highest among those who had farms with a positive market position. This pattern was visible in 1994, 1999, as well as in 2007. At the same time, the percentages of those who indicated lack of similarities were the lowest in that category. Exactly opposite was the situation of the farm owners with a negatively privileged market position. It is worth emphasizing that these relationships were statistically significant in each research period.

Observations made based on Table 3.24 led to another interesting conclusion. The dynamics of changes in perception of the differences mentioned above, in both discussed categories, were quite different in both analyzed periods of time; from

1994 to 1999 and from 1999 to 2007. In the case of farm owners with a positively privileged market position, the percentage of respondents indicating similarities went
down during the 1994 to 1999 period and then increased in the next period, from
1999 to 2007. In the case of farm owners with negatively privileged market position
the situation was similar, with the percentage of farmers noting differences declining
at first and then increasing. This makes another argument pointing to particularly
difficult times for farmers in the 1990s. What was occurring at that time sharpened
farmers' perception of their situation as very specific, not comparable with other
categories of owners.

Table 3.25. Perception of differences between the situation of the farm owner and the situation of the owner of a different type of enterprise, depending on the class position of the
respondents in 1994, 1999, and 2007

Perception of differences	Negatively privileged position			Averagely privileged position			Positively privileged position		
	1994*	1999*	2007	1994*	1999*	2007	1994*	1999*	2007
No differences	12.1	6.3	14.9	11.7	6.8	13.9	8.4	2.4	14.6
Responses referring to the specifics of agriculture	36.0	21.4	41.1	39.1	21.8	47.4	41.0	36.2	50.4
Sense of inferiority	48.5	70.8	41.1	44.1	70.2	37.0	47.3	58.9	31.4
Sense of superiority	3.5	1.3	2.9	5.0	1.2	1.7	3.2	2.4	3.6

* $p < 0.05$.

Source: Gorlach, 2009, p. 121.

The analyses of the variety of ways in which the representatives of various categories of the studied farms perceived differences between farmers and other enterprise
owners, provided additional material for the following reflection. This material was
presented in Table 3.25. It is worth noting that the perception of differences in connection with owners' class position was statistically significant only in 1994 and 1999.
This could mean that the respondents expressed a similar way of thinking regardless
of their class position. Was this the case in reality? To answer this question, one may
need to look into subsequent types of answers. The type asserting "no differences" in
2007 was almost the same in all three categories. In 1994 and 1999 these percentages
were different. This could confirm the above assumptions. However, the remaining
types of answers could not justify such a straightforward way of reasoning. Responses pointing to the specifics of agriculture were the most frequent among the
subjects with the positively privileged class position in all three research editions.
Such regularity could not be observed with the type of answers labeled as "sense of
inferiority." In 1994 the percentages of such answers were very similar in all three

categories of owners. The differences became more noticeable in 1999, when the percentage of such answers increased in all categories, which probably was the effect of the "difficult" 1990s decade. The increase was not the same across the board. It was much smaller among respondents with a positively privileged market situation than in the remaining two categories. A similar situation occurred in 2007. Here, the lowest percentage of such answers was also reported by the positively privileged. It should also be emphasized that the percentages of such answers in all three categories at this time were lower than in 1999. These declines were much smaller in the categories of negatively and averagely privileged than in the category of the positively privileged. Consequently, these differences were much smaller than in 1999. This could be treated as a certain, but not very strong, argument for the thesis on a similar way of thinking of various categories of respondents. At the same time, it should be noted that such a thesis is hard to support or weaken with the data related to the last type of responses, labelled as "sense of superiority." Such responses were rather rare and did not exceed the threshold of 5% in any research categories or editions.

The ways in which the respondents perceived conflicts relating to their socio-economic positions, characterizing various social categories could be seen as another aspect of the reconstruction of class consciousness of the studied farms. As already mentioned, Giddens (1973) treats the conflict awareness as another layer (level) of class consciousness, characteristic of the representatives of particular social categories. Following Giddens while simultaneously operationalizing this complicated issue, the concept proposed by Erik O. Wright (1990) seems quite appropriate. It treats class consciousness as a bundle of opinions on various aspects of the relations between the owners of the means of production and the employed workers. Such relations are the essence of the social structure of capitalism.

Three issues related to 1) compensation of the enterprise owner and the workers, 2) the ability to influence the destinies and action strategies of the enterprise, and 3) the possibility to hire other employees if currently employed workers go on strike or refuse to work, were chosen for the following study. The interviewed farmers were asked to express their opinions on these issues. The results were presented in tables 3.26, 3.27, and 3.28.

Table 3.26 presents the opinions of the respondents on issues related to work compensation for enterprise owners and workers in such enterprises. The opinions in the table only reflect the views of these participants whose farms survived the entire period of time of all three research editions, from 1994 to 2007. The analysis of the distribution of opinions provides some ground for one general statement. It turned out that the opinions in the studied sample population which could be described as "owner-friendly" or "capitalist friendly" were quite visible. The percentage of respondents who thought that enterprise owners should make more than their employees increased. Such an increase was particularly seen in the firm answer category of "decidedly yes."

Table 3.26. Distribution of answers to the question: "Should enterprise owners earn more than workers employed in those enterprises?" (in %)

Answers	1994	1999	2007
Decidedly yes	50.8	50.8	59.9
Rather yes	32.4	31.9	30.7
No opinion	7.7	6.9	4.5
Rather not	5.7	6.6	3.3
Decidedly not	3.4	3.8	1.6
Total	100.0	100.0	100.0

Source: Gorlach, 2009, p. 122.

Table 3.27. Distribution of answers to the question: "Should the enterprise employees have the same rights to decide on the future of the enterprise as the enterprise owners" (in %)

Answers	1994	1999	2007
Decidedly yes	11.5	14.5	5.9
Rather yes	28.5	28.6	24.2
No opinion	17.7	9.9	15.6
Rather no	26.4	29.9	34.6
Decidedly no	15.9	17.1	19.7
Total	100.0	100.0	100.0

Source: Gorlach, 2009, p. 123.

Table 3.28. Distribution of opinions to the question: "When a labor strike occurs, does the owner of the enterprise have the right to hire new people to replace striking employees?" (in %)

Answers	1994	1999	2007
Decidedly yes	10.4	3.4	7.1
Rather yes	12.3	8.3	12.3
No opinion	27.7	23.4	24.5
Rather no	34.4	34.5	36.9
Decidedly no	15.2	30.4	19.2
Total	100.0	100.0	100.0

Source: Gorlach, 2009, p. 123.

Could the observed tendency of the proliferation of "capitalist-friendly" opinions be confirmed in the case of statements regarding the influence of the owners and workers on the future of the enterprise? Data in Table 3.27 could confirm this hypothesis. In the studied period of time the percentage of opinions against workers having the same right as the enterprise owners to decide on the enterprise's future visibly increased. The "decidedly no" and "rather no" answers became more noticeable with every edition of the research study. At the same time, the group of those with opposite views on the issue decreased in the analyzed period with a small exception for the 1999 study.

What about the opinions related to the third issue, namely the right of the enterprise owner to terminate the employees on strike and hire new ones in their place? Would the answers of the respondents confirm the hypothesis about a growing trend of "capitalist-friendly" views among farmers? The analysis of the data in Table 3.28 did not provide an unequivocal outlook on that hypothesis. The matter was not as simple as in the previous two cases. Looking closer into the data in the analyzed table it should be stated that the category of respondents who had no opinion on this matter was more pronounced than in two previous cases. In the entire research time this category comprised less than one fourth of the population sample. This could indicate a relatively high controversy level for this question. It did not deal with the economic situation (compensation) or the situation of employees' influence (future of the enterprise), as was the case with the two previous questions. The issue of striking workers who can be replaced by "strike-breakers" was not only a problem related to structural relations between the enterprise owners and hired employees but it allowed for the possibility of depriving a certain category of people—namely, striking workers—of making their living through their work. This was the exclusion from economic activity of a certain category of people. It was not surprising that such a hypothetical issue could be seen as challenging. It was harder for respondents to have clear-cut opinions on this matter.

Considering the above data indicating that about one fourth of the interviewed farmers had no opinion on this issue in the three consecutive research editions, the opinions of the remaining three fourths of respondents in 1994, 1999, and 2007 should be analyzed more closely. The answers containing the term "decidedly" or "rather" were analyzed together, in both cases of positive and negative opinions. This allowed for better capturing of general tendencies in the respondents' attitudes on the studied issues. It turned out that in 1994 almost 23% of the interviewed farmers expressed the opinion that the enterprise owners would be completely right to get rid of employees on strike. It should be added, however, that almost 50% of respondents presented the opposite views. In 1999 the accents shifted. Less than 12% of the interviewed farm owners presented opinions that could be regarded as "capitalist-friendly." At the same time, as many as two thirds of the respondents (exactly 64.9%) expressed the opposite view. This indicated that in the difficult period of

1990s the atmosphere around this issue shared by farmers moved in the "anti-capitalist" direction. However, in 2007, the pendulum of opinion moved again, back in the other direction. Almost every fifth respondent (exactly 19.4%) took the side of enterprise owners in this case, and more or less confidently. In the same research edition, more than half (exactly 56.1%) presented the opposite view.

In conclusion, it should be emphasized that in all three research editions, i.e., in 1994, 1999, and 2007, the opinions of the majority of respondents were anti-capitalist. Secondly, intensification of such opinions could be observed in 1999. This should be treated as an effect of the difficult 1990s when the level of empathy towards all employees who were facing a tough situation was quite significant. Thirdly, the subsequent years of certain, moderate stabilization in agriculture, as well as Poland's accession to the European Union, could have prompted the increase in capitalist-friendly opinions and a better recognition of the situation and dilemmas faced by the owners. Noticeably, the scale of increase of these pro-capitalist opinions was much smaller here than in the cases of two previous aspects of relations between the enterprise owners and workers, manifested in the questions about compensation and the future of the enterprise. This was related to the specifics of family farms, which, on one hand, were becoming more similar to other economic enterprises and, on the other, fostered a more personal relationship between the enterprise owner and workers. It should be remembered here that farm workers were mostly family members. Neighbors and temporary hired workers were much less frequently employed on family farms. Having such an experience with employees made it harder to be anti-workers or have capitalist-friendly positions. However, despite such reservations, almost one fifth of respondents presented such attitudes.

The opinions of respondents presented above should be analyzed again but from a different perspective. In the next step, the analysis was conducted, not just to address the distribution of particular types of opinions but also their character, measured with the arithmetic average connected with the class position of the respondents. In order to emphasize this matter, the analysis was limited to two class categories of farmers; the positively privileged and the negatively privileged, in terms of their market positions. The average values presented in tables 3.29–3.32 were based on range values, attributed to particular types of answers from value 1 through value 5. The value 1 was allotted to the most capitalist-friendly (pro-owners, anti-workers) answers in all cases and reflected the "decidedly yes" answers, while value 5 was given to the most anti-capitalist (pro-workers, anti-owners) sentiments expressed by "decidedly no" answers. This allocation of values only applied to the first question related to compensation and the third one dealing with the strikes. However, the second question, referring to the enterprise's future, required an opposite allocation of values, due to its logic and wording. In this case the value 5 was used for decidedly pro-owner opinion (strong statement that workers do not have the right to decide on the future of the enterprise) and value 1 presenting completely opposite and pro-worker opinions that workers definitely have the right to decide on the future of the enterprise.

Table 3.29. Opinions on relations between the earnings of the enterprise owners and the earnings of workers in two opposite class categories

Class category	1994		1999		2007	
	average	standard deviation	average	standard deviation	average	standard deviation
Negatively privileged	1.90	1.138	1.88	1.070	1.64	0.934
Positively privileged	1.64	1.011	1.64	0.926	1.41	0.808

Source: Gorlach, 2009, p. 125.

Firstly, the opinions on earnings, addressing the economic aspects of relations between owners and hired employees (see Table 3.29) were analyzed. It turned out that generally both groups of interviewed farmers were very much owner-friendly (average values in both groups in all three research editions were close to the value of 1). Furthermore, both categories became even more owner-friendly over time (with the average value getting even closer to 1). These opinions in both studied categories became more homogenous and more focused, which was confirmed by the decreasing value of the standard deviation. In the category of the positively privileged, which was more understanding of owners, the opinions were even more concentrated around 1.

Table 3.30. Opinions on the right of owners and employees to decide on the future of the enterprise in both opposite categories

Class category	1994		1999		2007	
	average	standard deviation	average	standard deviation	average	standard deviation
Negatively privileged	2.95	1.220	2.99	1.359	3.17	1.216
Positively privileged	3.13	1.435	3.30	1.340	3.71	1.135

Source: Gorlach, 2009, p. 126.

Although, let's recall, Table 3.30 was constructed according to a different logic, it confirmed the rather owner-friendly opinions of the interviewed farmers related to the political dimension of the relationship between owners and hired workers.

Average values in both studied categories in all the selected years of research oscillated around the middle of the scale used, close to the value 3. Three tendencies were quite apparent. First of all, farmers of the positively privileged class position were more owner-friendly (values closer to 5) than representatives of the second studied category. Secondly, respondents with a positively privileged class position in 1994 expressed more significant divergence of opinions (higher value of the standard deviation) than in the two following research editions. In 1999 and 2007 the interviewed farmers from a positively privileged position made a more homogenous category than those of the negatively privileged market position. In 2007 they were owner-friendly and more focused in their opinions. Thirdly, general opinions regarding the "political" dimension of the relationship between the farm owners and workers were more subtle (not so close to the extreme values) than those related to the "economic" dimension (in this case, compensation).

Table 3.31. Opinions on the right of owners to hire "strike-breakers" as expressed in two opposite class categories

Class category	1994		1999		2007	
	average	standard deviation	average	standard deviation	average	standard deviation
Negatively privileged	3.36	1.090	3.84	1.017	3.59	1.049
Positively privileged	3.33	1.377	3.77	1.112	3.39	1.187

Source: Gorlach, 2009, p. 127.

The third dimension of the studied opinions dealt with the issue of "structural conflict" (see Table 3.31). It was focused on the very core of interclass relations, relating to the right of owners to lay off workers who were trying to execute their class interests. It is worth noting that generally the opinions from both categories of respondents were more anti-owner than the ones related to the two other questions of the economic and political dimension. It could be said that these opinions were rather ambivalent but with some noticeable stance that made the respondents take the side of striking employees, rather than the side of the owner who would like to get rid of them. It should also be stressed that the opinions on this matter expressed in 2007 (in both categories) were less focused than the opinions in 1999 but significantly more focused than in 1994, which was indicated by the values of standard deviation. In the case of categories of a positively privileged class position the 2007 opinions were somewhat more owner-friendly but more dispersed in comparison to the opinions of interviewed farmers with a negatively privileged position.

Table 3.32 contains collective opinions, also expressed with the use of average values. It should be emphasized that in this case, the rank value in the particular position of the already-mentioned second question (related to influence on the enterprise activities) was changed to ensure consistency of meaning with values related to answers in two remaining questions. In these two remaining questions the answers that were in favor of owners got a value of 1, but opinions that were decidedly anti-owner got a value of 5.

Table 3.32. Character of collective opinions of the respondents in two opposite class categories

Class category	1994		1999*		2007**	
	average	standard deviation	average	standard deviation	average	standard deviation
Negatively privileged	2.78	0.732	2.91	0.772	2.69	0.664
Positively privileged	2.62	0.849	2.70	0.764	2.38	0.653

* p < 0.01; ** p < 0.001.

Source: Gorlach, 2009, p. 127.

The analysis of data in Table 3.32 indicated that after a certain increase of anti-owner views in 1999, as compared to 1994, the new millennium and Poland's EU accession brought about the intensification of owner-friendly opinions. In general, it can be said that the category of family farm owners in Poland was, at the time, undergoing the process of becoming "bourgeoisie," especially in the consciousness aspect that was considered (see Manchin & Szelenyi, 1985; Szelenyi, 1988). Secondly, the process of becoming bourgeoisie was more advanced among those who had a positively privileged class position. Thirdly, the opinions in both categories were more pronounced (see the value of standard deviation) in 2007 than in the previous research editions. This process turned out to be consistent and systematic in the case of positively privileged respondents and rather non-homogenous (with the change in 1999) in the case of the negatively privileged category of respondents. Generally, the group of positively privileged tended to be more owner-friendly and, at the same time, more homogenous in terms of opinions expressed in 2007. The statement presenting this group as having clearer class consciousness is justified.

The differences between the averages in both studied categories deserve some attention, as well. In 1994 it was only 0.14 and it went up to 0.21 in 1999 and was statistically significant. By 2007 the difference became even more pronounced, at

0.31 and, at the same time, more statistically significant. It can be said that the polarization of class consciousness in the social category of the owners of family farms became more visible in the studied population sample in connection to the class position of the respondents.

The last level of class consciousness according to the concept proposed by Giddens (1973) is the level of "revolutionary" consciousness. In the current study the understanding of this "revolutionary" reflection is considered rather broadly and includes pointing to certain organizations, which, according to the respondents, were supposed to defend farmers' interests. The "revolutionary" character of class consciousness was understood as the preferred collective manner of fighting for and ensuring one's own interests. In the analysis, two organizations deserved special attention, namely the Polish Party of Peasants, the party traditionally connected with peasants in Poland, and "Self-Defense," a political formation of national and populist character that existed at the time.

Table 3.33. Percentage of respondents indicating the existence of organizations that defend farmers' interests

Category of answers	1994	1999	2007
Responses indicating at least one organization	13.6	41.1	17.5
Responses indicating the Polish Party of Peasants	7.4	9.1	4.5
Responses indicating the "Self-Defense" Party	0.6	27.0	2.7

Source: Gorlach, 2009, p. 128.

The data presented in Table 3.33 depict the dynamic of farmers' responses pointing to the organizations which, in their opinion, defend their interests. Comparison of these values could lead to several interesting conclusions. First of all, the responses reflected the tumultuous character of the 1990s. Intense processes of farm restructuring and waves of farmers' protests (see Gorlach & Mooney, 1998; Foryś & Gorlach, 2002, 2015; Foryś, 2008) could have resulted in 40% of respondents in 1999 pointing to at least one organization that defended their interests. This should be emphasized as a significant increase because in 1994 the analogous percentage was below 14%. This was particularly visible in the hopes that farmers had in collective actions and protests organized by the Samoobrona ("Self-Defense") Party. In 1994, less than 1% of respondents pointed to this party. However, in 1999 more than one fourth of respondents declared having interest in it. Those respondents who perceived the role of Samoobrona as crucial in the overall struggle for peasants' interests were three times as numerous in 1999 as those who pointed to the Polish Party of Peasants, which at the time was perceived as part of the political establishment.

Comparing the responses related to these two organizations was just as interesting. On one hand it should be noted that the Polish Party of Peasants had a more stable group of supporters, although it never exceeded 10% of our respondents. In 2007 their percentage was lower than in 1994. The case of Samoobrona showed a short-term popularity of the populist organization with a strong and very expressive leader. The party was involved in conflicts and protest actions during the mid to late 1990s, which was a challenging time for farmers, as was reflected in the research editions of 1994 and 1999. Samoobrona was mentioned by 0.6% respondents in 1994, and reached a popularity peak of 27% in the 1999 study, but by 2007 it had suffered a 10-fold drop. Lastly, regarding these parties, it should be noted that although in 2007 both the Polish Party of Peasants and Self-Defense were rarely named by respondents as organizations defending farmers' interests, the percentage of those naming at least one organization in that context was higher than in 1994. This could probably be explained by processes of diversification which impacted the category of family farm owners. The analysis conducted in reference to farmers' protests reflected the diversification of interests of the respondents. (see Gorlach, 2001, pp. 133–169; Foryś & Gorlach, 2002, 2015; Foryś, 2008).

Table 3.34 illustrates this problem in reference to the class position of respondents, but only within the 1999–2007 time frame.

Table 3.34. Responses indicating organizations defending farmers' interest, by class position of the farm operator in 1999 and 2007 (in %)

Category	Negatively privileged		Positively privileged	
	1999	2007	1999	2007
Responses indicating at least one organization	33.6	12.5	56.3	22.5
Responses indicating the Polish Party of Peasants	7.3	5.1	14.5	3.6
Responses indicating the Self-Defense party	22.1	2.8	33.8	3.6

Source: Gorlach, 2009, p. 129.

Considering the responses naming organizations defending farmers' interests, it was noticeable that the majority of such responses came from farm owners of a positively privileged class position. The same could also be seen at the general level (pointing to at least one organization), as well as in particular cases: Polish Party of Peasants and Self-Defense. The differences between the percentages of those naming organizations were noticeable between the positively and negatively privileged in 1999, as well as in 2007. Positively privileged respondents were more likely to express a revolutionary consciousness as understood in this very study. An explanation of this difference could certainly refer to ongoing processes of change in Poland's farms. The feeling of isolation, especially when facing market uncertainties, was more obvious on farms that had potentially lower chances to thrive in this type of economy and, thus, had a negatively privileged market position. On the other hand, the farm owners who were more involved in market processes also had a positively privileged market position. For that reason, they could be more aware about the various ways of fighting for farmers' interests, as well as more interested in collective efforts and institutional reassurances related to securing their interests.

One could ask whether these differences had to do with various preferences related to certain methods of fighting for farmers' interests. This issue is illustrated by the data presented in Table 3.35.

Table 3.35. Preferences for various forms of farmers' fighting for their own interests, by respondents' class position (1999 and 2007 in %)

General categories		Negatively privileged		Positively privileged	
1999	2007	1999	2007	1999	2007
Roadblock protests, demonstrations, building occupations		15.2	5.7	22.1	7.3
16.0	7.2				
Pressuring senators, MPs, and other elected officials		16.1	14.2	13.8	16.8
16.9	15.1				
Self-organization of farmers		30.0	46.0	46.2	50.4
35.2	48.0				
There is no point in doing anything		38.7	34.1	17.9	25.5
38.7	29.7				

Source: Gorlach, 2009, p. 130.

Looking into the values in Table 3.35, one can formulate two preliminary conclusions. Analyzing the first column of the table, it is worth noting that the increasing tendency could only be observed in one category of answers, which dealt with the method of farmers' organizing themselves in order to secure their interests. In 1999 this method was preferred by slightly over one third of respondents and in 2007 it was named by almost half of respondents. The drop in preferences for confrontational and political methods of fighting for their interests (blocking roads, demonstrations, and pressure on politicians) went together with an abandonment of the sense of powerlessness or alienation as well as the attitude that it does not make sense to do anything. Such interrelations were observed throughout the entire population sample.

The matter became more complicated when the categories of positively and negatively privileged were taken into consideration. The preferences for self-organizing were more pronounced in the category of positively privileged respondents. They were more visible in 1999 when juxtaposed with the responses of the negatively privileged research participants, as the percentages were 46.2% and 30%, respectively. In 2007 these percentages were 50.4% for respondents with a positively privileged market position and 46.0% for those with a negatively privileged position. It could be argued here that even the respondents with a negatively privileged market position became more convinced that the method of self-organizing was better for their interests.

The analysis of the remaining three categories allows for some general statements. The preferences related to the confrontational methods of securing farmers' interests in 2007 were somewhat stronger within the category of the positively privileged than in the category of the negatively privileged (7.3% and 5.7%, respectively). In the 1999 research edition, the difference in preferences was decidedly higher, with percentages of 22.1% for the positively privileged and 15.2% for those with a negatively privileged market position. Generally, this could serve as an argument indicating disappointment in the effects of securing farmers' interests this way. A quite different response was seen with the methods described as "political" (pressure on politicians, lobbying). Preferences for this form of action increased among the positively privileged respondent and decreased among those who were negatively privileged. Could this mean that such methods of fighting are more effective when carried out by farmers' circles with more economic and political weight? In this context, it was quite surprising that the respondents with a positively privileged market position, unlike the respondents with a negatively privileged market position, had the percentage spike in responses declaring a feeling of powerlessness, such as "It does not make sense to do anything" statements. Could this be seen as a result of the larger portion of disappointments, since the expectations drawn from a better market position were actually greater? However, it should also be added that in both the 1999 and 2007 research editions, the overall percentage of such declarations of disappointment was still smaller among the positively privileged than the negatively privileged. Was this some sort of expression of similarities in the way of thinking within the studied population sample or the effect of the processes of elimination on the weakest farms?

The last issue discussed in this chapter deals with various aspects of Poland's accession to the European Union and how they were reflected in the class consciousness of the respondents. One may wonder whether these issues should be discussed here in reference to other aspects of class consciousness. The answer should be a confident "Yes," as Poland's EU accession made the programs and instruments of Common Agricultural Policy available to Polish farmers, which consequently had a serious impact on agriculture and rural areas in Poland. In the late 1990s and at the beginning of the 21st century, certain political decisions caused a reaction among those who were the subjects of such decisions. They were political decisions that, at times, ignited protests. Such decisions, in one way or another, influenced farmers' acceptance of the system in which they operated.

Table 3.36. Knowledge of the possibilities of Polish farmers receiving aid from EU funds, by class position of the respondents (2007 in %)

Category	Generally	Negatively privileged	Positively privileged
Does not know anything	6.9*	11.4	0.7
Not interested	4.5	6.8	2.2
Knowledge, but not concrete	23.1	24.4	21.7
Knowledge, and concrete possibilities	17.0**	6.8	23.9
Remarks on bureaucracy, limited access to information	8.6**	2.8	17.4

* p < 0.01; ** p < 0.001.

Source: Gorlach, 2009, p. 131.

The data presented in Table 3.36 show the state of knowledge of the respondents in 2007 regarding various possibilities for use of EU funds. It should be emphasized that such a question was also asked in 1999, and the current study does not show all results obtained at the time. However, one aspect of that edition of the research should be mentioned here, namely, the case of respondents who were not interested in EU aid programs. In 1999 the percentage of those uninterested was 71% (see Gorlach, 2001, p. 261). By 2007 it had fallen to 4.5% (!) (see Table 3.36). The values in this table can indicate two things. On one hand, there was a high level of distrust towards the transformation outcomes and the expected EU accession among Polish farmers in 1999, as they were facing new and unknown solutions while still having memories of the threats to the peasant economy and the fears they experienced at the time of state socialism. On the other hand, the 2007 results reflected the effects of the EU aid information campaign, as well as the new experiences related to Poland's accession to the European Union in 2004.

In the 2007 research edition, the connection between knowledge of the EU programs and the class position of the respondents was particularly interesting. Some important interrelations can be observed in Table 3.36. For example, the percentage of respondents who did not know anything about the EU programs was particularly high in the category of the negatively privileged. Simultaneously, the percentage of those who displayed concrete knowledge or made specific remarks about the EU programs was particularly high among the interviewed positively privileged farm owners. It could clearly be seen that the operation of stronger, more prosperous farms, which were doing well in the reality of the market economy, and the possession of higher cultural capital (i.e., a positively privileged class position) went together with more extensive and more critical knowledge of the instruments of EU agricultural policy.

Another type of difference was visible in reference to the sources from which the respondents were getting their knowledge of the relevant EU programs (see Table 3.37). It was determined that in the case of traditional information sources, such as family, neighbors, representatives of local government (e.g., commune, head of the village) as well as conventional mass media (TV, radio) there were no differences between respondents. One might assume that such channels of communication were considered "natural" and typical for all rural residents. Using them did not require any special effort and was usually happening spontaneously. In the case of sources that required authentic effort to be put forth and a striving for certain knowledge (e.g., the purchase of press publications, surfing the Internet, or less typical sources, such as special workshops and training, meetings, etc.) the differences were visible and statistically significant. It is quite noticeable that the second type of sources was more widely used by farmers who had a positively privileged class position. Such farmers, who operated more modern and more productive farms, were more likely to seek out information on the pertinent EU programs.

Table 3.37. Sources of knowledge for the possibilities of accessing EU aid, by the class position of the respondents (2007 in %)

Category	All farmers	Negatively privileged	Positively privileged
Neighbors, family	28.5	29.0	23.9
TV	77.0	76.7	72.5
Press	36.3*	29.5	43.5
Radio	20.9	15.3	22.5
Internet	10.3**	3.4	21.0
Commune, head of the village	33.7	31.8	29.7
Others	15.0**	5.7	30.4

* $p < 0.05$; ** $p < 0.001$.

Source: Gorlach, 2009, p. 132.

The representatives of this category were also more likely to use the EU support programs. This is reflected in the numbers in Table 3.38. Almost three fourths of the respondents used various forms of EU aid after May 1, 2004, when Poland formally joined the European Union. This percentage was decidedly higher among respondents with a positively privileged class position. One might conclude that the EU programs mostly enhanced the position of larger farms which were already able to thrive better in the market economy. Contrary to earlier assumptions, the EU aid programs were not meant to improve the chances for success of all farmers who, in the context of the market economy, were in an unfavorable situation and owned farms with a negatively privileged market position.

The data contained in tables 3.38 and 3.39 provide additional support for the statement presented above. It turned out that the EU funds used to develop farm production and increase their potential also improved their market position. This was quite visible mostly in those farms which had a positively privileged market position. Such a statistically significant relation could not be observed in the case of the second purpose to which the EU funds were allotted, such as expenses to cover the current operating costs of a farm. It can be inferred that the EU funds significantly expanded the potential of farms which were already stronger.

Table 3.38. Using the various types of EU aid, starting from 2004, by class position of the respondents (2007 in %)

Usage of the EU aid—in general*	Negatively privileged	Positively privileged
Not using—26.7	40.9	15.2
Using—73.3	59.1	84.8

* $p < 0.001$.

Source: Gorlach, 2009, p. 133.

Table 3.39. Predestination of the EU funds, by class position of the respondents in 2007 (in %)

General category	Negatively privileged	Averagely privileged	Positively privileged
Development of production; 18.7*	8.0	24.3	25.4
Current needs; 59.8	51.7	63.0	65.9
Home investments; 2.1	1.1	2.3	2.9

* $p < 0.001$.

Source: Gorlach, 2009, p. 133.

To wrap up this part of the reflection, a reference should be made to the evaluative dimension of the consciousness of the respondents. This dealt with the evaluation of the accession of Poland to the European Union in 2004, from the perspective of benefits that it brought to various groups of farmers. The data relating to this topic are shown in Table 3.40.

Table 3.40. The evaluation of Poland's accession to the EU, by class position of the respondents (2007 in %)

Category of answers	In general*	Negatively privileged	Positively privileged
All farmers benefited	37.3	26.4	45.1
Large and modern farms benefited	57.2	68.4	46.6
Small and weak farms benefited	2.9	1.7	5.3
Everybody lost	2.5	3.4	3.0

* $p < 0.01$.

Source: Gorlach, 2009, p. 133.

First of all, it should be emphasized that the percentage of respondents who thought that all Polish farmers had experienced losses upon Poland's entrance into the European Union amounted to 2.5%. This rather marginal percentage combined with multiple statements made by Polish "Euro-skeptics," that joining the European Union would likely cause the annihilation of Polish agriculture, should be treated as an important fact, not only empirically, but also politically. However, it should also be stressed that the opposite opinion, that all farmers benefited from Poland's joining the European Union, was not a dominant one. This was stated by almost 38% of respondents and comprised the second largest viewpoint category. Interestingly, opinions of this kind were given by owners of farms with a positively privileged market position. They were the ones whose opinions correlated with the statements made by "Euro-enthusiasts," that the EU accession had resulted in beneficial consequences all across the board.

Respondents who expressed the opinion that Poland's accession to the European Union mostly benefited large and modern farms, made up the largest category in our research. Such an opinion was mostly shared by those who had a negatively privileged class position. They claimed that the EU accession brought prosperity to the "others," meaning farmers with a positively privileged market position. These statements, which could be described as opinions of the "Euro-realists," corresponded well with what could be observed in the case of using the relevant EU programs. This conviction was reinforced by another opinion expressed by the respondents, which mostly pointed to small and weak farms as the main beneficiaries of the integration process. The percentage of such opinions was slightly below 3%. Interestingly, such an opinion was decidedly more noticeable among these "others," that is, farm owners who had a positively privileged market position. It turned out that the opinions of the respondents did not confirm the predictions of those who, before Poland joined the European Union, had made statements indicating that the instruments of the EU agricultural policy would preserve a traditional agrarian structure in their country, and thus block the processes of farm restructuring. These opinions were more likely to be vocalized in 2007 by those respondents who represented "larger" and "more modern" farms (and were positively privileged). This could be, to some extent, explained by some socio-psychological mechanism, preventing the beneficiaries of a favorable situation from publicly admitting to its effects. This would also explain why the higher degree of benefits relating to Poland's accession to the European Union and experienced by "large" and "modern" farms, were reported by owners of those farms which could more accurately be described as rather "small" and "weak" (i.e., the negatively privileged).

3.9. Conclusions: Some results of major transformations

The decision made at the beginning of this chapter, on how far back in time to go for the best understanding of contemporary Polish agriculture and the changes it has been going through, was quite challenging to arrive at. However, in the view of the authors here, the historical perspective, encompassing Polish agriculture while taking into account a period of real socialism in Poland and the three decades of transformations which came after 1989, seems appropriate.

Socialism in Poland took the approach of repressive tolerance towards agriculture and the people involved in it. Despite lots of limitations that farmers and peasants had to face, such as attempts to impose state ownership over land, compulsory delivery of agricultural products, lack of rural modernization and difficulties with investments, they came away rather unscathed. The most significant reaffirmation of such a state was the fact that over 75% of agricultural land stayed in private hands, which was extremely rare among the countries of the Eastern Bloc. Later, it turned out that this fact had positive implications in regard to changes affecting agriculture and rural areas in Poland over the last 30 years. It allowed farmers to preserve the sense of ownership and relative ability to be in charge of their own life. Most of all, they were able to maintain entrepreneurial qualities and economic rationality. However, this is not to say that Polish agriculture did not inherit any notable burdens from the times of socialism, as well as the earlier problems of the interbellum period (1918–1939), which socialism could not solve. These problems included a fragmented farm structure, lack of modernization and "hunger of land." Several decades of neglecting these issues contributed to the significant scale of challenges that Polish agriculture had to face at the beginning of the transformation in 1989. At the time, agriculture received the same treatment as other branches of the economy requiring restructuring, and farmers were seen as no different as any other entrepreneurs. Yet, farmers, understood here as a social category, had to carry the costs of transformation almost by themselves. Not without reason did the changes of this period cause a great deal of fear and a sense of oppression, despite the constant presence of the freedom to farm. The changes activated the processes of farm selection, increased their average size, modernized them to a great extent, and accelerated the processes to change consciousness and the mentality of Polish farmers.

The process of farm selection, that caused the elimination of the smallest farms with utilized agricultural land of under 2 ha and between 2 and 5 ha, and increased the number of the largest farms, with over 20 ha of utilized agricultural area, was one of the most important consequences of the modernization processes in the last 30 years. This selection of farms, although not very thorough for small farms, mostly took place in the 1990s. It has also been going on to some extent in the last decade, but definitely with less intensity. The number of the smallest farms is decreasing, as well the number of medium-sized farms with utilized agricultural areas of 10–15 ha and 15–20 ha (GUS, 2017, p. 61). This reduction tendency does not apply to the largest farms, sized 20–50 ha and the ones with utilized agricultural area of over 50 ha, as their number is growing. The collective measure of these tendencies is the increased average farm size in Poland. In the beginning of the 1990s it was 6.3 ha, and now it is already 10.8 ha. It is still below the size of the average farm in the European Union, which is 18.7 ha.

Moreover, the quantitative changes are a significant indicator of modernization processes in the economic sphere. They are the foundational basis for the changes in consciousness that are manifested in attitudes and convictions. This evolutionary character of changes is the most interesting for the authors here, as it has explanatory potential for the question stated in the title of this chapter, regarding the peasant qualities in respondents' attitudes and mindset. It should be added, from the research perspective, that the studied issues illustrated in this chapter with data, indicate that the turning point for Poland was its accession to the European Union. The EU membership had a positive effect on the condition of farms in Poland. As indicated by the data, the opinions of respondents presented in the 1990s were much different than the ones expressed in the 21st century. While the attitudes towards hired labor did not change significantly and stayed at similar levels in the 1990s and in the first two decades of the 21st century, despite the increase of acceptance for such activities in the late 1990s, farmers' approach to taking credits and loans, or selling their land, went through notable changes. Farmers' readiness to take credits and loans was significantly higher in the 21st century than in the 1990s. On one hand, there was still the lingering memory of farmers taking credits and loans in the late 1980s and early 1990s, which made some of them fall into a spiral of debt and consequently face the threat of bankruptcy or even, in fact, go bankrupt. On the other hand, the increasing acceptance for using credits in the 21st century could also indicate a necessity that farmers intertwined in the economic market have to deal with. What appears more likely in the research context is that the growing tendency to take credits should be seen more like a chance for development than the necessity caused by a difficult economic situation. What confirms this is the declining trend to sell farms in the first decade of the 20th century in comparison to the situation in the 1990s when the market of agricultural products was in distress and most of all, farmers did not have the EU subsidies at their disposal. In the approach of respondents to the economic issues, it was quite clear that a positive outlook on the future, or a readiness to take some risk, grew with the level of capital that farm owners accumulated.

Third, the process that has clearly made an impression on farm owners was the diversification of their farms because of their economic position. Although this is not a radical tendency, it still confirms the process of farms transitioning from a less privileged to a more privileged position. In other words, the processes of advancement outnumber the process of degradation. This criterion is also a factor differentiating the ideas and attitudes of respondents and is reflected in how the farm owners perceive themselves. In the studied periods of time they increasingly perceived themselves as entrepreneurs and less as marginalized people. They noticed similarities between their own situation and the situation of entrepreneurs in general and the 1990s were the time caesura. In the 21st century, the processes of perceiving one's own position as an entrepreneur, as well as the processes of social advancements, were stronger than the sense of being marginalized or degraded, as was the case a decade earlier. It is worth emphasizing that even though farms had a more stable position in the 1990s than in the first two decades of the 21st century, the processes of their degradation were greater. The changes in farm status were more dynamic, more visible in the 21st century and they were more likely to bring advancement rather than degradation. Therefore, some positive effects of the modernization processes that started in the last decades of the 20th century can be noted.

Fourth, in the analyzed years the consciousness changes appear to become more visible over time. There is a noticeable growth in farmers self-identifying with the sense of being an entrepreneur, which is often coupled with searching for similarities between farmers' own situation and the situation of entrepreneurs in other fields and professions. This is also reflected in the approach towards employer-employee relations. There is also a rather visible polarization of opinions on these issues between farm owners, who are positively and negatively privileged in terms of market position of their farms. This kind of differentiation translates to differentiation of interests among farm owners, which can be reflected in their protest activities. Political changes allowed for legally sanctioned ways to fight for farmers' rights. To use the language of the theory of social movements, "the political opportunities" were expanded and this became one of the factors activating farmers' protests, besides the stimulating effect of the economic situation of farms.

Fifth, farmers protest activities, if analyzed closely, reveal to some extent phenomena and processes which are described in this chapter. This greatly applies to the ongoing process of "depeasantisation" of Polish agriculture, its professionalization and stronger ties to the market. The way the protests are organized and conducted, as well as their goals and methods, confirm the above statements.

Six, the organizing of protests provides the evidence of farmers' and agricultural producers' sensitivity to positive and negative signals coming from the market. In the broader perspective, two noticeable tendencies might be important for the issues discussed in this chapter. The first is the relation between the growing number and intensity of protest activities of this socio-professional category with the dire situations on the market of agricultural products and the consequently decreasing income of the owners. In such cases, the development of a protest cycle was a sure thing and sometimes there were even waves of protests such as at the beginning of the 1990s and the beginning of the 20th century. The second tendency meant professionalization of farmers' protests and their sectorial character, which became more apparent after Poland joined the European Union. It tied with the decreasing level of solidarity between the groups of protesting farmers and the narrowing particularism of their interests. Curiously, the afflictions of the economic system of the 1990s were mostly protested by the owners of large farms, strongly connected to the market but were initially supported by the group of owners of smaller farms, who did not have optimistic future perspectives and were doomed to be liquidated. The symbolic layers of these protests (i.e. slogans, banners, how the protesters express themselves) were manifested in the common denominator for various protest groups (not only various producer groups but also the owners of farms of various sizes), references to peasantry, and the peasant ethos. In the 21st century, and especially after Poland joined the European Union, this type of community had already become invisible. The owners of small farms did not show up for any protests (and not just because a large portion of such farms ceased to exist) and the lack of solidarity between the producers of various types of agricultural products became quite visible. It can be concluded that only the categories that at certain times experienced difficulties protested them, focusing on concrete difficulties such as wholesale buying prices, production quotas, or unfavorable laws and regulations. There was a rather obvious difference of agricultural interests among farmers, which shows a quite significant diversification of this category.

Therefore, farmers' protests should be treated as a validation of the processes described in this chapter. What first comes to mind here is the process of selection of farms in the first half of the 1990s and then gradual intensification of modernization processes and the changes reflected in farmers' consciousness and attitudes. The latter could be indicative of the professionalization and stratification of recent years with the notable expansion of these processes. This could be explained as peasantry making room for the professionalization of farmers, the owners of both the largest and—at the same time—the most modernized farms are better defined as agricultural producers rather than farmers. The remains of peasantry could still be found in the symbolic, and not the declarative or behavioral layer, but only as singled out, individual elements of the peasant ethos, treated by farmers in an instrumental fashion, which could be seen in farmers' protests over recent decades.

REFERENCES

Adamski, W. (1967). *Grupy interesów w społeczności wiejskiej* [Interest groups in rural community]. Ossolineum.

Adamski, W. & Turski, R. (1990). Interesy klasy chłopskiej jako źródło sytuacji kryzysowych [Peasant class interests as a source of crisis situations]. In: W. Adamski (ed.), *Interesy i konflikt. Studia nad dynamiką struktury społecznej w Polsce* [Interests and conflict. Studies on the dynamics of social structure in poland], pp. 127–150. Ossolineum.

Andrusz, G.D., Harloe, M., & Szelenyi, I. (2011). *Cities after Socialism: Urban and regional change and conflict in post-socialist societies*. Wiley-Blackwell.

CBOS (1993a). *Rolnicy o rolnictwie i o swojej pozycji w społeczeństwie* [Farmers on agriculture and their position in society].

CBOS (1993b). *Rolnicy o sytuacji rolnictwa i warunkach życia na wsi* [Farmers on the situation in agriculture and the living conditions in rural communities].

CBOS (1993c). *Rolnicy indywidualni o instytucjach życia społecznego i politycznego wsi polskiej* [Farmers on institutions of social and political life in the Polish countryside].

Davies, N. (2001). *Heart of Europe: A short history of Poland*. Oxford University Press.

Delanty, G. & Isin, E F. (eds.) (2003). *Handbook of Historical Sociology*. SAGE.

Foryś, G. (2008). *Dynamika sporu: Protesty rolników w III Rzeczpospolitej*. Wydawnictwo Naukowe Scholar.

Foryś, G. & Gorlach, K. (2002). The dynamics of Polish peasant protests under post-communism. *Eastern European Countryside* 8, pp. 47–66.

Foryś, G. & Gorlach, K. (2015). Defending interests: Polish farmers' protests under post-communism. In: B. Klandermans & C. van Stralen (eds.), *Movements in Times of Democratic Transition*, pp. 316–340. Temple University Press.

Gasson, R. & Errington, A. (1993). *The Farm Family Business*. CAB International.

Giddens, A. (1973). *The Class Structure of the Advanced Societies*. Harper and Row.

Gołębiowski, B. & Hemmerling, Z. (1982). Chłopi wobec kryzysów w Polsce Ludowej [Peasants in the situation of crises in people's Poland]. *Kultura i Społeczeństwo* 1–2, pp. 21–50.

Gorlach, K. (1989). On repressive tolerance: State and peasant farm in Poland. *Sociologia Ruralis* 29(1), pp. 23–33.

Gorlach, K. (1994). *Obronić ducha Ameryki. Kwestia rolna i socjologia wsi we współczesnych Stanach Zjednoczonych* [Defending the spirit of America: The agrarian question and rural sociology in the contemporary United States]. Wydawnictwo Uniwersytetu Jagiellońskiego.

Gorlach, K. (1995). *Chłopi, rolnicy, przedsiębiorcy. „Kłopotliwa klasa" w Polsce postkomunistycznej* [Peasants, farmers, entrepreneurs: "The awkward class" in postcommunist Poland]. Wydawnictwo Uniwersytetu Jagiellońskiego.

Gorlach, K. (1997). The class position of family farm owners in Poland. *Polish Sociological Review* 117(1), pp. 75–88.

Gorlach, K. (2001). *Świat na progu domu. Rodzinne gospodarstwa rolne w Polsce w obliczu globalizacji* [The world in my backyard: Polish family farms in the face of globalization]. Wydawnictwo Uniwersytetu Jagiellońskiego.

Gorlach, K. (2009). *W poszukiwaniu równowagi. Polskie rodzinne gospodarstwa rolne w Unii Europejskiej* [Searching for the balance: Polish family farms in the European Union]. Wydawnictwo Uniwersytetu Jagiellońskiego.

Gorlach, K. & Mooney, P.H. (1998). Defending class interests: Polish peasants in the years of transformation. In: J. Pickles & A. Smith (eds.), *Theorizing Transition: The political economy of post-communist transformations*, pp. 262–283. Routledge.

Gorlach, K. & Seręga, Z. (1991). *Chłopi we współczesnej Polsce. Przedmiot czy podmiot procesów społecznych?* [Peasants in contemporary Poland: Makers or outsiders of social processes?]. PWN.

Graham, A., Cohen, S., & Colton, T. (1992). Staggering toward democracy: Russia's future is far from certain. A roundtable discussion. *Harvard International Review*, 15 (2), pp. 14 – 17; 60–62.

GUS [Central Statistical Office] (2017). *Charakterystyka gospodarstw rolnych w 2016 r.* [Farm characteristics in 2016].

Halamska, M. (1981). Społeczne przesłanki powstania NSZZ RI "Solidarność" [The reasons for the emergence of rural "Solidarity"]. *Wieś Współczesna* 11, pp. 26–33.

Halamska, M. (1982). Rolnicy '80: Wpływ wykształcenia na zróżnicowanie postaw, poglądów i opinii [Farmers '80: The impact of education on diversification of attitudes, views and opinions]. *Wieś i Rolnictwo* [Village and agrictulture] 4, pp. 55–80.

Halamska, M. (1991a). *Chłopi polscy u progu XXI wieku* [Polish peasants on the verge of the 21st century]. *Wieś i Rolnictwo* [Village and agriculture] 4, pp. 42–55.

Halamska, M. (1991b). *Chłopi polscy na przełomie epok* [Polish peasants on the verge of the epochs]. IRWiR PAN.

Jagiełło-Łysiowa, E. (1967). *Od chłopa do rolnika* [From peasant to farmer], Ludowa Spółdzielnia Wydawnicza ("Młode Pokolenie Wsi Polski Ludowej" [Young peasant generation in People's Poland], 4).

Kochanowicz, J. (1988). Losy gospodarki chłopskiej w niekapitalistycznej próbie modernizacji. Spojrzenie historyczno-porównawcze [The history of peasant economy under non-capitalist modernization: A comparative history study]. In: J. Wilkin (ed.), *Gospodarka chłopska w systemie gospodarki socjalistycznej: Podstawy i skuteczność mechanizmu adaptacji* [Peasant economy in the system of socialist economy: A basis and effectiveness of the adaptation mechanism], pp. 61–92. Wydawnictwa Uniwersytetu Warszawskiego.

Kuczyński, W. (1981). *Po wielkim skoku* [After the great leap]. PWE

Lane, D. & Kolankiewicz, G. (eds.) (1973). *Social Groups in Polish Society*. MacMillan.

Leder, A. (2014). *Prześniona rewolucja: Ćwiczenia z logiki historycznej* [The dreamt revolution: Exercises in the logic of history]. Wydawnictwo Krytyki Politycznej.

Manchin, R. & Szelenyi, I. (1985). Theories of family agricultural production in collectivized economies. *Sociologia Ruralis* 3–4, pp. 248–268.

Marcuse, H. (1976). Repressive tolerance. In: P. Connerton (ed.), *Critical Sociology*, pp. 301–329. Penguin Books.

Mokrzycki, E. (1997). Od protokapitalizmu do posocjalizmu: makrostrukturalny wymiar dwukrotnej zmiany ustroju [From protocapitalism to postsocialism: A macrostructural dimension of the double change of political regime]. In: H. Domański & A. Rychard (eds.), *Elementy nowego ładu* [Some elements of the new order], pp. 33–46. IFiS PAN.

Mooney, P.H. (1988). *My Own Boss? Class, rationality, and the family farm*. Westview Press.

Ost, D. (2005). *The Defeat of Solidarity: Anger and politics in postcommunist Europe*. Cornell University Press.

Ploeg, J.D. van der (2008). *The New Peasantries: Struggles for autonomy and sustainability in an era of empire and globalization*. Earthscan.

Ploeg, J.D. van der (2013). *Peasants and the Art of Farming: The Chayanovian manifesto*. Fernwood Publishing.

Skocpol, T. (1990). *States and Social Revolutions: Comparative analysis of France, Russia, and China*. Cambridge University Press.

Skocpol, T. (1991). *Vision and Method in Historical Sociology*. Cambridge University Press.

Stark, D. (1992). Path dependence and privatization strategies in East Central Europe. *East European Politics and Society* 6(1, Winter), pp. 17–54.

Strange, M. (1988). *Family Farming: A new economic vision*. University of Nebraska Press.

Szczepański, J. (1973). *Zmiany społeczeństwa polskiego w procesie uprzemysłowienia* [Changes in Polish society under industrialization]. Instytut Wydawniczy CRZZ.

Szelenyi, I. (1988). *Socialist Entrepreneurs: Embourgeoisement in rural Hungary.* Polity Press.

Tilly, C. (1984). *Big Structures, Large Processes, Huge Comparisons.* Russell Sage Foundation.

Topolski, J. (1973). *Metodologia historii* [Methodology of history]. PWN.

The Warsaw Voice (1990). Issue of July 22.

Wielowieyska, D. (1997). Kim jesteś polski farmerze? [Who are you the Polish farmer?]. *Gazeta Wyborcza*, October 1.

Wilkin, J. (1988). Chłopski składnik losów gospodarki socjalistycznej (ze szczególnym uwzględnieniem polskich doświadczeń) [Peasant part of the fate of socialist economy (with some Polish peculiar experiences)]. In: J. Wilkin (ed.), *Gospodarka chłopska w systemie gospodarki socjalistycznej: Podstawy i skuteczność mechanizmu adaptacji* [Peasant economy in the system of socialist economy. Foundation and effectiveness of the adaptation mechanism], pp. 7–35. Wydawnictwa Uniwersytetu Warszawskiego.

Woś, A. (1978). Klasowo-warstwowa struktura ludności wiejskiej i jej ewolucja w procesie budowy rozwiniętego społeczeństwa socjalistycznego [Class and strata diversification of rural population and its evolution in the process of building of the developed socialist society]. *Studia Socjologiczne* 2, pp. 87–102.

Wright, E.O. (1990). *The Debate on Classes.* Verso.

W.T. (1997). Młodzi rolnicy inwestują i liczą [The young farmers invest and count]. *Wieści* August 3–10.

Zamorski, K. (2008). *Dziwna rzeczywistość. Wprowadzenie do ontologii historii* [The strange reality. An introduction to ontology of history]. Księgarnia Akademicka.

Think Locally, Act Globally: Polish farmers in the global era of sustainability and resilience, ed. by Krzysztof Gorlach and Zbigniew Drąg in collaboration with Anna Jastrzębiec-Witowska and David Ritter
Jagiellonian University Press, Kraków 2021, pp. 225–294
ISBN 978-83-233-4949-5
DOI: http://dxdoi.org/10.4467/K7195.199/20.20.12730

Chapter Four:
Family Farms in 2017: Drawing of the sociological portrait

Krzysztof Gorlach https://orcid.org/0000-0003-1578-7400/

Zbigniew Drąg https://orcid.org/0000-0002-9106-7758/

4.1. Introductory Remarks

The following chapter aims to present the characteristics of, and processes taking place in, the family farm owners and operators category in 2017 Poland, and starts the second part of this entire publication. As was already indicated in Chapter 2 describing the study's methodology, the research conducted in 2017 had a larger population sample and somewhat different selection criteria than that of previous editions. Therefore, direct, in-depth analyses and comparisons between the research editions might not be adequate or even possible. Nevertheless, on a more general level, the attempts to compare the findings from the 1994 and 2017 studies seem legitimate. In 1994 (see Gorlach, 1995) the research concept and presentation of family farms was elaborated in Poland for the first time after the political breakthrough of 1989. For research purposes, a stratified sample was randomly selected to represent a large poll of farms. Later analyses in 1999 and 2007 used the panel study methodology with the population sample chosen in advance in 1994. The trends, regularities and changes described in Chapter 3, which closes the first part of this publication, did not fully represent the entire population of family farms in Poland but only a specific category of farms selected in 1994, which were then studied consecutively, in 1999 and 2007.

It could be said that having the new population sample in the 2017 research has allowed for a different perspective. It is not a detailed analysis of the processes taking place in the selected category of farmers, as more research at various times should have been conducted to achieve such a goal. This new look could, however, be helpful in illustrating the change that occurred within the perspective of almost a quarter of a decade. Juxtaposing some research results from 1994 and 2017 could provide a valuable contribution to the reflection on family farms in Poland.

The main idea of this chapter is to offer an approach to the category of farm owners and operators that would help to identify the change that occurred in the time frame indicated above. To use the painting metaphor, it could be said that the second part of this publication, beginning with this chapter, is about presenting a family portrait of farms in Poland. Within that context, this particular chapter, which addresses the most fundamental characteristics of such farms, could be described as a draft portrait.

The draft consists of several parts. The first part (subchapter 4.2) contains the most essential characteristics, such as farm size and owners' ages and education. However, the draft of the portrait has been done according to a certain scheme that is crucial for this publication.

The subsequent part (subchapter 4.3) aims to show whether contemporary Polish agriculture and, in particular, the main portion of farms in Poland fits the picture of the typical family farm. In other words, the reflection in this subchapter concentrates on answering the question of whether agriculture in Poland still possesses a family character. Remarks on family agriculture appeared earlier in this publication and now it is essential to answer the question regarding family character. To answer this, one has to accept some particular definition of a farm which could describe its empirically verifiable characteristics. It seems that on the operational level, which is analytical, equipped with indicants that can be empirically verified, the most useful definition is that of two British authors, Gasson and Errington. They elaborated it as such:

> There are six elements to our own definition of a farm family business: 1. Business ownership is combined with managerial control in the hands of business principals, 2. These principals are related by kinship and marriage, 3. Family members (including these business principals) provide capital to the business, 4. Family members (including business principals) do farm work, 5. Business ownership and managerial control

are transferred between generations with the passage of time, 6. The family live on the farm (Gasson & Errington, 1993, p. 18).

The content of subchapter 4.3 was prepared in reference to these characteristics.

There is also another issue worth addressing. As the parameters of family farms change due to ongoing economic and social processes, the socio-cultural context and the awareness also change. What is meant here are the various ways of thinking, the opinions, and the preferences expressed by farm owners and operators regarding various aspects of how family farms function and how they change. For sociologists, this layer of social order is particularly important. It helps to explain various aspects of human actions and behaviors and, in this case, it is farmers who are being analyzed. The further reflection in subchapter 4.4 focuses on these issues.

4.2. Family Farms from 1994 to 2017: Essential characteristics

Firstly, one should take a close look at the changes in the area structure of the studied farms. These changes are illustrated by the data presented in tables 4.1 and 4.2 which reach the following conclusions. The first one relates to the category of the smallest farms, those with an area of under 2 ha. Their percentage in the research sample over the 1994–2017 time frame stabilized noticeably, while the minimal percentage increase of 0.5% could be considered meaningless. One could ask whether the stabilization trend was an effect of including Polish farmers in the programs of the Common Agricultural Policy. A positive answer could be formulated in reference to direct subsidies which, in the situation of small and weak farms, could be compared to an IV drip maintaining their life functions. However, this answer is insufficient in the case of the two other categories, namely, farms utilizing an agricultural area of 3 to 5 ha and those farms with an area of 6 to 7 ha. There is some visible downward trend which should be emphasized. Another category, of farms utilizing 8 to 10 ha of agricultural land, also presents an interesting picture. In Poland, the farms in this category are considered medium-sized and here a downward trend is also observed. The next category of farms, those with an area of 11 to 15 ha, also noted a decline, albeit very small. However, the category of farms with an area of 16 ha or more noted a significant increase. Currently, this category accounts for twice as many farms as it did in 1994.

Table 4.1. Farms according to the utilized agricultural land areas in the studied population sample (in %)

Size in hectares (ha)	Farms	
	in 1994	in 2017
Up to 2	23.0	23.5
3–5	31.2	29.0
6–7	15.0	10.9
8–10	14.0	11.5
11–15	9.9	9.5
16 and over	6.9	15.6*
Total	100.0	100.0

* Note: The category of "16 ha and over" includes farms of 16–30 ha (9.2%) as well as the farms of 31 ha and over (6.4%).

Source: Own research.

Table 4.2, with farms divided into three area categories, is also quite interesting. The purpose of this measure was to get a more pronounced picture of the family farm structure in Poland. In the mentioned period of time, a small drop in the percentage of farms smaller than 5 ha was noted, as well as a more dramatic drop in the category of farms situated on 6 to 10 ha. During the same time frame, there was a quite noticeable increase in the category of farms of 11 ha and more, which should not be ignored.

Table 4.2. Farms according to utilized agricultural land areas in the studied population sample (in %)

Area in hectares (ha)	Farms	
	in 1994	in 2017
Up to 5	54.2	52.5
6–10	29.0	23.4
11 and over	16.8	25.1
Total	100.0	100.0

Source: Own research.

Tables 4.1 and 4.2 show the tendencies that confirm the processes of concentration for utilized agricultural land in Poland. Table 4.1 provides evidence for stabilization in the percentage of the smallest farms, a drop in the middle categories, and a noticeable increase in the largest categories. The second table presents, even more clearly, the process of "moving the center of gravity" towards the categories that are larger in size. It should be stressed here that the percentage drops in the studied categories mostly occurred in the middle category of farms. The phenomenon is more pronounced in this category than in the category of smallest farms. This, combined with the increasing number of the largest farms (11 ha and more), could support the theory of the "disappearing middle," so popular in the specialized rural sociology literature (see Gorlach, 1995, 2009).

Tables 4.3 and 4.4 describe the characteristics of contemporary farms, such as the owner's age and educational attainment level. These variables are quite important, as they are the base characteristics which could be the crucial factors in explaining various actions and behaviors. They could also be part of aggregated variables, which are more complex characteristics of the studied farms. Table 4.3, containing data on the owners' ages, legitimizes statements that the category of family farms in Poland is going through the aging process. The drop in the percentage of farm owners under the age of 35 is particularly noticeable. In the remaining two categories, there are noticeable surges.

Table 4.3. Farms according to the age of the owner (in %)[1]

Age (in years completed)	Farms	
	in 1994	in 2017
18–34	31.5	10.7
35–54	42.8	56.2
55 and over	25.7	33.1
Total	100.0	100.0

Source: Own research.

Table 4.4. Farms according to the educational attainment of the owner (in %)

Educational level	Farms	
	in 1994	in 2007
Elementary school or less	45.7	13.9
Vocational education	36.4	42.4
High school	16.0	31.9
Post-high school education	1.9	11.8
Total	100.0	100.0

Source: Own research.

[1] Table 4.3. takes into account three age categories following the three phases of an individual's life that are described in the demographic literature (see Michałkowski, 1987, pp. 28–29). These phases are known as the exploration stage which ends at 24, the stabilization stage for ages 25–44, and the mature stage, combined with the reconciliation stage, for ages 45 and up. As can be seen, the age caesuras used in the following work are somewhat different. Due to specific characteristics of the farmer's profession, where the mature age connected with operating a farm is reached somewhat later in life compared to other types of professional activities, three age categories were adjusted accordingly. The first category consists of the exploration stage and the first part of the stabilization stage, up to age 34. The second stage includes the remaining years of the stabilization stage and the first phase of the maturity stage (ages 35–54). The third age category is connected with ages 55 and over, which is the furthest stage of maturity and reconciliation with life.

The structure of education among the studied population is improving. This is particularly visible in the decline of the percentage of people with only an elementary school education. Generally, the educational level of farm owners and operators in Poland indicates revolutionary processes. In the mid-1990s, farmers with an elementary education of 8th grade or less were the largest category of farm owners and comprised almost half of respondents. They did not even have a vocational education. Twenty-three years later, the percentage of respondents in this category dropped to less than 14% and was reduced by nearly four times. This category is no longer the most numerous, but it is apparently the third largest. The percentage difference between this category and the category of farm owners with the highest level of education (beyond high school level) is not very great when juxtaposing the 13.9% of farm owners with elementary school education with the 11.8% of those whose education is beyond high school. Interestingly, there is a significant increase in the categories of respondents with high school education and post high school education. In this last category, the number of farm owners increased five times over the studied period of time. Nevertheless, they are still the smallest category.

What is the conclusion of the above reflection? Two statements can be made without a doubt. Most importantly, the presence of farms that are large in size is becoming more noticeable in Polish agriculture. The farm owners are aging but they are also better educated. Such tendencies are also observed in other European countries.

4.3. Still a Family Farm?

The reflection on the family aspect of farms in Poland should start with a closer look at the evolution of two "entry" characteristics, presented in rural sociology literature, which are thought to determine the "family" nature of farms. It should be noted that the most widely recognized model of family farms is that in which the farm family owns the land that they use.

What can be concluded from Table 4.5? It appears that the category of the typical family farm, wherein the owner is also a user of the entire owned land, still dominates in Polish agriculture. In both research studies, such an ownership situation applied in over 80% of cases. However, the data compiled in the research showed some additional tendencies that should be analyzed here. The frequency of the most typical farm situation (that of owner-user) decreased over twenty-three years, from 85.6% to 80.2%. This presented slow, but noticeable, changes in the emergence process for relations typical of a market economy in the family farm model. The values in the category of owner-tenants, which described those people who did not operate their own land but rented land from other farmers, as well as owner-renters, who only used a portion of their owned land and rented the remainder out to others, confirmed more. These tendencies were not dominant, but still noticeable, which is well-reflected in Table 4.5.

Table 4.5. Farm ownership situation in 1994 and 2017 (%)

Owner's category	1994	2017
User	85.6	80.2
Tenant	12.6	17.7
Renter	1.8	2.1
Total	100.0	100.0

Source: Own research.

It should be recalled here that, in the time between the two studies, Poland joined the European Union (in 2004). This meant that from May 1 of that year the regulations of the EU's Common Agricultural Policy were applied to Polish agriculture, with some other adjustments being made later. Common Agricultural Policy is mostly oriented toward maintaining the status of family farms as the foundation of European agriculture. Could the data presented above confirm this thesis? This might be the case. Certainly, it can be stated that Poland's involvement in the globalization processes, after both the changes of 1989 and its admission into the European Union in 2004, did not cause any radical or rapid changes in the farm ownership structure or in the operation of farms.

The way of acquiring the farm is also important, with inheritance being the most typical means of farm acquisition. The data supporting this thesis can be seen in Table 4.6. It should be noted that, among the characteristics presented in the definition of family farm, Gasson and Errington indicated intergenerational transfer of the farm (i.e., land and other means of production) as being one of the core components of the family farm. This was still true for an overwhelming majority (well above 80%) of farmers, with a noticeable increase in farms acquired by rental. The percentage of farms acquired through outright purchase did not change at all over the last quarter of a century, and the data in Table 4.6 support that. It could be said that the processes for acquiring farms during the last quarter of a century were characterized by significant and visible stability.

Table 4.6. Ways of acquiring farms in 1994 and in 2017 (in %)

Way of acquiring the farm	1994	2017
Inheritance	88.5	86.4
Purchase	9.1	9.1
Rental	2.4	4.5
Total	100.0	100.0

Source: Own research.

The following characteristics of family farms present the role and the meaning of the labor type for the functioning of farms in both studies. Going back to the concept from Gasson and Errington, it should be recalled that the amount of hired labor was a significant factor in the de-familization process of family farms. It should be emphasized that the data presented in Table 4.7 point to the strengthening of the family character of the studied farms. The percentage of farms studied in 2017 that did not use hired labor was 85.3 and this was even higher than the analogous percentage in 1994 (80.6). The data on the number of people hired for farm work could also be considered valuable information. As indicated by the values in Table 4.7, the percentage of farms hiring one person increased over the last quarter of a century. The percentages of farms hiring two and three or more people decreased over the years. Over the analyzed period, the position of hired labor became somewhat stronger, but only as an additional factor in the process of de-familization.

Table 4.7. Number of hired workers in 1994 and in 2017

Number of hired workers	1994	2017
None	80.6	85.3
One person	6.4	9.9
Two people	6.3	2.3
Three or more people	6.7	2.5
Total	100.0	100.0

Source: Own research.

A similar conclusion might be drawn from the data on the number of days that the farms used hired labor (see Table 4.8). The percentage increase could only be seen in the category of farms that used hired labor up to 14 days per annum. This increase was so small as to be within the realm of statistical error. The other two durational categories (15–30 days and over 30 days) saw percentage decreases. This could confirm the thesis that the family factor of farm labor strengthened and solidified, while hired farm labor was a relatively small addition in the greater scheme of these relations.

Table 4.8. Hired labor: number of days utilized in 1994 and in 2017

Number of days utilized	1994	2017
0	80.6	85.3
14 or less	9.2	9.4
15–30	6.1	2.4
Over 30	4.1	2.9
Total	100.0	100.0

Source: Own research.

The tradition of the Polish countryside (and perhaps not only Polish) includes another type of farm labor, which goes beyond the simple dichotomy of family labor and hired farm labor. This involves the type of work which can be best described as "exchange of work with the farmer/neighbor." This type of work relation has a double context. In some way, it carries the dimension of the rural community functioning as a whole, where no family farm is truly independent and separated from others as an entity of its own. Thus, the use of help from neighbors, particularly during critical times for field work, such as grain and potato harvesting, played an important role in the traditional countryside's farm labor scheme. It should be emphasized here that the rules for the neighborly exchange of farm work with other farmers functioned according to principles that, in social science, are described as "the norms of reciprocation" (see, e.g., Gouldner, 1960; Kocik, 1976, 1986; Łukaszyński, 2013).

Table 4.9. The type of labor used on farms in 1994 and in 2017

Type of farm labor	1994	2017
Family	53.1	63.8
Family and work exchange with another farmer	27.5	21.3
Family and hired labor	11.7	10.8
Family, work exchange with another farmer, and hired labor	7.7	4.1
Total	100.0	100.0

Source: Own research.

Table 4.9 provides a summary of the labor aspects analyses. It contains the data that demonstrates the process of "familization" of farms. In 1994, slightly more than half of farms in Poland used family labor in order to operate properly. Almost twenty-five years later, the percentage in this category increased by 10% or more. At the same time, the percentages of farms where family labor was tied to other types of labor dropped: exchanging farm labor with another farmer declined, from 27.5% to 21.3%; hired labor decreased from 11.7% to 10.8%, and the combination of three types of labor, such as family labor, exchanging farm labor with another farmer, and hired labor, dropped from 7.7% to 4.1%.

What can be inferred from the observed tendencies? First of all, it is worth noting that development and transformations in agriculture within the process of its modernization, which spanned for nearly 30 years, starting after the political breakthrough of 1989, did not enhance capitalist relations (i.e., hired labor) on farms. Secondly, it should be emphasized that the percentage of traditional relations in farm work exchange shrank. Both of these categories in Table 4.9 decreased over the analyzed period of time. It could be said that farms in Poland, treated as a certain overall category, are moving towards familization, at least in terms of work relations.

The above observations can lead to the more general commentary that agricultural modernization and expansion of farms do not necessarily go hand in hand with the growth of capitalist relations (hired labor), although it has meant the abandonment of traditional forms of help between neighbors. It should be said that agricultural modernization is conducive to maintaining the family quality of farms, which was observed in other studies and presented by other authors (see Haan, 1994; Djurfeldt, 1996; Brandth & Haugen, 1997, 1998). This process relates to the fact that technical solutions and possibilities for performing various tasks significantly reduce the need for farm labor from outside the family that lives on the farm itself. Sometimes it is even possible for one person to operate the farm with minimal support from other members of the family.

Another important aspect in the process of the functioning and transformation of farms relates to using credits, mortgages, or replenishing farm resources with other financial means coming from outside of the family circle. It is worth remembering that the increased use of such financial sources is, according to Gasson and Errington (1993), an important factor in the "de-familization" of family farms and an indicator of family farms, generally, entering into such relations with an external economic system that is characteristic of the capitalist economy.

Earlier research (see Gorlach, 2009) has allowed for formulating observations about "leaving" the family farm model situation, which means resorting to external financing sources. This was a factor facilitating the function and continuous existence of these farms within the market economy, introduced in Poland after 1989. A thesis very popular in sociological literature points to the family economic model as a specific type of means of production organization (see Friedmann, 1981; Wright, 1989, 1997; Gasson & Errington, 1993; Gorlach, 2004) being so very different from that of a typical capitalist enterprise. Even when following this thesis, it should be stated that the use of typical instruments of the capitalist economic system facilitates the functioning of family farms within such a system. It turned out that the use of credits/loans by those family farms in Poland which remained in operation until the following research study (in 2007) had intensified over the time periods of 1990–1994 and 1995–1999 (see Gorlach, 2009, pp. 84–85). In the first of these periods, such farms comprised slightly more than one third of the studied sample population and in the second period they made up almost half of the studied sample. The use of credits/loans reached a statistically significant level and was connected with the continued operation of the studied farms.

Table 4.10 is worth some attention, as it presents the percentages of farms using credits/loans during various research editions. It should be recalled here that the studies of 1994, 1999, and 2007 took the form of a panel study and were conducted on the same original population sample, which diminished over 13 years, as some farms ceased to exist and some farmers abandoned farming altogether. The 2017 study was conducted on a new population sample which was also representative. Therefore, it is justified to compare the data presenting differences between the first and the last study, as can be seen in Table 4.11.

Table 4.10. Percentage of farmers using credits/loans during the time intervals 1990–2006 and 2015–2017 (in %)

Research	1990–1994	1995–1998	1999–2006	2015–2017
1994 (N = 800)	31.0	×	×	×
1999 (N = 685)	32.6	43.6	×	×
2007 (N = 510)	35.6	49.2	3.2	×
2017 (N = 3551)	×	×	×	19.4

Source: Own research.

Table 4.11. Farms according to sources of financing in 1994 and in 2017 (%)

Sources of financing	1994	2017
Own financing only	69.0	80.6
Own financing and credits	31.0	19.4
Total	100.0	100.0

Source: Own research.

When comparing the data from 1994 and 2017 it is clear that farms in Poland currently rely less on credits and mortgages that were once taken in order to fulfil the needs of the farms' everyday functioning and to foster farms' further development. In 1994, 30% of farms were in such a situation but, twenty-three years later, it was less than 20%. Several factors played a role here. Firstly, the situation is now more stable than it was in the 1990s, when Polish agriculture was experiencing the effects of rapid restructuring caused by the political breakthrough and the economic reforms implemented at the time and oriented toward the introduction of market economic principles. Secondly, with the effects of modernization and the increasing role of larger farms, the situation of Polish agriculture has become more favorable than it was more than 20 years prior. Thirdly, farms in Poland are currently fully impacted by the effects of the Common Agricultural Policy, which systematically provides a financial influx to them without the need for them to use risky financing sources like bank credits. It seems legitimate to say that the process of familization of farms can be observed in Poland, as described by the concept of Gasson and Errington. The researchers and readers of the following monograph, who might be looking for some more in-depth validation of this process, could be reminded that the Common Agriculture Policy is very much oriented towards maintaining and strengthening family farms. Thus, the Common Agricultural Policy appears to have had positive effects on family farms, at least in Poland.

The next issue deals with the farmers' attitude towards the farming profession, and their answers are presented in Table 4.12. Here (somewhat differently than that before) the data in the table contains the answers obtained in the 1999 and 2007 research studies. It should be noted that the reason for this is that in 1999 the "Would you become a farmer again?" question appeared in the surveys for the first time (see Table 4.12). Answers to the question on becoming a farmer again might be seen as validating the choice of such profession. They could also be indicative of a choice made consciously and autonomously. The "yes" answers could also be seen as statements of satisfaction related to the choice once made and stemming from the conviction that the farmer's profession was the right choice for the professional career and life strategy. This type of choice can be recognized as the manifestation of the attitude of an agricultural entrepreneur who is confident about their life choices and achievements.

The negative answers to the question on willingness to become a farmer again could be interpreted as disappointment in the particular results of choosing one's own life path in agriculture. They could also be seen as the result of coercion or necessity caused by pressure from the family and the need for someone to take over the farm. In confrontation with the reality of mediocre results in farming economic activity and the necessity for making various sacrifices and enduring certain hardships related to farming, such answers were the manifestation of negative attitudes towards the farmer's profession. The same can be said about the "Hard to say" answers, which only differ from the "no" answers in terms of the degree of negativity. Both categories can be seen as specific, negative manifestations of "peasantry," meaning, surrendering to the fate and enduring the consequence of negative experiences related to one's own choice and agricultural economic activity.

Table 4.12. Answers to the question: "Would you become a farmer again?" (in %)

Answers	1999	2017
"Yes"	16.8	36.2
"No" and "Hard to say"	83.2	63.8
Total	100.0	100.0

Source: Own research.

Here some generalizations might be tempting. The "peasant" complex, which could be reflected in the inability to answer the question or in negative answers to questions, was significantly reduced in the studied population. Considering that in 1999 (about 20 years prior) over 80% of respondents manifested the disappointment of their choice in one way or another, one should consider that in 1999 the researchers studied the farms of the earlier 1994 research which had survived the previous five years. In 2017, the percentage of negative answers dropped to less than two thirds of respondents. The percentage of positive responses expressed by farmers who felt positively convinced of their own life choices and accepting of the decisions related to engaging in farming almost doubled (growing from below 17% to over 36%). It could be said that the attitude of treating the farm as a family tradition and a family business became even stronger.

4.4. Changes in the Way of Thinking about Farming during the 1994–2017 Time Frame

This part of the reflection should start with the opinion on land rentals, which is one of the key elements in the modern architecture of agriculture's market relations. The issue of ownership is no longer the most important, but the utilization and operation of agricultural land has become crucial. The person using and operating agricultural land (who is not necessarily the owner of its entire area) is the subject of the game in market relations, and the one who, at the same time, must formulate a strategy for managing the farm. Positive opinions regarding land rentals could suggest an increasing "entrepreneur" consciousness, wherein the formal title of ownership is of secondary importance while the ability to use the resources and have the decision-making power over them is essential.

Some difficulties in analyzing the changes in the time period from 1994 to 2017 had to do with changing the way the questions on this issue were asked. In 1994 (see Gorlach 2009, pp. 88–89) respondents received the same set of seven possible opinions on renting land and were asked to mark those they agreed with or recognized as their own.[2] The first three of them contained a positive evaluation of renting land, seeing that it allowed for increased production, expansion of the farm area, and a better start for young farmers who might not necessarily be able to afford the outright purchase of land they would like to farm. The next four opinions referred to negative aspects of land rental, such as dependency on a contract with the land owner, the possibility of going into debt, as well the popular and widespread opinions in rural communities that tenants do not take good care of the land or that it is better to become the formal owner of the utilized land, as such a situation fosters confidence and stability (see Table 4.13). Highlighting the positive aspects of land rental could indicate the emergence of a way of thinking that is generally considered characteristic of entrepreneurs. Contrary to that, answers that emphasized the negative aspects of land rentals pointed to the traditional peasant way of thinking, marked by an uncertainty about the contracts and a sense of an incomplete and unfulfilled ownership status for the utilized farm land. The analysis of the data results of this survey provided fertile ground for some interesting conclusions (see Gorlach, 2009, pp. 88–89). The preference for statements in favor of land rentals was quite clear. Almost two thirds of respondents chose such statements in 1994. The negative aspects of renting land were acknowledged by less than 30% of respondents.

The opinions obtained in 1999 and 2007, when panel research was conducted to follow up on the 1994 study (see also Gorlach 2009, pp. 88–89), deserve a closer look. They illustrate the results of experience acquired by the interviewed farmers during the times of rapid change, after the political breakthrough of 1989 and Poland's accession to the European Union in 2004. Some interesting fluctuations and variations

---

[2]   It should be remembered that in that study (as well as in the 1999 and the 2007 editions) the respondents were allowed to mark as many positive and negative answers as they wished. However, in 2017 respondents received only one question containing four possible opinion responses, which they had to accept, negate or declare indifference toward.

could be observed here. In 1999, again, almost two thirds of respondents pointed to the positives of renting. At the same time, negatives were reported by 53.3% of respondents. In 2007, the situation was completely different with 90% of respondents recognizing rental as a good overall solution for farmers and just over 26% reported any negative opinions in this context. It could be underscored here that opinions on land rental among the respondents were equivocal, yet positive choices prevailed.

Table 4.13. Opinions of respondents toward land rental (in %)

Opinion	"Yes" answers		
	in 1994	in 1999	in 2007
Good solution as it makes it possible to increase production	30.8	28.6	43.1
Good solution because it makes aggregation of land possible	9.8	11.5	19.0
Good solution because it facilitates a start for young farmers	25.4	25.6	31.2
Bad solution because it causes contract dependency	3.3	7.4	6.1
Bad solution because one may go into debt	5.0	14.3	1.0
Bad solution because tenants do not take care of land	7.2	12.1	4.1
Bad solution because it is better to be an owner	13.4	19.5	15.1

Source: Gorlach, 2009, p. 88.

Table 4.14. Respondents' opinions on land rental (% of answers) in 2017

Opinion	Yes	No	Neither yes nor no / Hard to say
Renting is a good solution because it increases the possibilities for effective use of machines and allows for increased production	71.0	8.6	20.4
Thanks to land rental, it is easier for young farmers to start farming	77.3	6.0	16.7
Land rental is a good solution because it allows for land usage without the necessity to own it	63.2	13.2	23.6
Land rental is not a good solution, because it makes farmers dependent on contract conditions as well as prone to accumulating debt and there is no motivation to care for the land	36.2	35.7	28.1

Source: Own research.

The results of the research conducted twenty-three years later showed yet again the ambiguity of farmers' opinions on land rental (see Table 4.14). The opinions in the questionnaire, to which respondents had to declare a positive or negative attitude or decline the choice altogether, were formulated in reference to the results of the studies conducted in 1994 and in 2007. The majority of respondents, here meaning over 60% to almost 70%, identified positively with renting. It is worth noting that in the case of the last statement option, that renting is not a good solution, opinions were more equally divided. Slightly over one third of respondents (36.2%) agreed with this statement and nearly as many (35.7%) did not. Furthermore, slightly less than 30% (28.1%, to be precise) were not able to form a clear-cut opinion.

Table 4.15. Opinions on farmers owning the machines they use (% of answers) in 1994 and in 2017

Category of viewpoint answers	1994	2017
Opinion A: A farmer should own all the machines, even if they are only being used from time to time	55.2	46.9
Opinion B: A farmer should only own those machines which are being used on a daily basis and the rest of the machines should be purchased and used together by several farmers	28.9	26.6
Opinion C: A farmer should only own those machines that are being used on a daily basis and rent other machines from other farmers	12.8	20.7
Opinion D: Farmers should not own any machines and just commission machine services from others	3.0	2.3

Source: Own research.

This observation is confirmed by some of the data in Table 4.15 containing farm owners and operators' opinions on owning machines that are necessary for various types of farm work. Opinions taking a practical approach to owning machines could be seen here as indicative of "mental modernization." Such opinions express a preference for owning only those machines which are needed on a day-to-day basis. The remaining machines could be rented from others or farmers could commission external companies to perform services on the farm. These opinions mesh neatly with answers C and D. It is quite noticeable in Table 4.15 that the percentages of respondents agreeing with these types of statements changed—but it should be noted that respondents' preferences changed according to different patterns in the C and D answers, perhaps reflecting the contextual realities surrounding the answers. The C answers indicating acceptance for renting machines from other farmers rapidly gained acceptance in 2017 (20.7%) compared to the earlier study (12.8%). However, in the case of D type answers, which expressed preferences for commissioning services on the farms to other, outside entities, the percentage of favorable opinions for this approach was minimal in 1994 (below 3%) and had only diminished further, to 2.3%, by 2017.

The most "modernized," or rational type of thinking (the C answers) could only be noticeable in the studied population in 2017 when chosen by one fifth of the respondents. It should be highlighted here that the most numerous group of respondents still preferred answer A, which in some way could be regarded as a "psychosis of owning" (see Gorlach, 1995, 2001). It should not go unnoticed, however, that over the last 20 years this viewpoint's popularity decreased from 55.2 to 46.9%. Special attention should be given to the B answers, with communal ownership and cooperation between farmers presented as a solution to the problem of machine ownership. The percentage of this type of answer had a declining tendency in the years 1994–2017, with a noticeable drop from 28.9% to 26.6%.

The results of the analyses presented in tables 4.13–4.15 cannot be treated as the basis for a thesis regarding some radical reevaluation in the consciousness and mentality of the interviewed famers. Some decline in the opinions that represent the most traditional way of thinking, as indicated by the A answers, could be observed. Nevertheless, it should also be emphasized that the continued presence of such opinions still plays a crucial role in respondents' consciousness. The B answers, which were declining over time, might be worth a closer look, especially that such a tendency seems to run contrary to the observations of Western rural and agricultural sociologists (see Mooney, 2004), who reported an increased propensity among farmers in Western countries to apply collective or even communal solutions. To summarize this part of the reflection on machine ownership, it should be stated that the symptoms of "modernized" thinking, even though they're not yet dominant, are marking their presence (mostly through the C answers) in the consciousness of the interviewed farm owners and operators.

The last characteristic of the "modernization" process of respondents' consciousness related to a vision for the farm, wherein a farming family can lead a dignified life. In the analysis presented in Table 4.16, only one readily quantifiable indicator of the idea of an optimal farm—farm area—was taken into consideration. Many respondents in both studies made additional comments to their answers, pointing to various conditions that should be fulfilled on the optimal farm. It appears that the element which was easiest to imagine, and the most comparable over various periods of time, was the optimal farm size, as indicated by respondents.

Table 4.16. Optimal farm area; % of answers

Category of answers	1994	2017
Up to 10 ha	31.4	11.5
11–25 ha	40.3	25.6
26–50 ha	24.4	44.8
51–99 ha	2.0	5.9
100 and more ha	2.0	12.2

Source: Own research.

The characteristics in Table 4.16 could be a good illustration of the moderni-zation processes in the consciousness of the interviewed farmers. It is particularly noticeable in the two—as they could be described—extreme categories. On the one hand, there was an obvious percentage drop in the category of respondents who still thought in traditional terms and preferred relatively small farms (up to 10 ha) and, on the other, the category of respondents who thought of a farm of over 100 ha as the best solution was growing. All of the above confirmed the conviction that, for the majority of respondents, the size that was considered a "good" farm increased and this was the consequence of farmers having almost 30 years (1989–2017) of experience functioning in the market economy. It also turned out that the trans-formations taking place in the consciousness of Polish farmers were directed into a way of thinking that was typical of agricultural entrepreneurs. The entrepreneurial approach also emphasized the viewpoint regarding the favorable impact of the size of the farm on its operation. Similar tendencies were observed earlier in the United States (see Strange, 1988; Gorlach, 1995) and Western Europe (see Ploeg, 2003).

The dynamics of increased aspirations over the analyzed period of time, meas-ured by the "desired" farm size providing a dignified life for the farmer's family is well illustrated by the central tendency for 1999 (!) and 2017 (see Table 4.17). Al-though the average land area of such a farm had only increased by 13% during this period of time (from 38.3 to 43.2 ha) it should be noted that the median increased by 20% (from 25.0 to 30.0 ha), and the mode went up to 150% (from 20,0 to 50,0 ha). How to interpret these data, beside stating the obvious, that they reflect the increasing aspirations of farmers? First of all, the respondents' opinions on this matter (as well as aspirations) were decidedly more diverse in 1999 than they were in 2017. Lower values of the median and the mode than the value of the average in 1999 showed at the time a small group of farmers among respondents who, in comparison to other respondents, had "outrageous" expectations about the optimal size of the farm that would secure a dignity of life and work. Perhaps these were the farmers who were most afraid of the competition on the EU agricultural market which was about to open up for Poland? It is also possible that the respondents who estimated the "desired" farm size this way were aware of the reality of agriculture in Western countries. Diversification of farmers' aspirations was also noticeable in 2017 but was not as big as it had been two decades earlier. The average size of the "desired" farm is bigger than its median but smaller than its mode. It could be inferred that the majority of farmers expressing their opinions on the "desired" farm size are quite rational and to some extent their ideas and preferences stem from the experience of Polish agriculture's reality over the last two decades. It should be noted that, without a doubt, the most numerous category is that of farmers whose aspirations are higher (mode = 50 ha) than the aspirations of the regular farmer (average = 43.2 ha) as well as the aspirations of the other half of respondents (median = 30 ha). Are they Polish agricultural entrepreneurs and Polish farmers who have the chance to dominate agricultural production? These issues will be analyzed later in the text.

Table 4.17. Respondents' opinion on optimal farm area

Farm area in ha	1999	2017
Average	38.3	43.2
Median	25.0	30.0
Mode	20.0	50.0

Source: Own research.

Another issue related to farmers' behavior is the situation of the emergence of problems with the functioning of farms, such as lack of profitability. The question about the reactions of farm owners to the potential situation of problems with profitability seems interesting. The results of the study are in Table 4.18, and some remarks might be helpful. Only parts of the data from previous research editions were included here (see Gorlach, 2009, p. 94) and only those types of reactions were presented that carried respondents' exclusive agreement to certain measures in the research conducted in 2017.

Table 4.18. Answers to the question: "In a situation where profitability was lacking, would you be able to take the following measures?"

Type of measure	"Yes" answers	
	in 1994	in 2017
1) Selling or renting of part of the agricultural land	22.1	31.7
2) Selling of some farm machines and equipment	5.0	17.7
3) Resigning from further education of the children	2.8	0.6
4) Reducing the shopping expenses of the family to the necessary minimum	32.4	36.0
5) Selling part of the home amenities	5.0	2.2
6) Getting a loan	18.2	22.6
7) Increasing the volume of farmer's work and the work of his family	30.2	49.4
8) Taking additional job outside of agriculture	52.1	67.4

Source: Own research.

It should be recalled as well that these two research editions were conducted almost a quarter of a century apart (exactly twenty-three years). It is worth noting that some measures that were evaluated by the respondents related to three dimensions. The first one concentrated on the matters of the farm (compare points 1 and 2). The second one (compare points 3, 4, 5) was devoted to matters of the family. Finally, the third dimension (see points 6, 7, and 8) relate to issues connected with the larger context of market economy (loans, job market, etc.). What could be inferred from the presented analyses? In the first dimension (points 1 and 2 relating to the farm) the readiness of farmers to sell or rent part of the agricultural land and some machinery increased. This could mean the intensification of attitudes treating farms simply as business venues, to which principles of selling and purchasing, so characteristic of the market economy, could be applied. In the second dimension (points 3, 4 and 5, related to the family) the increase in opinions characteristic of thinking on the categories of the market economy could also be noted. The participants in the 2017 research were more likely to reduce their shopping purchases to the necessary minimum and, at the same time, much less likely to resign from educating children or selling household amenities. These tendencies intensified the market-oriented way of thinking, and were reflected in the opinions in the third dimension (points 6, 7 and 8), where the propensity for getting credits and loans increased significantly. The willingness to increase the volume of work of the farmer and his family, as well as the readiness to take additional jobs outside of agriculture, also went up. This could be interpreted as a progression of the modernization processes in the consciousness of the interviewed farmers.

The issue of hired labor provided for other farm owners and its evaluation could serve as an interesting addition to the reflection on the modernization of consciousness of the interviewed farmers (see Table 4.19). The initial assumption should be made that acceptance of such a situation could mean treating work as some kind of good or commodity, to which the principles of a market economy could apply. It could also serve as another argument for the modernization of farmers' consciousness in Poland.

Table 4.19. Opinions on hired labor in agriculture in 1994 and in 2017 (%)

Category of answers	1994	2017
Approval	43.8	51.7
Disapproval	52.7	25.5
Hard to say	3.5	22.8
Total	100.0	100.0

Source: Own research.

Working for money (not just in the form of work exchange with neighbors) for other farmers was, from the perspective of traditional peasant mentality, negatively evaluated. Providing such work was considered degrading, as it was the domain of the weakest farmers, not able to manage their own farms (Gorlach, 2001, p. 228). The data from the 1994 and 2017 research editions clearly indicated the gradual process of abandoning the peasant mentality and emergence of the pro-market attitude. In

1994, more than half of the respondents expressed reluctance to be hired by another farmer while less than 44.0% accepted this as a possibility. The 2017 research edition brought some significant changes. Farmers accepting the possibility of being hired by another farmer comprised more than half of all respondents while the percentage of those who didn't have a clear opinion went up to almost 23.0%. There was a radical reduction in the percentage of those who rejected such a possibility—down to only 25.0%. All the above can lead to the statement that the modernization process of the mentality of farm owners and operators in Poland is far from complete.

The last aspect of the reflection on the modernization of the mentality of the interviewed farmers relates to the perception of the category of people who operate farms and are the most capable in managing farms and dealing with difficulties or, in other words, have some success in the market economy. Table 4.20 contains the opinions presented by respondents in 1994. The categories of answers relating to the economy of scale, dealt with large and mechanized farms. This opinion was supported by answers related to entrepreneurial qualities, risk-taking, and conducting specialized production. Therefore, it can be stated that the dominant answers were connected, for the most part, with characteristics typical of the market economy.

Twenty-three years later, the question about these issues was differently formulated. The interviewed farmers received a list of factors that could determine a farmer's success and these factors were chosen on the basis of research results in 1994, 1999, and 2007. The farmers were asked to choose a maximum of 3 of them. The effects of the choices of research participants are presented in Table 4.21.

Table 4.20. Percentage of answers pointing to a particular category of farmers perceived as able to deal with difficulties in 1994 (in %)

Category	%
Owners of large and mechanized farms	34.1
Entrepreneurial and risk-taking	10.8
Conducting specialized production	9.3
Taking job outside of agriculture; processing and others	8.7
Young, healthy, and resilient	7.2
Working and producing more	6.7
Starting in the period of People's Republic of Poland (before the breakthrough of 1989)	5.9
Owning the good category/type of soil	5.6
"Wheeler-dealers" who know the right people	5.0
Having an appropriate number of workers	3.7
Having a family tradition	1.7
Hard to say	1.3
Total	100.0

Source: Gorlach, 2009, p. 91.

Table 4.21. Skills and abilities determining the success of the farmer on the 2017 market (intensity of choices in %, the percentages do not add up to 100.0, as the respondents were allowed to make 3 choices)

Category of skills and abilities	%
Diligence and willingness to take up hard work	63.7
Penchant for farm work	62.5
Providing appropriate quality of production	32.6
Orientation to the market of agricultural products and directing of production towards the market needs	31.0
Appropriate qualifications and agricultural education	27.2
Readiness to take on risk	26.6
Ability to "wheel and deal" and knowing the right people	12.9
Ability to combine agricultural production with environmental care	10.1
Orientation toward production of traditional local products	8.4

Source: Own research.

Table 4.22. Analysis of main components of the skills and abilities that determined farmers' success in 2017: Factor loading, Varimax rotation (loadings below 0.20 not included)

Skills and abilities	Component 1: entrepreneur	Component 2: sustainable (eco-farmer)	Component 3: risk-taker
Penchant for farm work	−0.698	−0.225	−0.238
Orientation to the market of agricultural products and directing of production towards the market needs	0.674		
Diligence and ability to work hard	−0.521	−0.342	−0.236
Appropriate qualifications and agricultural education	0.483	−0.395	−0.417
Providing appropriate quality of production		0.667	
Ability to combine agricultural production with environmental care		0.583	
Orientation to production of traditional local products		0.460	
Readiness to take risk			0.691
Ability to "wheel and deal" and knowing the right people			0.658
Percentage of variance explained by data component	16.30	15.20	14.20
Total percentage of explained variance		45.70	

Source: Own research.

It turned out that farmers' choices were significantly determined by designations that pointed to qualities and abilities usually considered to be traditionally peasant (i.e. penchant for farm work, as well as diligence and willingness to take on hard work) in determining success in agriculture. The analysis of these factors also revealed other dimensions of presented skills and abilities (see Table 4.22).

They can be named consecutively as: abilities of farmer-entrepreneur, abilities of sustainable farmer (eco-farmer), and abilities of risk-taking farmer. A farmer-entrepreneur is familiar with the market of agricultural products and able to direct production for market needs. He also possesses appropriate qualifications and agricultural education. A sustainable farmer (eco-farmer) focuses on production quality and traditional products. He is also able to combine agricultural production with environmental care. The risk-taking farmer is not afraid of endeavors that carry some risk, has the ability to "wheel and deal" and knows the right people (i.e., is in a position to, and has abilities to network). Such a farmer feels comfortable within the contemporary neoliberal economic model, mostly based on risk and flexibility.

For what groups of farmers are these particular types synonymous with the "successful farmer"? The entrepreneurial type could appear as the epitome of success in agriculture to those farmers who, when given three possible answer choices, picked those indicating familiarity with the market of agricultural products, ability to direct production towards the needs of the market and combining appropriate qualifications and agricultural education. The risk-taking farmer appeared likely to succeed with those respondents who pointed to the readiness to take on risks as well as the ability to "wheel and deal" and to know the right people. A successful farmer of the sustainable type appealed to those respondents who picked at least two out of three possible choices such as: providing appropriate quality of production, ability to combine agricultural production with environmental care, and orientation toward production of traditional local products. The "peasant" type was another example of the successful farmer and those who picked it named a penchant for farm work, as well as diligence and readiness to take on hard work as the characteristic features of this type. There was also a "mixed" type where three possible choices referred to characteristics and abilities describing three different types of "successful farmers" mentioned above. The percentages of choices of particular types of farmers as those who were able to become successful in the contemporary market are presented in Table 4.23.

The data in Table 4.23 confirmed that a significant portion of respondents representing Polish farmers (42.6%) shared the conviction that success in agriculture can by determined by possession of traditional peasant assets. Quite well represented were also those who did not have a precise vision of a farmer's success on today's market (the mixed type was chosen by 37.5% of respondents). Other types, alternative to the "peasant" type, were rarely constructed. This should be recognized as positive expressions of the changes in the mentality of Polish farmers who constructed the model of the "successful farmer" as an entrepreneurial farmer and most of all as a sustainable farmer (eco-farmer).

Table 4.23. Type of farmer who could be successful in today's market, as indicated by interviewed farmers (percentage of farmers referring to that particular type, in 2017)

Type of successful farmer	%
Peasant	42.6
Entrepreneur	8.2
Environmentalist/Ecologist	7.6
Risk-taker	4.1
Mixed type	37.5
Total	100.0

Source: Own research.

4.5. Conclusions

Summarizing the reflection on family farming presented thus far, a draft portrait of Polish family agriculture in 2017 can be sketched. The conclusions could be made about some objective dimensions characterizing farms, their owners/operators, as well as the changing ways of thinking about the most important issues relating to the managed farms. The processes taking place in Poland from 1994 to 2017 resulted in an increase in the average farm area while, at the same time, a continuation of the situation known in rural sociology literature as a "disappearing middle." Simultaneously, the processes of aging and the advancement of farmers' education were noticeable.

A different set of processes could be seen as confirming the general continuation of the family character of Polish agriculture. What should be mentioned here is the predominance of the situation where the land operator is also the land owner, although there was some small reduction in this category in the period of time mentioned above. The family character of agriculture was also confirmed by other characteristics such as: 1) inheritance as a main way of becoming an owner and operator of the family farm, 2) family labor as the main type of labor resource used on farms (and its presence intensified in the period of time mentioned above), 3) the diminishing role of external, risky sources of financing (credits), 4) increased conviction about making the right choice to become a farmer.

The above processes are also accompanied by a phenomenon called the modernization of farmers' consciousness. Such modernization was expressed through increased acceptance of renting the land, rationalization of ownership of machinery and equipment, as well as the significant growth of aspirations related to the optimal farm land that would allow the farmer's family to live with "dignity." It should be stated here that the processes of modernization of farmers' consciousness are far from being clear-cut and unambiguous. The respondents still had inclinations towards the "psychosis of ownership" when it came to machinery and their opinions on communal ownership of machines and equipment needed for production changed only slightly. The dominant conviction in the consciousness of respondents was that the penchant for farm work and readiness to work hard proved to be the best way to achieve success on today's market. It should also be noted that other types of thinking also appeared, specifically of the so-called eco-farmer type, which seemed to be playing an increasing role in the context of the popular concepts of sustainable agriculture.

REFERENCES

Brandth, B. & Haugen, M. (1997). Rural women, feminism, and the politics of identity. *Sociologia Ruralis* 37(3), pp. 325–344.

Brandth, B. & Haugen, M. (1998). Breaking into a masculine discourse on women and farm forestry. *Sociologia Ruralis* 38(3), pp. 427–442.

Djurfeldt, G. (1996). Defining and operationalizing family farming from a sociological perspective. *Sociologia Ruralis* 36(3) , pp. 340–351.

Friedmann, H. (1981). *The Family Farm in Advanced Capitalism: Outline of a theory of simple commodity production* (manuscript, University of Toronto).

Gasson, R. & Errington, A. (1993). *The Farm Family Business.* CAB International.

Gorlach, K. (1995). *Chłopi, rolnicy, przedsiębiorcy: „Kłopotliwa klasa" w Polsce postkomunistycznej* [Peasants, farmers, entrepreneurs: "The awkward class" in postcommunist Poland]. Wydawnictwo Uniwersytetu Jagiellońskiego.

Gorlach, K. (2001). *Świat na progu domu. Rodzinne gospodarstwa rolne w Polsce w obliczu globalizacji* [The world in my backyard. Polish family farms in the face of globalization]. Wydawnictwo Uniwersytetu Jagiellońskiego.

Gorlach, K. (2004). *Socjologia obszarów wiejskich. Problemy i perspektywy* [Sociology of rural areas: Problems and perspectives]. Wydawnictwo Naukowe Scholar.

Gorlach, K. (2009). *W poszukiwaniu równowagi. Polskie rodzinne gospodarstwa rolne w Unii Europejskiej* [In search of balance: Polish family farms in the European Union]. Wydawnictwo Uniwersytetu Jagiellońskiego.

Gouldner, A. (1960). The norm of reciprocity: A preliminary statement. *American Sociological Review* 25(2), pp. 161–178.

Haan, H. de (1994). *In the Shadow of the Tree: Kinship, property and inheritance among farm families.* Het Spinhuis.

Kocik, L. (1976). *Przeobrażenia funkcji współczesnej rodziny wiejskiej* [Changes of the functions of contemporary rural family]. Ossolineum.

Kocik. L. (1986). *Rodzina chłopska w procesie modernizowania się wsi polskiej* [The peasant family in the process of modernization of the Polish countryside]. Wydawnictwo Uniwersytetu Jagiellońskiego.

Łukaszyński, J. (2013). Inkontrologiczna koncepcja wzajemności [Incontrological concept of reciprocity]. *Nauki Społeczne* 7(1), pp. 112–113.

Michałkowski, M. (1987). Adjustment of immigrants in Canada: Methodological possibilities and its implications. *International Migration* 25(1), pp. 21–39.

Mooney, P.H. (2004). Democratizing rural economy: Institutional friction, sustainable struggle, and the cooperative movement. *Rural Sociology* 69(1), pp. 76–98.

Ploeg, J.D. van der (2003). *The Virtual Farmer: Past, present, and future of the Dutch peasantry*. Royal Van Gorcum.

Strange, M. (1988). *Family Farming: A new economic vision*. University of Nebraska Press.

Wright, E.O. (1989). *The Debate on Classes*. Verso.

Wright, E.O. (1997). *Class Counts: Comparative studies in class analysis*. Cambridge University Press.

Think Locally, Act Globally: Polish farmers in the global era of sustainability and resilience, ed. by Krzysztof Gorlach and Zbigniew Drąg in collaboration with Anna Jastrzębiec-Witowska and David Ritter
Jagiellonian University Press, Kraków 2021, pp. 251–294
ISBN 978-83-233-4949-5
DOI: http://dxdoi.org/10.4467/K7195.199/20.20.12731

Chapter Five: Large Farms and Big Farmers in Poland in 2017

Zbigniew Drąg   https://orcid.org/0000-0002-9106-7758/

Krzysztof Gorlach   https://orcid.org/0000-0003-1578-7400/

5.1. Introductory Remarks

The current chapter is devoted to analysis of so-called large farms. This matter is one of the most crucial aspects of the reflection in this publication and of equal importance with the analyses of various aspects of farms. In the rural sociology literature, large farms are usually juxtaposed with small farms, the act of which may have different purposes depending on the contexts and the specifics of the research. In the historical perspective it has been presented that large farms are usually under the ownership of dominant class representatives (aristocrats, magnates, landowners, etc.) unlike small peasant farms. Nowadays, the juxtaposition takes the form of relations between large corporate farms and relatively small family farms. Somewhat differently, in the analysis of modernization processes in contemporary societies, numerous authors notice the increasing average farm size, which goes together with the mechanization of farm work, where physical labor is being eliminated as large farms make it possible to conduct agricultural activities with modern machinery in a larger space. This fosters the use of agricultural machines and amps up the production scale. Such a reflection consequently leads to the concept of the economy of effect, which points to larger production volume and larger farm size, which together lower production costs per product unit and lead to greater profitability.

The issue of farm size reveals the relative nature of what can be recognized as a "large," "medium," or "small" farm. Common knowledge indicates that a 20–30 ha farm can be recognized in Poland as a large farm, while in the United States it would be rather small, as large farms there encompass hundreds of hectares. The average size of a farm in any given country can be a good indicator of the relativity of the terms "small farm" and "large farm." The data on average farm sizes in the European Union can provide some interesting information in this regard (Table 5.1.1).

Table 5.1.1. Average farmsize in selected member-states of the European Union in 2010

Country	Average farm size in ha
Czech Republic	152.4
France	53.9
Germany	55.8
Poland	9.6
Romania	3.5
Hungary	8.1
Great Britain	70.8
Italy	7.9
European Union (27 members)	13.8

Source: Own work, based on the data from Statistics Poland (Poczta, 2013, p. 23 [Table 7]).

By devoting one chapter of the monograph to an examination of large farms in Poland, the authors want to show how owning and operating a large piece of land might influence other characteristics of these farms, as well as the strategies and activities of those engaged in operating them. In this context, the demarcation line between large and small (remaining) farms in terms of area seems to be a problem. For the reflection in this publication—and this chapter in particular—the statistical line between large and small farm was set at 15 ha, which was 50% above the average farm size in Poland, as estimated by Statistics Poland (aka Central Office of Statistics) based on the 2010 Agricultural Census (see Table 5.1.1).

According to Statistics Poland, in 2016 there were 1,383,933 individual farms with areas exceeding 1 ha.[1] Taking into consideration ownership of utilized agricultural land, and distinguishing 7 area groups, the number of farms that qualified for the

---

[1] http://bdl.stat.gov.pl/BDL/dane/podgrupa/tablica, Kategoria: Rolnictwo, leśnictwo i łowiectwo;- Grupa: Gospodarstwa rolne; Podgrupa: Gospodarstwa rolne wg grup obszarowych użytków rolnych – nowa definicja gospodarstwa rolnego; Wymiary: grupy obszarowe, rodzaj własności (gospodarstwo indywidualne), lata (2016) (access: August 2, 2019).

three largest area groups and exceeded the size of 15 ha was estimated at 200,412.[2] These farms comprised almost 14.5% of all individual farms sized over 1 ha. In this chapter, attention will be given to such farms and the farmers managing them. They are not the most numerous among Polish family farms but, due to their size and production potential, they can play the same role as farms in Western countries. What characterizes large farms in Poland and Polish large-scale farmers? Are these farms more mechanized, more specialized, and more commodity-oriented, and thus more profitable than other family farms with a smaller area of utilized agricultural land? Is their employment structure different? Do their managers fall into the category of middle-class entrepreneurs or the category of wealthy peasants? Finding answers to these questions might be easier with an examination of the basic results of the 2017 study. The opinions of the interviewed presidents of agricultural chambers regarding the development prospects for Polish agriculture might be particularly helpful here. These opinions were acquired in the summer of 2019.

Considering the size of the utilized agricultural land, rather than the ownership (i.e., regardless of whether the land was owned or leased) in the study sample of 3551 farms, there were 624 farms (17.6%) operating farmland of at least 15 ha. These farms are the subject of analysis in this chapter. It should be added that in the study 76.6% of the managers of farms larger than 15 ha had at least 15 ha of owned land and the remaining 23.4% of study participants owned less than 15 ha. The fact that the latter study participants managed farms classified within the group of large farms was a consequence of farmers leasing farm land from (an)other owner(s), in addition to owning small pieces of land. The average size of the large farm in the studied sample was 36.5 ha, and the average size of the remaining farms was 5.1 ha. It is worth mentioning that only three farmers participating in the study with over 15 ha of owned land rented part of their farmland to others and, as a consequence, they used less than 15 ha of agricultural land and their farms were not classified as large farms, but as "other" farms (Table 5.1.2).

---

[2]   The seven farm area groups are the following: over 1 ha but smaller than 2 ha, over 2 ha up to 5 ha, over 5 ha up to 10 ha, over 10 ha up to 15 ha, over 15 ha up to 20 ha, over 20 up to 50 ha, and over 50 ha.

Table 5.1.2. Farms according to the area utilized and owned agricultural land (in %)

Farms	Total (N = 3551)	Farms with owned land comprising	
		15 ha and above (N = 480)	less than 15 ha (N = 3071)
Large farm, utilizing more than 15 ha of agricultural land (N = 624)	17.6	76.6	23.4
Remaining farms, utilizing less than 15 ha of agricultural land (N = 2927)	82.4	0.1	99.9
Total (N = 3551)	100.0	13.5	86.5

Source: Own research.

5.2. Users of Large Family Farms in Poland and Their Households

Before going into the characteristics of large family farms in Poland, it might be worthwhile to ask who is managing these farms, and whether their users are any different than the users of the remaining farms. A closer look at the territorial distribution of farms might be helpful here.

Almost half of large family farms can be found in two regions of Poland: Eastern Region and Central Region. The same regions also have the greatest number of the remaining farms. There is also a significant number of large farms in the Northern Region and Northwestern Region, where the number of small (remaining) farms is the lowest in the country. The Southwestern Region has the lowest number of large farms, as well as the lowest number of small farms, while in the Southern Region small farms prevail (Table 5.2.1).[3] Considering farm structure in particular regions, large farms are most common in the northwestern part of the country, in the so--called Recovered Territories that Poland was able to get back from Germany after World War II. In this area, state collective farms were established in the time of the People's Republic of Poland (1945–1989) and functioned in the state-run economy of the previous political system. The territorial structure of farms shaped during the time of the People's Republic of Poland still plays a decisive role in the territorial location of large farms in contemporary Poland. In the Northern Region, large farms make up 34.2% of all active farms over 1 ha, while in the Northwestern Region they comprise 30.3% of all farms, and in the Southwestern Region—22.7%. Their percentage is smaller in the Central Region (17.2%) and in the Eastern Region (12.6%).

---

[3] Detailed territorial characteristic of all regions at the NUTS 1 level was presented in the Chapter 2, footnote 2.

It is the smallest in the Southern Region (6.2%). In other words, farmers operating large farms are most common in the Eastern and Central Regions. However, from the perspective of particular regions, the largest farms are most common in the Northern and Northwestern regions. Graphic 5.1 addresses the first situation and Picture 5.2 addresses the second.

Table 5.2.1. Family farms and the characteristics of their users (in %)

Variable	Farms	
	large	other
Region*		
Eastern	24.2	35.9
Central	23.6	24.1
Northern	20.2	8.3
Northwestern	20.0	9.8
Southwestern	6.7	4.9
Southern	5.3	17.0
Total	100.0	100.0
Sex/gender*		
Women	18.6	37.0
Men	81.4	63.0
Total	100.0	100.0
Age*		
18–34	18.6	8.8
35–54	55.0	50.1
55 and above	26.4	41.1
Total	100.0	100.0
Average age*	46.3	51.3
Education*		
Grammar school	7.3	14.9
Vocational	41.5	42.3
High school	37.4	32.2
Beyond high school	13.9	10.6
Total	100.0	100.0
Percentage of respondents with agricultural education*	52.4	22.5
Percentage of respondents who participated in agricultural courses in the last three years*	47.7	14.1
Percentage of respondents using computers on the farm*	16.1	2.9
Percentage of respondents with knowledge of any foreign language	38.9	35.5

* $p < 0.01$.

Source: Own research.

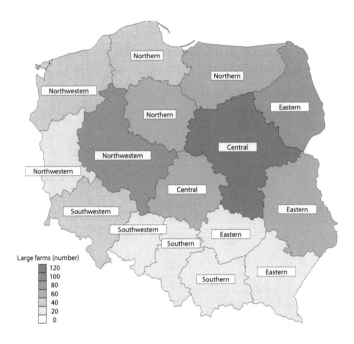

Picture 5.1. Number of large farms in provinces of particular regions. N = 624

Source: Own work.

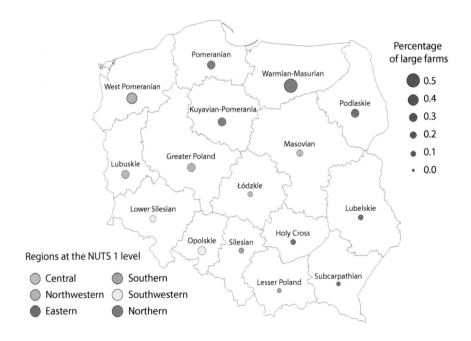

Picture 5.2. Percentage of large farms in provinces of particular regions. N = 624

Source: Own work.

Eight out of ten managers of large farms were male. For the small farms, the percentage of male managers was not quite two thirds. Generally, the managers of large farms were also younger and better educated than the rest. Their average age was 46.3 and more than half of them had at least a high school education. The average age for the rest of the surveyed farm users/operators was 51.3 and 43% of them were at least high school graduates. Over half of the managers of large farms had an agricultural education. Over half of the large farm managers participated in agricultural courses in the last three years to enhance their agricultural education. Furthermore, a notable percentage (16.1%) of managers of large farms used computers in their work. The managers of the remaining farms could not measure up to these characteristics. Less than one fourth of them had an agricultural education and only 14.1% of them participated in agricultural courses. Only 2.9% of study participants who managed non-large farms used computers in farm management (Table 5.2.1). Referring to the concept of the three forms of cultural capital by Pierre Bourdieu (1984, 1986), it can be stated that farmers managing large farms are much better equipped with institutional capital than other farmers. Are they different in term of embodied cultural capital and objectified capital? Analysing the situation in farm households and in their surrounding social context might help with answering this question.

Households of the study participants who managed large farms had more people than the households of small farm managers. The former were more likely to live in multigenerational families and rarely alone. They were more often married or had stable partners, as well as children, in comparison to the study participants who utilized and managed the remaining smaller farms (Table 5.2.2). Generally, it should be noted that farmers with large farms were closer to their families and more embedded in family networks than the farmers who operated smaller farms. In this case, the term family is being used in the traditional sense. Perhaps this has an impact on the development of embodied capital due to the influences of the older generation and spouses. Embodied capital can also be formed by the professional situation of household members. A lack of lasting and stable social and professional contacts with off-farm circles (which would be the easiest contacts to maintain through off--farm employment) as well as a lack of customs and habits transferred to the family farm environment from other social groups are responsible for the petrification of social relations and resistance from the respondents who managed the large farms to any modifications in that regard. Such petrification can also be linked to the lack of occasions to confront behaviors and the ways of thinking that are characteristic of being embedded on the farm with those of the outside world. Could the managers of large farms be less or more isolated from other professional and social circles than farmers managing the remaining farms?

Table 5.2.2. Situation of study participant households

Household situation	Farms	
	large	other
Family situation		
Average number of household members*	3.9	3.5
Percentage of respondents living in multigenerational families*	31.4	25.7
Percentage of respondents living in marriage/informal long-term stable relationship	81.9	76.3
Percentage of respondents having children	81.5	83.9
Percentage of respondents living alone*	4.2	6.7
Professional situation		
Percentage of study participants who utilized farmland and worked off-farm*	19.1	43.2
Percentage of households with at least one household member working off-farm*	31.7	55.7
Percentage of respondents having a farmer as a spouse*	64.4	35.4
Average percent of annual household income acquired from the farm including EU subsidies (and without subsidies)*	83.3 (52.9)	37.4 (22.9)
Material situation		
Number of rooms in the house*	4.90	4.16
Number of rooms per person in the house*	1.46	1.42
Percentage of households having a passenger car*	97.40	89.00
Percentage of households having two passenger cars*	61.60	40.40
Percentage of households having a delivery car or truck*	21.50	11.90
Percentage of households having a computer with Internet*	89.50	78.30
Percentage of households where, in the last 12 months, there had been a shortage of funds to buy needed food or clothing*	4.40	8.60
Percentage of households where, in the last 12 months, there had been a shortage of funds to cover living expenses (rent, electricity, water, gas, etc.)**	5.70	8.40
Percentage of households where, in the last 12 months, there had been a shortage of funds to cover health expenses and medical care of household members*	3.40	8.40

* $p < 0.01$; ** $p < 0.05$.

Source: Own research.

Among study participants who utilized large farms, one fifth worked off-farm. For the farmers participating in the study who managed smaller farms, as many as 43.2% worked outside of agriculture. The percentage of large farms with household members working off-farm was 31.7%, while for the small farms this percentage was 55.7%. Almost two thirds of the spouses of study participants managing large farms also worked on the farm. For the remaining farms participating in the study, just one third of the spouses of the managers also worked there. What seems to come to light is the stronger socio-professional isolation of families managing large farms in comparison to those managing smaller farms. Can this differentiate the managers of small and large farms to a significant extent, in terms of embodied capital? This is likely one of the factors influencing the resources of embodied capital. Before more detailed analysis, it should be noted, and it may even appear obvious, that the financial and material situation of households connected with a large farm is determined to a much greater extent by the income acquired from agricultural activities in comparison to households connected with small farms. Considering the annual income of the former, 83.3% comes from the farm (including the EU subsidies), on average, while, for the latter, only 37.4% of their income, on average, originated on the farm (Table 5.2.2). Generally, the overall situation of large farms appeared to be better than the situation of the remaining farms. In the households connected with large farms the average number of rooms in the house was 4.90. For the households connected with small farms, it was 4.16. The number of rooms per person in the home was still larger for households connected with large farms than for small farm households, but this disproportion was not very significant—1.46 to 1.42. Over 97% of large farm households had at least one car. Almost 62% of them had two cars and almost 22% of large farm households had a delivery car or truck. For the small farm households, the respective percentages were as follows: 89%, 40%, and 12%. Some differences were also noticeable in the farm households' access to a computer with Internet. In the case of the small farm household members, the percentage was 78%. Generally, on both types of farms, the scale of poverty is rather minor and the poverty is usually temporary. Still, it is somewhat smaller in households connected to large farms. In the last year (from the moment of conducting the study) the situation of having had insufficient funds for food, clothing, rent, living, and healthcare expenses applied to 3–6% of large farms and a bit over 8% of small farms (Table 5.2.2). If one wishes to consider characteristics of material situation as indicators of objectified capital, it should be stated that large farms had greater resources of such capital, although most of their income came from agriculture. The remaining farms got the majority of their income from off-farm work.

Table 5.2.3. Family traditions and ways to acquire farm by farm users (in %)

Variable	Large farm users	Users of other farms
Means of acquiring the farm		
Inheritance	83.5	82.2
Marriage	6.9	8.7
Purchase	7.1	6.8
Lease	2.5	2.3
Total	100.0	100.0
Family traditions*		
Both parents were farmers	80.0	67.6
One parent was a farmer	8.8	18.7
No parent was a farmer	11.2	13.7
Total	100.0	100.0

* $p < 0.01$.

Note: Inheritance means taking over the farm from parents or relatives.

Source: Own research.

Returning to the issue of embodied capital, and to the earlier statement that greater professional isolation of large farm managers, their spouses, and other household members could be a contributing factor to stronger petrification of their schemas of thinking and behaviors than is the case with smaller farm managers. However, the social circles where managers of large and small farms currently function might modify their cultural habits to some extent. These circles might not be able to constitute them as this is the domain of family circles, especially the ones in which the farms have been functioning for some time. This might also be the case with acquired institutional capital. A closer look at the origins of the managers of large and small farms participating in the study, and the analysis of the ways in which they acquired their farms, as well as the family farming traditions the managers came from, might also be helpful here.

There were no significant differences between both categories of farmers in the ways in which they acquired their farms. Over 80% of the study participants in both categories took possession of their farm via inheritance. However, there was an important difference in family professional traditions between the participants managing large farms and those who managed small farms. As many as 80% of large farm mangers came from families where both parents were farmers. Within the category of the remaining farm managers participating in the study, 68% came from families with two farmer parents (Table 5.2.3).

Could the strong roots in the family agricultural tradition of large farm operators contribute to their greater isolation from other social circles in the past and presently in comparison to users of smaller farms? Could such traditions differentiate these two categories in terms of embodied institutional capital at present? Could the fact that large farm managers have greater assets of institutional capital be meaningful here? It might be worth analyzing whether the uneven distribution of such capital deepened the differences between large and small farmers even more, or perhaps it had no bearing on their schema of thinking, conduct, and behaviors in that both categories of farm users were similar in this regard?

Analysis of several indicators of embodied capital resources might help with answering the above questions. The study participants' definitions of their own socioeconomic situation pointing to the character of their agricultural identities were chosen as a starting point of such analysis. A set of 10 statements was presented to the respondents with the request for their evaluation. The full texts of these questions as well as the distribution of answers evaluating them on a 5-point scale, from 1 = "I strongly agree" to 5 = "I strongly disagree" can be found in Table 5.2.4. Factor analysis conducted on the collected data distinguished three dimensions of farmer identities. For the purpose of this chapter these identities were defined as "farmer by choice," "farmer by coercion," and "farmer by tradition" (Table 5.2.5).

Table 5.2.4. Farmers' identities: Farmers' opinions on three conditions of their professional and social situation. Percentage of "yes" answers ("I strongly agree" answers combined with "I somewhat agree") and "no" answers ("I rather disagree" together with "I strongly disagree") as well as averages and level of significance (all issues were evaluated on a 5-point scale, from 1 = "I strongly agree" to 5 = "I strongly disagree")

Variable	Large farmers			Other farmers			Significance
	Yes	No	Average	Yes	No	Average	
Selling: Today, in the current situation, the best thing to do is sell the farm, put the transaction money in the bank or invest in some business and make a living from it	7.8	82.9	1.7	13.7	67.8	2.2	0.01
All over again: If I had the possibility to start my professional life all over, I would still choose to be a farmer	60.5	20.0	3.7	36.1	39.3	2.0	0.01
On one's own: In today's world being a farm owner means foremost the sense of working on one's own and having opportunities to realize one's vision. It is also a responsibility concerning the fate and fortunes of the farm and the necessity to plan ahead, to strive towards development and to increase the effort	84.6	3.1	4.2	80.3	8.9	3.8	0.01
Possession: Today, one can consider himself/ herself a farmer if the farm was inherited from parents. Having farm land, machines, and livestock can foster the sense of security and independence	65.5	18.0	3.7	63.2	14.6	3.7	none
Slave: Currently being a farm owner is tied with feeling of inferiority, hard and hopeless work, and being one's own slave	32.0	52.9	2.7	45.6	33.9	3.2	0.01
Inheritance: I became a farmer because I received the land as an inheritance from my parents	72.1	19.4	3.9	77.3	12.9	4.0	0.01
Affiliation: I wanted to be a farmer because I liked the work, and I had an affection for the farm land	74.2	12.1	3.9	50.4	26.9	3.3	0.01
Risk: I wanted to become a farmer because this profession provides opportunities to work on one's own and requires independence, decision-making, and risk-taking	72.1	9.4	3.9	48.1	26.5	3.3	0.01
Parents: I did not want to become a farmer but there was nobody else to work on our farm and my parents pushed for it	21.4	65.3	2.3	37.1	43.7	2.9	0.01
Unemployment: I did not want to become a farmer, but I could not get any other job, so I stayed on the farm	7.2	82.5	1.8	13.9	78.3	2.2	0.01

Source: Own research.

The identity of the "farmer by choice" consists first and foremost of an affiliation to the farm land and the conviction that the farming profession gives one the opportunity to work independently for one's self. The awareness that a farmer's job requires independence in decision-making and risk-taking could also be seen as an important part of this identity. No less important is the certitude that the person would choose farming again in the situation of having to start their professional life all over, without the parental pressure to choose farming. "Farmer by coercion" is defined as taking work in agriculture due to the inability to find work elsewhere, as well as parental pressure to stay on the farm. This identity is tied up with the conviction that being the farm owner could mean hard, hopeless work, being one's own slave, and having feelings of inferiority. There might be an accompanying thought that in the contemporary situation it would be best to sell the farm, put the money from the sale in the bank, or invest in some business that one could live off of. The conviction that working on one's own farm could foster opportunities for realization of one's potential which, in return, could bring about a sense of responsibility for the fate and fortunes of the farm, is absent. For the identity of "farmer by tradition," the most important components have to do with inheriting the farm from the parents and the conviction that ownership of farm land, farm animals, and agricultural machines can give one a sense of safety and independence, as well as the feeling that the realization of one's ideas is possible. Consequently, one feels responsible for the fate and fortunes of the farm, planning its development, and finding motivation for the greater effort. Although, these identities are independent from each other, some of their elements are common. Identities of "farmer by tradition" and "farmer by coercion" share the decision about taking up farm work because of parental pressure and receiving an inheritance from them. What "farmer by choice" and "farmer by tradition" have in common is the idea that to work on one's own allows for fulfilment of one's ideas and that consequently awakens a sense of responsibility for the fate and fortunes of the farm, as well as a necessity to plan farm development through greater effort.

The placement of both categories of farm managers/users on particular scales of identity is presented in Table 5.2.6. Managers of large farms are situated lower than the remaining farmers on the "farmer by coercion" scale and higher on the "farmer by choice" scale. There is no significant difference in placement of the two categories of farmers on the "farmer by tradition" scale. The distance between these categories is very small—0.046. This means that the identities of both these categories of farmers are well rooted in tradition, but the identities of large farmers are marked more positively as "farmers by choice" and less negatively as "farmers by coercion." Could this warrant the conclusion that, despite the equally strong mental roots in the traditional agricultural environment of both categories of farm managers/users, the managers of large farms have more of a propensity for modern thinking, characteristic of entrepreneurs, than that found among the remaining farm managers. Could the large farm managers also be less likely to display any characteristics of the classic peasant mentality?

Table 5.2.5. Factor loadings of main components of farmers' identities (Varimax rotation, loadings below 0.20 not considered)

Variable	Components		
	I: Farmer by choice	II: Farmer by coercion	III: Farmer by tradition
Affiliation	0.907		
Risk	0.884		
All over again	0.725		
Parents	−0.525	0.483	0.314
Unemployment		0.657	
Slave		0.647	
Selling		0.533	−0.288
Possession			0.702
Inheritance		0.220	0.643
On one's own	0.225	−0.446	0.564
Percentage of explained variance	25.1	16.6	14.7
Total percentage of explained variance	56.4		

Source: Own research.

Table 5.2.6. Average factor values of main farmers' identity components according to the category of farm manager/user

Category of farm manager/user	Identity scale								
	farmer by choice			farmer by coercion			farmer by tradition		
	average	significance	eta²	average	significance	eta²	average	significance	eta²
Large farmers	0.427	0.01	0.039	−0.450	0.01	0.043	0.036	none	0.000
Remaining farmers	−0.091			0.096			−0.008		

Source: Own research.

The opinions of both categories of farm managers/users on farm rentals and farm equipment are worth analyzing, as they can be treated as indicators of peasant thinking, as opposed to a modern approach to farming. The data on the study participants' attitudes to farm rentals are presented in Table 5.2.7. They confirm that large farm managers were characterized by a more modern way of thinking than the managers of the remaining farms. The opinion that renting of the farm could be a good solution was more common among the respondents in charge of bigger farms, as they saw it as providing more opportunities for using the already owned machines and increasing production. Unlike the small farm managers, the large farmers often rejected the conviction about farm renting not being a good solution due to a tenant's possible development of a strong dependency on the conditions of the contract and the increased risk of a tenant's falling into debt and subsequently lacking the motivation to take good care of the land.

Table 5.2.7. Attitudes towards farm rental. Percentage of "yes" answers ("I strongly agree" answers combined with "I somewhat agree") and "no" answers ("I rather disagree" together with "I strongly disagree") as well as averages and level of significance. All issues were evaluated on a 5-point scale, where 1 = "I strongly agree" to 5 = "I strongly disagree"

Statement	Large farmers			Other farmers		
	Yes	No	Average	Yes	No	Average
Renting the farm is a good solution as it provides greater opportunities for the effective use of the currently owned machines and to increase the production*	80.2	8.5	1.9	74.8	9.5	2.1
Renting the farm is not a good solution as the tenant may become too dependent on the conditions of the contracts, as well as easily fall into debt and then may have no motivation to take proper care of the land*	31.1	55.2	3.4	37.4	38.6	3.0

* $p < 0.01$.

Source: Own research.

Table 5.2.8. Farmers' opinions on machine ownership (in %)

Categories of answers	Large farmers	Other farmers
Farmer should own all the machines even if he only uses them from time to time	48.8	47.6
Farmer should only own those machines that he/she uses regularly in everyday work and the remaining machines should be purchased and used together collectively with other farmers	24.8	29.1
Farmer should only own those machines that he/she uses regularly in everyday work and other machines that are not used regularly should be rented from other farmers	24.5	21.1
Farmer should not own any machines and only contract out machine services	1.8	2.2
Total	100.0	100.0

Source: Own research.

The expected, more entrepreneurial, attitude of the study participants who managed large farms was not confirmed by data on the preferred model of supplying farms with agricultural machines. Almost half of the respondents from both categories of farmers were in favor of having their own machines, even if they reported using them only from time to time. This was clearly indicative of an irrational peasant mentality. It should be added, however, that slightly more than half of the respondents in both categories prefer more rational solutions supporting ownership of only those machines that are used regularly. As for the other machines, these respondents either proposed joint purchase and sharing with other farmers or renting the farm equipment from other farmers. There were also opinions that farmers should not own any machines and hire specialized firms to conduct farm work that requires use of the machines (Table 5.2.8).

Juxtaposing the peasant way of thinking with the modern approach to farming based on certain opinions of the study participants regarding paid labor of household members working with other farmers, as well as on borrowing agricultural machines from others when their own machines are defective, might be an interesting exercise. It is quite well known that the peasant tradition and a certain sense of peasant honor were against paid work on other people's farms but accepted the possibility of borrowing farm machines from them. Such a dichotomy might be considered contrary to the modern way of thinking.

Data on these issues did not provide an unambiguous answer to which category of farmers participating in the study displayed a more rational way of thinking and which was more rooted in tradition and the peasant mentality. In both categories, respondents who were not against the paid work of their household members on other farms were in the majority, but their percentage was higher in the small farm manager category. Therefore, considering only this one issue, it could be stated that the peasant way of thinking is more deeply rooted in the category of large farm owners. Similarly, in both categories of farmers, those who would rather borrow a machine from neighbors than use the services of a specialized company in the event of their own equipment malfunctioning were in the majority. However, the percentage of respondents with such attitudes was higher in the category of small farm managers, which could mean that solely considering this one particular issue, their approach to farming was more traditional. The managers of large farms displayed more rationality and more modern thinking in this regard (Table 5.2.9).

Table 5.2.9. Opinions on household members doing paid work with/for other farmers and on borrowing farm equipment from other farmers. Percentage of "yes" answers ("I strongly agree" answers combined with "I somewhat agree") and "no" answers ("I rather disagree" together with "I strongly disagree") as well as averages and level of significance. All issues were evaluated on a 5-point scale, from 1 = "I strongly agree" to 5 = "I strongly disagree"

Statement	Large farmers			Other farmers		
	Yes	No	Average	Yes	No	Average
I have nothing against my closest relatives working for money on someone else's farm*	43.5	37.4	3.0	54.2	27.4	2.7
If my farm equipment broke down, I would borrow it from another farmer, or I would use the service of some specialized company*	32.4	52.7	3.3	23.6	61.9	3.6

* p < 0.01.

Source: Own research.

Table 5.2.10. Farmer as an entrepreneur. Similarities and differences between farmers and other entrepreneurs. Percentage of "yes" answers ("I strongly agree" answers combined with "I somewhat agree") and "no" answers ("I rather disagree" answers together with "I strongly disagree") as well as averages and level of significance. All issues were evaluated on a 5-point scale, from 2 = "I strongly agree," to 5 = "I strongly disagree"

Statement	Large farmers			Other farmers		
	Yes	No	Average	Yes	No	Average
Currently the situation of the farmer is not different from that of any other entrepreneur. Each of them works for their own record, has to take risks and be able to sell their own products*	75.2	15.3	2.1	66.8	18.8	2.3
In today's reality the farmer has to work more than the owner of any other business and the farmer's income is still lower, as well as the social appreciation for the profession*	73.4	11.2	2.0	79.9	7.9	1.9

* p < 0.01.

Source: Own research.

The differences between the two categories of respondents in terms of institutional and objectified capital, family tradition, the extent of socio-professional isolation, or in regards to social identities, did not directly result in respective differences of embodied capital in the dimension of thought schemas. It turned out that in some matters the thought schemas characteristic of modern entrepreneurs were more common among large farm managers. The issues likely to bring about modern opinions were the following: leasing of the farm and the preferred solution when one's own machinery was in disrepair. Other issues, such as paid work of household members at someone else's farm, revealed a tendency for modern thinking among the users/managers of smaller farms. However, when it came to the preferred model for supplying a farm with agricultural machinery, both categories of farmers displayed modern thinking equally. In conclusion, it seems that the managers of large farms were more in opposition to the category of traditional peasants and more likely to identify themselves as modern entrepreneurs than the managers of the remaining farms. This observation is based on the higher percentage of large farm managers holding the opinion that a similar situation exists between the farmer and other entrepreneurs. Consequently, this category of farmers noted fewer differences between farmers and business managers. The conviction that the situation of the farmers and the situation of other types of entrepreneurs did not differ—as each of them work on their own, take risks, and must have an aptitude for selling their own products—was shared by three fourths of large farm managers, but just two thirds of the remaining research participants. The opinion that, in the current time, farmers have to work much harder than other business enterprise owners yet still earn lower incomes, was expressed by 75% of the large farm managers and 80% of the remaining farmers participating in the study (Table 5.2.10). Should it therefore be expected that the large farm managers, who—more than the remaining farmers—identified themselves with modern entrepreneurs, would also adopt their lifestyle? In other words, could any significant differences be seen in the schemas of behaviours reported by the two categories of respondents that would indicate that one of them was more rooted in peasant tradition, with its characteristic habits while the other was more open to modernity? Or perhaps, similarly to the schemas of thinking described above, both categories would present some ambiguity in their behaviors?

In the study, the participants were given a set of 16 activities and were asked to report how often they performed them and their answers were placed on a 7-point scale, from 1 = "I do it at least several times a week", to 7 = "I never do it." Table 5.2.11 presents the distribution of all activities, as well as factor analysis. Based on this scale, four dimensions of lifestyle were distinguished, and they were designated as "higher culture style," "Internet style," "traditional style," and "city style." The term "higher culture style" involved listening to classical music, reading books, going to theatre to see plays and performances, as well as recreational sport practice, and devoting time to hobbies. The components of "Internet style" included checking the prices of agricultural products on commodity stock exchanges, checking the stock exchange markets, as well as keeping track of local news via the Internet and using e-mail. "Traditional style" was strongly connected with hosting relatives and friends at home, talking to neighbors in front of the house, visiting family members, and attending church services. For the "city style" the most characteristic activities were dining at restaurants for business purposes or eating out with family, and not attending church services. Despite being independent from each other, the distinguished dimensions had common components. The "higher culture style," the "Internet style" and "city style" were more or less tied to each other through checking the local news on the Internet, using e-mail, and dining in restaurants for professional purposes. The "city style" and the "higher culture style" had recreational sports practices in common. Both of these styles, as well as the "traditional style" placed emphasis on spending time on one's hobby.

Which styles could be more characteristic of large farm managers and which ones might be appealing to small farm managers? Placing the four types of lifestyles on particular scales for both categories of research participants can be seen in Table 5.2.12. As indicated by the statistical results, the managers of large farms were situated lower on the "Internet style" and "city style" scales, as these styles characterized them better than they did other research participants. For the managers of the remaining farms, styles such as "higher culture style" and "traditional style" were more characteristic and these respondents were situated lower on these scales. However, the distances between the placements of both categories of farmers on particular scales were rather small and it made it difficult to determine whether the cultivation of any given style corresponded to a particular category of respondents. These distances were the following: 0.127 for "higher culture style," 0.101 for the "traditional style" and 0.102 for the "city style." The only exception was the "Internet style." The distance between these two categories on the "Internet style" scale was 0.599 indicating that using various Internet sites was the factor that most differentiated the managers of large farms from users of the remaining farms. This supported the assumption regarding their greater openness to modernity and the ascertainment that in the dimension of behavior schemas they are closer to entrepreneurs and further from traditional peasants than the small farm users are.

It is worth analyzing what following a particular lifestyle could mean in practice and to what extent each category of farm users realized the given lifestyle. It can be assumed that reading books at least once a year, going to the theatre at least once a year, listening to classical music at least once a month, as well as recreational practice of some sport at least once a year and spending time on one's hobby at least once a week constituted participation in the "higher culture." For the "Internet lifestyle" the indicators were the following: using e-mail at least once a week, checking the prices of agricultural products on commodity exchange markets at least once a week, checking local news at least once a week, and checking the stock market at least once a month. The frequency of engaging in activities constituting the "higher culture style" was about the same for both categories of farm users, which showed that this dimension was not differentiating for the farmers participating in the research. A quite different situation could be seen with the "Internet style." Large farm users stated that they checked prices on commodity market exchanges at least once a week and those on the stock exchange at least once a month, twice as often as users of small farms. It should be added that almost half of large farm users checked the local news on the Internet at least once a week, while only one third of the remaining farm managers did. A similar disproportion between the two groups of users could be seen in the use of electronic mail: 44% of large farm managers and not quite 29% of the managers of the remaining farms used e-mail at least once a week (Table 5.2.13). As was mentioned earlier, this type of lifestyle distinguishes the two categories of farm users the most.

Table 5.2.11. Factor loadings of main components of farmers' lifestyles (Varimax rotation, with loadings below 0.20 not considered). The study participants described the activities they engaged in (variable) on an incremental scale from 1 to 7, where 1 = doing something several times a week and 7 = never doing it

Variable (activities)	Components			
	I: Higher culture style	II: Internet style	III: Traditional style	IV: City style
Listening to classical music	0.733			
Reading books	0.723			
Attending theater plays	0.679			
Recreational participation in sports	0.526			0.317
Devoting time to hobbies	0.449		0.305	0.216
Checking the prices of agricultural products on commodity market exchanges		0.763		
Keeping track of the stock exchange		0.754		
Perusing local news on the Internet	0.310	0.679		0.236
Using electronic mail (e-mail)	0.377	0.576		0.324
Hosting friends and relatives at home			0.736	
Talking to neighbors in front of the house			0.696	
Visiting relatives			0.672	
Attending church services			0.417	−0.403
Dining at restaurants for social purposes				0.631
Going out to restaurants for a meal with a family				0.456
Dining at restaurants for professional purposes	0.339	0.221		0.402
Percentage of explained variance	15.000	12.900	11.400	8.200
Total percent of explained variance	47.5			

Source: Own research.

Table 5.2.12. Average factor values of main components of farmers' lifestyles, according to the category of farm user

Category of farm user	Lifestyles scale							
	I: Higher culture style		II: Internet style		III: Traditional style		IV: City style	
	average	significance /eta²	average	significance /eta²	average	significance /eta²	average	significance /eta²
Large farmers	0.105	0.01/0.002	−0.494	0.01/0.052	0.083	0.05/0.001	−0.084	0.05/0.002
Other farmers	−0.022		0.105		−0.018		0.018	

Source: Own research.

The weekly activities seen as constituting the "traditional style" included: talking to neighbors in front of the house, attending church services, visiting family members, and hosting friends at home. For the "city style" important activities entailed dining out with one's family at least once a year or having meetings in restaurants at least once a year for social and professional purposes. The activities characteristic of the "traditional style" were not significantly differentiating but small farm managers engaged in them a bit more frequently, with the exception of hosting friends in the home. Activities constituting the "city style" were more differentiating. Large farm users more often declared eating out with families in restaurants and meeting other people for social and professional purposes at least once a year than the small farm users did (Table 5.2.14). For purposes of the study, these types of activities, along with Internet use, could therefore be recognized as the most characteristic of modern entrepreneurs and as another factor determining their distinctiveness from the remaining farm users, in terms of lifestyle.

Table 5.2.13. Lifestyles' dimensions: "higher culture style" (I) and "Internet style" (II). Percentage of respondents engaged in the following activities

Activity	Large farmers	Other farmers
I. Listens to classical music at least once a month	11.7	12.2
I. Read books in the last year	57.3	56.1
I. Was in a theatre in the last year**	17.1	15.2
I. Practiced any sports in the last year	28.8	26.8
I. Spends time on a hobby at least once a week	25.2	26.2
II. Checks prices of agricultural products on commodity market exchanges at least once a week*	36.4	18.8
II. Checks stock market at least once a month*	22.8	11.4
II. Peruses local news on the Internet at least once a week*	49.3	33.4
II. Uses electronic mail (e-mail) at least once a week*	43.7	28.7

* $p < 0.01$; ** $p < 0.05$.

Source: Own research.

Table 5.2.14. Lifestyles' dimensions: "traditional style" (III) and "city style" (IV). Percentage of respondents engaged in the following activities

Activity	Large farmers	Other farmers
III. Hosts friends, relatives and acquaintances at least once a week	24.2	24.1
III. Talks to neighbors in front of the house at least once a week	62.6	65.1
III. Visits relatives at least once a week	25.1	28.1
III. Attends religious services at least once a month**	61.0	65.5
IV. Dined in restaurants for social purposes in the last year *	39.3	32.4
IV. Dined in a restaurant with one's family in the last year*	44.5	34.2
IV. Scheduled professional meetings in restaurants in the last year *	19.6	10.9

* p < 0.01; ** p < 0.05.

Source: Own research.

Besides analyzing the above four lifestyle dimensions, two other matters could be considered meaningful in the confrontation of behavioral schemas differentiating the modern entrepreneur and the traditional peasant. The first one pertains to attending to one's health and appearance and the second addresses social activity. Could any of these two issues or both of them be more important to one category of farm users than another, the way it would be more important to modern entrepreneurs compared to traditional peasant?

The attitudes toward a healthy lifestyle were studied with the application of a tool consisting of 6 questions presented in Table 5.2.15. In the case of the first two questions, the respondents were asked if they agreed with particular statements, using a 5-point scale, from 1 = "I strongly agree" to 5 = "I strongly disagree" (the table contains a certain percentage of answers "I strongly agree" and "I rather agree"). With the remaining questions the respondents were asked how often they engaged in a particular activity according to a 7-point scale, starting with 1 meaning "I do it at least several times a week" and ending with 7 meaning "never."

Table 5.2.15. Healthy lifestyle. Prevention aspects (I) and practical aspects (II). Percentage of respondents engaging in the following activities

Activities	Large farmers	Other farmers
I. Could give up tasty dishes for health reasons*	59.6	64.6
I. Pays attention to what he/she eats to avoid putting on weight*	42.5	50.7
I. Undergoes a periodic medical examination at least once every few years*	85.8	90.5
II. Goes to the dentist at least several times a year*	47.7	41.7
II. Eats fish for dinner/supper at least once a week*	39.0	34.3
II. Drinks juices and eats fresh raw fruit at least once a week*	87.1	81.7

* p < 0.01.

Source: Own research.

Generally, it should be stated that a significant percentage of respondents in each category of farm managers engaged in activities that were far removed from traditional peasant habits, and this could indicate that health and appearance were quite important to them. However, there was a difference in emphasizing certain types of activities between the two categories of respondents. Managers of small farms, more often than the managers of large farms, engaged in health-related activities of a preventive nature. These activities included: scheduling a periodical medical examination at least once every few years, a willingness to give up tasty dishes for health reasons and paying attention to their food intake in order to avoid putting on weight. However, the large farm owners were more likely to engage in health maintenance activities of practical kinds such as; drinking juices and eating raw fresh fruit, eating fish at least once a week, going to the dentist at least several times a year. Despite these differences, it should be noted that the healthy lifestyle cult, although not yet deeply rooted in the consciousness of Polish farmers, has already marked its presence in their mentality. Particular elements of the healthy lifestyle motivated farm users to take up certain activities, some of them related to prevention and some of them of a practical kind. Needless to say, the image of the rural family as shabby and prone to illness has become outdated.

The last of the analyzed behavior schemes relates to social activism. Could this study confirm the assumption that activism in Polish rural areas and among Polish farmers is usually limited to informal activities, and often connected with helping the neighbor and with participation in local elections, which leaves less room for participation in the more formalized structures? To some extent, yes.

Table 5.2.16. Social activities. Percentage of respondents participating in local elections in 2014, belonging to political parties or trade unions, belonging to associations of agricultural producers and breeders, currently fulfilling some social function, fulfilling functions in local associations, and fulfilling the function of the head of the village

Forms of social activity	Large farm managers	Other farm managers
Participated in the 2014 local elections	84.6	82.6
Belongs to association of agricultural producers or breeders*	10.9	1.9
Fulfils some social functions*	20.6	10.8
Fulfils social functions in local associations*	10.9	6.4
Fulfils the function of the head of the village*	5.2	1.8

* $p < 0.01$.

Source: Own research.

Both categories of farm users declared their participation in the 2014 local election and the percentage of respondents who claimed they voted in that election was over 80%. Notably, the activism in formalized structures was much lower, showing quite significant disproportions. Large farmers turned out to be more active than the remaining farmers. Almost 11% of large farmers belonged to producer groups and almost 21% of farmers in this category fulfilled some social functions, usually of a local character (10.9%) or as head of the village (5.2%). In the category of small farm users, less than 2% declared membership in producer groups and less than 11% fulfilled some local functions, usually on the local level (6.4%). Only 1.8% of respondents in this category served as the head of the village (Table 5.2.16). As indicated by the data, the spirit of the modern entrepreneur, who is engaged in social life and interested in things that happen outside his/her own business enterprise, has slowly entered the circles of large farmers. At the same time, the users of small farms remained in the sphere of informal activism the most.

Summarizing the results of all analyses presented above, two additional questions might deserve some attention. First, could there be any differences between large and small farms in terms of possession of cultural capital? Second, to what extent do the resources of cultural capital that are in farmers' possession make them more similar to modern entrepreneurs and to what extent do these resources still keep them comparable to traditional peasants?

Large farm managers had much larger resources of institutional and objectified capital than the small farm users. The former were younger, had higher social and cultural competencies, and more grounded agricultural knowledge. Their farms also had much greater material resources. The matters could be even more complicated with the embodied capital in the context of mental schemes and schemes of behaviors. Although, large farmers more often than small farmers constructed their identities in positive categories as entrepreneurs by choice they still referred to the peasant tradition in their self-definitions just as much as small farmers did. The main difference between the two categories of farmers in this regard had to do with the tendency of small farm managers to define themselves as farmers by coercion. It would not be an exaggeration to say that both categories of farmers displayed certain aspects of peasant mentality regarding various aspects of farm functioning, which might make them similar to each other or differentiate them. Consequently, the large farm owners expressed a more positive attitude towards leasing and displayed more rationality in the situations of the machines they owned being out of order. Nevertheless, similar to small farm users, they preferred the model of a "machine park" on their farm and personal ownership of machines. Additionally, the large farm managers displayed a sense of peasant honor even stronger than that of small farm managers, if judged by their attitude towards paid work of their family members at other people's farms.

As was mentioned earlier, the schemas of thought as well as the schemas of behaviors of both categories of farm users could not be interpreted unambiguously. It should be mentioned, however, that among the large farm managers the Internet and city lifestyles were more frequent than among the remaining farm managers. The large farm managers also engaged in practices related to healthy lifestyles, while the remaining respondents reported some prevention activities. Additionally, the large farm owners engaged more in social activities. Although, all the above factors made the large farm managers more similar to modern entrepreneurs in comparison to small farm managers, their participation in the higher culture was limited and the traditional lifestyle was dominant for large and small farm managers alike.

How to explain all these layers of the peasant mentality, especially among large farm users, considering that they did not only possess the larger resources of institutionalized capital but also, more often, that other farmers considered their own situation to be quite similar to the situation of other entrepreneurs? It appears that the socio-professional isolation of large farm users was greater than the isolation of the remaining farm users. Limited contacts with other socio-professional circles and deep embedment in the traditional agricultural setting could be quite meaningful here. The large farmers, their spouses, and the remaining household members were less likely than the remaining farm users to take up work outside of agriculture, they more often lived in multi-generational families, and usually both of their parents were farmers as well. Perhaps thanks to acquired resources of the institutionalized capital, they would want to be and considered themselves to be modern entrepreneurs, just like those who conduct their activity outside of agriculture. While they don't express it in the sphere of conscious attitudes and behaviors, in the sphere of unconscious habits of everyday experience, they still remained peasants, which was reported by those in their closest environment. Nevertheless, it should be admitted that generally in the resources of the embodied capital and cultural capital that large farmers possessed, there were more elements characteristic of an openness to modernity than in the resources of the small farm users. Could the resources of cultural capital of both categories of farmers to some extent determine their development strategies? Would, therefore, the development strategies of large farms—just like their managers—differ from the remaining farm managers in terms of greater openness to modernity? If so, could this bring measurable results? To get the answer, the development strategies of large and small farms need to be analyzed further.

5.3. Large farms in Poland: Strategies implemented and strategies planned

Before analyzing the main elements of strategy implemented on the farms of both categories it should be noted that large farm managers as well as small farm managers decidedly preferred those strategies fostering sustainable and gradual farm development. Only a few respondents preferred risky strategies and they were more likely to be found among the small farm users. Sustainable strategy, ensuring that the farm would not perish, even though at first only modest income could be noted, still granted a chance for gradual increase of income. It was preferred by 88.5% of large farm managers and 83.0% of the managers of small farms. The percentage of declared followers of risky strategies with the possibility of earning lots of money in a short time with no guarantee of continued market presence of the farm and gradual development in the future was barely 6.4% among the large farm managers and 10.9% among the remaining farm managers (Table 5.3.1). It could be concluded that Polish farmers, regardless of the type of farm they managed, resemble the ideal entrepreneur—a representative of the middle class with work ethics described in the literature as Protestant (Domański, 2001), even though the issues of denomination were not of importance in the described study.

Table 5.3.1. Farm development strategies preferred by farm users (in%)*

Type of development strategy	Large farmers	Other farmers
Sustainable: guaranteeing that the farm will not perish, even though it will only bring modest income initially and in the long-term perspective an increase of income is possible	88.5	83.0
Risky: possibility of making lots of money in a short time but with no guarantee of farm survival or gradual development in the future	6.4	10.9
Hard to say	5.1	6.1
Total	100.0	100.0

* $p < 0.01$.

Source: Own research.

The main determinant of the development strategy is obviously the size of utilized agricultural land, which in Poland, for the most part, stems from the ownership structure of the agricultural land. The prevailing dominance of small and medium farms (below 15 ha) combined with restrictive legal regulations on land trade and mental barriers such as reluctance to sell the "land of the fathers" makes Polish agriculture quite unique and influences its development directions. Consequently, this has an impact on the agricultural policy of the state as well as on the development opportunities for family farms regarding their production and profitability. In a situation of limited possibilities to increase the area of owned utilized land, renting the land from other farmers seems like the most rational solution, allowing for increased production abilities of the farm. Could such a solution be more present among the development strategies of large farms rather than small farms? The data in Table 5.3.2 are quite evident.

Table 5.3.2. Farms according to the farm users (in %). N = 3551*

Type of farm	Type of user			Total
	owner (owns the entire utilized agricultural land)	tenant (rents portion of the land from farm owner)	rentier (leases out portion of the owned land to others)	
Large (15 ha and more of UAA)	43.6	55.4	1.0	17.6
Other (less than 15 ha of UAA)	84.9	13.2	1.9	82.4
Total	77.6	20.6	1.8	100.0

* $p < 0.01$.

Source: Own research.

More than half (55.4%) of large farm managers rented farm land from other farmers, while among the remaining farm managers those who rented the farm land from others made up only 13.2%. On average, the farmers rented 19.8 ha, while the latter rented 2.7 ha from others. The category of rentiers who were active farmers but rented out a portion of their farms to others was quite small for both categories with: 1.0% of large farm managers and 1.9% of remaining farm mangers. On average, the former leased 12.7 ha of land to others and the latter 2.8 ha. As a result of applied mechanisms of leasing (taking the land in lease and leasing the land to others) the area of large farms increased by 10 ha on average and the area of the remaining ones increased barely by 0.3 ha. For large farm the average size of owned farm land was 25.7 ha and the average size of the utilized agricultural land was 36.5 ha. For the remaining farms these values were 4.8 ha and 5.1 ha, respectively.

The size and quality of the "machine park" that was connected with the farm and could be used on that farm was another significant determinant of the farm's production abilities. If the total value of the owned machines and equipment would be treated as an indicator of the size of the machine park and the value of the most expensive machine as its quality then the difference between large farms and the remaining ones could be considered gigantic (Table 5.3.3). According to farmers' declarations the average value of all machines and equipment on large farms was 343 thousand PLN and on the remaining ones was only 42.6 thousand PLN. Such large disproportions could also be noted with the most expensive machine. The average value of such a machine on a large farm oscillated around 99.3 thousand PLN and on the remaining farms around 16.6 thousand PLN. Nevertheless, these disproportions did not seem so extreme if the value of the owned "machine park" was calculated per 1 ha of utilized agricultural land. This value was 9.9 thousand PLN for large farms and 8.6 for the remaining farms.

Table 5.3.3. Farms according to the value of the owned machines and equipment, in thousands PLN

Farms	Does not own the machines (%)*	Declared value of agricultural machines and equipment, in thousands PLN		
		average value of all machines*	average value of the most expensive farm machine*	average value of machines per 1 ha of utilized agricultural land**
Large	1.9	343.0	99.3	9.9
Other	17.0	42.6	16.6	8.6

* $p < 0.01$; ** $p < 0.05$.
Note: For the calculation of averages the farms that did not own the machines were also included with the value of 0 PLN.

Source: Own research.

How could these data be interpreted? Could there be some limits of rational saturations with technics for various categories of farms, while certain proportions are observed? Perhaps both categories of farmers could share a similar desire to own machines but rationalize it depending on the utilized agricultural area? These issues will be addressed again in the upcoming analysis of profitability of both categories of farms. It should be added that around a dozen large farm owners (1.9%) and almost every sixth of the remaining farm users (17.0%) did not have any machines and used the services of external bodies to conduct the field work.

Another important aspect of the development strategies employed by farmers was the decision about choosing the sources of financing for agricultural production. The large farms differed from the remaining ones in that regard (Table 5.3.4).

Three times more often than the remaining farms, the large farms financed their agricultural production from their own financial means, including the EU subsidies and by acquiring loans. This strategy is characteristic of less than half of large farms (45.7%) and only every sixth of remaining farms (16.7%). The differences between the values of the loans taken should also be noted. In the last three years, large farms dispensed loans for an average sum of 162.7 thousand PLN, while the remaining farms took average loans of 32.8 thousand PLN value. The differences turned out to be much smaller when the values of loans were calculated per hectare of utilized agricultural area. Considering the totality of farms, each hectare on an average large farm was burdened by a loan of 1.8 thousand PLN and each hectare of the remaining farms was burdened with a loan of 1 thousand PLN. Nevertheless, in both categories, the majority of farms financed agricultural production from their own financial means exclusively. This was the case for 54.3% of large farms and 83.3% of the remaining ones.

Table 5.3.4. Farms and their sources of financing (in %)

Farms	Sources of financing*		
	Own sources only	Own sources and credits	Total
Large	54.3	45.7	17.6
Other	83.3	16.7	82.4
Total	78.2	21.8	100.0

* $p < 0.01$.

Source: Own research.

It could be assumed that expanding the machine park, increasing the mechanization of work on the farm, and consequently moving from extensive production based on significant use of labor resources towards intensive production with increased use of machines could be reflected in the farm employment structure. Could the larger and more mechanized farms differ in that regard from the remaining farms? Two issues are worth analyzing: involvement of household members in work on the farm and the use of non-family labor (Table 5.3.5).

It was much rarer for large farms than for small farms to have the manager doing all the work without the involvement of household members. The percentage of farms with such a work situation with an area of 15 ha or more was 16.9% and for the farms of less than 15 ha it was 25.2%. For large farms, 42.7% had at least two household members other than the main user engaged in farm work. For the remaining farms, 33.6% had such an employment scenario. On average, the main farm user on large farms could expect to get help from 1.6 household members, whereas on small farms of less than 15 ha, 1.3 household members could assist with the workload. The study also revealed that larger farms, more often than the small ones, utilized unpaid exchange of farm work between farmers and the respective percentages were: 28.8% and 24.9%. The large farms also used hired labor more often than the small ones, and the respective percentages were: 20.3% and 14.2%. In 2016, the large farms used, on average, 1.8 people for unpaid work exchange between farmers and this type of help was applied for a total of 66.0 workdays per annum. On the remaining farms it was 1.5 people and 38.2 days per annum. When it came to paid labor, large farms used, on average, 3.6 workers who put in a total of 112.3 workdays per annum and the remaining farms used 1.8 workers for a total of 32 days per annum.

Table 5.3.5. Farms according to the type of labor (in %)

	Farms' size	
	large	other
Number of household members working on the farm*		
Only respondent	16.9	25.2
Farm user + one person	40.4	41.2
Farm user + two people	23.4	19.0
Respondent + three people	11.9	9.7
Respondent + four people or more	7.4	4.9
Total	100.0	100.0
Non-family labor: forms of help/employment		
Exchange of work between farmers**	28.8	24.9
Hired labor*	20.3	14.2
Type of labor*		
Family	57.4	64.8
Family and work exchanges between farmers	22.3	21.0
Family and hired labor	13.8	10.3
Family, work exchanges between farmers, and hired labor	6.5	3.9
Total	100.0	100.0

* $p < 0.01$; ** $p < 0.05$.

Source: Own work.

The above data show that, despite the higher level of mechanization, large farms needed significantly more workers than the small farms, which was a consequence of the production scale but the data do not illustrate the intensity of the farm work on large farms in comparison to small farms. To measure the extensity/intensity level of production in both categories of farms, it might be helpful to consider the volume of labor calculated per 10 ha of utilized agricultural areas. When applying such a relative measure, it was revealed that, in 2016, for every 10 ha, the main user of a large farm used help from, on average, 0.6 household members and 0.1 of a non-family person working without pay, for a total of 24.7 working days per annum, as well as 0.08 of a hired worker who devoted 38.3 days to farm work per annum. For the remaining farms, the main user received, on average, help from 4.0 household members and 0.8 of a person who devoted unpaid work to the farm, for a total of 108.6 workdays per annum, as well as 0.5 of a hired worker who devoted a total of 70.3 workdays to the farm per annum. These data present, quite clearly, that agricultural production on large farms was decidedly more intense than production on the remaining farms. Before summarizing this matter, a closer look at the types of labor forces used on the farm could be helpful. The family type of labor force was dominant for both categories of farms, which meant that the farm did not use external help or work, be it paid or unpaid. All the work was conducted by the main farm user alone or, more likely, with the help of other family/household members. This model was characteristic of 57.4% of large farms and even more popular for the remaining farms, as 64.7% of them used family labor exclusively. The mixed model forms of farm labor were a bit more likely to appear on large farms. The family type of farm labor combined with an exchange of labor between farmers occurred on 22.3% of large farms and 21.0% of the remaining farms in the study. The family type of labor, supplemented with hired labor, was utilized on 13.8% of large farms and 10.3% of farms. A combination of family labor and the exchange of work between farmers and hired labor was used on 6.5% of large farms and 3.9% of the remaining farms. Large farms used various types of labor to a greater extent, but it should be noted here that the intensity of using these types of labor was lower in comparison to the remaining farms and this was connected to greater mechanization.

The size of utilized agricultural areas, the level of mechanization, labor resources, and available sources of financing for agricultural production constitute the fundamental elements of development strategies employed by farmers. They also could be seen as crucial determinants of success. However, even the most optimal decisions pertaining to the above issues cannot guarantee the adequate profitability of agricultural production. Even if the risk related to the unpredictability of nature is somehow omitted, success cannot be achieved without a proper assessment of the market situation informing the decision of what to produce and how, where to sell the products to incur profits, and where to buy the means of production and needed materials to minimize production costs. A closer analysis would answer the question of whether these issues could be significantly different between the two categories of farmers.

As indicated by Table 5.3.6 the agricultural production of large farms was decidedly more diverse than the production of remaining farms. Nearly three quarters of large farms (73%) combined plant production with animal production and half of large farms (51.0%) had multiple crops and multiple breeds of animals. One fifth of large farms had multiple crops and only one type of animal. Slightly over half (50.3%) of the remaining farms also combined plant and animal production and over one fourth of such farms (27.4%) had multiple crops and multiple animal breeds. Not quite one fifth of farms under 15 ha combined production of multiple plants with one breed of animal and over one third of them (34.5%) produced a variety of crops exclusively. 15.2% of such farms produced one crop only. Exclusive production of crops was also characteristic of slightly over one fourth (25.8%) of large farms with only a marginal percentage (1.1%) concentrating on just one crop.

Table 5.3.6. Farms according to types and focus of production (in %)

	Farm	
	large	other
Types of agricultural production*		
Plant and animal production	73.1	50.3
Plant production only	26.9	49.7
Total	100.0	100.0
Focus of agricultural production*		
Multiple breeds and multiple crops	51.0	27.4
One breed and multiple crops	20.2	18.6
One crop and at least one animal breed	1.9	4.3
Multiple crops	25.8	34.5
One crop	1.1	15.2
Total	100.0	100.0
Volume of plant production used for animal production*		
0% (lack of plant-animal production combination)	26.9	49.7
Up to 25%	10.9	8.4
26–50%	10.9	7.8
51–99%	13.9	10.2
100%	37.4	23.9
Total	100.0	100.0
Average	54.1	35.7

* $p < 0.01$. Note: For calculating the average volume of plant production used for animal production, farms not engaging in plant-animal production were also included, with a value of 0% assigned to them).

Source: Own research.

The data above show that the agricultural production of the large farms was more diverse than the production of the remaining farms. Considering the lack of market stability for agricultural products, this could be seen as a rational strategy for limiting risk. However, the data also indicated that large farms were likely to combine plant and animal production or grow multiple crops and have multiple breeds of animals. They were rarely oriented toward production of just one crop. Therefore, it can be assumed that their development was more sustainable than the development of the remaining farms. The scale of use of a farm's own plant production for animal production also makes an argument for the sustainable development of large farms that participated in the study. Over half of the large farms (51.3%) used at least half of their crop production for animal production. Over one third (37.4%) of large farms used their entire crop production to feed their livestock. For the remaining farms, the percentage of those using over half of their own plant production for their own animal production was 34.1%. The portion of small farms that devoted all of their crop production to animal production was 23.9%. Overall, large farms allotted, on average, 54.1% of their own plant production to animal production conducted on their farms. For the remaining farms, livestock animals were given 35.5% of the farm's plant production.

Table 5.3.7. Farms according to supply markets and sales markets (in %)

	Farm	
	large	other
Supply markets*		
Local markets	68.5	88.9
Local and supralocal markets	31.5	11.1
Total	100.0	100.0
Sales markets*		
Local markets	61.3	87.0
Local and supralocal markets	38.7	13.0
Total	100.0	100.0

* $p < 0.01$.

Source: Own research.

Decisions related to choosing supply and sales markets are also important to development strategies. Purchasing the needed means of production from local suppliers, local producers, and at local farmers' markets might, in the times of supply monopolization, significantly increase production costs. Similarly, sales of a farm's own products exclusively in local purchasing centers, to local processors or directly to consumers, might, in the context of strong competition on the agricultural production market, significantly decrease sales incomes. Relying exclusively on local supply and sales markets could be detrimental to a farm's development and even threaten its economic stability. Looking for alternative supply sources and alternative buyers on supralocal markets might be considered a more rational decision. Could there be any significant differences between the large and the remaining farms in this regard? What percentage of farms in the study operated exclusively within the local market and when, and to what extent, would they reach out to supralocal markets? The data regarding these matters are presented in Table 5.3.7.

In both categories of farms, those that functioned exclusively on the local market prevailed. They purchased the needed production means from local suppliers and sold their products to local buyers. However, the large farms, more often than the remaining ones, found some contractors on supralocal markets and therefore their strategies should be considered more development-oriented. Almost one third of the large farms (31.5%) in the study used suppliers from supralocal markets and 38.7% of large farms sold their products on supralocal markets. Such strategies were much rarer among the remaining farms. Only 11.1% of them purchased their needed means of production beyond local markets and 13% of such farms sold their products there.

Could a greater size of utilized agricultural areas, a higher level of mechanization, better rationalization of employment, more frequent farm loans designated for agricultural production, and greater openness to supralocal markets bring greater farm profitability? In other words, could large farms be more profitable than the smaller, remaining farms? The data in Table 5.3.8 should be given a closer look.

Table 5.3.8. Farms according to average annual income from agricultural activities in the years 2014–2016, in thousands PLN

Farms' size	Average annual income of the studied farms, in thousands PLN			Average annual income of the studied farms, in thousands PLN, per hectare of utilized agricultural land		
	2014*	2015*	2016*	2014	2015	2016**
Large	108.1	110.3	115.9	3.1	3.2	3.4
Other	15.3	15.6	15.7	3.0	3.0	3.0

* $p < 0.01$; ** $p < 0.05$.

Source: Own research.

When calculating the averages, farms whose managers declared no income from agricultural activity were also included, with an assigned income value of 0 PLN. From 2014 to 2016 they comprised between 0.9% to 1.6% of large farms and 3.7% of the smaller remaining ones.

Considering the absolute values of the income acquired by the two categories of farms in the years 2014–2016, the differences were enormous. The annual incomes of large farms were, on average, seven times greater than the incomes of the remaining farms. They went from 108.1 thousand PLN to 115.9 thousand PLN in three years and, per month, made a sum of over 9 thousand PLN. The incomes of the remaining farms oscillated around 15 thousand PLN, which meant around 1.3 thousand per month. The above data show quite clearly that only a large farm provided an opportunity for an average family to work on a farm, live off of its income, and have a decent standard of living on the average national level. The other category of farms did not create such opportunities, even in situations of a single person working on a farm.

Could the strategies employed by large farms be the only way to succeed on today's market? Is there less room on today's market for farms that are smaller in size?

It appears that the solution for maintaining the farm and developing agricultural activities, as well as providing for the family, is to increase the utilized agricultural area by leasing the land, continue the modernization of the "machine park," and to rationalize the employment, as well as to use loans and mortgages—just as other entrepreneurs do—to support business activities and to engage in expansion beyond local markets. Could this be the only way to go? The study revealed that such activities were the domain of large farms. Could they also be employed by smaller farms or were they doomed to be eliminated from the market in the very near future?

It should be noted that although the agricultural production of small, remainder farms is less diversified, more extensive, mostly oriented to local markets, and only to a minor extent financed by loans, the relative level of mechanization for such farms does not differ much from large farms. Let's recall that the difference between the value of agricultural machines on large and small farms, calculated per ha of utilized agricultural land, is only 1.3 thousand PLN in favor of the large farms. More significantly, large and small farms had relatively similar profitability per hectare: in the years 2014–2016 large farms reported between 3.1 thousand PLN and 3.4 thousand PLN and small farms reported profit of 3 thousand PLN. It should be emphasized again that similar values of profitability per hectare did not mean work on the farm for the entire family and the guarantee that working on the farm could ensure living at a decent level. It also did not mean that the small farms were completely out of date as contemporary market actors and that they should be eliminated from the market. Perhaps their future depends on the development of multifunctionality of rural families, for whom agricultural production could be one of several fields of professional activities and one of many sources of income ensuring a decent standard of living for such families. What seems vital in this process is the ability of small family farms to transform into a certain multifunctional family enterprise, where agricultural production is just one of the several lines of economic activities, but this is not only limited to farmers and their families. Appropriate activities of central authorities, responsible for macroeconomic agricultural policy, and the activities of regional authorities implementing concrete solutions in rural areas, could be quite helpful here. Similar opinions were presented by the representatives of farmers' self-government agricultural chambers. Before analyzing their opinions on general possibilities and the direction of family farm development in Poland, it might be helpful to take a look at the development plans for both categories of farms.

Analysis of particular elements of development strategies planned by users of the two farm categories during the next three years indicated that the differences in their situations would certainly deepen. This could apply to various aspects of farm functioning, as more large farms had concrete development plans than small farms (Table 5.3.9). To support this thesis, it was revealed in the study that over half of the large farm managers were planning to buy more agricultural machinery and equipment in the next three years. In the category of the remaining farms, not quite 14% reported such plans. Over one third of the former and only 11.2% of the latter planned to start construction of new farm buildings or the modernization of existing ones. Additionally, every fourth large farm manager wanted to take out new loans and expand the size of the utilized agricultural area and nearly every tenth farmer planned on increasing the use of fertilizers and plant protection products. Interestingly, every twentieth respondent was interested in starting new types of production. In the category of the other farm users such plans were rare, as only a small percentage of respondents expressed them. It was quite evident in the study that it was mostly the managers of large farms who were oriented toward further development. The remaining farmers who participated in the study were characterized by a willingness to survive and keeping the farm on the market in its current state.

Table 5.3.9. Elements of the planned strategies of farm development; percentage of respondents considering particular elements in the strategies over the next three years

Strategy elements	Farms	
	large	other
Purchase of agricultural machines and equipment*	50.9	13.9
Building of new farm buildings or modernization of the existing one*	37.0	11.2
Taking out new loans*	26.0	7.9
Increasing the size of utilized agricultural land*	25.2	4.7
Increasing the use of fertilizer and plant protection products*	9.4	3.5
Introduction of new types of production*	5.4	2.8

* $p < 0.01$.

Source: Own research.

Obviously, having development plans cannot guarantee actual farm development. Therefore, the dynamics of how large farms and, despite everything, small family farms can develop in Poland depends on—as indicated earlier in the text—social conditions on the macro scale. It might be helpful to find out what representatives of farmers' elected self-government body, specifically the Presidents of Regional Agricultural Chambers,[4] have to say on this matter. It should be added

---

[4] The study, involving free-form interviews with the presidents of six (out of an existing 16) Regional Agricultural Chambers and the President of The National Council of Agricultural Chambers, was conducted in July and August of 2019. All respondents were social activists with a long history of being involved in agricultural chambers, and all were active farmers with farms ranging in size from a dozen hectares to over 100 ha, with various production profiles (combining crop production with

that farmers in Poland generally perceive agricultural chambers as entities able to defend their interests, to the extent that they believe there are any organizations able to defend farmers' interests at all. To recap, in the study described above, only 12.2% of respondents recognized the existence of organizations defending farmers' interests, and 3.6% pointed to agricultural chambers.

All the respondents stated unambiguously that, due to ownership conditions characteristic of Polish family agriculture and the great prevalence of small farms, which they recognized as farms up to 30 ha, the best development model for Polish agriculture is the coexistence model. This was understood to mean having large farms (here described as exceeding 50 ha) that are commodity-oriented, as well as small ones. Both types were, in their opinion, necessary and potentially beneficial for Polish agriculture. As emphasized by one of the interviewees:

> . . . this is the best solution from the safety point of view. The big ones give quantity, the small ones create the brand, the quality of the products as they never use so much fertilizers as the big ones (R1).

In the interviews many benefits of having small farms were cited, not only their contribution to better food safety. In crisis situations and unfavorable market conditions they could play a crucial role in ensuring food security. Commodity farms provide an adequate quantity of food product to society, as stressed by one of the interviewees, but their production should not be taken for granted. To quote an interviewee (R2):

> . . . having commodity farms alone is quite dangerous as in a crisis situation, a big producer finishes production quickly.

Another representative of the agricultural chamber explained the meaning and importance of small farms for the continuity of food production by pointing to their greater flexibility and resistance to crisis:

> Things changed after the transformation that we went through with other countries of the European Union, such as Czech Republic, Slovakia, and Hungary, that used to have better agriculture. Today, they wonder what has happened that even though they have big farms, we are outplaying them on the European market. In Czech Republic,

animal breeding, crop production only, garden production, orchard production, and agricultural production combined with gardening), leasing from several to 20 ha of utilized agricultural areas.

animal production dropped by 50%. It affected all kinds of animal-related production, including poultry and meat. Our production increased because the peasants farms stayed afloat and they were more flexible economically. Say, I do not have income on my farm in a particular year, because of drought and bad prices. I do not invest at such time. I do not buy new shoes, I do not buy a car, I am trying to survive and stay in operation. On a big, commodity farm, affected by external forces, this might not be possible. Payments must be made to the Social Insurance Institution, workers must get paid (R3).

The importance of small farms for food security was emphasized even more by another interviewee. His observations were the following:

Who during the war fed the partisans? Large farms? It was the small farms. Large farms in case of Wars and military conflicts cease to feed the Poles. They start to feed the others (R4).

The interviewee (R2), claimed that it was the affiliation, devotion even, of the farmer to his land, unheard of in other countries that could be seen as a factor enhancing the resistance of Polish farmers to crises:

French farmers told us that two farmers quit milk production as the price went down. In our country it would be unthinkable to stop farming only because the price of milk went down.

Obviously, the notable role of small farms in food safety and food security, as well as their greater ability to withstand crises, do not guarantee their thriving on the market. Nevertheless, the interviewees argued that both categories of farms, large and small, can do well on the market, as both categories have distinct development potential and abilities. The great merit here could be attributed to the peasant character of Polish agriculture and the entrepreneurial capabilities that Polish farmers developed during the times of the People's Republic of Poland.

In 1989 there was only vinegar on the stores' shelves [in Poland]. But thanks to peasant farms, great peasant capital was released in rural areas. The peasants were able to sell their products in towns and cities, fill in all the market stands in urban areas. It started when the food showed up. If these people were wisely motivated and properly directed they would most likely develop processing . . . it could have been tremendous capital. Food stands in the cities would be well-stocked… (R3).

All of them agreed that, considering the reality of European agriculture, ecological food production was the best direction for small farm development. As one of the respondents stated:

> ... with such farm fragmentation, we should fully follow the path of ecology, become ecological basin as ecological products and commodities are three times more expensive than the conventional ones (R5),

while another one (R6) added:

> ... there is no other choice than production of crops and animals of good quality.

Multiple interviewees emphasized that such a type of agriculture could be a good future prospect for small farms in Poland, as the demand for healthy foods was on the rise in Europe, as well as in Poland:

> Ecological products are very in ... [stated (R7), who also added:] ... peasant farms might not be producing completely ecologically, but they are partially ecological, as they use less fertilizer. Their products are more expensive but there are more and more customers, who would be able to pay more.

Some stressed that even though the market for ecological products had great developmental potential, it was still not fully formed. It would require many legal regulations as well as advertising and marketing activities that would help to support access to the consumer.

> There is an increasing demand for ecological food in Poland as the middle class is emerging. There are also many aspects of ecological food, such as reliability of companies that certify foods, access to the means of ecological production, the issue of price. The consumer must know and accept that the ecological products are more expensive. Generally, the demand for ecological products is growing (R5).

The necessity to legally sanction ecological production was understood by the interviewee (R6) who expressed the hope "that nobody would pretend or cheat." He simultaneously admitted that the lobby of large producers was against legal regulations and sanctions. The interviewee (R1) brought up the issue of rationality regarding regulatory standards that ecological farms must meet:

. . . ecological products must meet very strict requirements and maintaining them is difficult as there is some pollution in the atmosphere and it permeates the products. Without a rational approach to these requirements it will be hard to develop the agriculture that should be developed.

According to some interviewees, the most effective way to develop ecological agriculture is to return to nature and sustainable development, which should be easier for Polish farmers than for the agricultural sectors of Western countries.

Nature can defend itself on its own. In the natural environment there are lots of insects that can clean the plants of pests. Nature maintains the balance, so if there is too much of something, there will always be a solution to it, another organism to take care of the surplus. We interfered with that balance . . . Now, we must return to natural agriculture. My father used to say that the three-field rotation of crops was not a good enough solution and a five-field rotation was a better solution. After five years the soil cleanses itself from weeds and diseases but if you plant corn 10 years in a row then the soil gets depleted completely and fertilizers must be used . . . Compared to the West . . . we still have better soils, water and chances for sustainable development. There, in the West, modernity was measured by the volume of fertilizers and now this is something to be ashamed of . . . Sustainable development means giving nature a chance, for example, through the rotation of crops, so nature could restore its power (R1).

Other interlocutors also pointed out the better situation of Polish agriculture in comparison to the Western countries in the context of ecological production development.

The idea [of sustainable development, return to natural agriculture] is what, for example, the Germans need as they mostly have corn and more corn for their biogas plants. Here, in the Podlasie, we do not have any problems with the crops, we have lots of green agricultural areas . . . Maintaining the quality is possible as our soils are not as depleted as in the West (R2).

According to (R6):

Compared to the West, Polish agriculture is sustainable, ecological even, especially in the south-east . . . I was in France on an ecological farm of 18 ha specializing in beef cattle. The farm was "ecological" probably because there was a donkey there. With the cows, I did not see if they were fed with ecological hay or if they did not get any enhancers.

The interviewee coded (R5) stated clearly that:

> The issues of sustainable development look better in Polish agriculture than in the West. Our food is not stuffed with chemicals as much as theirs.

The response of (R7) was quite similar:

> . . . We have better products, quality-wise, because of lower usage of plant protection products and the fertilizers than there is in the West.

Obviously, the future of Polish agriculture, and small farms (up to 30 ha) in particular, will not only be determined by the fact that these farms now engage in production of healthy food or may do so later. It should be stressed here that this direction of development is not yet very popular with Polish farmers. Among all the studied farms, only 5.9% were already producing healthy foods. This percentage was even lower for large farms of 15 ha or more of utilized agricultural areas (4.3%) and slightly higher for smaller farms up to 15 ha (6.3%). The interviewees also emphasized other factors influencing the developmental perspective of small and large family farms. They emphasized the role of farmers' activities as well as state policy towards agriculture and rural areas. When discussing state policy, they addressed the solutions that would facilitate increasing the area of utilized agricultural land for family farms. They noted, however, that buying land from the state by issuing a call for tenders would not be good for family farms, as big land owners, who can offer higher prices, usually win such bids. Some interviewees remarked that long-term lease of the state land to farmers would be a better solution:

> When talking about the family farm, one needs to think how this farm would develop in today's reality. Farmers who had large farms and were able to lease the land to others . . . this was all a friend making a deal with a friend. To thrive, family farms need more land and what should be proposed to farmers is not competitive bidding but 5-year leases, so the farms could develop . . . Not a tender, because the tender inflates the price, and there is no state agricultural policy then (R3).

However, not all agricultural chambers' representatives were in favor of farm leases, pointing to dangers leasing could carry.

> I rent from other farms half of the area that I farm and nobody wants to sell that land to me . . . so the farms are getting bigger but not through their own land. Farm leasing/

rentals can stunt development. For example, if you have animals, you also have to have hectares, where you can dump the manure. If there are no hectares you can't develop the animal production. If the owner decides suddenly to stop leasing the land, the farmer who rents has to wrap up production even though he is probably burdened with loans (R4).

Farmers who want to develop their production and overall economic activities have recently been encountering the increasing problem of conflicts with incoming residents. According to the interviewees, this problem might require a solution on the part of state institutions. As one of the respondents stated:

> We [the state authorities] must seriously approach the issue of designating agricultural production space. Farmers cannot develop their production as they are blocked by neighbors; some complain about the odor, others do not want the barn nearby. The countryside should be the area for agricultural production as it used to be. If somebody has had a farm in the family for 50 or even 100 years and then a doctor or a dentist, who has a lawyer in the family, builds a house close by, then there is a problem. Farmers get eliminated as they are not allowed to do this or do that . . . The manure cannot be transported because the gentlemen are complaining (R5).

Other issues that were perceived as needing some measures to be taken by the state involved regionalization of agricultural policy (particularly crucial for regions where small farms prevail), support for the development of direct marketing and domestic processing (which, according to the representatives of agricultural chambers, was almost completely in the hands of foreign and transnational entities), and development of rural infrastructure that would help farmers managing small farms to achieve dual professions, which would ensure survival of the farms and an improved financial situation for them. As noted by one of the respondents:

> Perhaps additional workplaces should be found for them [small farm users], so they could earn more money for their farms. They do need to make their own decision as to whether they enter direct marketing or maybe some small processing, which would allow them to support themselves and their families. If small farmers want to be involved in agriculture as people of two professions, they should be given a chance to do that. No one should be excluded from the countryside, as rural areas need people that would like to live there (R3).

The presidents of agricultural chambers thought that the most crucial activities that farmers could engage in for further development were the following: willingness to cooperate with other farmers; organizing farmers, especially the small farm users in various groups and communities; readiness to take up direct marketing and local processing; as well as rationalization of the possession of machines through limits placed on the purchases of one's own machinery and equipment, and using them together with other farmers or using the services of companies that specialize in providing agricultural machine work, which is a quite popular solution in the Western world.

Summarizing the opinions of the interlocutors from agricultural chambers, it is worth noting that they spoke with certainty that the development of Polish family farms, be they large or small, in the directions they point to, could be effective. However, would it be realistic to expect Polish farmers to follow in these directions? Referring to the earlier analyses, it is challenging to give a clear positive answer here. As one can recall from earlier parts of this chapter, many schemas of thinking, especially those displayed by large farm users, indicated their openness to modernity combined with a propensity for a peasant way of thinking, sanctioned by their immediate social environment. In the view of this chapter's authors, the remnants of peasantry can be seen as a potential obstacle to conducting farm activities and taking up endeavors recommended by farmers' self-government.

5.4. Large Farm Users: Peasants or entrepreneurs?

Over 80 years ago, Frank Chapin conducted a study which resulted in construction of the Living Room Scale, which was a scale of social position (Chapin, 1935). To place a particular family on such a scale, certain information on their material resources and the quality of these resources was used. It turned out that placement of particular families on the scale based on their resources was similar to the placement given by the interviewers on the basis of their own evaluation. In the study conducted for the purpose of this publication, the interviewers were also asked to place the interviewed farmer, his/her farm, and his/her household on particular scales based on the researcher's evaluation. All of these scales were 5-point scales with 1 described as "peasantry" and 5 as "modernity." The scale that the interviewers used had the following points: from 1 = traditional peasant (dressed rather carelessly, unkempt, having problems with articulating words, using dialect, etc.) to 5 = modern entrepreneur (well-groomed, nicely dressed, eloquent, not using dialect words in conversation, etc.). The interviewers placed their evaluation of the farm itself on the following scale: from 1 = old type farm (old barns, sheds, and warehouses; old and outdated farm machinery and equipment; unkempt and run-down farm, with messes around barns and sheds, etc.) to 5 = modern farm (modern farm buildings and machines, tidiness around farm buildings, etc.). The evaluation of the domicile considered the following scale: from 1 = farmstead (old residential building, close proximity between the farm house and barns, older and beat up furniture in the house as well as old equipment, mess inside and outside the house, etc.) to 5 = modern house of a high standard (new house of modern design and interior with well-manicured grass around a house and a garden, etc.). On all scales, the medium value was 3. Were the respondents more like traditional peasants or modern entrepreneurs in the eyes of the interviewers? Did the evaluators see more domiciles and farms of the old or modern type? Were the households more like peasant farmsteads or modern residences? Did the large farm users and their households receive better or worse evaluations? Were the evaluations of interviewers consistent with the results of analyses presented in the previous subchapters of this work or were they divergent?

It could be helpful to recap the results of the analyses previously described in the subchapters of this work, focusing on the differences between large and small farm users. The former had more institutionalized and objectified capital, but such an unambiguous interpretation is not possible for institutionalized and embodied capital. While in the dimension of behavioral schemes large farm users expressed more openness to modernity, it might be justified to say that, in both categories of respondents, elements of modern thinking coexisted with elements of the peasant mentality. A clearer and more explicit picture could be seen when resources of economic capital managed by both categories of farm users were analyzed. Large farms clearly towered over the small farms in every aspect: the area size, farm machines and equipment, rationality of employment, intensity of production, as well as profitability. Considering the above findings, it should be expected that the interviewers would place large farm users higher than small farm users on every one of the 3 scales. Would they be placed closer to the position of traditional peasants or closer to modern entrepreneurs? Were their households more like peasant farmsteads or modern residences, according to the interviewers? Were the farms considered old style farms by interviewers or did they more resemble modern farms?

The differences between average evaluations in both categories of farm users were statistically significant ($p < 0.01$), which confirmed that the interviewers distinguished both categories of farm users quite clearly. On all scales, average evaluations for large farm users were higher than the average evaluations for small farm users, and in all cases their values exceeded 3.0 quite notably. This means that, in the eyes of the interviewers, a propensity for modernity exceeded a propensity for peasant qualities in all evaluated dimensions of large farms (average evaluation of the farmer was 3.75, of the farm—3.54, and of the household—3.61). The average evaluations for small farms were not so high and only the evaluations of the respondents (3.24) and household (3.16) exceeded 3. On the scale evaluating the farms the average evaluation was below 3.0, only reaching the value of 2.85. Based on the results that the small farm users got in the evaluations of their households and themselves as farmers, it could be inferred that the interviewers saw them and their households as reaching beyond the boundaries of the peasant realm and expressing certain signs of modernity. However, their farms were perceived as still having dominant peasant characteristics. In conclusion, it can be stated that the interviewers' evaluations were quite close to the interpretation of empirical analyses of the resources of cultural capital and economic capital of family farm users. Therefore, the large farm managers can be called entrepreneurs more fairly than they can be called peasants.

REFERENCES

Bourdieu, P. (1984). *Distinction: A social critique of the judgement of taste*. Routledge.

Bourdieu, P. (1986). The forms of capital. In: J.G. Richardson (ed.), *Handbook of Theory and Research for the Sociology of Education*, pp. 241–258. Greenwood.

Chapin, F.S. (1935). *Contemporary American Institutions*. Harper.

Domański, H. (2001). Klasy średnie w Polsce, a wybrane elementy protestanckiego etosu. *Kultura i Społeczeństwo* 1, pp. 105–116.

Poczta, W. (ed.) (2013). *Gospodarstwa rolne w Polsce na tle gospodarstw Unii Europejskiej – wpływ WPR. Powszechny Spis Rolny 2010* [Farms in Poland as compared to farms in the European Union: The impact of the CAP: Agricultural Census 2010]. Główny Urząd Statystyczny.

Internet sources

http://bdl.stat.gov.pl/BDL/dane/podgrupa/tablica, Kategoria: Rolnictwo, leśnictwo i łowiectwo; Grupa: Gospodarstwa rolne; Podgrupa: Gospodarstwa rolne wg grup obszarowych użytków rolnych – nowa definicja gospodarstwa rolnego; Wymiary: grupy obszarowe, rodzaj własności (gospodarstwo indywidualne), lata (2016) (access: August 2, 2019).

Think Locally, Act Globally: Polish farmers in the global era of sustainability and resilience, ed. by Krzysztof Gorlach and Zbigniew Drąg in collaboration with Anna Jastrzębiec-Witowska and David Ritter
Jagiellonian University Press, Kraków 2021, pp. 295–325
ISBN 978-83-233-4949-5
DOI: http://dxdoi.org/10.4467/K7195.199/20.20.12732

Chapter Six: In Search of a Balance: Some aspects of sustainable farming

Krzysztof Gorlach https://orcid.org/0000-0003-1578-7400/

Zbigniew Drąg https://orcid.org/0000-0002-9106-7758/

Grzegorz Foryś https://orcid.org/0000-0002-9411-2681/

6.1. Introductory Remarks

This chapter reviews the issues of sustainable development, which are crucial to the content of the entire publication, and vital to the analysis of the selected results of the study to which the entire project refers. The concept of sustainable development has already been described in detail in the first chapter, which contains theoretical and analytical remarks and research findings. In this chapter, only references directly pertaining to the analyses of sustainability will be made. It should be noted, however, that this chapter will also attempt to approach the problems of sustainable development from the subjective perspective of individuals who are farmers and, because of that, have to make certain decisions and be engaged in ongoing social processes. This is in line with some of the messages that this work aspires to, such as "bringing farmers back in," which follow the concepts of the famous author George Homans (1964)—concepts well known in the history of sociology—who went against the macro-system perspective that dominated in sociology in the 1950s and early 1960s. Unlike his contemporaries, Homans called for more emphasis on the actions of individual people. Considering this, various aspects of sustainable development will be analyzed that relate to the various contexts and factors influencing the decisions taken by people operating farms in Poland.

When studying the actions of individuals it is always helpful to have some context variables or structural variables which, in one way or another, can influence the context, meaning, or character of decisions taken by individuals. Generally, in the tradition of sociological analyses of agriculture, such variables include farm area and production volume. They are usually, but not always (as is the case with animal breeding), connected with farm area. It is the production that maps out the direction of development and influences the way of thinking about the farm and its management. And even though the contrasts drawn to distinguish between farm types are described somewhat differently in the American sociological tradition (family versus capitalist/corporate farms; see, e.g., Strange, 1988), than in the European tradition (peasant-type versus entrepreneurial farms; see, e.g., Ploeg, 2020), it can be stated, judging by the years of these two publications, that categorical dichotomy of farms has been present in the sociological debate for decades, and is still present now.

This chapter also follows the typology of large and small farms, which could be helpful in analyzing various aspects of sustainable development and farmers' individual management styles, in particular. The differences between large and small farms, resulting in various approaches and methods of farming, are considered in the works of the previously mentioned Strange, and Ploeg, as well as Gorlach (2009), and Nelson and Stock (2018), who describe the processes of "repeasantization" in the United States. All of these authors stress that the implementation of sustainable development ideas is more noticeable on smaller farms. It can be argued that "small" and "large farms" represent a different reality in the Unites States than in Europe. The peculiar reality of the coexistence of so-called "small" and "large" farms can be found in Poland, with the presence of its traditional peasant farms still appearing to be quite strong (see also more detailed considerations: Gorlach & Starosta, 2016; Gorlach & Drąg, 2019).

The division of farms in Poland into large and small was extensively described in Chapter 5. Table 6.1.1. is only meant to recall this dichotomy as presented in the previous chapter and should be treated as an introduction to the reflection herein.

This division is very crucial for the analysis presented in this chapter. It will be possible to at least partially verify the thesis that it is smaller farms that create conditions for various activities which foster the type of development generally conceptualized as sustainable. Smaller farms can be seen as a counterbalance to large farms, the functioning of which is more in tune with the logic of the industrial development of agriculture, where most emphasis is being placed on mass production, cost-effectiveness, profitability, standardization of agricultural products, and strengthening of the tendencies for monoculture in agricultural production.

Table 6.1.1. Large farmers in the surveyed sample in 2017

User's category	Number	%
Large farmers (15 ha or more of operated agricultural land)	624	17.6
Other farmers (less than 15 ha of operated land)	2927	82.4
Total	3551	100.0

Source: Own research.

Before we focus on empirical analyses, it might be useful to look at several perspectives describing characteristics of sustainable development processes, which are well presented in the first chapter of this monograph.

These perspectives place emphasis on different aspects of sustainability and can be located in various analytical schemes. First, there is the mechanism of change described as neo-endogenous development, in which local initiatives are connected but with the use of powers and resources that supra-actors have at their disposal. The core of this approach is to point to the concrete mechanism of changes, where networks of locally cooperating actors are created, as well as the networks of supra-local actors and outside actors.

A different perspective for analyses of sustainable development issues stems from the reflection on certain general developmental goals or, to put it differently, certain values. This type of reflection has already appeared in Chapter 1, where Millennium Development Goals were addressed, as well as in other discussions on abandoning the neoliberal concepts of social development, mostly operating within the vision of market mechanisms and economic effectiveness, most typically measured by some quantitative parameters. In light of these discussions, development processes cannot be reduced to pure economic effectiveness or commodities created by the individual activities of entrepreneurs. There are also ways to define development to include various aspects of the natural and social environment. In the most general perspective, the concept of sustainable development is—to use the language of Terry Marsden (2003)—the abandonment of the productivist model of development. To some extent, the sustainable development concept draws from the idea of peasant agriculture (see Ploeg, 2008, 2013; Gorlach, 2009), where a certain balance between the demands of the farm enterprise and the needs of the family which operates it can be observed. The concept of multifunctionality also plays a role here, asserting that agriculture should fulfill various social functions, not only those connected with food production. In regard to the production of commodities, the safety of food and agricultural products matters as much as their quality and nutritional value.

Another approach to sustainable development highlights the role of actors directly involved in these processes, namely agricultural producers. Entrepreneurs operating in a sustainable manner should consider various factors when making decisions related to farming, and not just the profitability of production. The information formulated to influence farmers' decisions on agricultural production should go beyond considerations of production cost-effectiveness and include social context, use of traditions, opinions, and the experience of other producers, such as family members, friends, or neighbors.

This last perspective is particularly interesting due to our research interest in individuals operating farms. Special attention is given to sustainability and the resilience mindset connected with this concept, as well as their certain manifestations, which were analyzed quite broadly in the first chapter of this monograph.

Therefore, later in this chapter, attention will be redirected to certain matters which, in our view, can be seen as components of sustainability and the resilience mindset mentioned above. The first pillar of sustainability chosen to be analyzed is the issue of trust, with the assumption that trust is a factor influencing farmers' predilection to cooperate with others, which is often presented as an essential component of conducting economic activity in a sustainable manner. Subchapter 6.2 is devoted to analyzing this issue.

Such cooperation can take various forms. It can mean using various sources of information which can later be used in the decision-making processes on choosing farm strategy, or more broadly, be helpful to general economic activities of the respondents (Subchapter 6.3). Searching for information through various sources is meant to increase the knowledge of respondents and the decisions to use the advice (Subchapter 6.4. is devoted to this issue) are connected with more serious planning and the search for concrete suggestions and solutions that usually become convoluted in farm management strategies adopted by surveyed farmers.

In subsequent subchapters other contexts of sustainability will be studied. Subchapter 6.5. concentrates on matters pertinent to food safety, showing their importance in respondents' decision-making. Here, we will look into surveyed farm operators' contacts with the representatives of various institutions that ensure the quality of food products and safety of agricultural production processes and which can generally be seen as relevant to widely understood food safety. We will also analyze the attitudes of farmers toward the matters of food quality and food safety. The last issue deals with the respondents' attitudes toward potential use of illegal practices in production, which help agricultural producers to reach higher incomes and stay in business even if the vagaries of the market can be quite challenging. No less important are the attitudes of the surveyed farmers toward consumption of their own food products and their evaluation of the quality of such food products. The reflection of Subchapter 6.6 will close with the analysis of the role and meaning of various factors that farm operators take into consideration while making production decisions.

The subsequent subchapter containing the analysis of the study results expands the repertoire of contexts for respondents' production decisions (Subchapter 6.7). First, we examine farmers' opinions on their personal characteristics that could determine successful agricultural economic activity in the globalized market society. In Subchapter 6.8, we address the readiness of farm operators to make sacrifices and agree to self-limitations in a situation of continuously worsening profitability of agricultural production. Such planning for hard times can be seen as a strategy to defend the farm and the family.

6.2. Trust as a Component of Sustainable Development Context

We propose to start with a very special quality, which is trust. We should add that this is a very subtle variable describing relations between people and, in this case, between individuals operating farms, as well as those who form relations with them during the production process. It can be assumed that this exact variable itself shaping of interpersonal relations can have some impact on strategies of farm man-agement according to the principles of sustainable development. The sociological literature related to issues of trust is quite vast and diverse. Here a quotation from Piotr Sztompka might be quite appropriate (Sztompka, 1999, p. 17), as he stresses that: "There are also numerous empirical studies of trust in various settings, which do not have immediate theoretical relevance; there are erudite historical accounts of the genealogy of the idea, which illustrate twisted and convoluted roads of human thought." What makes trust essential for sustainable development are all the various roles it may play in a society and its transformations as a crucial characteristic of interpersonal and social relations. Trust, for the most part, is involved in all direct relations that subjects such as family farm operators enter into. Certain functions of trust can also be applied to more broadly understood communities and, in our case, rural communities, where the studied farms operate. They can also apply to representatives of various institutions that provide advice to farmers or help them with purchases of means of production and sales of their products. In the above examples, relativization of trust should also be noted, as the surveyed farmers only trust those individuals and institutions that they deem trustworthy. Thus, not all partners of social interactions are trusted (Sztompka, 1999, p. 107). Higher levels of trust seem to be conducive to higher levels of social capital, which—according to Piotr Sztompka (2016, p. 336), as quoted earlier—is a key to social welfare and happiness. This is particularly important for sustainable development, for which cooperation between various actors constitutes an essential mechanism.

Table 6.2.1. Opinions on trust (in %)

Categories of answers	Large farmers	Other farmers
Generally, the majority of people can be trusted	16.4	19.1
One should be very careful in contacts with other people	81.7	78.0
Hard to tell	1.9	2.9
Total	100.0	100.0

Source: Own research. Lack of statistical significance.

Table 6.2.1 presents two essential characteristics of the sampled population of farm owners and operators in Poland. As seen in the table, the percentage of respondents who are inclined to trust in other members of society is relatively low, not exceeding 20% in each category of farm users. In other words, not even every fifth respondent feels that trusting others is a good policy. Such research findings should not be particularly shocking if we compare them to the results of a nationwide study on social trust that was conducted in 2018 and revealed a low level of trust among Poles.

The study on social trust prepared by the Centrum Badań Opinii Społecznej— CBOS (Centre for Public Opinion Research) in 2018 showed that, even in recent years, Poles were still not very open to relying on others and preferred to keep their trust in others at a rather low level. According to the data collected by CBOS, various levels of social mistrust were declared by 68% of Poles and only 25% presented different opinions (CBOS, 2018, p. 4) indicating a willingness to trust in others. Table 6.2.2 presents data on social trust in Poland, from the perspective of a dozen years. The average value on the scale, from -3 to +3 for each year, was negative, which indicated that mistrust and caution dominated over trust and openness in Polish society. Moreover, in recent years the value of this general indicator has been decreasing, which demonstrates a growing mistrust.

Table 6.2.2. General level of trust in Poland in selected years (in %)

Values according to 7-point scale from –3 to +3	2006	2008	2010	2012	2014	2016	2018
Very significant mistrust, carefulness (–3)	26	18	21	19	25	22	25
Medium mistrust (–2)	15	13	13	15	14	15	17
Small mistrust (–1)	27	29	28	29	25	26	26
Ambivalent attitude (0)	10	10	9	10	8	9	7
Small trust (+1)	13	18	16	16	16	15	13
Medium trust (+2)	4	6	5	3	4	5	3
Very significant trust, openness (+3)	5	6	8	9	8	8	8
Average value on the –3 to +3 scale	–0.98	–0.62	–0.66	–0.64	–0.79	–0.72	–0.89

Source: CBOS, 2018, p. 4.

It might be interesting to view this indicator as a point of reference in comparison to other countries. As stated by Kasprowicz, Foryś, and Murzyn (2016, p. 124):

As regards the level of social trust, which many research studies treat as a synthetic measurement of social capital, Poland suffers from a deficit. In the results of research quoted in *Social Diagnosis 2007* [Czapiński & Panek, 2007, p. 258], the proportion of people declaring trust in others remained at a low level, amounting to 10.5% in 2003 and 2005, and 11.5% in 2007. According to data from the European Social Survey, as of 2014, Polish society was still suffering from a lack of trust. Research into the level of trust expressed, not in proportion, but with a scale of one to ten, revealed that, where social trust is concerned, the Polish people remain at the bottom of the ranking in Europe. The highest levels of trust are recorded in the Scandinavian countries (Denmark—6.95, Norway—6.68, Finland—6.58, Sweden—5.99), the lowest in Poland—4.11; Slovakia—3.83 and Portugal—3.61. It can thus be concluded that Poland has been characterised . . . by the phenomenon of lack of trust, which thus seems to have become a permanent feature. Despite the fact that the Polish people consider participation in organisations an important aspect of being a good citizen, actual participation remains at a relatively low level 2.6 . . . that places Poland at the bottom among European Union countries (Dzwończyk, 2006, p. 172).

When questioned about trust, surveyed farm users and operators expressed relatively low levels of social trust and this was true for both categories of farmers. There were no statistically significant differences between categories, as declared by farmers and presented in Table 6.2.1. This general lack of trust can be expected to have had a crucial influence on respondents' way of thinking regarding potential cooperation with other farm users, and comprises an important component of the mindset oriented toward sustainable development ideas. Interestingly, a variable like farm area, which determines the scale of farm production, does not correlate in any significant way with the respondents' general approach to other people. One might entertain the idea that users of smaller farms would display a higher level of trust towards others, especially those who are similar to them, as a measure to increase their own uncertain position in the agricultural market. A higher level of trust in that regard could indicate a greater propensity for cooperation, but in the studied sample such a phenomenon could not be seen. Therefore, it can be stated that Polish farmers, just like Polish society in general, treat others with mistrust and might not be inclined to engage in long-term and closer cooperation.

If one should form an opinion on cooperation between the surveyed farmers on the basis of the above research findings, it would be natural to say that they are more oriented toward autonomous actions rather than cooperation with others. It can be assumed that a generally lower level of trust towards others does not foster cooperation with them. Such a minimal predisposition for cooperation also means difficulties in moving towards sustainable development and increasing individual-istic tendencies in farmers, without looking into potential benefits of cooperation with others. Such an approach to farming could be attributed to farmers operating larger farms, who might be more oriented toward production outcomes, and not particularly interested in cooperation with others.

Table 6.2.3. Opinions on cooperation (in %)

Categories of answers	Large farmers	Other farmers
It is possible to achieve more by working together with other people than it is by working alone	66.5	62.8
Cooperation with other people rarely brings desired effects	23.6	27.7
It's hard to tell / hard to say	9.9	9.5
Total	100.0	100.0

Source: Own research. Lack of statistical significance.

The results collected in our study indicate something contrary to the above assumption (see Table 6.2.3). It turns out that the great majority of respondents, re-gardless of the farm area they use/operate, agree with the assertion that cooperating with others might bring better results than acting independently. Moreover, the differences in opinions between the owners of both types of farms are not statisti-cally significant. This means that the surveyed farmers in the prevailing majority are willing to cooperate with others and, according to the line of thinking of this chapter, ready to participate in the strategy of sustainable development.

6.3. Acquiring Information as a Pillar of Sustainable Farming

Information influencing farmers' decisions on farm development strategies constitutes, in our view, an essential part of the reflection on sustainable development. Sources of information are of particular importance. It can be assumed that their types indicate whether the respondents have an interest in a strategy for managing their farm in a sustainable manner, using local sources of information, or perhaps they should be included in the category of farmers who manage their farms with the help of information sources that are not linked to the local context.

Based on the study results presented below, it is worth highlighting that, when it comes to sources of information, farms in Poland are strongly connected with the local context. For the prevailing majority of the studied farmers, family members, friends, and acquaintances constitute a rather important collective source of information. Next in popularity are information sources such as local press, heads of the villages (representatives of local government), local announcements (leaflets, posters), municipal websites, other local Internet portals, priests, or even Facebook and other social media online. From this list only the last two examples can be viewed as sources which go beyond local context.

Admittedly, the data in Table 6.3.1 are quite surprising, as differences between the categories of farmers in regards to using the sources of information are, in most cases, statistically significant. Only with two types of sources—"other residents" and "priests"—are there no statistically significant differences between the categories of farmers. All the remaining types of information sources are more likely to be used by respondents operating larger farms, who seem to be more interested in local matters. It should also be noted that this category of farmer is more likely to use Facebook or other social networks, which are not necessarily limited to the local context. Therefore, general statements that individuals operating larger farms are more oriented toward using all the sources providing information on the context where they operate are quite justified. This could mean that large farmers, who are quite sensitive to various signals coming from the market, are also more attuned to various information sources from a local context such as town or municipality, where agricultural economic activity takes place. At the same time, they are also able and interested to use sources of information not necessarily linked to the local context.

Table 6.3.1. Sources of information on local matters pertinent to respondents' municipalities: frequency of acquiring information from particular sources is measured on a 1–7 scale, where 1 means "at least several times a week," 2—"less than once a week," 3—"2–3 times a month," 4—"less than once a month," 5—"several times a year," 6—"about once a year or less," and 7—"never." The table presents the percentage of answers "at least once a month" (sum of responses 1–4), "several times a year at most" (sum of answers 5–6), and the answer "never" (response 7), as well as the level of significance

Source of information	Large farmers			Other farmers			Significance level
	At least once a month	Several times a month, at most	Never	At least once a month	Several times a month, at most	Never	
Family, friends, and acquaintances	86.5	9.5	4.0	82.0	13.3	4.7	0.01
Other residents	75.5	18.4	6.1	74.1	17.6	8.3	none
Local press	62.2	16.0	21.8	50.7	18.5	30.8	0.01
Head of the village	50.0	32.7	17.3	42.2	36.6	21.2	0.01
Leaflets, posters, and announcements	49.2	31.6	19.2	43.3	33.7	23.0	0.01
Municipal website(s)	42.2	14.9	42.9	28.6	13.4	58.0	0.01
Other local Internet portals	35.2	10.0	54.8	19.1	9.5	71.4	0.01
Priests	24.7	30.7	44.6	25.6	26.9	47.5	none
Facebook or other social networks	20.7	5.3	74.0	12.8	5.9	81.3	0.01

Source: Own research.

The results of factor load analysis allow us to distinguish two types of information sources: traditional sources and modern sources (see Table 6.3.2). Traditional sources mostly include other residents, family, friends, and acquaintances, as well as heads of the village or town, and priests. Modern sources are the following: municipal websites, local Internet portals, Facebook and other Internet networks, as well as local press, leaflets, posters, and announcements. Although these sources can be treated as independent from each other they share some characteristics: local press, leaflets, posters and announcements may deal with both local matters and matters from outside the local context. Additionally, sources of information such as leaflets, posters, announcements, and local press can be treated as both traditional and modern information sources.

Table 6.3.3 illustrates location of information sources for both categories of farmers on particular scales. Owners and operators of larger farms are situated lower on the scales for both modern and traditional information sources. According to the data coding scheme, a lower position on the scale means more frequent use of particular sources (see Table 6.3.1). The data indicate however, that differences in the location of both categories are greater for modern sources (the distance on the scale between these two categories equals 0.395) than in the case of traditional sources (where the distance on the scale between both categories is small and equals 0.096). This means that both categories of farmers use traditional sources of information quite often but the owners of larger farms supplement their use of such sources with modern sources more often than other respondents.

Table 6.3.2. Factor loads of main components of local sources of information (Varimax rotation, loads below 0.20 were not considered)

Variables	Components	
	I: Traditional sources	II: Modern sources
Other respondents	0.735	
Family, friends, and acquaintances	0.690	
Head of the village	0.673	
Priests	0.654	
Leaflets, posters, and announcements	0.622	0.284
Local press	0.462	0.398
Other, local Internet portals		0.880
Municipal websites		0.817
Facebook and other Internet networks		0.780
Percentage of explained variation	28.200	25.600
Total percentage of explained variation	53.8	

Source: Own research.

Table 6.3.3. Average factor values of the main components of local information sources, depending on the user's category

User's category	The scale of information sources					
	Traditional sources			Modern sources		
	Average	Significance	eta²	Average	Significance	eta²
Large farmers	−0.079	0.05	0.001	−0.326	0.01	0.022
Others	0.017			0.069		

Source: Own research.

6.4. Sustainable Farming: Contexts of decisions on agricultural production (part 1)

Things look noticeably different when various sources of information are considered not just for simple acquisition of information mostly pertaining to local matters, but as sources of advice related to the process of managing the farm (see Table 6.4.1). Here, significant statistical differences between large farmers and all other farmers are visible in all cases. The differences are quite significant for the following sources:

TV and radio programs; professional magazines, press and books; Internet; expert advice and opinions; observing other farm operators; and using special computer programs (level of significance < 0.01). In only one case was the difference less statistically significant (level of significance < 0.05) and entailed the situation of using the advice and opinions of neighbors, friends, and acquaintances. However, even this source, understood as a source of advice and support, was more likely to be explored by farmers operating large farms. Generally, it can be stated that respondents managing larger farms and having more experience with market signals and trends are also more inclined to use other external sources of information that could be useful for their farm strategies. It can be argued that respondents operating larger farms are more likely to enlist some safeguards on their actions, also by checking various sources and external factors. It might be assumed that using and relying on various sources of information as well as considering multiple factors indicates farmers' emphasis on cooperation with various subjects. This could be treated as part of a more sustainable action strategy where it could be concluded that respondents operating larger farms are more likely to implement sustainable ideas to manage their farms.

Table 6.4.1. Using advice on farm management: frequency of using various sources of advice measured on a 1–7 scale, where 1 means "at least several times a week," 2—"less than once a week," 3—"2–3 times a month," 4—"less than once a month," 5—"several times a year," 6—"about once a year or less," and 7—"never." The table presents the percentage of answers "at least once a month" (sum of answers 1–4), percentage of answers "several times a year, at most" (sum of 5–6), percentage of "never" responses (answer 7), and level of significance

Using	Large farmers			Other farmers			Significance
	At least once a month	Several times a year, at most	Never	At least once a month	Several times a year, at most	Never	
TV and radio programs	68.1	18.2	13.6	55.9	20.2	23.9	0.01
Professional magazines, press, and books	59.7	25.8	14.5	29.9	30.8	39.3	0.01
Internet	57.6	12.6	29.8	31.8	15.4	52.8	0.01
Advice and expert opinions from agricultural and veterinary services and authorities	41.5	49.5	9.0	15.8	46.3	37.9	0.01
Observing other farmers	36.0	47.0	17.0	31.5	46.5	22.0	0.01
Advice and opinions from neighbors, friends and acquaintances	34.7	46.1	19.2	32.6	44.8	22.6	0.05
Computer programs helpful in managing the farm	12.8	5.8	81.4	4.1	4.3	91.6	0.01

Source: Own research.

In the analysis of factor loads, two types of sources of economic advice were distinguished: traditional sources and modern sources, according to the typology already utilized in this subchapter (see Table 6.4.2). Among the components of modern sources are the following: Internet, professional magazines, press and books, computer programs, TV and radio programs, as well as professionals representing agricultural services and authorities. The components of traditional sources include: neighbors, friends, and acquaintances; observing other farmers; experts from agricultural services and authorities; TV and radio programs; professional magazines, press, and books. As seen in Table 6.4.2, although these components are independent from each other, some of them, such as experts from agricultural services and authorities; TV and radio programs; professional magazines, press, and books, overlap.

Table 6.4.2. Factor loads of main components of sources of economic advice (Varimax rotation, loads below 0.20 were not considered)

Variable	Components	
	I: Modern sources	II: Traditional sources
Internet	0.753	
Professional magazines, press, and books	0.731	0.294
Computer programs	0.629	
Radio and TV programs	0.530	0.355
Neighbors, friends, and acquaintances		0.873
Observing other farmers		0.854
Experts from agricultural and veterinary services and authorities	0.487	0.532
Percentage of explained variation	29.100	28.500
Total percentage of explained variation	57.6	

Source: Own research.

Table 6.4.3 illustrates location, on particular scales of economic advice, sources for both categories of farmers. Owners and operators of larger farms are situated lower on the scales for both modern and traditional information sources. It should be remembered, however, that a lower position on the scale means more frequent use of particular sources. Data indicate that a significant difference in location of both categories of farm users can only be seen on the scale for modern sources (the distance between the two categories being 0.836). Such a difference for traditional sources of advice is quite small and the distance between the two categories of respondents equals 0.109. This means that, although modern sources of economic advice are the domain of large farmers, they also use traditional sources to an extent similar to that of the other category of respondents.

Table 6.4.3. Average values for factor loads of the main components of economic advice sources, according to user's category

User category	Scale of advice sources					
	Modern			Traditional		
	average	significance	eta²	average	significance	eta²
Large farmers	−0.690	0.01	0.101	−0.090	0.05	0.002
Other farmers	0.146			0.019		

Source: Own research.

6.5. Sustainable Farming: Contexts of decisions on agricultural production (part 2)

Tables 6.5.1 and 6.5.2 contain information on another important aspect of sustainable agricultural production, that of respondents' contacts with various institutions which are responsible for food safety (see Table 6.5.1), as well as the attitudes of respondents toward various opinions on this type of safety (see Table 6.5.2).

Table 6.5.1. Respondents' contacts with institutions responsible for food production safety (Veterinary Inspectorate, State Sanitary Inspectorate, Main Inspectorate of Plant Health and Seed Inspection, Trade Inspectorate) in %

User's category	Yes, many times	Yes, only once	No, never	Total
Large farmers	74.4	11.2	14.4	100.0
Other farmers	39.2	12.8	48.0	100.0

$p < 0.01$.

Source: Own research.

Table 6.5.2. Respondents' attitude towards the statement: "Farmers in Poland often use forbidden practices in food production to achieve higher income" (scale from 1—"I strongly agree" to 5—"I strongly disagree") in %

User's category	I agree ("I strongly agree" and "I somewhat agree" answers)	"I neither agree nor disagree"	I disagree ("I strongly disagree" and "I somewhat disagree" answers)	Total
Large farmers	22.6	23.9	53.6	100.0
Other farmers	26.7	28.8	44.6	100.0

$p < 0.01$.

Source: Own research.

Both tables clearly present two important tendencies. The first one reveals that farmers managing large farms have more contact with the representatives of various institutions responsible for the safety of food production. In this sense, they can be treated as agricultural producers implementing the principles of sustainable development. It is also the large farm users who, more often than the other category of farmers, disagree with the statement that farmers in Poland use forbidden agricultural practices to increase their incomes. This second tendency, revealed in the responses from large farm users, might suggest the influence of frequent contacts with the institutions responsible for food safety, as indicated by Table 6.5.1.

Certain aspects of the opinions presented earlier deserve further analysis, as questions can be asked as to what is more important to the studied farm users: profit levels from selling agricultural products or their quality and safety for the consumer. One might venture a conclusion that an emphasis on incomes and profits, at the expense of compromising food safety and quality, characterizes those respondents whose farms—due to their size—are more oriented toward the "industrial," mass production style. Is that the case?

Table 6.5.3. Respondents' attitudes towards the statement: "To stay on the market and have some income it is sometimes necessary to use forbidden practices in food production" (scale from 1—"I strongly agree" to 5—"I strongly disagree") in %

User's category	I agree ("I strongly agree" and "I somewhat agree" answers)	"I neither agree nor disagree"	I disagree ("I strongly disagree" and "I somewhat disagree" answers)	Total
Large farmers	13.6	16.2	70.2	100.0
Other farmers	18.4	22.9	58.6	100.0

$p < 0.01$.

Source: Own research.

The answers in Table 6.5.3 above reflect a rather negative overall response to the featured statement. It is worth emphasizing, however, that the differences between the responses of large farmers and other farmers are statistically significant. Most importantly, a smaller percentage of the large farmers agree with the statement on the necessity of forbidden agricultural practices compared to the other category of farmers (see percentages in column 2: 13.6% and 18.4%, respectively). This is even more visible in column 4, where as many as 70.2% of respondents operating large farms disagree with the positive response to the featured statement. In the remaining category of farmers, the percentage who disagree is 58.6%. It might be concluded that large farmers, who are more oriented towards industrial production and the maximizing of their profits, are less likely to accept forbidden practices in food production.

There are two possibilities for how these findings might be explained. First, it could be surmised that users of larger farms, who produce on a mass, industrial scale have, as already indicated in this chapter, more contacts and connections with various institutions that could spot and prevent forbidden practices in agricultural production processes. Second, a thesis could be formed that larger producers, due to the large number of purchasers of their products, might not want to jeopardize their reputation. It should be remembered, however, that these explanations are only hypotheses, which should be verified through additional studies in the future.

This matter might be additionally complicated by the data presented in Table 6.5.4, which connects food production on the farm with its partial consumption by household members. This is a situation that regularly occurs on a typical peasant farm and in the peasant household, where a certain part of agricultural production is allotted for "own use."

Table 6.5.4. Respondents' attitude to the statement: "To save some money on food shopping we often eat what we produce on our farm" (scale from 1—"I strongly agree" to 5—"I strongly disagree") in %

User's category	I agree ("I strongly agree" and "I rather agree" answers)	I neither agree nor disagree	I disagree ("I strongly disagree" and "I rather disagree" answers)	Total
Large farmers	71.2	12.9	15.9	100.0
Other farmers	75.6	11.7	12.6	100.0

$p < 0.05$.

Source: Own research.

It should be noted that the differences between the categories of farmers are statistically significant but their level of significance is not very high. Nevertheless, in this aspect of the analysis it is usually assumed that farmers others than those operating large farms are more likely to prefer activities and types of conduct which could be described as characteristically peasant in nature. In this study, this applies to consumption of at least part of the farm's production by the family who produced it. Farmers from the "other" category are more likely to agree with the statement on saving money and consuming the food produced on the farm. As many as 75.6% of respondents from this category agree with this statement compared to 71.2% of large farm respondents. The percentage of respondents disagreeing with this statement is 12.6% in the other farmers' category and 15.9% in the large farmers' category. It can be therefore stated that the respondents who were more like "peasants" were also more oriented toward this aspect of sustainable development that involved consumption of food produced on one's own farm. Respondents who operated large farms and were oriented toward industrial, large scale production of agricultural products were less likely to eat food produced on their farms, and save money this way.

Table 6.5.5. Farmers' approach to the statement: "We often eat what we produce on our farm and our products are of much better quality than the ones that are sold in the store" (scale from 1—"I strongly agree" to 5—"I strongly disagree") in %

User's category	I agree ("I strongly agree" and "I somewhat agree")	"I neither agree nor disagree"	I disagree ("I strongly disagree" and "I some-what disagree")	Total
Large farmers	83.8	8.8	8.4	100.0
Other farmers	81.4	9.5	9.2	100.0

Lack of statistical significance.

Source: Own research.

Table 6.5.5. seems to present another aspect of being "peasant," related to the conviction that products from one's own farm are of superior quality than those available in stores. Therefore, the respondents reported that "we often eat what we produce on our farm." Obviously, this statement is based exclusively on farmers' declarations. The lack of statistically significant differences between large farmers and other farm users in Poland is noticeable here. It might be assumed that the peasant tradition, with its message of "ours is better," is still quite prevalent among Polish farmers.

However, one might proceed with caution when formulating such statements as there is a lot of room for ambiguity. Appreciation for the quality of one's own product might not necessarily be related to tradition and peasantry. This might also be interpreted as part of the tendency that is quite typical in the most advanced type of market economy, with nearly every entrepreneur being wholly convinced, at least when making public declarations, which can be seen as advertising testimonials, that his or her product is the best. This interpretational dilemma will have to stay unresolved for now, leaving room for further, more detailed studies and analyses.

6.6. Sustainable Farming: Factors considered in making decisions on production

Table 6.6.1 presents the variety of factors that respondents take into consideration when making production decisions on their farms. It does not go unnoticed that there are statistically significant differences between the two categories of farm users. Large producers are more likely than other producers to consider factors such as: product price, costs of production, certainty of product's purchase (through contracts), requirements of purchasers relating to production quality, farm labor resources in one's own family, quality of farm labor (e.g., possibility of hiring suitable workers), possibility of getting a loan, and—what is quite interesting—factors related to environmental protection. Smaller producers, on the other hand, are more inclined than the larger ones to declare the importance of the following factors: the possibility of unhampered sales outside of contracts, local and family-related customs, as well as production traditions. These last two factors are often presented as characteristics of sustainable farming. Larger farmers appear to pay greater attention to the possibility of soil and water contamination with chemicals, which could be explained by their greater awareness and deeper understanding of this problem, possibly due to their production scale.

When analyzing factors considered by respondents, three dimensions of decisions have been distinguished, and described as: "quality with profit," "mass production," and "local harmony" (see Table 6.6.2). This matter might be somewhat complicated as only two categories of farmers are presented in the study.

In the decision strategy described as "quality with profit" the following factors are considered: market price of the product, production costs, requirements of purchasers related to food quality, possibility of sales outside of the contract, certainty of purchase confirmed by the contract or agreement, and number of people in the family able to do farm work. Within the decision strategy named "mass production" the following components can be distinguished: possibility to use credits and loans, certainty of purchase confirmed by contract or agreement, possibility of hiring suitable workers, market price of the product, production costs, purchasers' requirements related to food products, possibility of contaminating water and soil with fertilizers and pesticides, and lack of possibility/consideration to sell products outside of contracts. Factors recognized as components of the "local harmony" decision strategy primarily include: family and local production customs, number of people in farmer's family able to work on the farm, possibility of contamination of water and soil with farming chemicals, possibility to hire suitable workers, buyer requirements related to food quality, possibility to sell products outside of contracts and possibility to use credits and loans. Although these strategies are independent from each other, they share some components (see Table 6.6.2). For example, consider that requirements of purchasers related to food quality were present in all strategies but they played a different role in particular strategies. For strategies such as "quality with profit" and "mass production" components such as product price on the market, production costs, and certainty of purchase confirmed by contract are particularly important. Strategies described as "mass production" and "local harmony" overlap on components such as: use of credits/loans and possibility of contaminating soil and water with fertilizers. Strategies such as "quality with profit" and "local harmony" highlight the importance of factors like the number of family members able to work on the farm and the possibility to sell products outside of contract obligations.

Table 6.6.1. Factors considered in the process of making production decisions (measured on the standard 5-point scale, from 1—"definite yes" to 5—"definite no"). Percentage of "Yes" answers ("definite yes" and "rather yes") and "No" answers ("definite no" and "rather no") as well as the average and significance level are presented in the table. To show 100.0 % of responses, answers such as "Neither yes nor no" are also included

Type of factor	Large farmers			Other farmers			Significance
	Yes	No	Average	Yes	No	Average	
Price: market price of the product	89.8	5.6	1.6	77.2	15.3	2.0	0.01
Costs: production costs	93.6	4.2	1.5	84.0	10.1	1.9	0.01
Contract: certainty of purchase confirmed by contract or agreement	74.9	13.7	2.0	50.8	34.4	2.8	0.01
No contract: possibility to sell products outside of contracts	41.3	37.2	2.9	63.5	24.6	2.5	0.01
Quality: purchasers' requirement related to food quality	83.8	7.3	1.8	67.7	17.5	2.3	0.01
Resources: number of family members able to perform agricultural work	61.2	22.9	2.5	54.7	29.1	2.7	0.05
Employment: possibility of hiring well-suited workers	31.8	54.5	3.4	19.9	66.2	3.8	0.01
Customs: local and family customs reflected in production	35.2	42.6	3.2	46.4	34.3	2.9	0.01
Credits/loans: possibility of using credits/loans	46.2	39.4	3.0	19.6	67.3	3.9	0.01
Environment: possibility of polluting soil and water with artificial fertilizers, herbicides, etc.	50.7	31.2	2.7	39.7	39.9	3.1	0.01

Source: Own research.

Table 6.6.2. Factor loads for the main components of production decisions (Varimax rotation, loads below 0.20 were not considered)

Variable	Components		
	I: Quality with profit	II: Mass production	III: Local harmony
Price	0.813	0.283	
Costs	0.780	0.200	
Quality	0.694	0.215	0.253
No contract	0.558	−0.288	0.320
Credit/loan		0.737	0.243
Contract	0.411	0.704	
Employment		0.557	0.512
Customs			0.712
Resources	0.235		0.590
Environment		0.214	0.508
Percentage of explained variation	23.300	16.600	16.100
Total percentage of explained variation	56.0		

Source: Own research.

Table 6.6.3 illustrates the location of both categories of farm users on decision making scales. Farmers operating large farms are located much lower than the other studied farmers on the "mass production" scale and the distance between the two categories of farmers equals 0.846 indicating that they are more likely than other users to apply such a decision strategy. The differences between the two categories of farmers are much smaller for the two other scales. Farmers operating large farms are located lower on the "quality with profit" scale than other farmers, with the distance between the two categories being 0.207. On the "local harmony" strategy scale the situation is different, and the farmers operating farms in the "other" category are located lower on the scale, with the distance between these two categories being 0.129. This indicates that the "quality with profit" strategy is more appealing to respondents operating large farms and the strategy of "local harmony" has more adherents among the other farmers.

Table 6.6.3. Average values of factor loads for production decisions according to user's category

User's category	Scale of production decision								
	Quality and profit			Mass production			Local harmony		
	average	significance	eta²	average	significance	eta²	average	significance	eta²
Large farmers	−0.170	0.01	0.006	−0.695	0.01	0.105	0.106	0.01	0.002
Other farmers	0.037			0.151			−0.023		

Source: Own research.

6.7. Sustainable Farming: Selected characteristics of farmers according to their own opinions

Respondents' opinions on characteristics that might currently determine the success of the farm are presented in Table 6.7.1. Many similarities between the two categories of farmers are quite striking. These similarities are indicated by a lack of statistically significant differences between the percentages of opinions expressed by farmers operating large farms and those offered by the other farmers. Such observations can be applied to six out of nine farm success characteristics, those being: diligence and willingness to work hard, penchant for farm work, insight into the product market, attention given to the quality of products, ability to "wheel and deal" and to know the right people, as well as bringing together agricultural production and environmental care. Statistically significant differences between the two categories of farmers could only be observed in three cases of the listed farm success characteristics. This matter will be discussed further in the following subchapter.

Table 6.7.1. Opinions on farmers' characteristics determining their success in activities on today's market, with respondents given three choices out of nine possibilities (the percentage of responses indicating particular factors)

Characteristics	Category	
	Large farmers	Other farmers
Diligence and willingness to work hard	64.1	62.3
Penchant for farm work	62.8	64.7
Proper education and qualifications*	39.8	30.0
Readiness to take risk*	34.5	26.6
Insight into the product market and orientation towards the needs of the market	31.2	30.5
Attention to quality of products	29.8	30.3
Ability to "wheel and deal" and knowing the right people	12.2	12.4
Ability to bring together agricultural production and environmental care	9.2	10.0
Orientation towards the production of traditional and local products	3.9	8.2

* $p < 0.01$.

Source: Own research.

It seems that the differences noted in Table 6.7.1 are quite consistent with the popular assumption of the modern way of managing farms. Larger farms are expected to be managed according to the methods of production for which more professional and specialized education is essential, as well as certain qualifications to go with it. It can also be stated that larger farms, characterized by mass and standardized production, are more closely linked to the market and therefore more affected by how the market functions and how it fluctuates (see, e.g., Gorlach, 2009). This might be the reason why large farm users have such appreciation for characteristics such as readiness to take risks. A different tendency could be observed among farmers managing smaller farms (the "other farmers" category in Table 6.7.1), who were more positive toward the meaning and role of local and traditional products. It's quite possible that this approach was closely related to ways of thinking not so oriented towards mass and standardized production. These farmers are more likely to treat their products in an artisanal fashion, as something more than mere "commodities" intended for mass market and anonymous buyers.

6.8. Sustainable Farming: Defense strategies

Table 6.8.1 contains detailed information relating to the sustainable mindset and various factors that, in a more or less significant manner, determine farmers' action strategies in situations where their farms face uncertain market positions. Certain strategies may involve actions and solutions designed to keep the farm but deal with various matters related to the functioning of the family (increased engagement of family members in farm work, additional work outside of agriculture, cutting back on household expenses, sale of household conveniences, resignation from financing of children's university education). There are also quite opposite strategies that directly apply to farm matters, such as taking credits/loans, sale of agricultural machines and equipment, selling or renting out a certain part of the farm or in some cases even selling the entire farm. The first type of strategies described above aim to protect the farm by mobilizing the family, while the second type is oriented toward protecting the family situation and keeping the family functioning at a comfortable level. The first example illustrates a more entrepreneurial, capitalist way of thinking, where everything is aimed at continuation of business activities. The second example presents a more sustainable way of thinking with the welfare and well-being of the family being placed at the center, before the business, in the form of agricultural activity.

Foremost, no statistically significant differences have been noted between the way of thinking displayed by large farmers and other farmers and related to actions and solutions such as selling farms machinery and equipment and getting rid of household conveniences. In other words, it can be stated that in the studied sample, statistically significant differences in the situation of both categories of farms could not be observed when the action strategies in both categories of respondents involved limiting farm equipment and household conveniences.

Table 6.8.1. Activities and solutions that the respondents are ready to commit to in a situation of long-term worsening of agricultural production profitability[1]

Solution/activity	Large farmers			Other farmers			Significance
	Yes	No	Average	Yes	No	Average	
Farm work: increasing the volume of own farm work and the farm work of other family members	66.2	21.4	2.4	50.1	32.7	2.8	0.01
Additional work: taking additional work outside of agriculture	64.3	26.7	2.5	67.3	22.3	2.3	0.05
Home expenses: reducing the expenses to the necessary minimum	48.4	41.8	3.0	39.2	46.0	3.2	0.01
Loan: taking a loan	46.6	41.3	3.1	22.4	62.6	3.8	0.01
Sell the farm: selling or renting out part of the farm to others at market price	30.7	62.2	3.6	36.4	52.4	3.3	0.01
Sell machines: selling some farm machinery and/or equipment	22.5	66.7	3.8	19.4	65.3	3.8	none
Sell household conveniences: selling some furnishings and household conveniences	5.1	89.3	4.5	3.5	90.7	4.5	none
Education: resignation from providing university education to children	1.8	81.8	4.5	1.1	77.7	4.4	0.05

Source: Own research.

In any other situation, things are quite different. It turns out that large farm users are decidedly more inclined to increase the level of their own work and the work of family members when the farm is facing some crisis situation. The level of significance here is noteworthy—0.01. The percentage of large farm users who report such a possibility is around two thirds of all respondents in this category (66.2%), while

---

[1] These activities were measured on the standard 5-point scale, from 1—"definite 'yes'" to 5—"definite 'no'". The percentage of "Yes" answers included "definite 'yes'" answers and "somewhat 'yes'" answers, while the percentage of "No" answers included "definite 'no'" and "somewhat 'no.'" The table also provides the average values and the significance value. To maintain totals of 100.0 %, answers such as "Neither 'yes,' nor 'no'" and "Hard to tell" answers are also included. For the calculation of average values the "Hard to tell" answers are assigned the value of 3.

the percentage of other studied farmers who express such opinions only slightly exceed half in their category, at 50.1%. It can be stated that the respondents managing larger farms are more willing to increase the farm work load of family members in a situation of growing needs of the farm in comparison to the respondents who operate smaller farms. This shows that in the case of larger farms, the emphasis on farm needs is so strong that the needs of the family might be left behind. It can also be seen as a manifestation of the logic of unsustainable development, where everything is oriented towards the needs of production. Defending farm activity is seen as key to the well-being of the family.

Users of smaller farms were more inclined than large farm operators to apply the strategy of family members taking jobs outside of agriculture. The percentage of respondents with large farms supporting such a strategy is 64.3% and for the users of other farms it is 67.3%. The percentages of respondents who didn't identify with such strategies were the following: 26.7 (large farmers) and 22.3 (other farmers). The level of statistical significance is weaker here, with a value of 0.05. This is a far cry from the statistical significance level of 0.01 that was calculated for the strategy of family members engaging in more work on the farm.

The above declarations made by surveyed farmers indicate that, in a crisis situation, owners of various types of farms choose somewhat different action strategies. The user of larger farms seem to prefer the "all hands on deck" strategy, which means that the members of the farming family will not look for employment outside of agriculture and fully concentrate on the farm. Farmers assigned to the other category are more likely to try to support their families in crisis situations rather than devote all their efforts to farm work. The type of strategy they are likely to adopt means searching for employment outside of agriculture, which seems to be a more sustainable action strategy focused on the diversification of family incomes.

The three remaining situations presented in Table 6.8.1 confirm the above findings. It turns out that surveyed large farm operators are more likely to reduce household expenses (48.4 % confirming answers) than the respondents who operate smaller farms (39.2%). Respondents using and operating large farms are also more likely to take credits or loans to save the farm with 46.6% declaring so. Only 22.4% of respondents who use and operate smaller farms express such possibility. Interestingly, large farmers in the study are less likely to sell or rent out the part of the farm experiencing crisis (30.7% of responses) than the farm users managing smaller farms (36.4%). The differences in each of these cases are statistically significant at the level of 0.01. However, it is also the case that surveyed large farm users are less likely than other farmers to resign from educating their children in the situation of decreasing farm profitability with 81.8 % of answers indicating unwillingness to resign. Users of other farms also declare a similar stance on their children's education, but to a lesser degree—77.7% of answers. The difference between the two categories of respondents is less statistically significant and its level is 0.05).

Factor analysis of main components of anti-crisis activities and measures allows us to distinguish three configurations of particular action strategies applied in situations where profitability is threatened, described here as: "physical and financial efforts," "measures limiting production," and "household-related austerity measures" (see Table 6.8.2).

The anti-crisis strategy described as "physical and financial effort" includes such measures and activities as: increasing the volume of one's own farm work and the farm work of other family members, taking credits/loans, taking additional work outside of agriculture, reducing household expenses to the necessary minimum, as well as selling some furnishings and household conveniences. The activities and measures of the "limiting production" strategy mostly involve: selling or renting out part of the farm, selling off some farm machinery, taking up additional work outside of agriculture, and selling some household conveniences. The strategy of "household-related austerity measures" consists of the following components: resignation from providing university education to children, selling some furnishings and household conveniences, as well as reducing the household expenses to the necessary minimum, selling some of the farm machinery, and not taking any additional work outside of agriculture.

Table 6.8.2. Factor loads for main components of anti-crisis solutions and activities (Varimax rotation, loads below 0.20 were not considered)

Variable	Components		
	I: Physical and financial efforts and measures	II: Measures limiting production	III: Household-related austerity measures
Farm work	0.819		
Loan	0.685		
Additional work	0.576	0.332	−0.424
Home expenses	0.570		0.293
Sell farm		0.856	
Sell farm machinery		0.790	0.238
Education			0.799
Sell household conveniences	0.307	0.299	0.628
Percentage of explained variation	23.800	19.700	17.300
Total percentage of explained variation	60.8		

Source: Own research.

Table 6.8.3 illustrates the location of both categories of farmers on particular scales for anti-crisis strategies. The biggest difference can be seen in the "physical and financial effort" category. Large farm users are located well below the other farmers, which means that they implement this strategy much more often. The distance between the two categories is 0.385. The strategy of "limiting the production" is more often applied by farmers who do not manage large farms and they are located on the scale well below large farm users, while the distance between the two categories is 0.230. The strategy involving "household-related austerity measures" is the least differentiating between the two categories of respondents. Large farmers are located lower on the scale than the other farmers but the distance between the two categories is 0.075.

Table 6.8.3. Average factor values of main components of anti-crisis solutions and activities, according to user's category

User's category	Scale of anti-crisis solutions and activities								
	Physical and financial efforts and measures			Measures limiting productions			Household-related measures		
	average	signifi-cance	eta²	average	signifi-cance	eta²	average	signifi-cance	eta²
Large farmers	−0.317	0.01	0.022	0.189	0.01	0.008	−0.062	none	0.001
Other farmers	0.068			−0.041			0.013		

Source: Own research.

6.9. Some Concluding Remarks

It's time to draw some conclusions to analyses presented in this chapter. It is quite evident that the studied farmers generally declare a low level of trust towards other people. They are not different in that regard than Polish society in general. There are no significant differences between the two categories of farmers when it comes to trust. However, what is important here, lack of trust does not necessarily mean an unwillingness to cooperate with other people. Interestingly, a significant majority of respondents representing both categories of farmers declared the conviction that cooperation with other people can be quite beneficial. Does it negate any earlier declaration about lack of trust towards others? Or does it indicate lack of consistency or even some contradiction in how the respondents perceive their situation? An affirmative answer should not be surprising, as in various social situations contradictions and even a lack of logic occur with how individuals and various social categories think and act. However, in this concrete case of farmers' approach to trust and cooperation, the contradiction noted at first can be explained in another way. The low level of trust towards other people might be seen as an effect of the historical insufficiency of democratic traditions, as well as a related lack of trust towards the state. There is also the phenomenon of the "sociological vacuum" (see, e.g., Nowak, 1979), which, according to Polish sociologists who coined and described this term, has not completely disappeared even after radical social changes took place as a result of the political breakthrough of 1989 and despite the emergence of civil society in Poland (see, e.g., Czapiński, 2006; Cześnik, 2008; Pawlak, 2015; Jacobsson & Korolczuk, 2017). It should also be mentioned that the farmers' opinions on the benefits of working with others could be the product of the everyday experience of the studied farmers. Thus, the opinions presented above do not necessarily have to be contradictory. The respondents seem to appreciate the benefits of cooperation with others, which does not prevent them from applying the principle of limited trust towards those with whom they do cooperate.

The differences between large farm users and other farmers are noticeable in how both of these categories use various sources of information. Generally, large farm operators use these sources to a greater extent than other farmers. This should not be particularly surprising, as the logic of managing large farms that are more dependent on the global market and more prone to be affected by various external factors causing market volatility requires the farm operators to gather information from a variety of sources. Interesting relationships have been discovered through factor analysis, leading to categorization of information sources as either "traditional" or "modern." While the former are used by respondents of both categories to a similar extent, the operators of large farms are more likely to supplement them with modern sources of information. This last finding seems to go well with the assumption that it is the larger farms, with intensive production based on more complicated technological processes that, in a way, "force" their owners/operators to be familiar with all of the aspects of the "modern" ways of conducting economic activity.

The analysis of the frequency of use of various sources of advice for farm management show a greater intensity of such activities being performed by large farm users. It might indicate that this category of farmers is more likely to base their decisions and production-related activities on the consideration of various sources and external factors. As noted earlier, entering into various relationships, even those involving the use of advice, is an important element of cooperation. In this sense, farmers operating large farms are more inclined to implement this aspect of the concept of sustainable farming. The matter of using various sources of advice is also quite interesting. Similar to the use of sources of information, the factor analysis has led to distinguishing two types of sources: modern and traditional. As revealed in the study, farmers operating both types of farms use traditional sources to a similar extent. However, modern sources of economic advice are the domain of large farmers, who—as can be assumed—have to follow the logic of more intense and more market-oriented agricultural production.

Farmers who operate larger farms declare contacts with representatives of various institutions in charge of ensuring the quality and safety of food products more often than other farmers do. Furthermore, this category of farmers, more than other respondents, disagrees with the opinion that agricultural producers use forbidden practices in production in order to have higher incomes. How to explain these relations that in the study are statistically significant? It seems that the mechanism of denial is applied here. In the situation when large farm operators usually acquire greater incomes, statements on the use of forbidden practices to increase income, might appear to them as hurtful hearsay directed at them. On the other hand, it should be remembered that larger farms, whose owners declare quite frequent contacts with the representatives of institutions monitoring food safety and food quality, are under more visible scrutiny related to the quality of the production process and safety of produced commodities. This correlates with the more frequent declarations of large farm users in comparison to other respondents about disagreeing with statements on applying illegal practices in order to increase incomes.

Consumption of food produced on one's own farm can also provide a look into the characteristics of surveyed farmers. Those managing smaller farms—more often than farmers operating large farms—declare consumption of food produced on their own farm. This might lead to a formulation of a thesis that these respondents are more "peasant" in that regard than the farmers operating large farms who are, therefore, more oriented towards mass and more industrial production, which could be seen as an impersonal model of agricultural commodities production. However, both categories of respondents do not differ much in their declarations that they consume food produced on their farms because the quality of these products is better than the quality of foods in the stores. It can, therefore, be assumed that the peasant tradition with its message of "ours is better" is still thriving among Polish farmers. However, the popularity of this conviction can also be attributed to every contemporary producer who claims his or her product is the best. Such self-promotion of products is expected due to the hard rules of the market economy.

The study has revealed that larger farmers making certain production decisions related to their farms pay attention to factors such as commodity prices, production costs, credits and contract agreements, as well as possibilities for potential contamination of the natural environment. At the same time, the respondents who operated smaller farms are more likely to sell their products outside of the contract system and focus on various local products. These last two factors are recognized in sociological literature as the building blocks of sustainable farming. The factor analysis pertinent to this matter presents a more complex and more nuanced picture of the decision processes for both categories of farmers. Factors identified by farmers as essential to making production decisions can be arranged into three configurations, described as: "mass production," "quality with profit," and "local harmony." It turns out that respondents operating large farms preferred the first two decision strategies. The other farmers identify more with the strategy of "local harmony." Its name seems to reflect the ideas presented in sociological literature as crucial to the concept of sustainable development.

The opinions of large farm operators and other surveyed farmers on the characteristics that currently determine farm success are quite similar. This applies to characteristics that are traditionally linked to farming and rural areas, such as diligence and dedication to work, willingness to work hard, having a penchant for farm work, market insight, attention to the quality of products, ability to "wheel and deal" and know the right people, as well as paying attention to environmental concerns. Such similarities can also be found in the Dutch study results presented by Jan Douwe van der Ploeg (2020) which compare two types of thinking about the crisis experienced by farmers representing two categories of producers, namely: peasant-type farmers and large entrepreneurial ones. As Ploeg writes: "What farmers have in common is that they love their jobs, cherish their independence and are proud of the farm they develop. They equally share a feeling that things are getting worse as they experience ongoing regression and crisis" (Ploeg, 2020).

However, the above similarities, including those presented by Ploeg, do not exhaust all the characteristics of the categories of farmers distinguished in the study. It turns out that large farm operators are more likely than other farmers to point to the role of the characteristics connected with suitable education, qualifications, and willingness to take a risk. Other respondents prefer a different characteristic such as orientation towards traditional, local products, which in their opinions could foster market success for farmers. This characteristic is compliant with the concept of sustainable development.

As a final note, some attention should be given to different declarations of both categories of respondents regarding the preferable action strategies in situations of worsening profitability for agricultural production. Large farm operators tend to declare a preference for the "physical and financial effort" strategy more often than other farmers, which means that they choose more work and additional financial support to maintain farm profitability. Owners of smaller farms tend to pick a quite different set of activities and solutions in such situations. The core of their strategy might be "limiting the production," namely reducing the production potential of the farm. What is seen as a remedy to the crisis situation involves the shrinking of production rather than an increase in the effort. The least divergent between the two categories is the propensity for the "household-related austerity measures" strategy, meaning sacrificing some family needs, conveniences, and household expenses for farm profitability.

REFERENCES

CBOS [Centre for Public Opinion Research] (2018). *O nieufności i zaufaniu*. Study release No. 35/2018.

Czapiński, J. (2006). Polska – państwo bez społeczeństwa [Poland: The state without society], *Nauka* 1, pp. 7–26.

Czapiński, J. & Panek T. (2007). *Diagnoza społeczna* [Social diagnosis]. Rada Monitoringu Społecznego.

Cześnik, M. (2008). Próżnia socjologiczna a demokracja [Sociological vacuum and democracy], *Kultura i Społeczeństwo* 52(4), pp. 19–50.

Dzwończyk, J. (2006). Rozwój społeczeństwa obywatelskiego w Polsce po 1989 roku [The development of civil society in Poland after 1989]. In: B. Krauz-Mozer & P. Borowiec (eds.), *Czas społeczeństwa obywatelskiego. Między teorią a praktyką* [The time for civil society: Between theory and practice], pp. 161–181. Wydawnictwo Uniwersytetu Jagiellońskiego.

Gorlach, K. (2009). *W poszukiwaniu równowagi. Polskie rodzinne gospodarstwa rolne w Unii Europejskiej* [Searching for the balance: Polish family farms in the European Union]. Wydawnictwo Uniwersytetu Jagiellońskiego.

Gorlach, K. & Drąg, Z. (2019). Od gospodarstwa chłopskiego do przedsiębiorstwa rodzinnegod [From the peasant farm to the family business]. In: M. Halamska, M. Stanny, & J. Wilkin (eds.), *Ciągłość i zmiana. Sto lat rozwoju polskiej wsi* [Continuity and change: A hundred years of rural development in Poland], pp. 323–350. IRWiR PAN, Wydawnictwo Naukowe Scholar.

Gorlach, K., & Starosta, P. (2016). *Farming Families in Rural Communities: Changing Rural Social Organization in a Modern and Postmodern World*. In: M. Shucksmith & D.L. Brown (eds.), *Routledge International Handbook of Rural Studies*, pp. 518–530. Routledge.

Homans, G.C. (1964). Bringing men back in. *American Sociological Review* 29(5), pp. 809–818.

Jacobsson, K. & Korolczuk, E. (2017). *Civil Society Revisited: Lessons from Poland*. Berghahn.

Kasprowicz, D., Foryś, G., & Murzyn, D. (2016). *Politics, Society and the Economy in Contemporary Poland: An introduction*. Wydawnictwo Naukowe Scholar.

Marsden, T. (2003). *The Condition of Rural Sustainability*. Van Gorcum.

Nelson, J. & Stock, P. (2018). Repeasantisation in the United States. *Sociologia Ruralis* 58(1), pp. 83–103.

Nowak, S. (1979). System wartości społeczeństwa polskiego [Value system in Polish society], *Studia Socjologiczne* 4(75), pp. 155–173.

Pawlak, M. (2015). From sociological vacuum to horror vacuity: How Stefan Nowak's thesis is used in analyses of Polish society. *Polish Sociological Review* 1(89), pp. 5–27.

Ploeg, J.D. van der (2008). *The New Peasantries: Struggles for autonomy and sustaiability in an era of empire and globalization.* Earthscan.

Ploeg, J.D. van der (2013). *Peasants and the Art of Farming: A Chayanovian manifesoro.* Fernwood Publishing.

Ploeg, J.D. van der (2020). Farmers upheaval, climate crisis and populism. *Journal of Peasant Studies* 47(3), pp. 589–605.

Strange, M. (1988). *Family Farm: A new economic vision.* University of Nebraska Press.

Sztompka, P. (1999). *Trust: A sociological theory.* Cambridge University Press.

Sztompka, P. (2016). *Kapitał społeczny. Teoria przestrzeni międzyludzkiej* [Social capital: A theory of inter-human space]. Znak.

Think Locally, Act Globally: Polish farmers in the global era of sustainability and resilience, ed. by Krzysztof Gorlach and Zbigniew Drąg in collaboration with Anna Jastrzębiec-Witowska and David Ritter
Jagiellonian University Press, Kraków 2021, pp. 327–354
ISBN 978-83-233-4949-5
DOI: http://dxdoi.org/10.4467/K7195.199/20.20.12733

Chapter Seven: Polish Farm Women as Managers*

Zbigniew Drąg https://orcid.org/0000-0002-9106-7758/

Krzysztof Gorlach https://orcid.org/0000-0003-1578-7400/

7.1. Introductory Remarks

Sex- and gender-related social inequalities in the job market were of interest to numerous sociologists for nearly the last half of the 20th century but in the last two decades analyses of these issues have become more intense and more substantive. Undoubtedly, this is strongly connected with the growing popularity of feminist theories and the increased social activity of women, especially in the form of feminist movements. Analyzing gender inequalities in the job market—and they are reflected upon in this publication—some researchers (see Domański, 1999, pp. 29–37) refer to such mechanisms and processes as segregation of professions (relating to divisions made between male and female professions), ghettoization (division of female and male tasks within a given profession) or professional re-segregation (feminization of professions previously dominated by men), all of which impede professional integration aimed at achieving gender balance in a particular profession. Although direct workplace and profession-related discrimination of women mostly happens on a micro- -scale at the level of companies, business enterprises, and offices, these structural

---

*   The Polish version of this text was published in the leading Polish academic journal *Studia Socjologiczne* (Sociological Studies), No. 4/2019, pp. 129–156.

mechanisms indirectly enhance the discrimination. It is a fact that women, despite abilities and qualifications similar to those of men, have lower income and lower professional positions, usually reaching "the glass ceiling" at some point in their career. It is also a fact that in the more holistic approach to economics professions traditionally recognized as female, women are considered less prestigious and have fewer power resources (see Domański, 1992, 1999).

Without a doubt, agriculture is one of the economic sectors where gender-related inequalities are externalized and a traditional Polish farm is a special type of enterprise that petrifies these inequalities. As Polish researcher Barbara Tryfan notes, for many years the traditional type of peasant economy has determined the order of prestige in the family group according to economic usefulness. As the man performed works recognized as important and had decision-making powers, his position in the family was privileged. Without permission from her husband, a wife could not sell any livestock or equipment, even though it came from her dowry. Upon entering a peasant home, guests and family members greeted the man of the household first. He was also getting the best parts of the family meal (Tryfan, 1987, p. 160). While there are acknowledgements of changes affecting rural woman, as seen, for example, in the work of British researcher Sally Shortall, there are also arguments that a traditional peasant model of everyday life still prevails in rural areas. To quote Shortall: "Agriculture and the position of women in agriculture and society generally, has transformed over the last 40 years" (Shortall, 2017, p. 89). In order to get a better picture, it is helpful to juxtapose this generally accepted statement with the writings of Tiina Silvasti, a researcher from Finland, whose views are less optimistic. Silvasti states that "many powerful features of the traditional peasant script are still influential in everyday, rural life, especially on farms" (Silvasti, 2003, p. 162). As the cultural model of favoring male successors is still quite powerful, the situation of a farmer's daughter taking over the farm is not easy, even if she has the blessing of the family as the farm's successor. Her life as a female farm successor is determined more by her sex than by changes taking place in agriculture and in rural areas (Silvasti, 2003, pp. 162–163). Similar conclusions are drawn by Barbara Heather, Lynn Skillen, Jennifer Young, and Teresa Vladicka, who state: "Rural gender discourses portray women as responsible for the reproduction of the family farm as a social form" (Heather et al., 2005, p. 93). The discussion on the changes of the position of women in contemporary farming can be summarized with the opinion of German researcher Elizabeth Prugl, who noted the following: "Gender orders embedded in these regimes have produced distinctive outcomes for women and men working on farms. Under both regimes, gender has operated as a key regulatory means to divide labor, assign rewards, and produce power" (Prugl, 2004, p. 366). In other words, it is not the character of women's work on the family farm that changes much, but the work done by women becomes more meaningful and valuable and is no longer treated as unpaid labor.

The functioning of women in agriculture and their situation still receive significant attention from researchers as a study subject. Nowadays, rural areas are no longer treated as a selected segment in contemporary society. Thanks to the elaborate network of electronic connections (e.g., the Internet), these areas are more included in the mainstream of social changes. In describing how and where the information is unbound, Bettina B. Bock and Sally Shortall write: "In theory, rural peripheral locations may be as included in the network society as central and urban places" (Bock & Shortall, 2017, p. 2). However, they add: "In practice, the so-called 'connectivity' of rural residents is often lower . . ." (Bock & Shortall, 2017, p. 2). While taking that into account, it should be stated that the current situation of rural women does not justify perceiving them as a marginal social category. However, it should not be forgotten that the situation of rural areas and, at the same time, the situation of rural women in various communities, societies, or regions are quite diverse as civilizational changes do not exhibit the same dynamics everywhere. Bock and Shortall argue that the perspective of studies conducted nowadays should be changed to include "theoretical and methodological insight gains, and of the relevance of mobility and globalization" (Bock & Shortall, 2017, p. 3). Mobility and globalization are the two processes that undoubtedly have significant meaning and far-reaching consequences. These two processes allow for identifying and defining the differences and specifics of particular regions, places, and local communities. It is essential to understand how these differences influence the situation of women in agriculture and rural areas.

The reflections in this chapter focus on analysis of the situation of the rather unusual category of women working in Polish agriculture, namely the female owners of the farms or the main farm operators. In order to observe the evolutionary tendencies related to various parameters of their situation in the last quarter of a century, the data compiled from 1994 to 2007 is referred to, as well as the newest data collected in 2017 (detailed analysis of the situation of female farm owners and operators from 1994 through 2007 can be found in Gorlach, 2009; Gorlach & Drąg, 2011; Gorlach, Drąg, & Nowak, 2012). Determining whether the observed tendencies indicate an ongoing process of abandoning professional ghettoization in agriculture, as characterized by the secondary roles of women is one of the main goals of the scholarly reflection in this chapter. It seems also valuable to explore to what extent the process of multilevel professional integration, connected with balancing out the sex/gender proportions among the owners (main users and operators) of farms, has been taking place. Are the farms operated by men and women becoming similar in their economic dimensions? Do men and women share socio-demographic characteristics and do they become more similar in the cultural dimension? Perhaps these tendencies signal the process of a certain economic re-segregation, meaning women taking management of only specific types of farms, (namely, the smallest and the weakest economically) or becoming farm managers and operators only in very specific circumstances, when, for example, men are employed off-farm? Before delving into the data analysis and formulating answers to the above question, a closer look needs to be taken at the crucial elements of the picture of female farmers, as found in the international literature on the subject. The reflection of this chapter should refer to the broader, international context of how the roles of women are perceived in the functioning of farms.

7.2. The Role of Women in the Functioning of Farms and in Farm Changes: Comments based on international analyses

In contemporary sociological literature it is often emphasized that the situation of women in agriculture and rural areas has gone through significant changes over the last few decades. However, these changes did not go unambiguously in the direction of eliminating all regionally specific characteristics and nullifying, at least to a certain extent, some traditional elements of social order. In other words, it can be said that, despite advanced globalization processes and the tendencies towards convergence of the social landscape that stem from them, regional and traditional characteristics are still visibly intact. They remain significant features characterizing societies in various regions of the contemporary world. This is highlighted by British researcher Sally Shortall, who states, "Gender segregation is still a pervasive feature of all labor markets and there are differences in men's and women's employment by sector, workplace, and occupation" (Shortall, 2017, p. 89). She also adds that "social patterns persist and within agriculture the perception of farming as a male industry is hard to break . . ." (Shortall, 2017, p. 89). This particularly applies to family farms where, as Shortall points out, the work of women seems to be treated as invisible. Attention is given to those activities which mostly relate to field work and are the domain of farm men. Women's work in agriculture is still, in fact, overlooked but the extent of this omission depends on the region or country. It should be noted that it is women who mostly tend to livestock on family farms and they do participate in field work when such work accumulates to a point beyond the abilities of one person, or even a few people for example, during harvests and haymaking in particular. It is true that women's involvement in the family farm is concentrated on reproduction of the functioning of the farm. Nevertheless, it would not be accurate to say that women are not engaged in production itself. Therefore, it needs to be emphasized that the "invisibility" and "secondary character" of women's activities do not stem from objective premises and these activities do not have less value for the overall functioning of the family farm. However, due to the immersion of farms in the patriarchal system of social relations and cultural values, the presence of a woman in agricultural work and activities is only noticed when the man of the house gets employment outside of agriculture, and earns higher income than the revenues from the farm that the woman is then managing. The male domination is so obvious that even in the academic literature on the subject of rural relations one can find statements that women are not authentic farmers. They only become them when something happens to the men, who are seen as naturally predestined to be farmers (Bock & Shortall, 2017). As a general rule, women enter agriculture not by inheriting the farm but through marriage. They are also underrepresented in agricultural organizations. Finally, the programs that are meant to prepare would-be farmers for agricultural work are usually designed for men. For that reason, the sex and gender division persists in agriculture more so than the advanced social and cultural changes within the farming world would indicate is to be expected and standard, and this also applies to highly developed societies. Within this context Sally Shortall formulates the quite explicit conclusion that the ". . . patriarchal nature of farming and farming industry persists despite resistance and a changing society" (Shortall, 2017, p. 92). Despite many years of feminist influences, patriarchal relations continue to have significant influence on the agriculture of Western Europe. The stereotypes and convictions related to sex and gender stay remarkably intact despite ongoing economic and political transformations. In this context, Ralph Dahrendorf's metaphor for changes in Europe, especially those in Central and Eastern Europe, seems quite fitting (see Dahrendorf, 1991, pp. 68–100). This metaphor—worked out in detail by Piotr Sztompka (2016, p. 13)—alludes to three different rhythms (clocks) that social changes tend to follow. Certainly, the fastest clocks are political and legal changes and solutions creating frames for new institutions and the modes of organizing social and individual life. Somewhat slower is the clock of economic changes. The slowest clocks are the changes of social life, especially those taking place in the cultural layer and related to value systems, lifestyles, civil society, etc. Therefore, Dahrendorf's metaphor (see also, 1990) explains well the persistence of patriarchal relationship patterns on family farms in societies that have moved well beyond the traditional, or pre- and early capitalist, forms.

Obviously, the situation is quite dynamic. The researchers point out that women have become increasingly more active in social life and they are more consciously planning their social participation or, to use the language of sociology, their fulfilment of social roles. They are still not able to break the dominance of men in the farm context, and more broadly—in the context of the food production system. They gain empowerment and independence only in those situations where they find employment outside of agriculture. In other words: women need to leave the patriarchal system to become visible and meaningful actors in the given society. Women's employment outside of agriculture is an effect of better education and putting a certain pressure in this direction. It turns out that women earn more outside of agriculture, and they are more connected to the job market, than men who are involved in farming. Also, within the non-agricultural job market, women are more dynamic than men who stayed on the farm. Furthermore, the incomes of women working outside of agriculture have a substantial value which greatly supports the functioning of farms. Women sometimes play the role of rescuers, especially in situations where the farm is not bringing in sufficient income (see Bock & Shortall, 2017). Interestingly, even women who work outside of agriculture and earn substantial pay from such work are still considered to be an element of the farm, unless they break from it and move elsewhere. As emphasized by Sally Shortall (2017, p. 93): "Women's off farm labour is often part of a farm household survival strategy to maintain the farm and men's occupation as the farmers . . . In addition, women's off-farm work can have an implication of men's sense of identity." There is no ambiguity in this statement about the strength of traditional ideologies and how they relate to the place of women in agriculture.

It should not come as a surprise, considering the general principle of women holding subordinate positions on farms, that the percentage of women managing farms is fairly marginal. In Sweden it is 25% (Stenbacka, 2017, p. 114) and in Greece and Croatia this percentage is quite similar (Shortall, 2017, p. 94). The situation in Poland will be discussed later in this publication. In all these countries, including Poland, the secondary position of women is beyond dispute and this is not just about the above percentage values falling well short of 50%. It is also the mindset that does not question patriarchy. Here, Shortall (2017, p. 94) provides the example of Ireland where women generally accept the patrilineal tradition of inheriting farms. It happens even in those situations where the male descendants of the current owner of the farm are not particularly interested in working in agriculture and females on the farm have an adequate level of education and a certain level of financial independence. These characteristics—as was previously noted—are more likely to put them on a path to leave the family farm rather than take up managing it.

When analyzing the role of women in the functioning of farms it might be worthwhile to go beyond the findings of Sally Shortall (2017, pp. 89–99), presented above. In Australia, Margaret Alston (2017, pp. 100–113) places the role of women on farms in the context of climate change (!). The researcher asserts that Australian agriculture is subjected to two regimes. On the one hand, there is a natural regime (even though it is influenced by human activities) of global warming, which results in repeating periods of drought, as experienced by Australian farmers. On the other, there is a sociopolitical regime regulating the functioning of the economy (including agriculture) within the framework of neoliberal market principles and individual responsibility of entrepreneurs/farmers, who are mostly men. In situations of unstable farm incomes, the off-farm earnings become meaningful and, to a large extent, they are the domain of women. Alston (2017, p. 102) states that the work of farm wives has broadened to also include the off-farm work. She adds: "A critical farm survival strategy has been the adoption of off-farm work as a major source of income support and the women are usually responsible for this income generation" (Alston, 2017, p. 110). This becomes a problem for rural family dynamics that the off-farm work of farm wives fosters the emancipation of women and the neoliberal ideology of individual responsibility undermines the family character of the farm. What happens on the farm is treated more like a personal success or personal failure of the main operator/user, with no reflection on the entire family. It can be stated that off-farm, non-agricultural work, especially that performed by women, and neoliberal ideology oriented around the individual are the factors that break the family ties constituting the most important elements of a family farm, where close relation and blending of agricultural household with family business used to be fundamental on farm. Alston goes even further, claiming that climate changes are indirectly caused by the chain of causes that have a lot to do with severing the strategies and life paths of men and women who have constituted farming families. The author emphasizes that, "While men remain deeply entrenched in the ideology of family farming, women have moved along the continuum and many have mentally, if not physically, exited farming" (Alston, 2017, p. 110). This statement shows the lack of appropriate agricultural policies which could meet the needs of women, as well as the needs of agricultural families. This also explains the lack of mobilization of Australian rural (agricultural) policies in the face of climate change. The author recognizes and highlights "a move away from a joint commitment to a farm family ideology towards a more individualistic, de-traditionalized view" (Alston, 2017, p. 111).

It seems, however, that individualistic attitudes can also lead women directly to agriculture. This is confirmed by the results of a Swedish study indicating that in Sweden, when women share the entrepreneurial ideas, linked to various individual or social (local) contexts, the percentage of female farmers/managers increases (see Stenbacka, 2017, p. 114). It should be added that the average woman engaged in agriculture is somewhat younger than the average man involved in it. Could this possibly be a challenge to the traditional vision of agriculture as a masculine reality? More important might be the fact that values shared by female farmers (see Stenbacka, 2017, pp. 124 and others) do not only apply to quality of life and financial security. Contrary to what is recognized as traditional female values, focused on safety and security of family life, the goals of these female farmers/managers go beyond the traditional treatment of women as secondary and only supporting shareholders of the entire system. Female farmers and the values they share allude to the financial success of the farm, which is tied up with the welfare of the family and being environmentally friendly (Stenbacka, 2017, p. 125). Could this mean that women mitigate baser tendencies in agriculture, preferring linking market results (such as financial success) to quality of life, equally in the social sense (quality of social life) and the natural sense, with an emphasis on protecting natural resources?

For the academic reflection of this chapter, studies that show the willingness of women to acquire or improve their formal qualifications as farm managers or farm operators seem particularly valuable. Comparative studies conducted in Greece and Croatia related to women's motivation to participate in agriculture-related educational courses show interesting tendencies. As indicated by Černič Istenič and Charatsari (2017, pp. 129–146) the motivations of women in both countries are similar, despite their different historical paths. It's the "invisibility of women" stated at the beginning of this chapter that appears to be a dominant factor driving women to look for knowledge in the framework of the offered educational courses. This desire to gain specialist knowledge which allows women to use any opportunities to manage a farm enterprise is an indicator of a growing motivation level among women to explore innovative farm development (see Černič Istenič & Charatsari, 2017, p. 147). The study results show that women actively try to break tradition and customary regulations relating to the function of farms. This means getting rid of existing stereotypes regarding the role of women in farming. However, such stereotypes do not die easily and here Ireland can be a prime example, as described by Anne Cassidy (2017, pp. 148–161). To highlight this, the researcher refers to the paradox of empowerment. Farmers' daughters believe that their position is equal to the position of men, as they can advance their careers thanks to education and life opportunities. However, they do so outside of agriculture. At the same time, they are aware that in competition with their brothers they do not have a fair chance in taking over the family farm. Farmers' sons have no problem believing that their chances for a career outside of agriculture are open, due to various professional opportunities, but they simultaneously feel an overwhelming pressure to take over the farm from their parents. This shows, without much ambiguity, that sons are guided in the direction of the individualistic and dominant position of the farmer, while daughters are motivated—what comes coincidentally in line with the feminist position—to seek further education. However, this manner of raising daughters is still subject to the dominant matter of the farm and its succession. The message that one could get from this situation is the following: the farm is passed to a male (son) and women (daughters) are encouraged to study and have a professional life off-farm. This way, they do not disturb the natural and traditional path of the family farm succession. In this sense, the life choices of young Irish people connected with agriculture and rural life cannot be fully considered to be simple expressions of individual free choice or empowerment but: "are often choreographed from childhood through decisions made at an early age around the intensity of their involvement in farming and engagement in education" (Cassidy, 2017, p. 159).

7.3. Female Farmers: The context of the research problem

The above analysis uses the example of highly developed countries such as Australia, Sweden, and Ireland, or relatively developed ones (Greece, Croatia), to show that women's access to the farming profession and—to be more precise—to the position of primary farm manager/operator is not without significant difficulties. What is important here is that the difficulties of that access are mostly caused by cultural factors, namely by the patrilineal character of farm succession and entrance into the farming profession. It should be noted that the efforts made by women to acquire better education and competencies foster women's exodus from farm operations and, more generally, from the countryside. Education that is formulated to serve operating the farm into the future is, for the most part, aligned with the needs of men. The masculinization of the farmer's profession and the patrilineal character of farm successions provide strong enough reasons for academic interest in the situation of women who—despite life challenges and unfavorable circumstances—play the roles of farm owners and farm managers. This specific category is the subject of this work.

Berit Brandth and Marit S. Haugen (1997), two women researchers from Norway, identify three perspectives in studies on women engaged in change in rural areas and agriculture. The first perspective focuses on practices that discriminate against women and are not necessarily connected with rural areas and agriculture and are more relevant to the patriarchal tradition characteristic of European culture. In this approach, the dominance of men over women in rural families is the effect of a more general tendency that does not have any specific conditioning from the local setting. The second view can be described as a certain "female perspective" pointing out essential differences between men and women. The prime example is agriculture where, for many years, physical labor played a major role, as it was a very necessary and integral aspect of agricultural work. The effects of these dependencies are reflected in some types of work being treated as men's work and other types as women's work. In English language publications there are so-called "his and her jobs" (see Brandth & Haugen, 1997). The third perspective described by the Norwegian authors concentrates on differentiation among rural women in terms of education, competencies, abilities, and the variety of tasks that women perform on a regular basis, both on the farm and in its domestic household. It should be remembered, however, that rural households and farms make a certain union within the family farm.

Considering Polish rural tradition, the second perspective would seem to be the most fitting for analytical purposes. In Polish agriculture, men's work was mostly connected to farming the land and using the agricultural machinery. These tasks were external to the rural domestic household. Women's work was mostly concentrated on the internal part of the farm and household and involved tending to animals and growing crops for family needs (vegetable gardens, flowerbeds, etc.). Obviously, household duties are not even discussed here, as in the traditional peasant family they were exclusively the domain of women. This perspective, however, has very little use in the context of changes connected with the modernization of farms. The third perspective appears to be a more suitable one, as technological advancement plays a crucial role in the transformational processes of farms and changes in the roles of women. What is meant here by transformation are the processes of mechanization which, from a contemporary vantage point, could be treated as a traditional type of technological change, as well as the much more recent changes connected with the digitalization of life, including the functioning of farms. Both types of technological changes can bring about more visibility and more importance for women in agricultural households. As stated in the earlier work of Krzysztof Gorlach, "It is worth remembering that the modernization of agriculture changed, first and foremost, the typically male areas of work, namely the field activities. Here, the burden and nuisance of work were eliminated primarily by progressing mechanization" (Gorlach, 2009, p. 186). Sociological literature usually presents this process as supportive of the relative dominance—not to be confused with traditional or absolute dominance—of men due to the masculinization of work done in the field (see, e.g., Brandth, 1994, pp. 144–147). But male dominance is no longer absolute and does not stem exclusively from the simple advantage of physical strength, which was necessary to perform such tasks as harvesting or sowing in the time before mechanization. Nowadays, men stress their relative dominance, pointing out higher competencies and abilities to use farm machinery. However, this type of dominance might well be questioned by women, who can acquire this type of knowledge with only the necessary motivation to do so. Therefore, this male dominance could be seen as more of an effect of the processes of social construction, rather than a result of natural factors connected with the respective, typical physical strengths of the sexes. Additionally, the process of achieving a certain balance between the roles of men and the roles of women in the farming context can be augmented by the increasing presence of digital technologies in farm management, which allows farmers to accomplish goals based on their level of knowledge and competencies, not their sex/gender. Another aspect of the same process of change deals with mechanization and digitalization, and consequently decreases the engagement of family members, neighbors who help, or hired workers.

These processes create conditions that change the situation of women from rural families in two ways. On one hand, the requirement for physical strength connected with field work and some work related to tending to animals no longer applies. Ability to use the farm equipment becomes a decisive factor. It means that women using a tractor or being at ease with the computerized aspects of managing a barn are no longer a rarity. On the other hand, changes in assignation of work to people also bring about organizational changes in the functioning of the family which operates the farm. Mechanization or computerization, as well as the potential to hire specialist service companies or subcontractors (i.e., a kind of outsourcing) to perform various tasks, can lead to a situation in which the farm might be operated just by one person (male or female) and the rest of the family is free to focus on other activities. Such situations enable women to become full decision-makers in farm matters or, alternately, completely concentrate on professional activities in another realm. It is worth remembering that social changes occur rather slowly compared to material factors such as technological or economical changes. Therefore, it should be emphasized that changes related to the position, role, and importance of women in changing agriculture and modernizing and globalizing rural areas can be viewed, ultimately, as the result of sociocultural changes. They are the effects of changes in habits, customs, convictions, and perceptions. These are much harder to change in comparison to technological, institutional, or organizational changes and they do occur much later.

To recap, one might ask what image of the contemporary woman on a family farm could be inferred from the findings presented above, and whether this image is dominant in the global sociological literature. What elements of that image should be accentuated in empirical studies related to the processes of acquiring gender balance in agriculture, which is often considered nothing less than a form of professional integration? Generally, it is said that, despite the processes of mechanization and digitalization of agriculture, which indeed limit the significance of physical strength, and which have *de facto* broken the monopoly that men had on performing the central tasks of operating a farm, professional integration—understood as the establishment of an equal footing between men and women in managerial positions on farms—is still lacking the expected dynamics. The state of sex/gender segregation is still preserved: men maintain the primary positions, including the position of authority and the position of lead person doing the field work, while women are left to conduct work of secondary importance, mostly those related to household duties. Such a situation is determined by well-preserved cultural habits that still have decisive meaning in the sphere of farm succession and in the preferred model of education for children and for adult members of the family farm. It is the man who is recognized as the legitimate successor of the farm and, in order to fulfil that role he is educated in both his youth and adulthood to increase his agricultural qualifications by taking professional courses. The woman is more likely to become a farmer through marriage. Most of all, she is prepared for leaving the farm and motivated to improve her qualifications in such way as to facilitate a professional career outside of agriculture. This is conducive to casting her in the role of "main rescuer" of the farm in situations of financial crisis. It is a woman's task to support the family with earnings acquired outside of agriculture during such periods.

Is this state of affairs adequate to the reality of Polish family farming? To answer this question, several matters should be analyzed: a) the extent to which Polish women farmers take over farm management, b) the processes of farm mechanization and digitalization, c) farm economic condition, d) manners of succession, e) the education of farm owners and managers and f) the off-farm employment that household members take up.

In the Polish academic literature, more general and more synthetic analyses devoted to women who manage farms are rather lacking. One of the most recent works on this subject addresses only one region in Poland, the Mazovian Province. The results of empirical studies show that women in this province manage farms that are rather small and quite fragmented. Women take less land in lease and rent out less land than men do. With the attempts to generalize the particular situation of female farmers in the Mazovian Province, it should be emphasized that they are less active in the agricultural real estate (property) market. Women are also less inclined to make decisions about expanding the farm as this ties in with the increased expenses related to mechanization and a greater production workload. Women, less frequently than men, operate farms that are specialized, although—and this is worth emphasizing—the tendency does not apply to college-educated individuals who manage farms (Pięta, Skierska-Badura, & Skierska-Pięta, 2013, pp. 226–233). In this category it is the women who make management decisions on specialized farms (see Pięta, Skierska-Badura, & Skierska-Pięta, 2013, p. 234). At the same time, women are less likely to change their situation, especially if doing so would add to their workload. The situation of women is embedded in the stereotypes that proliferate in the rural community. Stereotypes designate women as responsible for the smoothness of family life and raising children but rarely tie their work to managerial positions, so their work input in the functioning of farms is perceived as being of lesser value. The authors state the need for a new mechanism, as well as courses and information campaigns, which would help in breaking psychological barriers and stereotypes, as well as foster and promote successes achieved by rural women (see Pięta, Skierska-Badura, & Skierska-Pięta, 2013, pp. 244–245).

Within this context it is worth recalling that the main issues discussed here refer to the thesis on women in agriculture that was presented in the earlier works of the main author of this publication (see Gorlach, 2009, pp. 179–208). First, the dynamics and the extent to which women take managerial positions on family farms will be analyzed here. This will allow us to determine the level of gender imbalance in regards to primary and secondary positions on farms and also show the extent of professional ghettoization. Second, the comparisons between male and female farms will be made in the economic dimension. This will be helpful in exploring whether women taking over the management of farms can be treated as a demonstration of steps taken towards addressing the sex/gender imbalance in the farming profession. What if this is only a simple—be it lasting or fleeting—consequence of men being employed off-farm? In other words, is professional integration really occurring thanks to the processes of mechanization and digitalization of farms, with physical labor losing its importance and meaning? Or perhaps these are the processes of a certain professional re-segregation and a consequence of men shifting to better paying jobs and professions. Third, focus will be placed on analysis of socio-demographic and cultural characteristics of farmers of both sexes, and the similarities and differences between them will be addressed. The results of this will help us to pinpoint whether socio-cultural factors can contribute to achieving full professional integration or to weakening of this dynamic. The answers to all the questions above will address and describe in detail the issue of sex/gender segregation in Polish agriculture. In particular, the question of whether the profession of farmer is typically male, as well the manifestations of such an assumption will be analyzed here. The factors fostering or limiting the processes of professional integration (achieving the sex/gender balance) will be identified.

Analyzing the role of women in the functioning of family farms in Poland after 1989 and wanting to grasp the extent of changes, studies covering the period of time from 1994 to 2007, and those conducted later in 2017, are often referred to. Both studies were conducted using the nationwide sample of individual farms sized over 1 ha, representative of the space and area structures. The studies conducted over the period of time from 1994 to 2007 took the form of research panels and served to illustrate the fates and fortunes of family farms in the period of adjustment to market conditions and the requirements of the Common Agricultural Policy of the European Union after Poland's accession in 2004. In 1994, a sample of 800 farms was drawn, representative of macro-regions (there were 9 macro-regions outlined in reference to the 1975 administrative divisions of the country) and the size of farm land (6 area groups were established: 1–2 ha, 2–5 ha, 5–7 ha, 7–10 ha, 10–15 ha and 15 ha and over). It was based on data acquired from the Central Statistical Office. Their continued fates and fortunes were analyzed in 1999 and then in 2007. This showed that out of 800 surveyed farms 685 lasted to 1999. In 2007 the number of surviving farms from the initial 800 dropped to 510. Sample selection and empirical studies were conducted by the Research Center of the Polish Sociological Association (see Gorlach, 1995, pp. 9–21). In 2017, a research sample was constructed on the basis of the 2010 agricultural census data. The size of the sample was apportioned in such a way as to facilitate the use of multilevel models in the analysis of the impact of territorial differentiation on farm characteristics. The sample maintains the representative character for regions (16 voivodeships (provinces) were considered in the study according to the present administrative divisions of Poland) and farm size (four area groups were set up: 1–1.99 ha, 2–4.99 ha, 5–9.99 ha, and 10 ha and over). Due to obligatory principles regarding the confidentiality of statistical data, direct sampling of individual farms through the Central Statistical Office was not possible and, therefore, development of the research sample in the 2017 study was a two-stage process. First, the sample of households was drawn from the database of the Central Statistical Office. Then the selected households were verified by the Agency for Restructuring and Modernization of Agriculture as to whether they utilized their agricultural land. As a result, 3551 farms were studied and the study results illustrate the current situation of family farms in Poland. Sample selection and empirical studies were conducted by the Center of Sociological Research in the Institute of Philosophy and Sociology at the Polish Academy of Sciences.

7.4. The Dynamics of Changes in the Role of Women in the Functioning of Family Farms from 1994 to 2017

The profession of farmer, similar to that of a miner, is considered to be typically male. This does not mean that both professions experience sex and gender segregation equally. While it is very hard for women to be employed as miners, the situation in farming is quite different and the professional integration, understood as sex/gender balance, has reached a rather high level in the farming profession. According to the 2010 census data, women made up 45.7% of farm family labor.[1] Perceiving the farming profession as masculine does not make it unattainable for women. It is mostly the already mentioned division of male and female work within the profession and the ghettoization of the profession that makes it harder for women to be accomplished farmers.

Has the last quarter of a century changed anything in this regard? What was the nature and the course of the process of women achieving managerial positions on Polish farms, which were traditionally recognized as reserved for men? Could this be indicative of processes of professional integration understood as processes for achieving sex/gender balance in managerial positions?

Table 7.1. Farm owners/operators within the sample (in %)

Sex/gender	Farm owners/operators according to the year		
	1994	2007	2017
Women	28.2	30.2	33.7
Men	71.8	69.8	66.3
Total	N = 800	N = 510	N = 3551

Source: Own research.

---

[1] See https://bdl.stat.gov.pl/BDL/dane/podgrup/tablica (access: March 4, 2019).

Data on sex/gender of the main operators of farms suggest a slow increase in the share of women managing farm work. In 1994, women made up 28.2% of managers of family farms, and by 2017 this percentage had grown to 33.7% (Table 7.1). It is hard to state with certainty that these numbers are an indicator of an approaching meaningful sex/gender balance on farms. The question remains whether this is a lasting tendency, connected with limiting burdensome types of work, similar to processes occurring in highly developed countries. There is also a possibility—already discussed—that women's takeover of farms might be of a situational nature and a consequence of certain other factors, such as improvement of the job market outside of agriculture, which give men the opportunity to find stable off-farm employment. If this last factor turns out to be decisive, it could indicate growing integrational tendencies and, even more so, professional re-segregation understood as women taking over professions dominated, until recently, by men as the men move to other, better paying, jobs and professions. This should be explained further through analysis of the basic economic characteristics of farms operated by women and men.

Based on studies covering the period of time from 1994 to 2007, a general statement is justified regarding the fact that farms which are operated by women have a much worse market position than farms operated by men (Gorlach, 2009, pp. 179–222). Considering the results of the 2017 study, and taking into account the dynamics of changes in the last decade with reference to the 2007[2] study, the objective determinants of the financial situation of farms operated by women and by men should be analyzed. These determinants are: size of the operated agricultural land, value of the machines and equipment in one's possession, as well as revenues from the agricultural business.

Table 7.2. Utilized agricultural area in ha (in %)

Sex/gender	2007*			2017*		
	up to 5 ha	6–10 ha	over 10 ha	up to 5 ha	6–10 ha	over 10 ha
Women	62.3	20.8	16.9	73.8	16.2	10.0
Men	38.9	29.6	31.5	55.0	21.2	23.8

* p < 0.01.

Source: Own research.

---

[2] Households surveyed in 2007 were part of the 1994 research sample that initially included 800 farms. The 510 farms in the 2007 sample were those that remained operational out of the original 800 selected in 1994. Considering how carefully they were selected in the initial study it could be argued that they were the best, most successful farms and, for that reason, did not reflect the actual size structure of farms in 2007. However, the data from 2007 fulfill the criteria of our comparative analysis of farms operated by women or men.

Both in 2007 and in 2017 women operated much smaller farms than men (Table 7.2). The average size of a farm operated by a woman was 7.4 ha in 2007 and 5.5 ha in 2017, while the average size of a farm operated by a man was 12.2 ha in 2007 and 10.3 ha in 2017. These data show that the differences in size between farms operated by men and women were becoming greater: in 2007 the farms operated by women were 39% smaller than the ones operated by men and by 2017 this difference had reached 47%. It could therefore be inferred that women generally take over the management of farms with the smallest land area.

Could the farms operated by women be more mechanized with the limited use of physical labor? Looking into data acquired in the 2017 study, presenting the self-declared value of machines and equipment that were at farmers' disposal might give some clue to this question (Table 7.3).

Table 7.3. Declared value of agricultural machines and farm equipment in 2017, in thousands PLN

Sex/gender	No ownership of agricultural machines and equipment (%)*	Declared average value of agricultural machines and farm equipment, in thousands PLN*	
		All machines	The most expensive machine
Women	26.1	46.5	16.0
Men	13.3	93.5	31.1

* $p < 0.01$. When calculating the averages the farms that did not have agricultural machines or farm equipment were also included, with the assigned value of 0 PLN.

Source: Own research.

Farms operated by women farmers were much less mechanized than those operated by men. Women were twice as likely as men not to have any machines and the average value of the machines and farm equipment on the farms operated by women was 46.5 thousand PLN. On the farms operated by men it was 93.5 thousand PLN. This could lead to the conclusion that farms operated by women had half the mechanization level value of farms operated by men. Similar disparities could be seen when the average values of the most expensive piece of farm equipment were compared. Such machines on the farms operated by women had typically half the average value of the most expensive machines on the farms operated by men. Considering this, it is rather hard to form a thesis that the mechanization of agriculture fosters farmers' professional integration because of sex/gender. The opposite situation appears more likely. The most mechanized farms remain in the hands of men and women take the management of the least mechanized farms.

The analysis of the use of information technologies in agricultural households could also shed some light on the types of farms that women manage. However, according to the 2017 study, the extent to which such technologies are used on farms overall is so small that it does not seem to have any meaningful influence on women taking the management of farms. To be more precise, computers were used on 4% of farms operated by women and on 5.3% of farms operated by men, and that percentage included even minimal usage.

Are size-related discrepancies between farms operated by women and those operated by men reflected in farm profitability? Taking a look at farm incomes declared in 2007 and in 2017 might be useful for this analysis. In 2007, study participants estimated their revenues for the years 2004–2006 and in 2017 the revenues for 2014–2016 (Table 7.4). Before starting a more detailed interpretation of the data one general remark should be made. Considering the significant risks of conducting agricultural activities connected with unpredictable natural phenomena that remain out of the control of agricultural producers (for example, droughts, hail, flooding, heavy rain, etc.) it could be expected that the income farmers make could vary dramatically from year to year. However, the declarations of farmers participating in the study did not confirm this. The income structure of all producers in particular years showed very small fluctuation: from year to year the percentage of particular income groups increased or decreased by only several percent. On one hand, agricultural production turned out not to be as risky as initially assumed and extraordinary losses were not so frequent. On the other hand, there were no incredible profits. Looking into the data from years 2004–2006 and 2014–2016 it can be inferred that in the last decade the profitability of family farms increased significantly (at least nominally), yet those increases did not take the form of surges (for example connected with Poland's access to the European Union), and were slow and gradual. Thus, Polish farmers could not be counted among the most risk-taking market players but should be seen as subscribers to regular, stable development and a safety-oriented model of conducting business activities.

The declared profitability of farms operated by women was much lower than the profitability of farms operated by men. This was true in the years 2004–2006 and later in the years 2014–2016, as well. Within one decade the nominal income of farms operated by women increased, on average, by 8–9 thousand PLN, while this same increase for the farms operated by men was 12.5–13 thousand PLN. This indicated that farms operated by women had lower growth abilities in the last decade than the farms operated by men. It should also be noted that in the years 2004–2006 only farms managed by men had a regular annual increase of incomes by 5–7%. Such a regular increase in incomes was also reported in the farms operated by women only in the years 2014–2016. However, this annual increase was very modest and only slightly exceeded 2%. It was somewhat lower than the income noted in farms operated by men, which was 2–4%. This might be indicative of the stabilizing perspectives of growth for farms operated by women in recent years. To summarize, it should be stated that the difference in profitability between the farms operated by women and the ones operated by men—measured in absolute categories—increased regularly. In the years 2004–2006 this difference was, on average, 8.5–11 thousand PLN, and in the years 2014–2016 it was 13–14 thousand PLN. Obviously, it should be remembered that the income data described above has only a declarative character and only includes the income from agricultural activity conducted on the farm. In most cases, such income comprises only part of all revenue of the study participants' households.

Table 7.4. Average annual incomes* from farming, in thousands PLN, according to farmers' declarations during the 2007 and 2017 studies**

Sex/gender	Average annual income of farms surveyed in 2007; income for years 2004–2006, in thousands PLN			Average annual income of farms surveyed in 2017; income for years 2014–2016, in thousands PLN		
	2004	2005	2006	2014	2015	2016
Women	8.33	8.81	8.13	16.42	16.82	17.18
Men	16.89	17.69	19.01	29.67	30.29	31.54

* for all years: p < 0.01.

** When calculating average income values, farms with no income from agricultural activity were also included, based on declarations made by research participants. The value of such income was assigned as 0 PLN. In the years 2004–2006 these farms with zero income comprised about 25–27% of all farms, and in the years 2014–2016 they accounted for only about 4%.

Source: Own research.

Taking into account farm size, the level of mechanization (measured by the value of agricultural machines and farm equipment), and the income from farming, should confirm that the economic condition of farms operated by women is much weaker than that of farms managed by men. In 2017, farms operated by women had incomes that were 46% lower than the incomes of farms operated by men. The female-operated farms were also smaller by 46% and the value of the machines in their possession was half the value of machines on the farms operated by men. Would this validate the thesis that women are taking managerial functions mostly on family farms that are the weakest economically? Could this be something different than the expressions of the process oriented toward professional integration, and indicate some other form of sex/gender inequality or perhaps even a gradual withdrawal from ghettoization and moving towards professional re-segregation? If that were the case, what other phenomena would accompany this process? If women took over the management of farms from men, who found better employment off-farm, and not as legitimate successors, then they should be the ones who entered farming through marriage more often than men. Receiving a farm as inheritance should be less frequent among female farm managers than among male farm managers. Furthermore, playing along with this line of thinking, women should combine farm work with other paid job activities less frequently than men. These types of jobs should be the domain of other household members. Could the research data confirm that female farmers were more likely than their male counterparts to acquire farms through marriage and less likely to take additional jobs off-farm? In comparison to farms operated by men, were the household members connected with farms operated by women more willing to help with farming and, at the same time, more involved in off-farm employment?

Table 7.5. Circumstances of taking over the farm in 1994 and in 2017 (in %)*

Sex/gender	Inheritance**		Marriage		Purchase		Lease	
	1994	2017	1994	2017	1994	2017	1994	2017
Women	73.0	76.4	15.9	16.2	8.4	5.5	2.7	1.9
Men	82.1	87.3	6.3	7.1	9.4	3.6	2.3	2.0

* For 1994 and 2017 statistical significance was at the level −0.01.

** Taking over the farm after parents or other relatives.

Source: Own research.

The data in Table 7.5 show that agricultural family traditions are more often being respected by men than women. Male farmers who acquired the farm as a result of inheritance made up four fifths of the total number of men surveyed. This ratio for female farmers was three-fourths. Women entered farming through marriage more frequently than men, and that made them similar to rural women in Western Europe. The data in Table 7.5 confirms that, in Poland, similar to the countries of Western Europe, a patrilineal character of land succession still applies. This could serve as an argument for the interpretation of the process of women taking over the managerial functions on farms as re-segregation rather than professional integration.

Now it seems appropriate to take a look at the employment of farm operators and other household members. While in 2007 14.3% of female farmers and 23.9% of male farmers worked off-farm, a decade later those respective values were 35.9% and 46%. Such a significant increase in additional employment could be seen as related to the improvement of the market situation outside of agriculture. It could also mean rising possibilities for finding jobs by farm owners/operators as well as other household members. Besides the lower participation of women in supporting their households with additional off-farm work, the data show that in 2017 female farmers used the help of other household members in performing farm work more than their male counterparts did. Furthermore, other members of farm households managed by women more frequently took work outside of agriculture than the household members on the farms managed by men. The members of female-run farm households worked as contract workers for domestic companies or abroad. Some of them ran their own businesses outside of agriculture.

Table 7.6. Work/employment of the members of family farms, excluding the main user/operator in 2017 (in %)

Jobs held by members of the family farm household, other than the main user/operator	Farms	
	Operated by women	Operated by men
Providing help on the farm*	86.3	80.7
Contract work in Poland*	57.5	47.8
Own business outside of agriculture*	8.8	4.9
Work abroad*	8.0	4.7
Sales of own agricultural products, services for farmers, agritourism*	0.7	1.4

* p < 0.01.

Source: Own research.

What conclusions can be drawn from the data on off-farm employment of other household members of female-run farms? First, the data reveal different developmental strategies for farms operated by women and those operated by men. Male farmers, more frequently than female farmers, decided for themselves to take additional work off-farm. In female-run farms such work was more frequently taken by other household members. Could this be linked to the economic conditions of farms? It should be recalled here that the farms operated by men had better financial conditions than those operated by women. Could it be assumed that the farmer would not be inclined to pass the prosperous farm over to his wife or another woman from the family, when he enjoys good, stable employment off-farm but he would be willing to do so with a less prosperous farm? Second, family farms in Poland are different than farms in more developed countries in terms of being supported by financial means acquired off-farm. In Poland, contrary to the earlier example of Australia, this duty is known to fall more on men, which can be seen as a certain paradox. Although women are educated, are prepared to leave the farm for an off-farm career, and have competencies superior to men to work outside of agriculture, they still stay on the farm, even when the job market creates real opportunities for finding good non-farm employment. The members of their households do exactly the opposite. This strategy of family farms in Poland might be considered irrational, and this would be the case if the work of better educated women would pay better than the work of men with lower qualifications. However, the market today is not predictable and there is no guarantee that a female with a white-collar job or a pink-collar job (services) could earn more than a man with a blue-collar job doing physical labor. Third, such a significant increase in off-farm employment of farm managers and other members of their household in recent years has had a rather positive impact on the change in structure of Polish family farms and their financial state. On the flip side, this has also led to greater dependence of their financial situation and their further development opportunities on the general condition of the non-agricultural job market and, more broadly, the state of the entire economy. This is one of the most important changes in recent years affecting family farms. This is not just a symptom of leaving the 'peasant autarchy' but also a sign that family farms are truly entering global market processes. Could these conditions foster the process of professional re-segregation related to family farming? The answer could be positive as, already mentioned earlier and as confirmed by studies in Western countries, women from agricultural families are educated and prepared to leave farms and agricultural education is oriented towards men. The analysis of Polish data is therefore in order (Table 7.7).

Table 7.7. Education of farm owners/operators in 2007 and in 2017 (in %)

Year	Sex/ gender	Education of farm owners/operator			
		No more than grammar school	Vocational	High school	Beyond high school
2007*	Women	27.9	32.4	31.8	7.9
	Men	27.2	51.2	16.3	5.3
2017*	Women	13.9	33.1	36.1	16.9
	Men	13.9	47.9	29.4	8.8

* $p < 0.01$.

Source: Own research.

The data in Table 7.7 show that Polish women farmers, like women farmers of Western Europe, are better educated than men farmers. Comparison of the data from 2007 and 2017 suggest that the situation is rather stable. The percentage of male farmers who had no less education than high school in 2017 barely reached the percentage of female farmers with no less than high school education in 2007. In 2017 female farmers with no less than a high school diploma comprised 53.0% of all women participating in the study. It could be stated that in terms of formal education Polish female farmers are ahead of their male counterparts by nearly

a decade. To use the terminology of Pierre Bourdieu (1986), women have greater resources of institutional capital than men do, in the form of formal college/university
educations. However, this is not the only indicator of the conviction that lingers in
agricultural circles (not just in Poland and Ireland) that a woman getting an education and improving her general civilizational competencies should be prepared
to leave the farm and leave the farm management to a man from her family circle.
The cues of such a way of thinking could be noted in the data on participation in
instructional courses, courses on computer-based farm management, courses on
usage of e-mail, or knowledge of foreign languages. As indicated by the 2017 studies,
in the last 3 years such courses were attended by 25% of male respondents and 11.6%
of female respondents. Courses directly linked to agricultural production, including
the use of chemical products (e.g., courses on how to use the sprayers, fertilizers,
plant protection products, etc.), were completed by 12% of male farmers participating
in the study and 2.8% of female farmers, while courses on the methods of agricultural production saw 11.7% of the male and 5.3% of the female study participants,
and courses on farm management methods were attended by 4.9% of the male
and 3.0% of the female respondents. Female farmers were more likely than male
farmers to finish computer courses—4.9% and 3%, respectively. The same could be
said about foreign language classes, courses in agritourism and rural tourism, and
accounting or tutorial classes for completing EU subsidies applications. The percentages for female and male study participants were, respectively, 3.5% and 1.6%.
Interestingly, more male farmers than female farmers used computer programs for
farm management and the percentages were, respectively, 3.3% and 1.5%. Female
farmers participating in the study used e-mail more often than male farmers (41.0%
to 36.4%) and declared an ability to communicate in a foreign language (40.1% and
36.7% respectively). Certainly, such a large percentage of participants declaring the
ability to communicate in another language is rather surprising. It should be remembered, however, that in the 2017 study, 44.3% of participants had no less than
high school education. Among them, 11.5% had bachelors or master's degrees. All
these respondents took at least one foreign language course in school. The older ones
took Russian in school and knowledge of that language was declared by 21.9% of
respondents. The younger respondents were more familiar with Western languages;
12.3% of all respondents stated that they could communicate in English and 7.2%
said they had some knowledge of German. Additionally, over a dozen respondents
declared some knowledge of French, Italian, or Ukrainian language. There were
also a few respondents who stated knowledge of one of the 17 languages listed in the
study. Such a large spectrum of languages might be connected with seasonal work
abroad and its various destinations.

Table 7.8. Age of farm owners/operators in 2007 and in 2017 (in %)

Year	Sex/gender	Average age	The age of owners/operators		
			18–34 years	35–54 years	55 and over
2007	Women	46.8	13.7	58.8	27.5
	Men	46.8	16.0	62.1	21.9
2017*	Women	49.7	8.6	58.6	32.8
	Men	49.1	11.9	54.9	33.2

* $p < 0.01$.

Source: Own research.

Knowing that acquisition of institutionalized capital is largely determined by age, just as it is easier for young people to assimilate new information, the structure of farm users/operators is worth analyzing (Table 7.8). The data show unambiguously that the process of aging is exacerbated among farmers just as intensely as it is in the entire society. Male-operated farms and female-operated farms alike face the successive decline in the involvement of young people, while the percentage of the oldest farmers is growing. Whether the farm is managed by a man or a woman, the dominant age group is 35–54 years, comprising over half of the studied population. However, if the pace of the aging process is measured by the drop in changes in average age, it could be expected that in 20–25 years farm management will be dominated by people from the largest age group—55+. In the 2007 study the average farmer's age was 46.8 years, for both men and women, and in the 2017 study it was 49.1 for men and 49.7 for women. It turns out that female farmers age slightly faster than male farmers. This could mean that increasingly older women are forced to take over farm management due to the lack of a male manager, who might be more absorbed with off-farm work. The possibilities for these women to multiply their institutional capital, whether in general terms or in such a way that could be useful in effective farm management, are rather limited due to their age. This could have negative consequences for the development prospects of the farms they manage. It should be noted here that the case of Polish female farmers is much different than the case of Swedish farmers. As already mentioned, in Poland and Sweden alike, the slow process of women taking over the management of farms is taking place but in Sweden female farmers are younger than male farmers. Summarizing this part, it should be stated that female farmers are ahead of male farmers in terms of their institutionalized capital of a general character. However, their assets of institutional capital which are specifically related to agriculture are smaller than those of men. This could indicate the persistence of significant cultural barriers in the processes of acquiring sex/gender balance in the farming profession. Considering that female farmers are aging somewhat faster than men, it is hard to think that their motivation to improve their agricultural qualifications, especially with no support from their social circle, which has rather traditional convictions regarding the role of women on farms, will be greater than the motivation of men, or at least equal to it. Obviously, better agricultural qualifications for women could contribute to a reduction in the disparities in economic conditions of male- and female-run farms. The way things appear to be now does not allow for such optimism. It should be expected that the gap in economic assets between farms managed by women and those managed by men will be deepening.

7.5. Conclusions

Researchers analyzing the topic of women functioning in agriculture point to the sex/gender segregation that still persists. Despite technological changes such as progressive mechanization and digitalization and considering the significant reduction in the need for physical labor on the farm, the farming profession is still perceived as a male profession. Even though the peripheral character of rural areas has been largely vanquished, thanks to the development of the Internet among other things, and rural areas are being included in the bloodstream of social life, the secondary role of women in agriculture remains the case. It is emphasized that the main barrier in overcoming sex/gender inequalities and moving in the direction of sex/gender balance in agriculture and rural life no longer lies in technology and mostly has a cultural character. The conviction that a man is the only legitimate manager of a family farm is deeply rooted not only among the male population of a rural community but also among women themselves, even in situations when, due to various circumstances, they have experience in fulfilling the role of farm manager. Such a way of thinking is legitimized through the existing educational system, which shapes the opportunity of agricultural education in such a way that it appeals mostly to men. At the same time, it proposes that women from farm families acquire civilizational competencies that would enable them to find employment outside of agriculture. This often prompts them to leave the farm. In the best-case scenario, it puts on women the duty of being a farm "rescuer" in crisis situations, when women finance the farm production needs from earnings acquired off-farm. Despite the general similarities in the functioning of women in agriculture in Western countries the country or region related specifics still apply. One of them could be the frequency of women fulfilling the function of farm manager. Such specifics can also be noted in the case of family farms in Poland.

The percentage of women who fulfill managerial functions on Polish farms, compared to the analogous data in the majority of European countries, is relatively large. This applies even when Poland is compared to Sweden, one of the leading countries in this regard. However, women manage farms that are decidedly weaker economically than those managed by men. They are also smaller in size, less mechanized, and less profitable. The production on the farms managed by women could be characterized as extensive, requiring a significant amount of work from other household members. This is not so much the case with male-run farms. Female farm managers take additional off-farm work less frequently than men, but other members of their households do so more often in comparison to male-run farms. Although Polish female farmers, like female farmers in other Western countries, have much better general education and civilizational competencies than male farmers, which makes them more predisposed to look for good employment opportunities off-farm, it is the men who are obliged to ensure proper financial functioning of the farm, even if it means working outside of agriculture. This distinguishes Polish family farms from the farms in some other Western countries. What makes them similar is female farm managers having weaker agricultural education and skills necessary to manage the farm than the male farmers. This shows that the Polish educational system has the same limitations of a cultural character that exist in other developed countries fostering petrification of sex/gender inequalities. Another difference, characteristic of farms in Poland, is that Polish female farm managers are somewhat older than men in farm managerial positions. Similar to other countries, women more frequently than men, take over farms through marriage, and much less frequently through inheritance. This is a sign of the persistence of the patrilineal character of farm succession.

What general conclusions can be drawn from the interpretations presented above? Could the extent to which women take the managerial functions within Polish family agriculture validate the thesis regarding the process of limiting gender inequalities and moving towards professional integration or is it rather an expression of moving away from professional ghettoization in the direction of re-segregation in the agricultural profession? It seems that the processes taking place in recent years make it hard to provide a straight answer. It is not yet a situation of openness of farmers' profession to women that would enable them to bid for managerial positions on the farm on an equal footing with men. It is more likely a situation which allows women to fulfil these functions when the "legitimate" farmer—a man—decides to take a job in another, better paid profession. However, there is not strong enough ground, considering the extent of this phenomenon, to see this as an expression of a lasting process of moving away from ghettoization and in the direction of professional re-segregation. Perhaps the answer to this question needs to wait until the processes of mechanization and digitalization encompass Polish farms more thoroughly and female farmers gain the conviction that agricultural knowledge is available to them just as much as the general/non-farm knowledge and general modern life competencies. They already possess the latter, which gives them a certain advantage over male farmers.

REFERENCES

Alston, M. (2017). The genderness of climate change, Australia. In: B.B. Bock & S. Shortall (eds.), *Gender and Rural Globalization: International perspectives on gender and rural development*, pp. 100–113. CABI.

Bock, B.B. & Shortall, S. (2017). Gender and rural globalization: An introduction to international perspectives on gender and rural development. In: B.B. Bock & S. Shortall (eds.), *Gender and Rural Globalization: International perspectives on gender and rural development*, pp. 1–7. CABI.

Bourdieu, P. (1986). The forms of capital. In: J. Richardson (ed.), *Handbook of Theory and Research for the Sociology of Education*, pp. 241–258. Greenwood.

Brandth, B. (1994). Changing femininity: The social construction of women farmers in Norway. *Sociologia Ruralis* 34(2–3), pp. 127–149.

Brandth, B. & Haugen, M.S. (1997). Rural women, feminism and the politics of identity. *Sociologia Ruralis* 37(3), pp. 325–344.

Cassidy, A. (2017). The agency paradox: The impact of gender(ed) frameworks on Irish farm youth. In: B.B. Bock & S. Shortall (eds.), *Gender and Rural Globalization: International perspectives on gender and rural development*, pp. 148–161. CABI.

Černič Istenič, M. & Charatsari, Ch. (2017). Women farmers and agricultural extension/education in Slovenia and Greece. In: B.B. Bock & S. Shortall (eds.), *Gender and Rural Globalization: International perspectives on gender and rural development*, pp. 129–146. CABI.

Dahrendorf, R. (1990). *Life Chances: Approaches to social and political theory*. The University of Chicago Press.

Dahrendorf, R. (1991). *Rozważania nad rewolucją w Europie* [Reflections on the revolution in Europe]. Niezależna Oficyna Wydawnicza.

Domański, H. (1992). *Zadowolony niewolnik? Studium o nierównościach między kobietami i mężczyznami w Polsce* [A satisfied slave? Studies on inequalities between men and women in Poland]. Wydawnictwo IFiS PAN.

Domański, H. (1999). *Zadowolony niewolnik idzie do pracy. Postawy wobec aktywności zawodowej kobiet w 23 krajach* [A satisfied slave goes to work? Attitudes towards occupational activities of women in 23 countries]. Wydawnictwo IFiS PAN.

Gorlach, K. (1995). *Chłopi, rolnicy, przedsiębiorcy. „Kłopotliwa klasa" w Polsce postkomunistycznej* [Peasants, Farmers, Entrepreneurs: The Awkward Class in Post-communist Poland]. Kwadrat.

Gorlach, K. (2009). *W poszukiwaniu równowagi. Polskie rodzinne gospodarstwa rolne w Unii Europejskiej* [Searching for a balance: Polish family farms in the European Union]. Wydawnictwo Uniwersytetu Jagiellońskiego.

Gorlach, K. & Drąg, Z. (2011). Kobiety na kombajnach. Właścicielki gospodarstw rolnych w Polsce współczesnej [Women on combine harvesters: Women farm owners in contemporary Poland]. In: K. Slany, J. Struzik, & K. Wojnicka (eds.), *Gender w społeczeństwie polskim*, pp. 280–297. Nomos.

Gorlach, K., Drąg, Z., & Nowak, P. (2012). Women on… combine harvesters? Women as farm operators in contemporary Poland. *Eastern European Countryside* 18, pp. 5–26.

Heather, B., Skillen, L., Young, J., & Vladicka, T. (2005). Women's gendered identities and the restructuring of rural Alberta. *Sociologia Ruralis* 45(1–2), pp. 86–97.

Pięta, P., Skierska-Badura, A., & Skierska-Pięta, K. (2013). Identyfikacja i charakterystyki warunków życia i pracy kobiet-rolniczek [Identification and characteristices of life and work conditions among female farmers]. In: J. Sawicka (ed.), *Rynek pracy na obszarach wiejskich Mazowsza – perspektywa gender* [Labor market in rural Mazovia: The gender perspective], pp. 226–245. Wydawnictwo SGGW.

Prugl, E. (2004). Gender orders in German agriculture: From the patriarchal welfare state to liberal environmentalism. *Sociologia Ruralis* 44(4), pp. 349–372.

Shortall, S. (2017). Gender and agriculture. In: B.B. Bock & S. Shortall (eds.), *Gender and Rural Globalization: International perspectives on gender and rural development*, pp. 89–99. CABI.

Silvasti, T. (2003). Bending borders of gendered divisions on farms: The case of Finland. *Sociologia Ruralis* 43(2), pp. 154–166.

Stenbacka, S. (2017). Where family, farm and society intersect: Values of women farmers in Sweden. In: B.B. Bock & S. Shortall (eds.), *Gender and Rural Globalization: International perspectives on gender and rural development*, pp. 114–128. CABI.

Sztompka, P. (2016). *Kapitał społeczny. Teoria przestrzeni międzyludzkiej* [Social capital: The theory of inter-human space]. Wydawnictwo Znak.

Tryfan, B. (1987). *Kwestia kobieca na wsi* [Women's issue in rural areas]. IRWiR PAN.

Internet sources

https://bdl.stat.gov.pl/BDL/dane/podgrup/tablica (access: March 4, 2019).

Think Locally, Act Globally: Polish farmers in the global era of sustainability and resilience, ed. by Krzysztof Gorlach and Zbigniew Drąg in collaboration with Anna Jastrzębiec-Witowska and David Ritter
Jagiellonian University Press, Kraków 2021, pp. 355–372
ISBN 978-83-233-4949-5
DOI: http://dxdoi.org/10.4467/K7195.199/20.20.12734

Chapter Eight: Farm Women as Participants in Social Life*

Zbigniew Drąg https://orcid.org/0000-0002-9106-7758/

Krzysztof Gorlach https://orcid.org/0000-0003-1578-7400/

8.1. Women's Issues in the Post-modern Countryside

Currently, women's issues in rural areas are, from our perspective, determined by two sets of factors which take the form of multi-layered and overlapping social processes. They are the processes of rural development, which can be seen as transformations happening simultaneously on many levels (economic, social, political, cultural, and personal). Various entities participate in these processes and the direction of these changes indicates attempts to have a multifunctional countryside. As indicated in the concluding remarks of the extensive work by Henri Goverde, Henk de Haan, and Mireya Baylina, which included extensive research in various European countries and regions, the new contract between agriculture and society in rural areas was being created with a multifaceted and innovative approach to rural development problems (2004, p. 176). Such an approach, however, creates a certain problem, as the presented vision of rural development is mostly a vision shared by social elites, the representatives of authority and administration, as well as economic and academic institutions. However, the people actually residing in

---

* The earlier draft of this chapter has been published in Polish as "Rolniczki jako uczestniczki życia społecznego," *Acta Universitas Lodziensis. Folia Sociologica*, Vol. 68/2019, pp. 47–66.

rural areas, who are directly caught in the rhythm of rural functioning, still see changes as an interplay of market forces and requirements. This discrepancy in points of view regarding the surrounding reality and various factors responsible for that were noted by the authors, but it might be a symptom of a phenomenon much stronger than indicated in the analyzed literature. This chapter supports the notion of the second combination of factors as a social and cultural legacy, in the form of a specific tradition functioning in rural communities, mapping out the ways their members act when confronted with new phenomena, processes, and new initiatives taken towards rural development. This is not just a direct and open protest against the new programs, or an effort to change the countryside's image. This is more a case of some well-established habits, or schemes of thoughts and actions, which make it difficult for rural residents to establish themselves in the new reality. Among such existing habits, patterns, and schemes are: patriarchy of agricultural communities, as manifested by the treatment of women's activities as "hidden" or "unpaid" de-spite the obvious changes in women's position in European agricultural and rural communities of the second half of the 20th century. To emphasize the importance of this problem, the authors state that rural families became embedded in the economy, trying to adjust to the new paradigm of rural development. Attention switched to the issues of the engagement of women in this process (Goverde, Haan, & Baylina, 2004, p. 178; see also Brandth & Haugen, 1997).

The economic engagement of rural women increases their empowerment. Within this context Liv Toril Pettersen and Hilde Solbakken stressed that such "empower-ment can be suitable as a strategy for change" (Pettersen & Solbakken, 1998, p. 327). This pertains to the change in the situation of women seen as an element of the wider changes within the rural area development strategy that is being applied. What needs to be noted is the fact that such development is largely the result of actions of particular subjects, actors, and participants in public life who take part in various projects and development strategies. In this sense, the changes in women's situation may be, for the most part, the result of actions taken by women themselves or the actions of other social categories and other participants in social life who cooperate with women in that realm. Women's increasing level of confidence and trust in their own abilities, as well as the possibility to control the resources that are crucial to the functioning and changes in rural communities create a foundation for taking up effective actions aimed at improving the situation of women. Quite rightfully, the women authors mentioned above state that, "Empowerment permits grassroots women to work for change on their own terms" (Pettersen & Solbakken, 1998, p. 327).

Rural women do not comprise one homogenous social category. Australian authors Margaret Grace and June Lennie state, "The diversity of rural women's personal identities, skills and knowledge, and in terms of the wide range of issues they bring to public forums is one of the greatest strengths and needs greater recognition" (Grace & Lennie, 1998, p. 366) and this is characteristic of the category of rural women. The American author Betty L. Wells (1998) shares this viewpoint and emphasizes the necessity for women's interests to be represented in the public space. According to Wells, such representation is only possible if women gain decision-making power regarding the structure and processes that bring about rural life transformation. Moreover, such a full-fledged, empowered presence of women could be an important factor in the proper development of the agricultural sector and the rural economy outside of agriculture. In other words, empowered women who, in this case, have the ability to control and make decisions related to crucial resources, "may develop their collaborative potential" (Wells, 1998, p. 387).

Empowerment of rural women still remains an unfinished process. Even in societies that are considered the most advanced, like those of Scandinavia, this process and its effects were, and still are, judged with some ambiguity. For example, Berit Verstad, in analyzing the election process of the Norwegian Farmers' Association in the second half of the 1990s, emphasized that "women's range of opportunities in the Farmers Union has increased" (Verstad, 1998, p. 424). However, other female authors from Norway presented different opinions. In their research, Berit Brandth and Marit S. Haugen (1998) used a longer historical perspective among communities engaged in forest farming, which is an important part of Norwegian agriculture, and pointed out the slow process of women breaking the male discourse in farm circles. In the mid-1970s women were nearly absent from this discourse, and they remained "invisible" as farmers, as was already mentioned in this text. About 10 years later, in the mid-1980s, the process of women becoming more active could be noted in agriculture and rural areas. Their work started to be noticed and this was somewhat reflected in the discourse. After another decade, in the mid-1990s, women appeared in the discourse, and not just in the role of workers. Nevertheless, as indicated by the same authors, despite these changes, the influence of female farmers in the rural and agricultural realm remained rather minor. Similar processes could be noticed in the Spanish countryside, as described by Clemente J. Navarro Yañez. The women there still experienced the process of social closure, although it was not as intense as in previous decades. As the author noted, "Nothing new has been stated in pointing out that modern societies, and particularly Spanish society, are marked by processes of social closure with respect to gender, although the exclusion of women has been declining recently" (Yañez, 1999, p. 231). Such a decrease could mostly be seen as the effect of the strategy of usurpation employed by women, who took up various actions aimed at improving their situation. Such activities might be an effect of the

stated that on the level of normative solutions, sex and gender equality are taken into consideration but, in practice, the actions of the institutions mentioned above did not adequately reflect that consideration. The situation, as concluded by Shortall, required vigilance in order to prevent the introduction of an unwanted relationship between the representatives of both sexes in the programs described above.

The power of tradition was still the factor that had an impact on the situation of rural women in the late 1990s, even in highly developed countries where agriculture was clearly entering the postmodern stage of development. The tradition was particularly present in the lives of the older generation of women. In the words of German researcher Ilse Modelmog, "the interviewed farmers' wives, primarily the older ones, still have an awareness of tradition. Yet, traditions no longer have an obligatory character" (Modelmog, 1998, p. 121). In other words, tradition was no longer a factor that determined, without hindrance, the functioning of rural areas and the changes taking place there. The countryside at the turn of the 20th century, similar to the entire society of the time, was to some extent influenced by the processes of de-traditionalization, characteristic of the postmodern and globalized society. The effects of these last processes were most visible in the middle-aged and youngest generations and this particularly applied to the new image of rural women created by this age category. Pointing to research conducted in Lower Saxonia, Modelmog explained:

> This image is based on self-confidence and respect. Since these women are able to transcend narrow boundaries of gender, they can "conquer the world" for themselves. They try to organize their lives in such a way that they not only fulfill their duties but can also consciously follow their inclinations and desires (Modelmog, 1998, p. 121).

The source of underachievement in the process of women's empowerment to participate in proper rural development, consistent with its intended goals, could be found in traditional male domination of the society's rural sector. This domination was possible mostly as a result of indecisiveness among women in fighting for their rights. This argument, made by Australian researchers Margaret Alston and Jane Wilkinson, stressed that "Women must strengthen their resistance to the male-dominated discourse and break down the attitude that women's exclusion is somehow their fault" (Alston & Wilkinson, 1998, p. 405). In this case, there were different aspects of empowerment, pointing to the need for building better self-confidence in women and their higher self-esteem.

Freeing the discourse from traditional male domination, including the perception of the agriculture of growing crops and raising animals as typically male activities, is only possible by overcoming the causes of such a status quo, which still exists in rural communities with a strong patriarchal structure (see Bennett, 2004). As indicated by Katy Bennett from United Kingdom, rural women also experienced discrimination outside of farm work. On one hand, they had gained independence, but the aforementioned discrimination took place when they were forced to transfer their income acquired outside of farming to farm needs. The author noted such situations and saw them as subordinate to patriarchal structures. She also stated that women were not without blame, especially if they raised their daughters in a manner subordinate to the patriarchy.

Another author who analyzed these issues, but in a somewhat different context, was Margaret Alston (2003) from Australia. She argued for the necessity of greater and more pronounced representation of women in agricultural organizations, where the dominant perspective for perceiving agriculture was generally masculine, or dominated by men, due to the character of the work. A similar opinion was also expressed in an article written by another Australian author, Barbara Pini (2002), describing the situation of women who were members of sugar producer organizations. It turned out that the newly meritocratic criteria for activists' advancement in the organizational structure, contrary to expectations, did not lead to the end of the discrimination of women. According to the author, the meritocratic criteria merely became another method for masking the problem of women's exclusion (see Pini, 2002, p. 75). Concluding her remarks, Pini took the rather extreme position that terms such as "meritocratic principles," "election systems," or "democratic procedures" must change the content which they carry out. They need to be taken out of the dominant discourse, so they could gain a new character advocating equality and fostering a context of real sex and gender equality.

8.2. Women's Issues in the Contemporary Polish Countryside

The general characteristics of rural women in Poland during the special period of the 1990s are worth a closer look. This was the time when the first results of basic socio-economic transformations related to the political breakthrough of 1989 could be observed. Based on a study conducted on a random, nationwide sample of rural residents over the age 15, Barbara Perepeczko (1996, pp. 7–11) identified three essential elements characterizing the category of rural women. First of all, the undercount of women in younger age categories was noticeable. Rural women were better educated than rural men, although many of them did not have a profession they were trained for. These last characteristics gained special meaning in the face of multifunctional rural development. On one hand, the better education could be an asset, yet on the other, the lack of a trained profession was a factor that makes adaptation to the changing reality quite difficult and taking action in the direction of changes even more difficult. In Poland, similar to the Western European countryside, statements that a multifunctional rural development could be a factor in increasing women's social and professional activities consequently would lead to more symmetrical relations between the two sexes in rural communities. Polish author Grażyna Kaczor-Pańków formulated a similar thesis, assuming that "with the foreseen multifunctional development of rural communities, professional heterogeneity and the development of various forms of political and economic pluralism in the process of shaping a particular frame of mind, the role of women will be quite serious" (Kaczor-Pańków, 1996, p. 21). The author added that rural women displayed several types of thinking which influence their forms of activity. First of all, women were critical towards changes that took place after 1989 and judged rather critically the material situation of rural families (see Kaczor-Pańków, 1996, p. 20). However, the ongoing process of professional emancipation of women and their ability to follow—and use to their advantage—the instructional broadcasting in mass media, particularly television, also got the attention of the above author (see Perepeczko, 1996, p. 28). In this context, women's ties to entrepreneurship had special meaning, both in the sphere of activities as well as in the way of thinking, which was no less important for "the spirit of entrepreneurship."

What "entrepreneurial" abilities could be particularly valued in rural women? As it turned out, very specific ones. While discussing this part of the research, Maria Mydlak (1996, p. 75) noted that these abilities included: maintaining good relations with neighbors, skills to do a variety of work ("handyperson," "a jack of all trades," "golden hands," etc.), skills in animal husbandry, vegetable gardening, etc. The role of "entrepreneurial" women on farms was seen from the perspective of a specific division of labor, shaped into traditional "male" and "female" types of work, which was already discussed earlier. Here, a woman should be a team member servicing the farm and mostly operating within certain sections of the farm (livestock breeding and gardening). Female labor was, and still is, treated as "emergency labor," a "jack-of-all-trades" situation, and women, in particular, are entrusted with stewardship over the sphere of social capital, meaning the maintenance of good interpersonal relationships.

Other research on changes in the situation of women during the regime change after 1989 brought attention to the crucial change in the social position of Polish women, including women in rural areas. These changes applied not only to very general contexts, such as increased educational levels, improved rural living conditions, the fostering of labor market competitiveness, and the disintegrating social safety net, but also to more specific phenomena such as the weakening of the traditional value system, and atrophy of functions fulfilled by families (Witkowska, 2013, p. 43). This consequently led to scenarios where "partner behavior" became a more important factor shaping the situation of women and increasing their level of empowerment, possibly leading to increased engagement in public life. Such activities mostly encompassed membership and work in farmers' wives associations and rural women's organizations, as well as fulfilling the mandate of the village head (Żak, 2013, p. 178). Comparing these types of activities between women and men brought some surprising results. As emphasized by Ilona Matysiak (2013, p. 89), who conducted comparative research among female and male village heads (mayors), noted that the interviewed women more often declared cooperation with the representatives of local government, local institutions, and school structures, as well as family members and acquaintances. Quite surprisingly, female village heads, somewhat more often than their male counterparts, cooperated with parish priests or the parish council. The author also stressed that ". . . women effectively expanded their networks of social capital with new actors of cooperation, who were essential to local community" (Matysiak, 2013, p. 89). Quite important here was the role of the culture, which became less and less encumbered by tradition and thus fostered increased activity among rural women (Matysiak, 2017, p. 240).

8.3. Towards the Empowerment of Women

Implementation of a strategy for multifunctional rural development, which could create favorable conditions for empowerment, appears to be the solution preferred by many authors for the issues of rural women. They also stress that a necessary factor for empowerment is intensification of the social activities of the women themselves. The control of essential community resources and the decision-making powers regarding the direction of transformations can only be acquired by women who take interest in the life of the rural community and actively participate in it. The extent of social activity of women in the Polish countryside is worth analysis. A very specific category of farm women, those who fulfill the role of farm owners or farm managers, appears to be particularly intriguing. This is one of those categories of rural women who have achieved some success on their way to empowerment. They control the resources of their farms and have the ability to make independent and autonomous decisions regarding the direction of farm development.

Generally, it should be stated that even a decade ago farmers were rather reluctant to engage in farmers' unions, political parties, national politics and matters beyond the local. They limited their interests to issues of their immediate social environment. As indicated by the 2007 study, farm women with managerial positions were homebound even as recently as ten years ago (see Gorlach, 2009). In 2007, only 3.2% of them had some roles and positions in social organization and a total of 7.7% of them could claim any experience in this regard. To compare, 22.2% of male farm managers had ever played any social role, and those who held them in 2007 stood at 12.4%. Female farmers were also less likely to take up communal and joint action in agricultural production than their male counterparts. Only 2.0% of female farmers were members of producer organizations while the percentage for male farmers was 8.5%. However, female farmers were more critical of the Polish political scene than male farmers were, as they were not able to discern their political representation there. Only 11.7 % of women farmers recognized the existence of organizations defending farmers' interests, while the percentage for male farmers was 19.9% (Table 8.1). Perhaps this increased criticism towards the political life on the national level, and an aversion to politics in general that was more pronounced than it was in male farmers, had to do with a lack of free time and an overload of farm and household duties. Female farmers generally did not abandon their household when they took up farm management.

Table 8.1. Farmers' social activity in 2007 (in %)

Variables	Women	Men
Percentage of respondents fulfilling notable roles in social organizations*	7.7 (3.2; 4.5)	22.2 (12.4; 9.8)
Percentage of respondents who are currently members of agricultural producer organizations*	2.0	8.5
Percentage of respondents stating that there are organizations defending farmers' interests**	11.7	19.9

* $p < 0.01$, ** $p < 0.05$.

Source: Gorlach, 2009, pp. 210, 220.

Has the last decade brought any changes? Have there been any signs of empowerment of female farmers in managerial positions, their increased engagement in social matters, better ability to influence the decisions shaping their close and more extended social space? Data from the 2017 study conducted on a nationwide representative sample of farm owners and managers might be helpful in answering these questions. The research covered 3551 farms and was part of a project entitled "Think Locally, Act Globally: Polish farmers in the world of sustainable development and crisis resistance," research project MAESTRO No. 2015/18/A/HS6/00114/, financed by the National Centre of Science.

The data from the study show that the engagement in social life by female farmers in managerial positions increased significantly, but it still remained smaller than the engagement of male farm managers (Table 8.2).

Table 8.2. Social activity of farmers in 2017 (in %)

Variables	Women	Men
Respondents with notable roles in social organizations (present, past)*	21.3 (9.8; 11.5)	27.2 (14.6; 12.6)
Respondents who currently are members of agricultural producer organizations	2.2	4.0
Respondents stating that organizations defending farmers' interests exist*	8.4	15.1
Respondents who have ever been members of unions and political parties*	3.7	7.9
Respondents participating in local elections in 2014**	84.6	87.1

* $p < 0.01$, ** $p < 0.05$.

Source: Own research.

As indicated by the collected data, the propensity for communal and jointly conducted professional activities, with 2.2 % of female respondents and 4.0 % of male respondents being members in producer organizations, did not increase in any category of farm owners or managers. At the same time, it was noticeable that the experience of having a function in social organizations (local associations, local social committees, parish organizations, supralocal organizations, etc.) showed an increasing trend. 21.3% of female farmers and 27.2% of male farmers admitted to be fulfilling such roles or had such roles presently or in the past. At the time of the 2017 study, 9.8% of female farmers and 14.6% of male farmers reported performing such functions. With such a significant increase in social activity, especially by women, the notion of the lack of their political representation also increased. In 2017, only 8.4% of female farmers and 15.1% of male farmers acknowledged the existence of organizations (parties, trade unions, etc.) that helped farmers' interests. It seems that this was closely related to the persistent reluctance of farmers to actively engage in the work of political parties and associations or in politics on the local, supralocal, or national level.

Only 3.7% of women farmers and 7.9% of men farmers have ever been members of some political party or trade union. No more than 0.4% of the former and 1.8% of the latter fulfilled some notable functions in parties or unions (Table 8.3).

Similar trends could be seen in organizations that operated beyond the local level. Experience with fulfilling functions in such organizations was reported by 1.6% of the female respondents and 1.4% of the male respondents. Both women and men had more significant involvement in local associations, with 13.1% of female farmers and 15.7% of male farmers holding significant roles. Moreover, female farmers were somewhat more likely to fulfill important functions on local social committees (5.1%) and parish organizations (4.3%). Men were more likely to be council members (6.0%) in local governments or heads of the village (5.3%).

Table 8.3. Farmers' fulfillment of notable social roles in 2017 (in %)

Variables: Roles fulfilled presently or in the past	Women	Men
Mandate of head of the village*	2.6	5.3
Mandate of councilperson*	2.5	6.0
Executive roles in local government (mayor, head of local administration, chief of staff at municipal office, member of executive board for the municipality or the county, etc.)	0.6	0.5
Notable roles in agricultural organizations* (president, member of executive board of agricultural circle, cooperative, etc.)	0.9	2.7
Notable roles in local associations* (volunteer fire department, farm wives association, etc.)	13.1	15.7
Notable roles in parish organizations	4.3	3.6
Notable roles on social committees* (school, roads, etc.)	5.1	2.9
Notable roles in political parties and trade unions*	0.4	1.8
Notable roles in supralocal organizations (e.g., Polish Red Cross)	1.6	1.4

* $p < 0.01$.

Source: Own research.

Generally, the factor analysis pointed to three dimensions of fulfilled functions: functions in a social organization of local character, functions in executive bodies, and functions in organizations operating beyond the local level (Table 8.4). All of them were somewhat more often the attributes of men and the most significant disparity appeared in functions in executive bodies. However, the sex/gender factor was not a meaningful determinant for any of the listed kinds of functions (Table 8.5).

It should be emphasized here that the interest in local matters from female and male farmers who managed their farms was confirmed by their above average participation in local elections. In the 2014 local elections, the participation rate of female respondents was reportedly 84.6% while the participation rate of male respondents was even higher, at 87.1%.

Table 8.4. Analysis of main components of social roles fulfilled by farmers in 2017: Factor loads, rotation Varimax (loads below 0.20 not included)

Farmers' social roles	Components		
	I: Local functions	II: Executive functions	III: Functions beyond local level
Parish organizations	0.650		
Social committees (for building roads, schools, etc.)	0.641		
Local associations	0.522		
Agricultural organizations	0.472	0.319	
Institutions of local executive bodies		0.735	
Head of the village	0.215	0.557	−0.268
Councilperson of local government	0.303	0.389	
Political parties and trade unions		0.247	0.732
Associations of supralocal character			0.641
Percentage variance explained by particular component	16.90	13.67	12.49
Total percentage variance explained	43.06		

Source: Own research.

Table 8.5. Types of social functions fulfilled by farmers in 2017: average factor values of scales for local, executive, and supralocal functions

Sex/gender	Scale of local functions			Scale of executive functions			Scale of supralocal functions		
	average	significance	eta²	average	significance	eta²	average	significance	eta²
Women	−0.006	no	0.000	−0.097	0.01	0.005	−0.041	no	0.001
Men	0.004			0.056			0.024		

Source: Own research.

Table 8.6. Farmers' sources of information regarding local matters and events in 2017. Scale from 1 to 7: 1—"never," 2—"about once a year or less," 3—"several times a year," 4—"less than once a month," 5—"2–3 times a year," 6—"about once a week," 7—"at least several times a week"

How often does the respondent get information on local matters and events (taking place in a particular village) from:	Never (answer 1)	A few times a year, at most (answers 2 and 3)	2–3 times a month at most (answers 4 and 5)	At least once a week (answers 6 and 7)
Family members, acquaintances	3.9	12.1	28.4	55.6
Other residents	7.5	16.4	39.3	36.8
Head of the village	17.9	37.5	32.1	12.5
Leaflets, announcements	20.7	34.0	35.8	9.5
Local press	30.0	19.3	32.2	18.5
Priest	42.9	26.2	17.0	13.9
Municipal website	53.6	15.4	19.2	11.8
Other local Internet sites	68.3	12.2	11.0	8.5
Facebook and other social media networks	78.3	7.5	6.1	8.1

Source: Own research.

Any type of social activity was conducted by taking up various forms of contacts, gaining and transmitting information on interesting matters to various people, organizations, and institutions. It was justified to address the sources of information on local matters, events, and the forms of social contact preferred by farmer-managers.

Table 8.7. Farmers' sources of information on local matters and events in 2017: averages by sexes/genders (scale 1–7)

How often does the respondent get information on local matters and events (taking place in particular village) from:	Averages		Significance	eta²
	Women	Men		
Family members, acquaintances	5.19	5.32	0.05	0.002
Other residents	4.65	4.66	no	0.000
Head of the village	3.38	3.40	no	0.000
Leaflets, announcements	3.41	3.26	0.01	0.002
Local Press	3.37	3.39	no	0.000
Priest	2.93	2.54	0.01	0.010
Municipal website	2.66	2.49	0.05	0.002
Other local Internet sites	2.08	2.01	no	0.000
Facebook and other social networks	1.84	1.74	no	0.001

Source: Own research.

Table 8.8. Analysis of main components of sources of information on local matters and events used by farmers in 2017: factor loads, rotation Varimax (loads below 0.20 not included)

Information sources	Components	
	I: Traditional sources	II: Modern sources
Other residents	0.760	
Family members, acquaintances	0.686	
Head of the village	0.672	
Priest	0.666	
Leaflets, announcements	0.639	0.244
Local press	0.440	0.386
Other local Internet portals		0.868
Municipal websites		0.807
Facebook and other social media networks		0.805
Percentage variance explained by particular component	28.860	25.410
Total percentage variance explained	**54.27**	

Source: Own research.

Table 8.9. Farmers' sources for information on local matters in 2017: average factor values for the scales of traditional and modern sources, depending on sex/gender

Sex/gender	Scale of traditional sources			Scale of modern sources		
	average	significance	eta^2	average	significance	eta^2
Women	0.024	no	0.000	0.042	no	0.000
Men	−0.014			−0.024		

Source: Own research.

Participation in religious services gave farmers the best opportunity for contact with other residents of the community. As many as 69.6% of respondents attended these services weekly and another 20.6% did so at least once a month. Among other frequently reported forms of social contact were conversations with neighbors in front of the house. Such conversations were conducted at least once a week by 68.2% of respondents and at least once a month by another 23.6% of respondents. A frequent form of direct contact involved hosting friends and acquaintances at home. Over one fourth of respondents (26.2%) did so at least once a week and over half (52.2%) of them at least once a month. Besides direct forms of social contact, farmers also engaged in indirect forms of contact. Perusing local news on the Internet could be considered one of them. Over one third of the interviewed farmers (35.1%) used this form of involvement in local matters at least once a week and another 16.0 did so at least once a month. A significantly lower percentage of farmers took up social activities to benefit the community. Only 4.9% of respondents engaged in such activities once a month and another 11.4% did so at least once a month. A small percentage of farmers engaged in the types of social contacts preferred by urban dwellers, for instance, dining in restaurants for social or professional purposes. Only 4.7% of the interviewed farmers met in restaurants for social purposes and less than 2% (1.8%) dined in restaurants for professional purposes. The frequency of such contacts was categorized as at least once a month (Table 8.10).

According to the collected data, both female and male farmers strongly preferred direct social contacts of a traditional character. They tended to talk with neighbors in front of the house, participate in religious services, and host friends and acquaintances at home. Having friends over and participating in religious services were the activities preferred by women rather than men. Modern forms of social contacts were relatively rare. These modern forms included meeting in restaurants for social and professional purposes and male farmers engaged in these forms slightly more often than female farmers (Table 8.11). Two dimensions of social contacts—modern and traditional—are confirmed by factor analysis in Table 8.12. Generally, the social contacts of farm women had a somewhat more traditional character than that which male farmers engaged in. Men showed a slightly higher preference for a more modern style of social communication (Table 8.13).

Table 8.10. Various forms of farmers' social contacts in 2017. Scale from 1 to 7: 1—"never," 2—"about once a year or less," 3—"several times a year," 4—"less than once a month," 5—"2–3 times a year," 6—"about once a week," 7—"at least several times a week"

How often does the respondent:	Never (answer 1)	A few times a year, at most (answers 2 and 3)	2–3 times a month, at most (answers 4 and 5)	At least once a week (answers 6 and 7)
Dine in restaurants for professional purposes	87.8	10.4	1.3	0.5
Dine in restaurants for social purposes	67.0	28.3	4.2	0.5
Do some things for the community	48.2	35.5	11.4	4.9
Peruse local news on the Internet	40.6	8.3	16.0	35.1
Talk to neighbors in front of the house	3.0	5.2	23.6	68.2
Participate in religious services	2.7	7.1	20.6	69.6
Host friends/acquaintances at home	2.2	19.4	52.2	26.2

Source: Own research.

Table 8.11. Forms of farmers' contacts in 2017: averages depending on sex/gender (scale 1–7)

How often does the respondent:	Averages		Significance	eta^2
	Women	Men		
Dine in restaurants for professional purposes	1.14	1.24	0.01	0.005
Dine in restaurants for social purposes	1.53	1.62	0.01	0.002
Do some things for the community	2.23	2.14	no	0.001
Peruse local news on the Internet	3.71	3.60	no	0.000
Talk to neighbors in front of the house	5.75	5.75	no	0.000
Participate in religious services	5.60	5.31	0.01	0.013
Host friends/acquaintances at home	4.69	4.55	0.01	0.003

Source: Own research.

Table 8.12. Analysis of the main components of farmers' social contacts in 2017: Varimax rotation (loads below 0.20 not included)

Form of farmers' social contacts	Components	
	I: Modern form	II: Traditional form
Dine in restaurants for professional purposes	0.831	
Dine in restaurants for social purposes	0.733	
Peruse local news on the Internet	0.650	
Do things for the community	0.409	0.335
Talk to neighbors in front of the house		0.723
Host friends/acquaintances at home		0.668
Participate in religious services		0.611
Percentage variance explained by particular component	26.49	21.16
Total percentage of variance explained	47.65	

Source: Own research.

Table 8.13. Farmers' forms of social contacts in 2017: average factor values for the scales of modern and traditional forms, depending on sex/gender

Sex/gender	Modern form scale			Traditional form scale		
	average	significance	eta²	average	significance	eta²
Women	−0.058	0.01	0.002	0.127	0.01	0.009
Men	0.034			−0.074		

Source: Own research.

8.4. Concluding Remarks

In recent years, despite farmers sharing the notion of a lack of proper political representation, as well as a general unwillingness to actively engage in the life of political parties and trade unions operating on the national and supralocal level, there was a noticeable increase in interest in local social matters. This was more the case with female farmers than male farmers. Their active engagement in local matters went up almost three times, approaching the level of involvement reported by male farmers. Female farmers were particularly active in working on solutions to local problems, mostly by fulfilling functions in local associations, local social committees, and parish organizations. At the same time, they were less likely than men to fulfill the functions on the local government council or as heads of the villages. It was also noteworthy that female farmers were more active than male farmers in obtaining local news and information on local municipal matters from various sources, traditional and modern alike. They reported obtaining such information from local priests or by reading leaflets and announcements but also from municipal websites and through social media networks online. They preferred the more traditional forms of social contacts, such as inviting friends and acquaintances over to their homes and participating in religious services, and therefore were less likely than men to engage in the modern form of contacts such as social and professional meetings in restaurants. However, they perused the Internet in search of local information more proactively than their male counterparts.

Generally, it can be stated that female farmers in recent years "got out of the house" more often than they used to and started to engage in matters of their local community as much as their male counterparts do. Such engagement has a real influence on decisions regarding their immediate social space. The ongoing, progressive process of the empowerment of female farmers can be observed here. At the same time, female farmers are "traditional participants" in social life more than male farmers are, especially in regard to preferred social contacts. This does not prevent them from being more open than men to getting information from more modern sources. They are also better at recognizing crucial problems in the local community, as well as civilizational problems caused by globalization. One could ask: What is the cause for such significant dynamics of increasing interest from female farmers in local community matters and de facto global problems? Is it only an attempt to transfer the skills they have acquired while managing their farms into a more open social space? Certainly, such aspirations are not unique to female farmers. However, it seems that freeing these aspirations and placing them out in the open stems from the ongoing worldwide discourse on women's empowerment and their equal rights in public life. Such a debate is also occurring in Poland.

REFERENCES

Alston, M. & Wilkinson, J. (1998). Australian farm women: Shut out or fenced in? The lack of women in agricultural leadership. *Sociologia Ruralis* 38(3), pp. 391–408.

Alston, M. (2003). Women's representation in an Australian rural context. *Sociologia Ruralis* 43(4), pp. 474–487.

Bennet, K. (2004). A time for change? Patriarchy, the former coalfields and family farming. *Sociologia Ruralis* 44(2), pp. 147–166.

Brandth, B. & Haugen, M.S. (1997). Rural women, feminism and the politics of identity. *Sociologia Ruralis* 37(3), pp. 325–344.

Brandth, B. & Haugen, M.S. (1998). Breaking into a masculine discourse on women and farm forestry. *Sociologia Ruralis* 38(3), pp. 427–442.

Gorlach, K. (2009). *W poszukiwaniu równowagi. Polskie rodzinne gospodarstwa rolne w Unii Europejskiej* [In search of balance: Polish family farms in the European Union]. Wydawnictwo Uniwersytetu Jagiellońskiego.

Goverde, H., Haan, H. de, & Baylina, M. (eds.) (2004). *Power and Gender in European Rural Development*. Ashgate.

Grace, M., & Lennie, J. (1998). Constructing and reconstructing rural women in Australia: The politics of change, diversity and identity. *Sociologia Ruralis* 38(3).

Kaczor-Pańków, G. (1996). Aktywność społeczno-zawodowa kobiet i ich rola w społeczności wiejskiej [Social and professional activity of women and their role in the rural community]. *Wieś i Rolnictwo* [Village and agriculture] 4, pp. 12–21.

Matysiak, I. (2013). Źródła i zasoby kapitału społecznego sołtysów i sołtysek w wybranych kontekstach lokalnych [Sources and resources of social capital of village leaders and village leaders in selected local contexts]. In: J. Sawicka (ed.), *Rynek pracy na obszarach wiejskich Mazowsza – perspektywa gender*, pp. 71–91. Wydawnictwo SGGW.

Matysiak, I. (2017). Gender desegregation among village representatives in Poland: Towards breaking the male domination in local politics? In B.B. Bock & S. Shortall (eds.), *Gender and Rural Globalization: International perspectives on gender and rural development*, pp. 222–245. CABI.

Modelmog, I. (1998). "Nature" as a promise of happiness: Farmers' wives in the area of Ammerland, Germany. *Sociologia Ruralis* 38(1), pp. 109–122.

Mydlak, M. (1996). Społeczne uwarunkowania przedsiębiorczości kobiet wiejskich [Social determinants of entrepreneurship of rural women]. *Wieś i Rolnictwo* [Village and agriculture] 4, pp. 70–84.

Navarro Yañez, C.J. (1999). Women and social mobility in rural Spain. *Sociologia Ruralis* 39(2), pp. 222–235.

Perepeczko, B. (1996). Rola środków masowego przekazu w procesie edukacji i emancypacji kobiet wiejskich [The role of mass media in the process of education and emancipation of rural women]. *Wieś i Rolnictwo* [Village and agriculture] 4, pp. 22–29.

Pettersen, L.T. & Solbakken, H. (1998). Empowerment as a strategy for change for farm women in western industrialized countries. *Sociologia Ruralis* 38(3), pp. 318–330.

Pini, B. (2002). The exclusion of women from agri-political leadership: A case study of the Australian sugar industry. *Sociologia Ruralis* 42(1), pp. 65–76.

Shortall, S. (1996). Training to be farmers or wives? Agricultural training for women in Northern Ireland. *Sociologia Ruralis* 36(3), pp. 269–285.

Shortall, S. (2002). Gendered agricultural and rural restructuring: A case study of Northern Ireland. *Sociologia Ruralis* 42(2), pp. 160–175.

Verstad, B. (1998). Cracking the glass ceiling: The story of the election process in the Norwegian Farmers Union in 1997. *Sociologia Ruralis* 38(3), pp. 409–426.

Wells, B. (1998). Creating a public space for women in US agriculture: Empowerment, organization and social change. *Sociologia Ruralis* 38(3), pp. 371–390.

Witkowska, D. (2013). Zmiana sytuacji kobiet w okresie transformacji [Change in the situation of women in transition]. In: J. Sawicka (ed.), *Rynek pracy na obszarach wiejskich Mazowsza – perspektywa gender*, pp. 27–45. Wydawnictwo SGGW.

Żak, M. (2013). Więzi społeczne na obszarach wiejskich oraz aktywność obywatelska [Social ties in rural areas and civic activity]. In: J. Sawicka (ed.), *Rynek pracy na obszarach wiejskich Mazowsza: perspektywa gender*, pp. 164–179. Wydawnictwo SGGW.

Think Locally, Act Globally: Polish farmers in the global era of sustainability and resilience, ed. by Krzysztof Gorlach and Zbigniew Drąg in collaboration with Anna Jastrzębiec-Witowska and David Ritter
Jagiellonian University Press, Kraków 2021, pp. 373–409
ISBN 978-83-233-4949-5
DOI: http://dxdoi.org/10.4467/K7195.199/20.20.12735

Chapter Nine: Class Diversification among Polish Farmers in 2017

Krzysztof Gorlach https://orcid.org/0000-0003-1578-7400/

Zbigniew Drąg https://orcid.org/0000-0002-9106-7758/

9.1. Some Theoretical Considerations

9.1.1. General Considerations about the Issues of Classes

The thesis about the death of classes appeared in sociological literature some twenty years ago (see Pakulski & Waters, 1996), and its authors made numerous arguments to support it. The most important arguments were related to the socioeconomic role of the state in interfering with market processes which, according to the traditional sociological view, previously served the purpose of creating the effect of structure. The taming of these processes weakened the class structure. Other arguments stressed the increasing social role of various organizational structures and the authority of these bodies. At the same time, educational qualifications, as well as professional knowledge and skills, also gained importance while other, more complex socio-professional, ethnic, nationality, and race-related divisions became more noticeable. Authors such as Pakulski and Waters (1996, p. 280) pointed to the weakening of patriarchal relations and strengthening of the role of cultural dimensions in lifestyles and personal tastes.

In our view, this thesis stretches much too far. Admittedly, the factors mentioned above diminish the visibility of economic factors and dependencies, such as the availability of production means or ownership of economic capital, and consequently obscure the importance of class structure; still, they do not make classes irrelevant. Quite to the contrary, references to economic and class interests, as well as various social and economic factors, may well lead to a more comprehensive approach to class issues, especially in conceptualization of class positions. Such a strategy of analytical construct is highly preferred in this chapter, mainly in the conceptualization of class positions. The analytical construct strategy will be applied to the reflection of this chapter.

On the surface, it would seem that the subject of class position in farmers' circles has been addressed quite extensively in the fifth chapter devoted to "big" farmers. As presented in that chapter, farm area was the main factor distinguishing between the large and the remaining (smaller) farms that constituted the only two categories in our analysis. In our view, leaving the analysis at that would be insufficient. Obviously, the social position of farm users depended on farm size, but this was not the only variable playing a significant role here. Therefore, getting to the core of social divisions that could be observed among farmers, we have decided to apply the perspective of class divisions, which would address the full spectrum of variables determining social position of particular farm user.

It might be helpful to start with a look at how the problems of the socio-professional categories, such as peasants/farmers, are addressed within the concept of class analysis. These issues referring to farmers are treated as having only marginal importance. Suffice it to say that in a voluminous publication on social structure, such as Grusky's work of 2001, the category of farmers did not appear at all, and the category of peasants was only mentioned once, in reference to Karl Marx's iconic reflection on classes in early capitalism and pre-capitalist society (see Grusky, 2001, pp. 100–101). Peasants were treated there as a rather homogenous category of small farm owners, separated from each other due to the character of their economic activity. This isolation was additionally enhanced by peasants' poverty and lack of communication networks at the time, meaning the first half of the 19th century in France, as Marx's reflection was clearly applied to that period of time and place. The peasants described by Marx were engaged in a certain exchange with nature and did not participate further in a wider network of social exchange. They were the proverbial "sacks of potatoes," unable to represent their group (class) interests. In this context, peasants constitute a social category typical of the fading pre-capitalist and pre-modern society. In this view, peasants (i.e., farmers) were a rather marginal social category, in that they lacked diversification and were connected with a traditional, pre-modern type of social structure that was becoming a thing of the past.

This is an overly simplistic picture. In the synthetic history of class analysis published in the same year as a collection of sociological work edited by Grusky, the topic of peasants (farmers) appears far more often and in various contexts (see Day, 2001). Within pre-modern (feudal) society, the peasant was presented as a representative of the exploited and oppressed social category. As Day put it, "... the specific form of exploitation under feudalism was the lord's power to extract 'surplus value' from the peasant by making him work on the lord's land without pay. The class struggle in feudalism was therefore over the individual liberty of the peasant" (Day, 2001, p. 14). In the later period, when capitalism emerged, as well as economic exchange relations (related to market and modern society) as basic principles of social relations, the situation of peasantry changed significantly but this category was still treated by researchers (including Polish ones) as a rather homogenous category.

The authors of this chapter strongly oppose the simplistic views which treat the entire population of farmers as such a monolithic category with a certain place in the system of social stratification (see Gorlach & Klekotko, 2010).

> It might be illustrated by recent work published by the leading team of Polish sociologists [see Domański, 2008]. Farmers are treated as a homogenous category, a homogenous element of the stratification system. In this concept of social classes, patterned after the international EGP model, farmers are treated as a single social category (Gorlach & Klekotko, 2010, p. 110).

This statement, unfortunately, is reflected rather closely in several other sociological publications, with farmers as an entire socio-professional category being generally placed in the lower levels of the stratification hierarchy. For example, in his analysis of social changes and inequalities related to access to education, Zbigniew Sawiński (2008, p. 38) presented farmers as a social category located between qualified workers and workers performing very simple tasks. Similarly, Domański and Tomescu-Dubrow (2008a, p. 65), in their analysis of educational inequalities in Poland from the historical perspective, also referred to farmers as a uniform category positioned below qualified and unqualified workers. In another publication, the same two authors (see Domański & Tomescu-Dubrow, 2008b, p. 148) examined the problems of entering the employment market in relation to attained educational level and, while doing so, presented professional career farmers as a socio-professional category which could be lumped together with farmers lacking qualifications. Another author, Dariusz Przybysz (2008, p. 225), in his analysis of Poles' election activities, also lumped all farmers together into a rather uniform, monolithic socio-professional category and, just like the other authors mentioned above, positioned them below workers (qualified and unqualified alike) in the stratification system.

The above could not be applied to more general statements as well as more concrete indications, related to particular characteristics of class and stratification groupings. In the sociological reflection on consciousness and recognition of political interests, farmers were often presented in their entirety as a large socio-professional category, distinct from other categories such as the upper-class or workers (see, e.g., Dubrow, 2008, pp. 271–292). This, however, did not mean that the authors who followed this concept were not aware of the internal divisions within the broadly defined category of farmers. The following quote about Polish farmers might illustrate this quite well:

> Considering qualifications and wealth [farmers] are a diverse category, but what they have in common is possession of lands and farming. In the socialist era they constituted a peasantry class dependent on the state in the scope of purchase of equipment and other means of production as well as entering into contracts for food production. The post-communist transformation and international competition forced professionalization of Polish farms so significant that part of them today represent a farmer type (Słomczyński & Tomescu-Dubrow, 2008, p. 95).

Nevertheless, despite such statements, analyses of social mobility are still conducted as if farmers constituted a homogenous group.

It might appear that such an approach could be an outcome of thinking about farming as a very specific economic activity, of homogenizing the situation of people who take up farming. In some sense, this is a continuation of the way of thinking which is quite widespread in contemporary sociological literature, particularly in the works of British authors.

Therefore, a classic concept of Goldthorpe (see Goldthorpe et al., 1969; Goldthorpe & Hope, 2001) should be mentioned along with its two main premises:
1) the perspective of social structure is viewed through the lens of professional groups,
2) the main factor determining the hierarchical position of these professional groups is the amount of prestige they receive.

As a result, the social structure in this view appears to be an arrangement of certain elements (see Goldthorpe, 1987, pp. 40–43) divided into seven categories. The system consists of: class I, class II, class III, class IV, class V, class VI and class VII.

It is important, however, who belongs to these classes. Class I includes: upper service class, large proprietors, higher professionals, higher administrators, and managers. Class II consists of: lower service class, lower professionals, technicians, lower administrators, small business managers, and supervisors of non-manual workers. Class III is composed of so-called routine, non-manual workers. Class IV is represented by the petty bourgeoisie and farmers. In Class V there are lower technicians, foremen, and shop supervisors. Class VI includes skilled manual workers, while Class VII consists of semi-skilled and unskilled manual workers. It turns out that farmers, along with the petty bourgeoisie, make up only one category and are generally placed into the middle class. In the quoted work there is not even a trace of reflection on any potential differentiation of farmers' positions depending on farm parameters, their personal characteristics (i.e., age, education, or gender), or even their financial situation, land ownership, etc. It can be assumed that in his analyse, Goldthorpe operated almost exclusively with a concept of a rather small family farm.

The matters look quite similar in publications by those following in Goldthorpe's footsteps. In two important works published by Mike Savage (2000)—the renowned, contemporary representative of this tradition—there were significant references to the matters pertinent to farmers as a specific socio-professional group in British society, or even—to put it more broadly—in developed Western societies. It should be remembered, though, that Savage primarily referred to British society. Moreover, in the index of names and terms used in the publication, not a single reference to "farmers," or "family farms" could be found. In the later work of Savage (2015), which was supported by many other authors, a similar trend could be noted with the terms "farmers" and "family farms" not being used at all. Just as in the previous publication, Savage mostly focused on British society. He and the other authors devoted their interests to changes in the meaning of "class" and the strengthening of the role of cultural factors in this context. Interestingly, among the several dozen interviews that supplemented the qualitative results of the conducted survey there was not a single instance of the word "farmers."

The problem of farmers' class is not noticeable in the mainstream reflection on transformations of class structure. It can also be viewed as a controversial issue, approached in a multitude of ways in the context of various traditions of social structure analysis in sociology. The issues appearing here relate to disputes, not just at the level of academic analysis, but also in the areas of ideology and political practice (see Gorlach, 1995, p. 125 and subsequent publications). These issues could be turned into the following questions:

a) Are farmers a social class?

b) In what circumstances can they be referred to as a social class?

c) What is more appropriate: speaking of one peasant class or many peasant classes (classes of farmers), due to the many different characteristics of farms in farmers' possession as well as nation-state, regional, or local traditions and patterns of behavior displayed in certain situations, etc.?

d) Are farmers more a class of "owners and entrepreneurs" or are they a "working class"?

The ambiguity of the described problem was well illustrated by the phrase "awkward class," being applied to peasants by Teodor Shanin (1972), a well-known researcher of peasantry. It should be added here that this term was introduced by the author in the analytical context debating the situation of Russian peasants in the late 19th and early 20th century. The ideological frames of analytical and political dispute at the time involved debates between so-called Narodniks and Marxists. The former perceived Russian peasants as a certain people's mass and entity, while the latter paid attention to the ripening class conflict within the peasantry, between the owners of large farms, who could be seen as a rural version of the bourgeoisie, and small farm owners, as well as landless peasants, who were the rural equivalent of the proletariat (see also Mitrany, 1972). Remembering that such discussions were taking place at the turn of the 20th century, and were part of wider deliberations on modernity, peasants' place in society was interpreted in the Russian discourse of the time in two ways. The peasants were either presented by Narodniks as a specific social category, based on their shared destinies or, by Marxists, as a group with a specific logic of the family farm within the community, differentiated by economic characteristics, which made the owners of small farms likely to form alliances with workers and their political representation (Marxists). This was the area of dispute, between the supporters of Chayanov's concept of similarities of destinies, and the ideas of agrarian Marxists (see more in Gorlach, 1995, pp. 125–128).

Another path of this discussion may lead to the statement that, within agriculture, various arrangements of class relations may appear. The most well-known concept addressing the variety of class arrangements is the proposal by Stinchcombe (1961), today considered to be classic, which refers to at least five elements of such arrangements. They are: manorial, family-size tenancy, family smallholding, plantation, and ranch. It should be stressed here that only smallholding family farm peasants (farmers) have status as the owners' class. In the remaining cases they are no more than a subordinate class, barely holding the status of workers or tenants, at best. Class here (see Shanin, 1973, p. 253) is a gradable quality depending on socio-historical context. Low in everyday life, it increases rapidly in times of crises as well as during social and political tensions. This matter is further discussed in chapter fifteen of this monograph, written by Grzegorz Foryś, and devoted to the political mobilization of farmers.

One might ask what concept of class division would be of optimal use when farmers are considered. In such an approach to social analyses, two traditions are usually referred to, as they map out the field of possible analyses. The first one is the tradition connected with the concept of Karl Marx and the other—that appeared about a generation later—is linked to the tradition of Max Weber. It might create a dilemma as to which of these two perspectives should be applied to the analysis of issues that interest the authors of this chapter of the monograph.

The answer to this question is unexpectedly simple. As it turns out, in the contemporary sociological literature, both of these perspectives are integrated to some extent, combining economic and cultural issues. This is manifested in the American and European tradition of rural sociology and, particularly, in the reflection on agriculture's dominant form in North America and Western Europe, that of the family farm.

9.1.2. The American Tradition in Family Farms Analysis

The authors of this chapter propose to start the reflection on the family farm from the American tradition. Its first author, Harriet Friedmann, refers to Chayanov's concept, already mentioned above, stating that a certain coexistence of a peasant farm and a capitalist enterprise comprise the core of a family farm. At first glance, this message might appear weak in both logic and coherence. One might recall the statement by Chayanov that peasant and capitalist modes of production are based on completely different logics of functioning. However, Friedmann makes the point that this combination could be possible within the framework of a simple commodity production system and may be crucial to it. Within this system, demographic factors characteristic of peasant farms and pressures from market competition on capitalist farms occur in combination and have consequences for the entire system (see Friedmann, 1978a, p. 85). The labor resources supplied by the family contribute greatly to the flexibility and market adjustment of the surveyed farms. This actually enables family farms to function on the basis of a pre-capitalist economic logic, readily adjusting to the principles of a capitalist marketplace.

Friedmann's line of argumentation does not end here. There are other factors that seem to support the durability of the family form of production in agriculture. One of them is the mechanization of agricultural tasks as the emergence of new machines and equipment made it possible to replace hired workers (one of the fundamental characteristics of a capitalist economy) with the work of one or two members of the farming family. Besides mechanization, there are certain characteristics of the simple commodity production system that play a role in the process of strengthening competitiveness. Among them are the following: a flexible consumption level in the household (tightly connected with the farm), and the possibility of additional employment and, consequently, extra income for family members working outside of the farm. It should be remembered that these factors were also presented in Chayanov's work, as characteristics of a peasant farm (Friedmann, 1978b).

Noticing certain similarities between the model of a family farm (as described by herself) and the peasant farm, most accurately described in Chayanov's conceptualization, Friedmann emphasizes the specifics and differences of the family farm, that it is—to use a Marxist category—not a mode of production but only a specific form of production. It means that the family farm can only exist in a certain context mapped out by capitalist mode of production (capitalist economy). The functioning of the farm in the simple commodity production system is quite similar to the functioning of a capitalist enterprise, with the only difference being the structure of family relations, which remains intact (Friedmann, 1981, p. 3).

Three other concepts elaborated by American rural sociologists may be an important and valuable addition to this consideration. The authors of the first concept are Susan Mann and James Dickinson (1978), who attempted to explain why capitalist relations struggle to completely subordinate the system of farm production. The core of the problem lies in the process of agricultural production (unlike in industrial production), as natural processes, such as the maturation of plants and animals to the form that can be treated as a final product or commodity for sale takes time. The cereal must grow and later be harvested and prepared for sale to processors or directly to consumers. A similar situation applies to animals raised for meat as their bodies must grow to the point that they can be offered to processors or consumers. In other words, according to the authors quoted here, in agriculture there is an obvious difference between the time of production and the time of work. The time of production contains both the time of work and the time needed for natural processes. Obviously, it can be shortened or sped up with the use of various chemicals, genetically modified animal feed, and through other genetic manipulations, but the natural process of changes that a plant or an animal must go through will, to some extent, always remain part of agricultural production. And this—according to Mann and Dickinson—is a factor in both slowing down capitalist relations' introductory process in agricultural production and, at the same time, maintaining the nature of family farms.

The second of the selected concepts is a dialectic concept of the family farm (see Gorlach, 2004, pp. 104–106) and it was coined by the renowned American rural sociologist Frederick H. Buttel (1980). This proposal contains the factors already mentioned in the previous which relate to the flexibility of work relations and the difference between the time of work and the time of production. External factors, such as state policy, which is generally rather supportive of family farms, are also quite important here. In effect—as stressed by Buttel and his corroborator Flinn (1980, p. 951)—the position of farm owner becomes a certain combination of the role of capitalist, who employs both family and hired labor, and the role of worker, who operates their own farm and works on it but quite possibly also works outside of agriculture. The farm owner loses the clarity of his role on the farm.

The third of the presented concepts elaborated by American sociologists was elaborated by Patrick H. Mooney (1988). It is the most complex out of the three concepts discussed here, as it quite directly refers to the concept of class locations by Erik O. Wright (1989, 1997). The complexity of Mooney's concept lies in acknowledgement (after Wright) that, besides the relation concerning the means of production, there are many other factors determining class position.

According to Wright, class position is determined by four main factors, of which only one is the standard Marxist relation to the means of production. The three remaining are relation to authority, relation to scarce skills, and number of employees. As a result of combining these characteristics, twelve categories of class position emerge (see Wright, 1997, p. 25), of which two, namely capitalists and non-skilled workers (proletariat) can be found in the classic Marxist bipolar scheme of the capitalist society class structure. The remaining categories can be treated as displays—according to Wright—of various combinations of contradictory class locations, indicating that ten out of twelve class categories are examples of various contradictory class locations.

Mooney's reasoning follows this path. In his view, there is not just one class position like the one described by Harriet Friedmann as a simple commodity production. It cannot be just one class position as family farms, according to Mooney's analyses, have to face the pressure of various factors which differentiate their location in social class structure. They are the following:
a) land rentals,
b) processes related to farmers' debts,
c) non-agricultural employment engaged in by the farmer and/or their family members,
d) employment of hired workers on the family farm,

e) the system of contract farming being carried out on the basis of an agreement between a farmer and a potential buyer of farm products which, on one hand, facilitates the certainty of the purchase but, on the other, limits the future freedom of the farmer selling his or her agricultural product.

As a result, at least several various class positions can be distinguished among the owners of family farms (see Mooney, 1988, p. 50; Gorlach, 2004, p. 112). They are the consequences of simple commodity production, as well as the capitalist mode of production. Therefore, in these contexts, the extreme categories are the types of class position, such as: capitalists and the representatives of simple commodity production, already described in Friedmann's concept. The four remaining categories represent the contradictory class locations. These four categories are:
   a) salaried managers and supervisors of capitalist firms,
   b) landlords and small employers,
   c) new petty bourgeoisie,
   d) tenants, debtors, contract producers, and those employed off-farm.

To summarize the above reflection, it can be stated that the five factors listed above are the sources of various forms of social relations on farms and the ambiguous positions of farmers in the class sense. Each of these factors can have various consequences depending on the intensity or regularity of their influence. As a result, with the situation of a farmer in the position of being a practitioner of simple commodity production, one can speak of the processes of moving towards the position of capitalist (bourgeoisification) or towards the position of hired worker (proletarization). In effect, the position of farm owner/user moves in the system of class relations from the "clean" position of simple commodity production towards two opposite class positions typical of the capitalist mode of production, namely the position of capitalists and the position of hired workers. Consequently, the position of farmers (family farm owners/users) can be seen as many variations of positions of contradictory class locations. These variations include, for example, on the side of capitalists: the positions of landlords/small employers, and salaried managers/supervisors of capitalist firms, and on the side of workers: new petty bourgeoisie as well as tenants, debtors, and contract producers employed off-farm (Mooney, 1988, p. 50).

As presented above, Mooney's concept definitely takes inspiration from Wright's reflection, but it is not its direct continuation. It is just borrowing from a certain analytical concept or ideas related to theoretical perspective and it would not serve any good purpose to install the proposal of Mooney directly into the concept of Wright. The concept of various class positions of family farm owners/users would be very hard to place in the last scheme. This is because Mooney introduces more specific variables that are more suitable to describe class positions of family farm owners/users than that which are elaborated in Wright's concept.

9.1.3. European Tradition in the Analysis of Family Farms

The European tradition of family farm analysis is more differentiated than the American tradition described above. Its presentation should start with an introduction of the concepts of French sociologists, who consider the evolution of family organization (Lacombe, 1991, p. 62) as an important aspect of such analysis. The typical model of the family farm is therefore increasingly questioned. First, it should be noted that the particular members of the family use their economic capital, their activities (cultural capital), as well as connections and relations (social capital) to strengthen the farm within the economic system in a variety of ways. A farm operation can be conducted even by one person, as this is currently possible due to technical and technological advancement of agricultural production but other family members, as well as the influence of the local community and tradition, can play an important role in farming. From this perspective, the family farm appears to present a rather uniform class form, for which the existing environment is quite important.

Another French author, Marcel Jollivet (1994, p. 46), takes a different approach. In his view, family farms comprise a rather homogenous—in the class sense—form of production that can exist in symbiotic relationship with large agricultural enterprises in the class system. These large farms intercept the added value from family farms, which farmers and their families produce, but in doing so leave the social relations on the farm unchanged.

The perspective of German researchers and authors with an emphasis on cultural tradition seems just as important and valuable as the French perspective. One of the German authors, Hans Pongratz (1990, p. 6), claims that the strategies of adjustment of family agriculture to the reality of industrial society (i.e., modern capitalist society) stem from relations between cultural tradition and the process of agricultural modernization. As the modernization results in larger farm size, it also increases farmers' debts in the banks. At the same time, farmers do not want to adjust to the changes occurring in industrial society. They do not fully emerge in the city/urban lifestyle and prefer to concentrate on the farm as a value in and of itself. Two types of rationality described by Max Weber, such as substantial and formal rationality, can be seen here. The former is focused on maintaining the farm as a value that is not just economic, and the latter ensures maintaining the profitability of production, so that the farm brings profits and not losses. When these two types of rationality clash, along with the modes of conduct they entail, the first option seems to be winning, with its effective use of traditional methods of organizing work and management, closely related to the idea of the peasant family farm (see Pongratz, 1990, p. 8).

The German tradition of analyzing the family farm can be summarized by the work of Andreas Bodenstedt (1990), whose name was already mentioned in the first chapter of this monograph. In his concept, Bodenstedt stresses the loss of traditional peasant autonomy in the modernization process. What determines their structural position in this process—as indicated by the author—is the growing dependence on the state and its policies. Farms begin to differ from each other in terms of their modernization progress and their degree of dependence on the state policy. In that sense, Bodenstedt's concept shares some similarities with ideas proposed by Mooney or Buttel, although obviously the German author does not refer to the principles of class analysis in this context. It seems that, in his perspective, the peasant farms undergoing modernization, namely family farms, still remain a rather homogenous category. The differences between them lie in detailed characteristics. They are mostly related to culture and tradition, where the significant social potential of farms can be found (see Bodenstedt, 1990, p. 47).

Another important aspect of the discussion on European tradition in analyzing family farms is the approach presented by the Dutch authors. The status of the owner working on his own, which is the epitome of the entrepreneur connected with the modern capitalist economy, is at the core of the Dutch perspective. The family farm is therefore treated as an individual enterprise, based on the rationality and entrepreneurial abilities of the owner, while the family connected with the farm is only a consumption unit, dependent on the farm as a source of income (Haan, 1993, p. 164). This is not to say that there is some absolute unification of farm model happening in Dutch agriculture or that there is only one type of class position. Sam de Haan also notices here the intergenerational transfer of particular farms as a factor differentiating their situations and impacting the course of their evolution.

Another author, Jan Douwe van der Ploeg (1995), adds some finishing touches to the way of thinking described above, using a model of the family farm that, due to its structural characteristics, may be treated as an example of one specific production type. The author introduces clear-cut factors differentiating farms due to certain cultural variables. The style of farming is such a variable, which is aggregated, presenting certain types of rationality among farmers and encompassing three component variables. These are:

    a) the set of strategic terms, values and perceptions by the group of farmers on how agriculture should be organized,

b) the specific character of agricultural activity, corresponding with the above terms, and collectively described as the so-called cultural repertoire of the surveyed farmers,

c) the specific set of relations between the farm and the surrounding market, stemming from the character of the agricultural activity, consisting of not just institutions but also state policy and certain technological variables.

The core of this approach is the subjective method of conceptualizing factors determining the position of the farm from the perspective of the cultural apparat that a farmer is equipped with (Ploeg, 1995, p. 122).[1]

Van der Ploeg's concept, which is related to cultural differences regarding family farms, has two limitations. One of them is its narrow analysis to just one region of contemporary Netherlands, namely Friesland, and the other is the focus on just one sector of agricultural production, which is milk production. Still, the author distinguishes five main types of farmers. The first type is defined as economic farmers, encompassing farmers who are focused on ensuring the functioning of their farms by supplying the external sources of financing. At the same time, these farmers care about the balance between external supply of finances and internal assets and resources of the farm. The main goal of these efforts is, on one hand, the minimization of production costs and, on the other, treating human labor as the factor that is decisive for the success of the farm. The second type of farmers in van der Ploeg's typology are intensive farmers. In this category, there are large farms which aggressively use all the technological novelties, are oriented toward constant expansion, and have strong, multifaceted ties with the market. They have a lot in common with large-scale industrial farming, described by American economist Marty Strange (1988) in his reflection on agriculture in the United States as being in opposition to the model of family farming. The third type distinguished by van der Ploeg are the machine-farmers. The strategy of operating the farm in this case is about minimizing the need for human labour with the goal of having a farm that would be completely mechanized. For the fourth type of farmers, farm animals are the most important. As the reader may recall, van der Ploeg's typology relates to farms engaging in milk production, so for this type of farmer he coined the term "cowmen." In this type of farm each animal receives highly individualized care, and farm land is devoted to producing crops for the animal feed that the livestock could consume later. Finally, there is the group of big farmers. In this case, similar to the

---

[1] This matter is discussed more broadly in Chapter 1.

category of intensive farmers, the most important aspect of the strategy (style) for operating the farm is its expansion. Each farm activity is evaluated from this point of view. However, these farmers conduct agricultural operations on an even bigger production scale than those belonging to the intensive farmer category.

9.1.4. Short Summary and Transition to Empirical Considerations

The content of this chapter can be seen in two perspectives regarding family farm analysis. In the first perspective, classic concepts of Marx and Weber can be mentioned as well as their more contemporary representation which can be found—as indicated earlier—in the works of Mooney and van der Ploeg. The second perspective is more about technical matters, relating to the construction of particular variables, determining class position (using Marx's terminology) or market position (to use the language of Weber). It should be emphasized that in this second case the difference between these two approaches is of a nuanced character. In both cases there are references to class and market, which can be treated as a scene where the classes play out their roles.

The first perspective imposes at least two messages if the presented class analysis is to be based on some synthesis of classic approaches (Marx and Weber) and their contemporary continuations (Mooney and van der Ploeg). It is consistent with the conviction that discernment of classes should not be derived merely from one single factor alone, for example just economic characteristics (type of ownership of production means, size of the owned capital, etc.), or just non-economic characteristics (cultural, related to educational level, style of conducting agricultural activity, etc.). Second, both types of factors should be present together, at least economic and cultural factors.

It should be emphasized that, although the concept of class analysis of family farms proposed below by the authors of this chapter draws upon the ideas elaborated by Mooney and van der Ploeg, it is not just their simple application. It is more the use of ideas present in these concepts that refer to equipping farmers with various types of capital, including economic and cultural capital. The conceptualization of these two types of capital is a rather technical matter.

9.2. Economic and Cultural Capital and the Market Position of the Surveyed Farmers

The technical matters mentioned above require more careful analysis and orientation to the details. Therefore, detailed construction of the two most important variables, namely "economic capital" and "cultural capital," is of great importance. For the reflection of this chapter it is essential to note that the 2017 study constitutes the bases for the analyses here. For the construction of two aggregated variables ("economic capital" and "cultural capital"), somewhat different "output" variables were assumed than in the previous research editions, the results of which are described in publications by Krzysztof Gorlach (1995, 2001, 2009) and already mentioned in this monograph. Therefore, all comparisons made between the 2017 study results and the results of earlier research editions should be treated with caution. It should be remembered that comparative references would be likely to apply some more general tendencies characteristic of the category of family farm owners in contemporary Poland but would not be very useful to support statements on more precise tendencies for change in regards to class characteristics of the studied category. In other words, the results of the 1994, 1999 and 2007 studies (see Gorlach, 2009, pp. 107–136) only provide some general background for the analyses related to the 2017 study.

It seems appropriate to start with the construction of both indicators, characterizing the class position of the farms surveyed in 2017.

In the construction of the indicator of economic capital two elements, namely, farm size and the value of farm equipment, were considered. Both of these elements had assigned values from 0 to 2. The 0 value was assigned to any farm of 5 ha or smaller, the value of 1 to farms sized 6–10 ha, and the value of 2 to farms exceeding 10 ha. In regard to the value of farm equipment, the 0 value was assigned to farms whose machine park was worth 30 thousand PLN or less, the value of 1 to farms with the equipment worth 31–80 thousand PLN and the value of 2 to farms where the value of the total equipment exceeded 80 thousand PLN. The values of both elements were summarized in such a way that each farm could receive value points within the 0–4 range. The economic capital of farms with the combined value of 0 was described as low. The combined value of economic capital between 1 and 2 was considered to be medium and farms with that result were assigned a value of 1, while farms with a combined value of economic capital between 3 and 4 were assigned the value of 2 and were considered to have high economic capital.

The construction of the indicator of cultural capital involved consideration of three factors such as age, education of the main user, as well as the type of rationality this user displayed. Regarding age, the 0 value was assigned to all farm users aged 44 and over and the value of 2 to users under the age of 44. The values assigned to educational level were the following: 0 was given to farm users with grammar school education, 1 to farm users with vocational school education, and 2 to the respondents with at least a high school education. The characteristic described as "type of rationality" was coined in regards to three issues that could serve as indicators of the way of thinking described as "peasant" and opposed to "modern." These issues were the following:

1) attitude towards renting the land,
2) definition of the role of farm owner,
3) preferred model of equipping the farm with the necessary machinery.

The first issues, such as the attitude on renting the land was measured on a 5-point scale that the respondents were to follow when giving their evaluations. The value of 1 was assigned to the "I strongly agree" answer and the value of 5 was given to the "I strongly disagree" statement. "Renting of the land is not a good solution because one is dependent on the contract conditions, it is easy to fall into debts, and there is no motivation to take good care of the land." The answers "I somewhat disagree" and "I strongly disagree" were recognized as characteristic of the "modern" way of thinking and given a value of 1. The remaining answers were identified as the "peasant" way of thinking and given a value of 0. On the same scale, the second issue, which involved the way to define the role of the farm owner was measured by respondents' evaluation of the statement: "Currently being a farm owner is connected with the feeling of inferiority, hard and hopeless work, and being one's own slave." In this case the answers "I somewhat disagree" and "I strongly disagree" were perceived as indicative of the "modern" type of thinking and assigned the value of 1, and the remaining answers as the "peasant" way of thinking with the assigned value of 0. Preferences regarding the third issue, which is the model of farm equipment that farm users preferred, were measured through the answer to the question on how farmers should solve the problem of equipping their own farms with the machines. The opinion that the farmer should own all the machines, even if he only used them from time to time, was recognized as indicative of the "peasant" way of thinking and was allotted the value of 0. The value 1 was assigned to the remaining answers, which stated that the farmer should only own the machines which are used daily or the opinion that he should not own any machines and order all services from others. The value of 1 was considered as representative of the "modern" way of thinking.

When the sum of the three issues above was equal to 0, the characteristic described as "type of rationality" was 0, when it was 1 the "type of rationality" was 1, and when the value of the sum was 2 or 3, the "type of rationality" received the value of 2. The total sum of points, considering age, education and the "type of rationality" could fall anywhere between 0 and 6. For a sum of points between 0 and 1, the level of cultural capital was described as low and assigned a value of 0. A sum of points such as, 2, 3 or 4 was considered medium, with a value of 1. Finally, a sum of points between 5 and 6 was given a value of 2.

Finally, the construction of the synthesis-related indicator defined as "market position" and the combination of two previously described aggregated indicators deserves a closer look. Referring to class concepts described earlier, and the concept from Mooney, the term "class position" can be used interchangeably. The value of this indicator is the sum of the values of two indicators, such as "economic capital" and "cultural capital." As both of these indicators can assume values from 0 to 2, their sum can be found in the 0–4 range. The sum of points from 0 to 1 in market/class position was defined as negatively privileged and assigned the value of 0. The market/class position of 2 points was categorized as averagely privileged and given the value of 1. The market/class position with the number of 3 or 4 points was rec-ognized as positively privileged and assigned the value of 2.

The effects of these procedures and analyses are presented in Table 9.2.1.

Table 9.2.1. Level of economic capital, cultural capital, and the market position of surveyed farms (in %)*

Level of capital/market and class position	Economic capital	Cultural capital	Market position
Low/negatively privileged	41.6	15.7	40.6
Average/averagely privileged	31.2	64.2	28.4
High/positively privileged	27.2	20.1	31.0
Total (N = 3948)	100.0	100.0	100.0

* Data come only from those farms for which it was possible to calculate all three characteristics, such as level of economic capital, level of cultural capital and—as a result of those two data points—the market/class position.

Source: Own research.

Taking into account the reservations related to the somewhat different mode for constructing the basic variables in the 2017 study than that used in the earlier stud-ies, it might still be worthwhile to compare their results. Comparing the characteris-tics of family farms in Poland in 2017 and 1994, during the first edition of the studies conducted from 1994 to 2007 (see Gorlach, 2009, p. 112) it can be stated that the studied category is somewhat more stratified in the class sense. Although currently similarly to 1994, when they made up 47.9% of the whole, farms that are negatively privileged still dominate, constituting 40.6 % of all farms, but the situation of the two other farm categories looks quite different. In 1994 over one third (34.5 %) of the surveyed farms were considered to be in an averagely privileged position, then in 2017 only 28.4 % (slightly over one fourth) held such a position. Over the years the percentage of farms that were positively privileged increased from less than one fifth (exactly 17.6%) in 1994, to almost one third (31.0% to be precise) in 2017.

9.3. Is a Farmer an Entrepreneur?

A farmer was traditionally presented as both: a farm owner and a farm user. Later, during the time of modernization, when capitalist relations started encompassing rural areas and agriculture, bringing about market economic principles, the situation began changing. A farmer started to be perceived as a certain type of entrepreneur and the farm treated as a special kind of enterprise. It is worth emphasizing that, in Poland, the idea of a farmer's profession is, first and foremost, deeply rooted in the peasant tradition but, to some extent, also connected with the landlord tradition of large farm ownership. This tradition proved to be quite effective in suppressing the way of thinking about agriculture as a specific area of economic activities, quite different from other areas of economic activities such as industry or services. For that reason, a large part of the reflection of this monograph is devoted to the profession of farmer as an entrepreneur. A traditional slogan placed on the flags and pennants of various Polish peasant political movements reads "With Life and Arms" ("Żywią i Bronią") and refers to the special role of small entrepreneurs who, by producing food, fulfill the needs of other social categories.

Tables 9.3.1–9.3.4 show various aspects of this matter. It was determined that all opinions presented by farmers should be matched with their class position and the market position of the operated farm. Elaborate and extensive headings of particular tables in this subchapter allow for more detailed discussion on particular aspects of the surveyed farmers who perceive their role as, more generally, that of an entrepreneur.

The first table (9.3.1) presents a general perception of the role of the owner and eventually the sense of security and independence that go with that.

The values presented in Table 9.3.1 might serve as a basis for formulating at least two important statements. Firstly and most importantly, a farmer's class position and the market position of the farm do not have a significant influence on the opinions of the surveyed farmers. Therefore, one might risk a statement based on the collected data that a general sense of security and independence stemming from land ownership is reported by a solid majority of respondents—two thirds of respondents in the study. This is also confirmed by the average value of the evaluations, which are located in the column next to the positive impressions for each category (below the indicator of 3). Nevertheless, it is worth noting that at least 30% of the surveyed farmers are not inclined to express such a positive opinion as they think that being farmers does not give them a sense of security and independence. It can be assumed—at least initially—that this sense depends on some other factors, perhaps connected with some more individual, family-related or local determinants.

Table 9.3.1. Market position and perception of the role of the owner in terms of farm possession, based on the respondents' evaluations of the statements: "Today one feels like a farmer if they took over the farm as an inheritance from their parents, and the ownership of land, machines, or animals gives one a sense of security and independence"—percentage of "I agree" answers (combining the answers "I strongly agree" and "I somewhat agree") and the average measured on the 1–5 scale, where 1 means "I strongly agree" and 5 means "I strongly disagree"

Market/class position	Percentage of "I agree" answers	Average on the 1–5 scale
Negatively privileged	69.7	2.24
Averagely privileged	66.5	2.32
Positively privileged	68.9	2.22
Total	68.5	2.26

Source: Own research.

Is a farmer simply just a type of owner/entrepreneur? The subsequent questions could bring some concrete answers to this topic. Table 9.3.2 contains information on the attributes of the farm owner's role with respect to their class position and the market position of the farm in their possession.

It is worth noting here that the matters discussed in this subchapter relate to the very meaningful layer of class consciousness (see Giddens, 1973) which is closely linked to the concept of individual identity or the identity of social category. It is important for the reflection of this chapter to determine to what extent farmers identify themselves with the broadly defined category of entrepreneurs and to what extent they have a sense of their own, collective distinctiveness. Data on these matters can be found in tables 9.3.2, 9.3.3, and 9.3.4.

In this case, the opinions of surveyed farmers are no longer significantly connected with their class positions or the market position of their farms (see Table 9.3.1). As can be observed, the majority of the respondents (over two thirds in every category, with the average below the value 3) agree with the statement that farm ownership gives one a sense of agency, expressed through the possibility to realize one's vision and linked to the sense of responsibility and increased effort, etc. However, in the class category of positively privileged farmers these indicators are notably stronger. It turns out that such farmers, more than negatively and averagely privileged farmers, felt more like entrepreneurs. This statement can be verified through the study results in tables 9.3.3 and 9.3.4.

Table 9.3.2. Market position and perception of the role of farm owner as an entrepreneur based on the respondents' evaluations of the statement: "In today's world, being a farm owner means, first and foremost, the sense of working on one's own and having opportunities to realize one's vision. It is also a responsibility concerning the fate and fortunes of the farm and the necessity to plan ahead, to strive towards development and to increase the efforts"—percentage of "I agree" answers (combining answers "I strongly agree" and "I somewhat agree") and the average, measured on the 1–5 scale, where 1 means "I strongly agree" and 5 mean "I somewhat disagree"*

Market/class position	Percentage of "I agree" answers	Average on the 1–5 scale
Negatively privileged	68.5	2.24
Averagely privileged	73.0	2.12
Positively privileged	84.4	1.81
Total	74.7	2.07

* p < 0.01.

Source: Own research.

Table 9.3.3. Market position and perception of the similarities between the situation of a farmer and any other entrepreneur based on the respondents' evaluations of the statement: "Currently the situation of the farmer and the situation of any other entrepreneur do not differ. Each of them works autonomously and has to take risks and be able to sell their own products"—percentage of "I agree" answers (combining answers "I strongly agree" and "I somewhat agree") and the average measured on the 1–5 scale, where 1 means "I somewhat agree" and 5 means "I strongly disagree"*

Market/class position	Percentage of "I agree" answers	Average on the 1–5 scale
Negatively privileged	66.0	2.32
Averagely privileged	65.0	2.36
Positively privileged	74.2	2.12
Total	68.3	2.27

* p < 0.01.

Source: Own research.

Similar to the previous case, the majority of respondents indicated that the situation of a farmer is not different in any way from the situation of an entrepreneur conducting any other economic activity. However, the differences between the positively privileged farmers and the two remaining categories are quite significant. It could be said that the feeling among farmers of having similarities with the other categories of entrepreneurs—due to working on one's own, having a willingness to take risks, and possessing an ability to sell their own products—is significantly higher among the representatives of this category of farmers who have a positively privileged position on the market.

How to approach the conviction still present in the social consciousness, that farmers are entrepreneurs who are vulnerable to various unfavourable situations since agricultural production still depends on barely controlled natural conditions? It should be remembered here that nowadays, in the countries of the European Union, the program of direct agricultural subsidies is still ongoing and there are other financial instruments which support the functioning of farms. The fundamental idea behind the subsidies has been, for decades, concentrated on the distinctiveness of agriculture as a production area, with a relative lack of flexibility, which—for what it's worth—may mean serious difficulties for the activities of the producers. Therefore, it is necessary—according to those who support such an approach—to subsidize the functioning of farms and, at a minimum, to stabilize their incomes to some extent (see, e.g., Sheingate, 2001).

The respondents' evaluation of the "lower social appreciation" given to agricultural producers in comparison with other categories of entrepreneurs is also quite important. This is related mostly to the characteristics of the feudal system in Poland and its tradition of the serfdom of peasants who, at the time and throughout most of the 19th century, were treated as a totally subordinate social category and forced into slave-like labour (see also Gorlach, 1990).

Table 9.3.4. Market position and perception of the differences between the situation of a farmer and any other entrepreneur based on the respondents' evaluations of the statement that "In today's reality a farmer has to work more than the owner of any other business and the farmer's income is still lower, and they receive a lower social appreciation for the profession"—percentage of "I agree" answers (combining answers "I strongly agree" and "I somewhat agree") and the average measured on the 1–5 scale, where 1 means "I strongly agree" and 5 means "I strongly disagree"*

Market/class position	Percentage of "I agree" answers	Average on the 1–5 scale
Negatively privileged	80.6	1.85
Averagely privileged	80.6	1.88
Positively privileged	75.2	1.97
Total	78.9	1.89

* $p < 0.05$.

Source: Own research.

The data in Table 9.3.4 indicate a quite explicit feeling of discrimination among agricultural entrepreneurs, who "work more than the owners of any other business and the farmer's income is still lower, and they receive a lower social appreciation for the profession." Almost 80% of respondents generally agreed with this view and the average on the scale indicated high level of acceptance for such a statement. However, in the surveyed sample certain differences could be observed. The percentage of respondents agreeing with this statement is somewhat lower among those with a positively privileged position and the numerical average of their answers is situated further from a value 1, which means full acknowledgement of the farmer's situation being worse than the situation of other entrepreneurs. It can be stated that the sense of farmers' situation being worse than the situation of others is mostly shared by respondents with negatively privileged market/class positions.

9.4. Peasant Classes (Farmer Classes)?

It might be valuable to get into the core of class perception analysis of the surveyed farmers. Similar to earlier studies (see Gorlach, 2009, p. 122), it was determined that the class would be measured from the perspective of how the respondents perceived the conflicts connected with the various aspects of their socio-economic position. Here the concept of Anthony Giddens (1973) might be referred to, as for him the perception or awareness of conflict was a special characteristic of class position. This particular aspect Giddens treated as an important layer of class consciousness. Following the proposal of Giddens, and at the same time attempting to operationalize this particular matter, the concept of Erik Olin Wright (1989, 1997) was applied. The class consciousness is seen here as a collection of opinions related to various aspects of relations between the owners of means of production and the workers employed by them. These relations can be treated as a core of the class structure in a capitalist society.

Similarly to previous studies (see Gorlach, 1995, 2001, 2009), three crucial issues were distinguished and they respectively dealt with: remuneration of farm owner (entrepreneur), remuneration and his/her workers, abilities of the farms and the workers to have an influence on the functioning and transformation of the farm enterprise, and the possibilities for the owner to hire other people in case of strike. The results of the analyses are presented in tables 9.4.1, 9.4.2, and 9.4.3.

The analysis of the values in Table 9.4.1 indicates that the preferences of the surveyed farmers were biased towards the entrepreneurs (owners). As many as 70% of the respondents with a negatively privileged position agreed that enterprise owners should earn more than the workers hired by owners. This tendency was significantly stronger among the representatives of the positively privileged class (over 80%). The average values on the scale (generally under 3) confirmed the location of the surveyed farmers on the side of owners. Here, the average values indicate the highest intensity of the "pro-owner" attitude among the representatives of the positively privileged class.

Table 9.4.1. Market/class position and the opinions on the statement: "Enterprise owners should earn more than workers employed in these enterprises"—percentage of "I agree" answers (combining answers "I strongly agree" and "I somewhat agree"), percentage of "I disagree" answers (combining answers "I strongly disagree" and "I somewhat disagree") and the average measured on the 1–5 scale, where 1 means "I somewhat agree" and 5 means "I strongly disagree"*

Market/class position	"I agree" answers	"I disagree" answers	Average on the 1–5 scale
Negatively privileged	70.8	10.2	2.10
Averagely privileged	76.5	10.8	2.01
Positively privileged	82.4	8.2	1.85
Total	75.9	9.7	2.00

* $p < 0.01$.

Source: Own research.

Table 9.4.2. Market/class position and opinions on the statement: "Enterprise employees should have the same rights to decide on the future of the enterprise as the enterprise owners"—percentage of "I agree" answers (combining the answers "I strongly agree" and "I somewhat agree"), percentage of "I disagree" answers (combining the answers "I strongly disagree" and "I somewhat disagree") and the average measured on the 1–5 scale, where 1 means "I somewhat agree" and 5 means "I strongly disagree"*

Market/class position	"I agree" answers	"I disagree" answers	Average on the 1–5 scale
Negatively privileged	37.1	38.0	3.06
Averagely privileged	34.8	47.0	3.22
Positively privileged	30.6	47.5	3.30
Total	34.4	43.6	3.18

* $p < 0.01$.

Source: Own research.

Interestingly, the distribution of opinions looked somewhat different when the respondents evaluated the theoretical issue of farm owners and workers deciding on the future of the enterprise. Here, the distribution of answers was more balanced. The percentages of "I agree" and "I disagree" answers assumed values, respectively, from 37.1% to 30.6% and from 38.0% to 47.5%. This aspect of class consciousness was not as clear-cut as the issues of differences in earnings, discussed in the previous case. It should be emphasized that there are significant differences between farmers with positively and negatively privileged class positions. The former did not quite accept the situation of workers having the same rights as owners on deciding the future of the enterprise. They presented quite strong pro-owner (pro-capitalist) attitudes. In their case, the average value on the scale was 3.30 and it was the closest to the extreme value of 5 from the value 3, meaning pro-owner—or, to put it differently, pro-capitalist—attitudes (i.e., anti-worker). Generally, it can be stated that all farmers (including the representatives of the two remaining class categories) presented the pro-owner (pro-capitalist) attitudes in that regard which, at the same time, could be defined as anti-worker.

Table 9.4.3. Market/class position and opinion on the statement: "The owner of the enterprise where the strike is going on should have the right to hire new people to replace the employees on strike"—percentage of "I agree" answers (combining the answers "I strongly agree" and "I somewhat agree"), percentage of "I disagree" answers (combining answers "I strongly disagree" and "I somewhat disagree") and the average measured on the 1–5 scale, where 1 means "I strongly agree" and 5 means "I strongly disagree"

Market/class position	"I agree" answers	"I disagree" answers	Average on the 1–5 scale
Negatively privileged	20.0	56.2	3.56
Averagely privileged	21.9	53.5	3.48
Positively privileged	22.7	55.3	3.50
Total	21.4	55.2	3.52

Source: Own research.

The situation looks much different in the last aspect of the surveyed farmers' class consciousness that was analysed. No significant differences were observed in coexistence of opinions of the representatives of various class categories of farm owners. Generally, it can be said that they were rather against the owner hiring new workers to replace the ones who went on strike. This was confirmed by the opinion results (see Table 9.4.3), with over 50% of respondents in each of the three categories being against such practices, as well as by the average values on the scale which, in every case, were located above the middle value of 3, on the side of those who expressed a lack of acceptance for such practices. In this aspect of class consciousness, anti-owner (anti-capitalist) and, at the same time, pro-worker attitudes dominated.

The last part of this subchapter relates to the core of class positions of the surveyed farmers. The concept reflected in the table is the continuation of the idea of Erik O. Wright (1989, 1997), according to which the state of consciousness is essential to reporting the existence of class divisions. The state of this consciousness is, according to the premise of this work, expressed with the help of an indicator, which is an integrated variable based on three previously analyzed output variables (presented in tables 9.4.1, 9.4.2, 9.4.3). The paragraph below contains the explanation of the construction of this aggregated indicator.

This indicator is the average of the sum of the three variables, described by the opinions on three statements below, that were measured on the 5-point scale (where 1 means "I strongly agree," and 5—"I strongly disagree"). These statements are: "Enterprise owners should earn more than workers employed in these enterprises," "The owner of the enterprise where the strike is going on should have the right to hire new people to replace the employees on strike" and "Enterprise employees should have the same rights to decide on the future of the enterprise as the enterprise owners" (as a statement of opposite meaning, it was appropriately recoded during the construction of the summary indicator). The indicator can have values from 1 to 5, where 1 means the highest level of class consciousness (pro-owner/pro-capitalist awareness), and 5 means the lowest level of class consciousness (anti–owner/pro-worker awareness).

Table 9.4.4. Market/class position and class consciousness: average and standard deviation on the 1–5 scale, where 1 means "the highest level of class consciousness," and 5—"the lowest level of class consciousness"*

Marker/class position	Average	Standard deviation
Negatively privileged	2.85	0.66
Averagely privileged	2.74	0.70
Negatively privileged	2.68	0.72
Total	2.77	0.69

* $p < 0.01$.

Source: Own research.

The values in Table 9.4.4. show statistically significant levels of class consciousness among the surveyed farm owners/users. It was observed quite clearly that the representatives of the categories of farmers with a positively privileged position within the farmer/peasant class had a higher level of pro-capitalist (pro-owners, anti-workers) attitudes than those respondents who could be perceived as the representatives of the negatively privileged class. It is worth noting that the opinions from the representatives of all surveyed categories, were located more on the side of the owner (pro-capitalist) and below the average value of 3, which mean ambivalent attitudes. At the same time, it is worth noting that the representatives of the positively privileged class category have more varied opinions (standard deviation: 0.72), than the respondents of the negatively privileged class category (standard deviation: 0.66).

9.5. Political Perspectives for Peasant Classes

Table 9.5.1. Market/class position and opinion on the existence of organizations defending farmers' interests—percentage of respondents indicating the existence of such organizations

Market/class position	Indicating at least one organization*	Indicating Agricultural Chamber*	Indicating trade unions ("Solidarność RI"— Solidarity of Individual Farmers)**	Indicating Polish Party of Peasants	Indicating producers/ breeders unions*	Indicating agri-cultural circles
Negatively privileged	8.9	1.7	1.5	2.0	0.7	1.1
Averagely privileged	10.9	2.8	1.8	2.1	1.5	1.8
Positively privileged	17.8	6.7	3.0	2.3	2.6	2.2
Total	12.2	3.6	2.1	2.1	1.5	1.5

* $p < 0.01$, ** $p < 0.05$.

Source: Total—own research.

The data in Table 9.5.1 illustrate several interesting facts and relations. Firstly, roughly the same percentages of surveyed farmers in all categories, and regardless of their class position, named two organizations defending their interests that were traditionally connected with the agricultural sector. They were the Polish Party of Peasants and "farmers' circles." Therefore, it can be concluded that organizations which had existed in the public realm for many years had somehow become automatically ingrained in the respondents' consciousness. However, it is quite obvious that the percentage of respondents admitting to an affiliation with these organizations (as seen in the fifth and seventh columns in Table 9.5.1) is—to put it mildly—very modest.

The situation looks somewhat different among more contemporary organizations, such as agricultural chambers, trade unions (e.g., "Solidarność Rolników Indywidualnych"—Solidarity of Individual Farmers), and producer/breeder associations (as presented in the third, fourth, and sixth columns in the table, respectively). Here, significant differences, even in the statistical sense, can be observed. Farmers with positively privileged positions—decidedly more often than the representatives of other categories—named agricultural chambers, agricultural associations, and unions as organizations defending their interests. In each case, farmers with positively privileged class positions were more likely to identify groups and institutions that they believe secure their interests.

This statement is supported by the data in the second column in Table 9.5.1, which contains the answers pointing to at least one organization defending farmers' interests. Here, the differences between farmers with positively privileged class positions, who identified more such organizations, and the representatives of the other, less privileged, class categories within this socio-professional category were quite clear.

It should be stressed that generally the percentage of respondents pointing to at least one organization defending farmers' interests was relatively small, 12.2% to be precise. This meant a slight decrease in comparison to the 1994 data, where this percentage was 13.6% (see Gorlach, 2009, p. 128 [Table 3.33]).

Table 9.5.2. Market/class position and farmers' preferred forms of fighting for their own interests (in %)

Market/class position	Road blocks, demonstrations, occupation of buildings	Pressure on elected officials (representatives and senators)	Self-organization of farmers	There is no sense in doing anything
Negatively privileged	10.6	17.9	37.0	34.5
Averagely privileged	11.1	17.9	38.1	32.9
Positively privileged	11.8	17.3	42.3	28.6
Total	11.1	17.7	39.0	32.2

Source: Own research.

The class position did not differentiate the respondents (as indicated in the data in Table 9.5.2) in terms of the preferred forms of fighting for their interests. Not quite one third of them presented a certain fatalist attitude that "there is no sense in doing anything." The remaining majority presented the opinion that it is better to choose one of the forms of fighting presented in the second, third, or fourth column. Some interesting observations can be made here. Self-organizing was the most popular among farmers, and this opinion was expressed by 40% of respondents. Proponents of lobbying numbered far fewer (not quite 18%), but aggressive and spectacular forms of protests such as roadblocks, demonstrations, and building occupations generated the least interest (around 11%).

Social activism of the surveyed farmers is quite varied, depending on their class position (see Table 9.5.3). Interestingly, this issue did not apply to participation in the local elections of 2014, which were the most recent elections in Poland prior to the 2017 study. As indicated in Table 9.5.3, over 80% of respondents claimed to have voted in the 2014 local elections, confirming their civic engagement. This voter participation rate was much higher than that of Polish society overall, as the indices for the general population oscillated around 50% or slightly below.[2] One might reasonably conclude that farmers are more focused on local politics than the typical members of Polish society.

Table 9.5.3. Market/class position and social activism. Percentage of respondents participating in the local elections of 2014 that were members of political parties and/or trade unions, or members of producer groups and cooperatives

Market/class position	Respondents participating in 2014 local elections	Respondents who at the time of study were the members of political parties and trade unions**	Respondents who at the time of study were the members of producer groups, cooperatives, etc.*
Negatively privileged	83.0	1.7 (6.9)	0.6
Averagely privileged	81.3	2.0 (5.1)	2.4
Positively privileged	84.7	2.7 (4.1)	7.8
Total	83.1	2.1 (5.5)	3.4

* $p < 0.01$, ** $p < 0.05$. Numbers in parentheses are percentages of respondents who were members of parties and trade unions in the past but were not at the time of the study.

Source: Own research.

At the same time, farmers with a positively privileged class position were more engrossed in the activities of political parties and other social institutions (see the third and fourth column in Table 9.5.3). As presented in Table 9.5.3, class position might roughly correspond with membership in various institutional forms of engagement in public life, although the percentage of these respondents who belong to the privileged categories and were engaged with political parties and trade unions was on a trajectory of very gradual decline.

The situation looks somewhat better with respondents' fulfilment of social functions (see Table 9.5.4) in local institutions and, in particular, the function of the head of the village. It can be observed quite clearly that having a positively privileged class position facilitates social activism, understood as fulfilling functions in local institutions, whether before or during the year of the study, 2017.

Quite valuable information can be found in tables 9.5.5 and 9.5.6, and related to the opinions of the surveyed farmers in regards to various issues connected to Poland's membership in the European Union. Table 9.5.5 contains data on the evaluation of Poland's accession to the European Union, 13 years prior to 2017. It can be

---

[2] See https://samorzad2014.pkw.gov.pl/356_Frekwencja.html (access: December 11, 2019).

inferred from the analysis of Table 9.5.5 that only a marginal number of respondents believed that all farmers in Poland lost upon Poland's admission to the European Union. In none of the categories did this value exceed 5%, although it should be stressed that the respondents from the positively privileged class position were the most critical about the effects of Poland entering the European Union. The correlation of opinions with class positions is statistically significant here and, therefore, this issue deserves some more attention. It will be discussed further in the conclusion of this chapter.

Table 9.5.4. Class/market position and social activism. The percentage of people currently fulfilling some social functions in an association of local character or currently fulfilling the function of the head of the village

Market/class position	Respondents fulfilling some public and social functions*	Respondents currently fulfilling the functions in local associations*	Respondents presently fulfilling the function of the head of the village*
Negatively privileged	9.5 (13.9)	5.4 (8.2)	1.5 (2.0)
Averagely privileged	11.8 (14.0)	6.5 (7.4)	2.5 (2.7)
Positively privileged	17.3 (13.4)	10.2 (6.4)	3.4 (3.7)
Total	12.6 (13.8)	7.2 (7.4)	2.4 (2.7)

* $p < 0.01$. Shown in parentheses are the percentages of respondents who previously fulfilled the described functions but no longer do so.

Source: Own research.

Before this more in-depth analysis, a closer look at other relations between the data in Table 9.5.5 might be in order. The opinions that all farmers benefited due to Poland's European Union accession were mostly expressed by those farmers (64.9%) who, according to the concept assumed in this work, had positively privileged class position. The most likely to agree with the thesis that the European Union accession mostly benefited the owners of large and modern farms (37.5%) were those with a negatively privileged class position. Generally, around 60% of respondents thought that all farmers benefited from joining the European Union.

To some extent this could be explained through the fact that the great majority of farmers, almost 95% of them, have been receiving subsidies and direct payments offered by the European Union (see Table 9.5.6). It is quite clear that farmers with a positively privileged class position use this source of support for their farms quite significantly.

Table 9.5.5. Market/class position and evaluation of Poland's accession to the European Union (in %)*

Market/class position	All farmers benefited	Large, modern farms benefited	Weak and small farms benefited	Everybody lost
Negatively privileged	57.6	37.5	3.3	1.6
Averagely privileged	59.6	34.8	3.4	2.2
Positively privileged	64.9	28.4	2.1	4.6
Total	60.5	33.9	2.9	2.7

* $p < 0.01$.

Source: Own research.

Table 9.5.6. Market/class position and using the subsidies and direct payments offered to farmers by the European Union (in %)*

Market/class position	Using direct payments	Not using direct payments
Negatively privileged	92.1	7.9
Averagely privileged	95.3	4.7
Positively privileged	97.9	2.1
Total	94.8	5.2

* $p < 0.01$.

Source: Own research.

9.6. Developmental Strategies of Peasant Classes

The last aspect of the respondents' class consciousness is, according to the concept of this work, the strategy of farm management declared by farmers. This aspect was operationalized by use of the question on the main forms of investments preferred by farmers. The respondents could choose from six options, which encompassed various situations, ranging from actual farm development to financial help for children (see first column in Table 9.6.1). Firstly, it should be noted that the opinions in Table 9.6.1 coexist with the type of respondents' class (market) positions, with one exception, which is the opinion on the fulfillment of day to day pleasures through acquired financial means. Only 6% of the surveyed farmers referred to the category of pleasures and there was no differentiation between the respondents in regard to the class and market positions of their farms. The opinions in the remaining categories represented a different logic.

According to the data presented in Table 9.6.1, around 45% of respondents with farms, and who have a positively privileged market position wanted to invest in further farm development. Among the respondents operating farms with a negatively privileged market position, only 15% expressed a willingness to invest in farm development. These respondents were three times less likely to invest in their farms than their positively privileged counterparts. This meant that the strategy of positively privileged farms concentrated on continuing to strengthen their position. Such a strategy was rather marginally applied by farmers with a negatively privileged farm position. Among the respondents who operated farms, two other strategies, such as putting savings in the bank and providing financial help for children, were given. Both of these strategies were twice as likely to be applied for by respondents with a negatively privileged farm position than by those who were positively privileged.

Table 9.6.1. Market and class position and preferred forms of investment: average amount of money, in thousands PLN, that farmers would invest in particular areas if they won a lottery prize of 100 thousand PLN

Areas of investment	Market/class position			
	Negatively privileged	Averagely privileged	Positively privileged	Total
Farm development*	15.2	27.4	45.6	28.1
Business outside of agriculture*	8.8	11.5	7.8	9.2
Savings (e.g., in a bank)*	14.7	10.9	7.1	11.3
Building or modernization of the residential home *	23.4	21.0	17.6	20.9
Day-to-day pleasures	6.1	5.7	5.1	5.7
Financial aid to children*	31.8	23.5	16.8	24.8

* $p < 0.01$.

Source: Own research.

Two other situations appear to be quite interesting. On the one hand, it should be mentioned that the relatively strongest preferences to develop business outside of agriculture could be observed among farmers existing within the "middle position" (averagely privileged class position) but preferences for building or modernizing residential homes were again expressed among those respondents who had a negatively privileged market position.

9.7. Some Concluding Remarks

To summarize the reflection in this chapter, several crucial issues should be addressed. First, it is worth noting that the class location of family farms in Poland went through changes from 1994 to 2017. During this period of time, for nearly a quarter of a century (23 years to be precise), changes did take place, but analyzing them—as indicated earlier—should be done with caution, due to a somewhat different way of constructing the variables during the two editions of the study.

It should be noted that, in 1994, the negatively privileged farms comprised 47.9% of all the surveyed farms, and this was the largest category. Averagely privileged farms made up 34.5% of all surveyed farms and were the second largest category. The smallest category was that of positively privileged farms, being only 17.6% of all surveyed farms.

In 2017 the situation looked quite different. The farms of average privilege were the least numerous category, making up 28.4% of all the surveyed farms. Farms with positively privileged market position were the second largest category and made up 31.0% of all the surveyed farms. The farms with negatively privileged farm positions were the largest category, comprising 40.6% of all surveyed farms. As the farms with averagely privileged market position numbered the fewest in the study, this could serve as an argument supporting the thesis on the farm dualism process and the "disappearing middle" phenomenon. This process is widely described in sociological literature. Attributed to the influences of capitalist market principles it is also occurring in Poland (see Gorlach, 2004, pp. 50–85). Consequently, the surveyed sample appeared to be more stratified in 2017 than a quarter of a century earlier, which shows that the logic of the capitalist market economy has had an increasing impact on the family farms situation in Poland.

Another matter that was of interest to the authors of this chapter is whether farmers should be defined as some socio-professional category, or should they be located in a more general category of entrepreneurs? It can be stated that after almost thirty years (period of time from 1989 to 2017) of the ongoing and quite intensive impact of market forces and the global capitalist system on socio-economic relations in Poland, it is to be expected that a specific sense of identity will develop. Farmers have become more inclined to identify with entrepreneurs. Therefore, the specific sense of identity connected with the peasant tradition is increasingly less important.

To address this matter, some juxtaposing of the situations from 2017 and 1994 might be in order, with all of the reservations that should be mentioned yet again, as in each study edition there was a different manner of constructing the class position of the respondents and somewhat different way of asking the questions related to similarities and differences between the situations of farmers and other entrepreneurs. In 2017, over 68% of respondents agreed with the statement that "Currently the situation of the farmer and the situation of any other entrepreneur do not differ." Back in 1994, only 45% of respondents reported the existence of similarities between the situations of farmers and other entrepreneurs (see Gorlach, 2009, p. 119 [Table 3.22]).

It should also be noted that differences in opinions on this matter were among farmers of various class positions in 1994 as well as in 2017. The respondents with a positively privileged class position were more inclined to the self-perception of being just like other entrepreneurs in other sectors (see Table 9.3.3. in this chapter and Gorlach, 2009, p. 120 [Table 3.24]).

It is also vital to take a look at the general level of class consciousness, while still keeping in mind the somewhat different way of constructing the variable of "class position" in the study from 1994 and from 2017. It should be noted that in 1994 (see Gorlach, 2009, p. 127 [Table 3.32]) the average values in the study indicated that the respondents were more likely to display pro-owner attitudes than the respondents in 2017 (see Table 9.4.4 in this chapter). The average values oscillated between 2.78 and 2.83 among those with a negatively privileged market position and stayed within the 2.62–2.68 range for those with a positively privileged market position. At both times that marked the two studies, pro-owner attitudes were more intense among the respondents with positively privileged class position.

Things look much different in regards to variety of attitudes in the studied categories of farm owners, measured by the value of standard deviation. Currently, as presented in the above reflection, more diversification could be observed compared to the 1994 study (see Gorlach, 2009, p. 127 [Table 3.32]). At the time, the values of standard deviation were, respectively, 0.732 for the interviewed farmers with a negatively privileged class position and 0.849 for farmers with a positively privileged class position. As might be inferred from these values, the attitudes of the respondents with a positively privileged farm position were more diverse than the attitudes of respondents with a negatively privileged farm position.

After 23 years, it became quite evident (in the 2017 study) that the attitudes in both extreme categories of farm owners are a bit more homogenous. However, other tendencies observed in 1994 still continued well into the 21st century. In 2017 these attitudes were more diverse among the respondents with positively privileged class position as compared to the farmers with negatively privileged class position. In 1994, the differences between these two categories were not statistically significant. They became so in the 2017 study as indicated in Table 9.4.4, and this was one of the most important changes noted between the two study editions. According to the authors of this chapter this was one of the most important changes. This could mean that the class differences, measured by level of class consciousness, have deepened among farmers.

What could be inferred from these analyses? It should be emphasized that farmers with positively privileged class positions appeared to be taking better care of their own interests. As indicated by the data presented in tables 9.5.1, 9.5.2, 9.5.3, and 9.5.4, these farmers were more likely to identify organizations and institutions defending farmers' interests, were more socially active, and more likely to be members of various organizations. They also declared a willingness to self-organize as a method to defend their own interests.

The farmers with a positively privileged farm position differed from others in terms of evaluation of Poland's accession to the European Union (see tables 9.5.5 and 9.5.6). First of all, they were more likely to receive European Union support in the form of direct subsidies and other programs. More than any other categories of farmers, they were likely to express opinions that Poland's joining the European Union benefited all farmers and all types of farms. Furthermore, such farmers were more inclined to invest in their farms, as presented in Table 9.6.1. This was confirmed by other opinions that emerged during other studies conducted among farmers, and described below.

The impact of Poland's formally joining the European Union on May 1, 2004, and its previous eligibility for European Union funds thanks to the pre-accession programs, seems like the most interesting issue in the context of class and market position differentiation among the surveyed farmers and the farms. It should be remembered that due to the construction of the "class position" variable, it is the characteristics that divide large scale commercial farms and their owners from farms less equipped with economic and cultural capital which counterbalance the big farms. In an interview that Edyta Bryła (2019, p. 9), a journalist specializing in agricultural issues, conducted with Professor Monika Stanny, the Director of the Institute of Rural and Agricultural Development of The Polish Academy of Sciences and then the expert on the situation of Polish rural areas, there were remarks on farmers, who were able to break even financially thanks to agricultural economic activity and then used the European Union subsidies for other purposes. Professor Stanny was quoted saying that this usually meant investment in children, in their education, "studies and apartments in Warsaw." The interview also contained information that the great majority of farms treated the direct subsidy payments as some kind of "social pension." As indicated by Stanny, smaller farmers lived off the European Union subsidies but the bigger ones saved the funds for other purposes or to make investments. Sometimes they invested in agricultural production and equipment but more often they had other investments goals, unrelated to agriculture (Bryła, 2019, p. 9).

It should be made clear that the thesis presented in the aforementioned interview is quite one-sided and cannot be supported by empirical evidence. There are some other outrageous statements in the interview that farms of at least 30 ha or even 50 ha are truly the ones that produce food. As described by Stanny, "Agricultural production in Poland can be summarized with a general statement that 20% of the largest farms produce as much as 80% of food." This would mean that out of 1.3 million farmers receiving subsidies, who are signed up with the Agency for Restructuring and Modernization of Agriculture, only 300 thousand farmers produce food for the market. The remaining million farmers defined by the typology above would just be the beneficiaries of agricultural subsidies in the form of the so-called "social pension" (Bryła, 2019, p. 9). Is this really the case?

Quite different pictures have emerged from the study that served as the empirical groundwork for this work. It was quite noticeable that farmers with positively privileged class and market positions were benefiting from various programs of the European Union, and direct subsidies in particular, to a greater degree than other categories of farmers. The difference between these categories was significant in a statistical sense (see Table 9.5.6). This is in stark opposition to the statements of Professor Stanny, made during the aforementioned interview with Edyta Bryła (2019, p. 9), indicating that the owners of the largest farms would gladly give up the direct payment subsidies from the European Union, because they only inflate the prices of land and costs of farm equipment and fertilizer.

REFERENCES

Bodenstedt, A. (1990). Rural culture: A new concept. *Sociologia Ruralis* 30(1), pp. 34–47.

Bryła, E. (2019, December 8). Rozmowa z prof. Moniką Stanny z PAN: Rolnicy z największych gospodarstw są przeciwni dotacjom unijnym. *Gazeta Wyborcza*, p. 9.

Buttel, F.H. (1980). W(h)ither family farm? Towards a sociological perspective on independent commodity production in US agriculture. *Cornell Journal of Social Relations* 15, pp. 10–37.

Buttel, F.H. & Flinn, W.L. (1980). Sociological aspects of farm size: Ideological and social consequences of scale in agriculture. *American Journal of Agricultural Economics* 5(62), pp. 946–953.

Day, G. (2001). *Class*. Routledge.

Domański, H. (ed.) (2008). *Zmiany stratyfikacji społecznej w Polsce* [Changes in social stratification in Poland]. IFiS PAN.

Domański, H. & Tomescu-Dubrow, I. (2008a). Nierównosci edukacyjne przed i po zmianie systemu [Educational inequalities before and after the systemic change]. In: H. Domański (ed.), *Zmiany stratyfikacji społecznej w Polsce* [Changes in social stratification in Poland], pp. 45–74. IFiS PAN.

Domański, H. & Tomescu-Dubrow, I. (2008b). Wejście na rynek pracy a poziom wykształcenia [Entrance to the labor market and a level of education]. In: H. Domański (ed.), *Zmiany stratyfikacji społecznej w Polsce* [Changes in social stratification in Poland], pp. 133–152. IFiS PAN.

Dubrow, J.K. (2008). Świadomość klasowa interesów politycznych. Pojęcie i pomiar [Class consciousness of political interests. A concept and its measurement]. In: H. Domański (ed.), *Zmiany stratyfikacji społecznej w Polsce* [Changes in social stratification in Poland], pp. 271–292. IFiS PAN.

Friedmann, H. (1978a). Simple commodity production and wage labor in the American Plains. *Journal of Peasant Studies* 1(6), pp. 71–100.

Friedmann, H. (1978b). World market, state, family farm: Social bases of household production in the era of wage labor. *Comparative Studies in Society and History* 4(20), pp. 545–586.

Friedmann, H. (1981). *The Family Farm in Advanced Capitalism: Outline of a theory of simple commodity production*. University of Toronto, unpublished manuscript.

Giddens, A. (1973). *The Class Structure of the Advanced Societies*. Harper and Row.

Goldthorpe, J.H. (1987). *Social Mobility and Class Structure in Modern Britain*. Clarendon.

Goldthorpe, J.H. & Hope, K. (2001). Occupational grading and occupational prestige. In: D. Grusky (ed.), *Class, Race and Gender: Social stratification in sociological perspective*, pp. 264–271. Westview Press.

Goldthorpe, J.H., Lockwood, D., Bechhofer, F., & Platt, J. (1969). *The Affluent Worker in the Class Structure*. Cambridge University Press.

Gorlach, K. (1990). *Socjologia polska wobec kwestii chlopskiej* [Peasant question in Polish sociology]. Universitas.

Gorlach, K. (1995). *Chłopi, rolnicy, przedsiębiorcy. „Kłopotliwa klasa" w Polsce postkomunistycznej* [Peasants, farmers, enterpreneurs: "The awkward class" in post-communist Poland]. Wydawnictwo Uniwersytetu Jagiellońskiego.

Gorlach, K. (2001). *Świat na progu domu. Polskie rodzinne gospodarstwa rolne w obliczu globalizacji* [The world in my backyard: Polish family farms in the face of globalization]. Wydawnictwo Uniwersytetu Jagiellońskiego.

Gorlach, K. (2004). *Socjologia obszarów wiejskich. Problemy i perspektywy* [Sociology of rural areas: Problems and perspectives]. Wydawnictwo Naukowe Scholar.

Gorlach, K. (2009). *W poszukiwaniu równowagi. Polskie rodzinne gospodarstwa rolne w Unii Europejskiej* [Searching for the balance: Polish family farms in the European Union]. Wydawnictwo Uniwersytetu Jagiellońskiego.

Gorlach, K. & Klekotko, M. (2010). Together but separately: An attempt at the proces of class diversification among Polish peasantry. *Przegląd Socjologiczny* 59(2), pp. 109–126.

Grusky, D. (ed.) (2001). *Social Stratification: Class, race & gender in sociological perspective (2nd edition)*. Westview Press.

Haan, H. de (1993). Images of family farming in the Netherlands. *Sociologia Ruralis* 33(2), pp. 147–166.

Jollivet, M. (1994). Kapitalizm i rolnictwo [Capitalism and agriculture]. In: P. Rambaud & Z.T. Wierzbicki (eds.), *Socjologia wsi we Francji* [Rural sociology in France], pp. 36–47. Wydawnictwo Uniwersytetu Mikołaja Kopernika.

Lacombe, P. (1991). Farming, farms and families. In: M. Tracy (ed.), *Farmers and Politics in France*, pp. 49–64. The Arkleton Trust.

Mann, S.A. & Dickinson, J. (1978). Obstacles to the development of a capitalist agriculture. *Journal of Peasant Studies* 5, pp. 466–481.

Mitrany, D. (1972). *Marx against the Peasant: A study in social dogmatism*. Collier Books.

Mooney, P. (1988). *My Own Boss: Class, rationality and the family farm*. Westview Press.

Pakulski, J. & Waters, M. (1996). *The Death of Class*. Sage Publications.

Pongratz, H. (1990). Cultural tradition and social change in agriculture. *Sociologia Ruralis* 30(1), pp. 5–17.

Ploeg, J.D. van der (1995). From structural development to structural involution: The impact of new development in Dutch agriculture. In: J.D. van der Ploeg & G. van Dijk (eds.), *Beyond Modernisation: The impact of endogenous development*, pp. 109–146. Van Gorcum.

Przybysz, D. (2008). Pozycja społeczno-zawodowa a zachowania wyborcze Polaków [Socio-occupational positions and electoral behavior of Poles]. In: H. Domański (ed.), *Zmiany stratyfikacji społecznej w Polsce* [Changes of social stratification in Poland], pp. 211–246. IFiS PAN.

Sawiński, Z. (2008). Zmiany systemowe a nierówności w dostępie do wykształcenia [Systemic changes and inequalities in the access to education]. In: H. Domański (ed.), *Zmiany stratyfikacji społecznej w Polsce* [Changes in social stratification in Poland], pp. 13–44. IFiS PAN.

Savage, M. (2000). *Class Analysis and Social Transformation*. Open University Press.

Savage, M. (2015). *Social Class in the 21ˢᵗ Century*. Penguin Books (A Pelican Introduction).

Shanin, T. (1972). *The Awkward Class*. Clarendon Press.

Shanin, T. (ed.) (1973). *Peasants and Peasant Societies*. Penguin Books.

Sheingate, A.D. (2001). *The Rise of the Agricultural Welfare State: Institutions and interest group power in the United States, France, and Japan*. Princeton University Press.

Słomczyński, K.M. & Tomescu-Dubrow, I. (2008). Systemowe zmiany w strukturze społecznej a ruchliwość społeczna [Systemic changes in social structure and social mobility]. In: H. Domański (ed.), *Zmiany stratyfikacji społecznej w Polsce* [Changes in social stratification in Poland], pp. 75–96. IFiS PAN.

Stinchcombe, A. (1961). Agricultural enterprise and rural class relations. *American Journal of Sociology* 67, pp. 165–176.

Strange, M. (1988). *Family Farm: A new economic vision*. University of Nebraska Press.

Wright, E.O. (ed.) (1989). *The Debate on Classes*. Verso.

Wright, E.O. (1997). *Class Counts: Comparative studies in class analysis*. Cambridge University Press.

Internet source

https://samorzad2014.pkw.gov.pl/356_Frekwencja.html (access: December 11, 2019). Poland local government elections 2014—Voter turnout statistics (in Polish).

PART THREE

DIVERSIFICATION
OF FARMERS' STRATEGIES

Think Locally, Act Globally: Polish farmers in the global era of sustainability and resilience, ed. by Krzysztof Gorlach and Zbigniew Drąg in collaboration with Anna Jastrzębiec-Witowska and David Ritter
Jagiellonian University Press, Kraków 2021, pp. 413–415
ISBN 978-83-233-4949-5
DOI: http://dxdoi.org/10.4467/K7195.199/20.20.12736

Some Introductory Remarks by the Editors to Part Three

Zbigniew Drąg https://orcid.org/0000-0002-9106-7758/

Krzysztof Gorlach https://orcid.org/0000-0003-1578-7400/

What is a contemporary Polish farmer like? What specific characteristics can be found in Polish farmers? Previous chapters presented attempts to answer these questions while treating surveyed farmers as one category and concentrating analyses primarily on determining the extent of similarities and differences between them. The analyses were conducted at the level of farm units and involved simple, one-level modeling. No less interesting was the issue of differentiation of farmers' characteristics according to place of residence, as presented in this part of the book. Such characteristics can be perceived as local specifics of farms and their users. This matter is also essential in the context of the leading thought of this work: think locally, act globally.

Considering the above, single-level modeling appeared to be insufficient to learn about the impact of territorial differentiation on matters such as: the functioning of farms, decision-making strategies, and the course of farm users' everyday lives. Multilevel analysis seems more appropriate here. It is applied in the form of a two-level model and presented in two consecutive chapters of the monograph; these chapters are then distinguished as a separate part to emphasize the specifics of their content. Chapter 10 discusses the impact of territorial differentiation on the choice of farm management strategy and Chapter 11 addresses the relationship between the place of residence and life strategies of surveyed farm users/operators.

Application of multilevel analysis is determined by the structure of collected data. The data are only to be used if they qualify as hierarchically nested data, meaning that the observations from the lower level are included in the higher level. In the case of two-level analysis, territorial units such as counties might constitute a higher (second) level of analysis and this is reflected in the described study. Level one consists of farmers and their farm units nested within these counties. What also might be discouraging researchers from applying the multilevel model is the number of observations necessary on particular levels. To fully utilize the potential of the two--level model, one should accumulate data from at least 100 units of level two, and in each of the level two units there should be a minimum of 30 nested units from level one. Therefore, to conduct such a study it is necessary to compile data from 3000 appropriately structured observations. This requirement was met in the presented study and analyses were based on the data collected from 3543 farmers (and farm units) from 101 counties. It should be noted that, in each county, at least 30 farmers (and 30 farm units) were identified as active. The data were collected in the studies which serve as the groundwork for all quantitative analyses presented in this monograph and the methodology for these studies was described in detail in Chapter 2.

The main asset of multilevel analysis—in comparison to the classic approach—is that with analysis of the differentiation of the dependent variable from level one, the predictors (independent variables) can also be considered, including variables from level one, as well as variables from the higher level. Importantly, this kind of modeling also provides the possibility to specify four models, from the simplest to the most complex, with every more complex model being a development from a simpler, previous model. The most basic model is a null model, which allows us to determine the relative influence on differentiation of particular characteristics of farm users (dependent variable of level one), as well as other characteristics of farmers (on level one) and qualities describing the counties (level two), but without indicating which particular characteristics they are. Further development of the null model is the model with independent variables, which allows one to determine if differentiation is also present in a situation where independent variables that are characteristics of farmers or counties are fully controlled. A model in which the independent variable is a variable from level one is called a random intercept model and its development is a random slope model. Application of this model allows one to determine whether the interdependency between the dependent and the independent variables from level one is similar in all counties or if there are significant differences between counties. Further explication of this model and its most complex form is the hierarchical model, also known as the Intercepts-and-Slopes-as-Outcomes model, which allows one to identify the county characteristics differentiating the relationship between dependent and independent variables. It is a situation in which differentiated variables from level one (characteristics of farmers) are explained with variables from level two (characteristics of counties). In this publication it would be impossible to use multilevel analysis to its fullest potential. Chapter 10, describing the strategies for farm management, also contains analyses conducted with the use of the null model with independent variables, while Chapter 11, describing the life strategies of farm operators, presents analyses using the applied random slope model.

Think Locally, Act Globally: Polish farmers in the global era of sustainability and resilience, ed. by Krzysztof Gorlach and Zbigniew Drąg in collaboration with Anna Jastrzębiec-Witowska and David Ritter
Jagiellonian University Press, Kraków 2021, pp. 417–457
ISBN 978-83-233-4949-5
DOI: http://dxdoi.org/10.4467/K7195.199/20.20.12737

Chapter Ten: Regional Farming Strategies in Poland of 2017

Zbigniew Drąg https://orcid.org/0000-0002-9106-7758/

10.1. Introductory Remarks

The results of the analyses presented in previous chapters pointed quite directly to differentiation of the family farms situation in Poland, particularly addressing the economic dimension of farming. Although farms with an annual income of up to 30 thousand PLN (in 2016) were the dominant category, comprising around three quarters of all surveyed farms, there were also farms with no reported income. Such farms made a marginal group of less than 5%. Farms with an annual income of over 30 thousand PLN constituted one fifth of the studied farms and those with income over 60 thousand PLN per year made up one tenth of the studied sample. Even a casual look at the data could lead to the realization that the area of utilized agricultural land was the main factor in determining income level. The average annual income of large farms of at least 15 ha exceeded 100 thousand PLN, while farms with areas smaller than 15 ha had an average income below 15 thousand PLN. Additionally, the analyses considering both categories of farms demonstrated that they differed significantly in terms of applied development strategies, including matters such as level of farm mechanization (measured by the value of machines and equipment), renting land from others, using loans and bank credits to finance the production, preferences in divisions of production and preferences for supply and outlet markets. It can be stated that the utilized agricultural area influences to a significant extent the development strategy adopted by a particular farm and, in effect, determines its profitability.

In Poland, there is a territorial diversification in the area of family farms, which is a consequence of historical factors and conditions that will not be analyzed here. In our studies, this diversification was reflected at the regional level (according to the NUTS classification: level-1), at the levels of provinces/voivodeships (level NUTS 2 and at the county level. The 2017 study results indicated that the average farm area in the Southern Region was 4 ha, while in the Northern Region it was around 15 ha. At the level of provinces/voivodeships, the average farm area in Lesser Poland and Subcarpathian did not exceed 4 ha, but in Warmian-Masurian, and Western Pomerania, this figure oscillated around 20 ha. Even greater differences could be seen at the level of counties/districts. In some counties of Lesser Poland and Subcarpathian, such as Limanowski, Myślenicki, Niżański, and Krośnieński, the average farm size was around 3 ha, while in the counties located in the northern part of Poland, such as Gryficki, Szamotulski, Wąbrzeski, and Łomżyński, it exceeded 23 ha. Could it be expected that the developmental strategies applied by farmers would be similar in those regions, provinces/voivodeships, and counties where large farms dominated and different in the regions, provinces/voivodeships, and counties with a prevalence of small farms? Would, consequently, farms in some regions, provinces/voivodeships, and districts be more profitable than in others? At the regional level, such relations between strategies, farm area and profitability could mostly be observed in the Southern region, where the smallest farms prevailed. Farm strategies of this region differed from the strategies of other regions and their profitability was the lowest in the country. Machinery on farms in the southern regions were of the lowest value in the country and, on average, it was 45 thousand PLN. Only 12% of farmers in the southern region took credits and loans and 10% sold their products beyond local markets. As many as 62% of farmers in that region conducted mixed production of animals and plants on a very small scale. Consequently, the farms of this region had the lowest profitability; in 2016, their incomes averaged somewhat under 10 thousand PLN. Therefore, it could be stated that farms with a small area of utilized agricultural land, which was determined territorially, had a very limited repertoire of development strategies and, as a result, a low profitability level. Such a clear relationship between farm area, adopted farming strategies, and also profitability could not be observed in the remaining regions, where the average farm area was larger than in the Southern Region. In the Northern Region, where the average farm area was highest, investment indicators were also highest with the average value of farm equipment exceeding 110 thousand PLN. This, however, was not accompanied by a greater than average propensity for renting (18%), taking lines of credits and loans

(22%), specialization of production (55% of farms conducted mixed production of plants and animals), expansion beyond local markets (15% of farms bought production outside of local markets and 19% of farms sold their products beyond local markets). Such characteristics were more often displayed by farms in the Southwestern and Northwestern Regions where the average farm area was smaller and oscillated around 13 ha and the value of farm equipment fell between 83 thousand PLN and 93 thousand PLN. In these regions the land was rented by over 20% of farmers. The Northwestern Region has the highest percentage of taking credits and loans (29%), and the Southwestern Region had the largest expansion beyond local markets with 35% of supply and 29% of sales, as well as the highest level of specialization, with only 29% of farms combining plant and animal production. It was revealed in the study that the most profitable were not the farms of the Northern Region, with an average annual income of 32 thousand PLN, or the Southwestern Region, with slightly over 25 thousand PLN in annual income, but the farms of the Northwestern Region, with slightly over 41 thousand PLN in annual income. It should also be mentioned that farms of the central region, with 33 thousand PLN in annual income had similar profitability level to the farms of the Northern region. It should also be worth noting that farms in the Eastern Region, with an annual income of 26 thousand PLN, had a profitability level similar to that of farms in the Southwestern Region. The average area of farms in the Central Region was only slightly larger than the average area of farms in the Eastern Region, with an area exceeding 7 ha.

Similar relations between the farm area and the applied development strategy and profitability could also be observed at the province level. They were quite obvious in the provinces with the smallest average farm area and more ambiguous in provinces with the largest average farm area. The smallest average farm area (under 4 ha) could be found in provinces: Lesser Poland and Subcarpathian). These farms could be characterized as having the lowest technical potential, as the average value of farm equipment was 36 thousand PLN in Lesser Poland and 32 thousand PLN in Subcarpathian provinces. The surveyed farm operators were also not very likely to use farm credit and loans with only 11% of respondents doing so. The farms in provinces Lesser Poland and Subcarpathian had mixed production (66% and 74% of farms, respectively) and were closely connected with the local market. Only 7% of the farms surveyed in Lesser Poland and 5% of the farms surveyed in Subcarpathian sold their products beyond local markets. The farms surveyed in these two provinces declared the lowest profitability, of 8 thousand PLN per year. The largest average farm area could be found in Western Pomerania (almost 23 ha) and in Warmian-Masurian (close to 20 ha). Farms in these provinces were the most likely to take credits and loans and 36% and 32% of respondents, respectively, confirmed doing so. The average value of the machine park in these provinces was among the highest in Poland: 154 thousand PLN in Western Pomerania and 158 thousand PLN in Warmian-Masurian Province. The highest average value of machine park of almost 174 thousand PLN was reported in the Podlaskie Province, where the average farm area was 15 ha. Farmers in Western Pomerania and Opolskie Province, as well as in Warmian-Masurian Province, were the most likely to rent land in order to add area to the owned and operated farm land. The rental percentages were: 29% for Western Pomerania, 26% for Opolskie, and 22% to Warmian-Masurian. Furthermore, even though the farms of West Pomerania and Warmian-Masurian Provinces had above average expansion beyond local markets in terms of supply and sales, it was not as great as in Opolskie and Pomerania Provinces, where 32% of farms sold their products beyond local markets. As many as 39% of farms in Opolskie Province and 30% of farms in Pomerania were purchasing equipment and other supplies outside of local markets. The farms of Western Pomerania and Warmian-Masurian Province were not specialized as the farms of Lubuskie, Łódzkie, and Opolskie Provinces, where mixed animal and plant production was under one third. Additionally, in Western Pomerania, farms with plant production comprised 55%, while in Warmian-Masurian Province, the farms with animal production made up 63% of all the surveyed farms. However, the farms of these two provinces differed in terms of profitability. The farms of Western Pomerania had the highest level of profitability—88 thousand PLN in 2016. The farms in Podlaskie Province came close with an annual profitability of 86 thousand PLN in 2016. Interestingly, the average profitability of farms in Warmian-Masurian Province was also around the national average, at approximately 27 thousand PLN. This was comparable to the profitability in Pomerania Province (which averaged 25 thousand PLN), Lower Silesia (26 thousand PLN), and Łódzkie (28 thousand PLN). Farms of Mazovia, Kuyavian-Pomerania, and Greater Poland reported greater profitability, with 36 thousand PLN in the first of these provinces and 37 thousand PLN in the other two. The average farm area in these three provinces was significantly smaller than in the Warmian-Masurian Province; it was around 9 ha in Łódzkie and Mazovia provinces and not quite 14 ha in Kuyavian-Pomerania.

What were the relations between farm area, development strategies and profitability at the county level? In the counties with the smallest farm area (about 3 ha), applied strategies of the development and profitability level were similar. In counties such as Limanowski, Myślenicki, Niżański, and Krośnieński, farms with mixed production of plants and animals prevailed, from 69% of farms with mixed production in Krośnieński County to 79% in Niżański County. The farmers mostly operated the land they owned (renting additional land was reported by 3% in Krośnieński

County to 18% in Niżański County), self-financed the agricultural production (as the farm loans and credits were only used by a small percent of farmers, from 6% of farms in Krośnieński County to 11% in Myślenicki County). The sales were mostly conducted in local markets, from only 3% of farms in Niżański County to 14% in Myślenicki County sold their products beyond local markets. Greater differences could be observed among counties in terms of supplying equipment and materials needed for production. The farms of Niżański County were supplying exclusively to the local market. In Krośnieński County, such farms made up 94% of the whole. Supplying outside local markets to an extent greater than the national average was reported in Myślenicki County (20%) and in Limanowski County (18%). Perhaps this could be a consequence of the close proximity of these counties to the Kraków metropolitan area. Moreover, the farms of Limanowski County were equipped with more expensive machines (with an average value of 37 thousand PLN) than the farms of the remaining three counties (with an average value of 20–22 thousand PLN), but they also reported better profitability in 2016 than the other three counties. The respective values were 7 thousand PLN and 4–5 thousand PLN. In both of these dimensions, the farms of these counties were well below the national average. Nevertheless, this could still indicate that the only strategy aimed at increasing the profitability of small farms was to increase their level of mechanization.

The strategies and profitability of farms in the counties with the largest farm area, such as Gryficki County, Wąbrzeski County, Szamotulski County and Łomżyński County are worth closer examination. The farms of Gryficki County had the largest average farm area of 32 ha, while the other counties had an average farm size of 23–25 ha. The farms of these counties were definitely above average, not just in terms of farm area but also due to their profitability, technical equipment, use of credits and loans, or renting of agricultural land. Nevertheless, there were still noticeable differences between them. The farms of Łomża County were the most mechanized with the average value of their machines and equipment close to 350 thousand PLN. They also noted in 2016 significantly higher profitability level (on average 220 thousand PLN) than the farms of the other three counties. When analyzing counties with the largest farm area, the lowest values in these dimensions were reported in Wąbrzeski County, where the average value of farm equipment was 160 thousand PLN and the incomes exceeded 60 thousand PLN. Out of the four counties listed above, it was also Wąbrzeski County, where the farmers were the least likely to use credits and loans (27%) and to rent the land (23%), which situated them only slightly above the average for the entire sample of farms. The farms of the remaining three counties reported higher values in these dimensions with 43% of respondents of Gryficki County taking credits and loans and 53% of surveyed farmers in the

Szamotulski County renting farm land to increase the area of they operated. However, what differentiated the farms of these four counties the most was the extent of expansion beyond local markets. Four different situations could be presented here: 1) intensive activities beyond local markets in the supplying of farms and in sales, 2) limited activities beyond local markets in supplying and sales, 3) marginal activities of both types beyond local markets, 4) marginal activities beyond local markets in farm supplying and intensive activities in sales. The first situation was characteristic of Szamotulski County, where 43% of farms supplied beyond local markets and 57% of farms sold their products there. Less intense activities could be seen in Wąbrzeski County, where 10% of farms were supplying beyond local markets and 23% of them sold their products beyond local markets. Marginal activity beyond the local market was reported for Gryficki County—3% for supplying and 7% for selling farm products). The situation of marginal supplying and significant sales beyond local markets could be observed in Łomżyński County, where only 3% of the surveyed farms supplied outside of local market and 39% of the farms sold their products there. The farms in these counties had a predominantly mixed production of plants and animals, which was the characteristic they shared with the farms of the counties with the smallest average of farm area. To the smallest extent this was the case with the farms of Gryficki County (53%), and to the greatest for the Łomżyński County (85%).

What conclusions can be made on the basis of the presented data? Farm profitability depends most of all on the area of the farm land, owned and rented for farming and this is obviously determined territorially. However the relation between the profitability of the farm and its area is not unconditional. The strategy of farm development appears to be an important variable that includes several components such as: level of farm mechanization, chosen direction of production, decision on sources of financing, supply, and sales market. The importance of the applied development strategy for profitability increases with operated farm area. Differentiation of development strategies can be observed at the county level and then consequently leads to identification of significant differences in functioning of farms in particular provinces and regions. Here, one might ask what factors determine the application of certain strategies by particular farms and consequently reach various levels of profitability (these issues—as was already mentioned earlier—mostly relate to counties with the smallest farms, as their situation appears to be homogenous regardless of the territorial context. Could the personal qualities of farm operators and the accumulation of such operators in certain territories be of key importance? Or, perhaps, the territorial context is essential here, with the specific characteristics of local community at the county level, which "impose" on farmers or enable them to accept and apply certain strategies of farm development? Which group of factors and to what extent affect the economic situation of family farms? To answer these questions the method of multilevel analysis will be applied here and, more specifically, due to the size and structure of the surveyed sample, the two-level model, where the level-1 model is represented by farmers "rooted" in the counties and the level-2 is represented by counties (Radkiewicz & Zieliński, 2010; Domański & Pokropek, 2011; Raudenbush et al., 2011; Nezlek, 2012).

10.2. Territorial Differentiation due to Farm Profitability

Could the above reflection on relations between territorial and development strategies applied by the farm users be confirmed by the results of suitable statistical analyses? Usually in sociological studies the analysis of territorial differentiation due to a certain variable is conducted through the analysis of variance (ANOVA). It serves to establish whether the differentiation due to a certain variable between the territorial units (e.g., municipalities or counties) is statistically significant. It also allows one to calculate intraclass correlation coefficient and estimate to what extent the observations from one territorial unit are more similar to each other than similar observations from various territorial units. In other words, the value of the correlation ratio indicates what percentage of total changeability of a certain characteristic (dependent variable) corresponds with territorial differentiation (independent variable). The value of correlation ratio falls into the 0–1 interval and the values closer to 1 indicate greater territorial differentiation due to a certain characteristic. In a situation where this value equals 0, it can be concluded that the analyzed territorial units do not differ due to the analyzed dependent variable and, therefore, territorial differentiation does not have any impact on the changeability of the variable. In such a case it is typically sufficient to just use classical statistics such as multivariate regression analysis to explain the changeability of a certain characteristic. However, whenever the correlation ratio is higher than 0 it is more useful to apply multilevel analysis if the structure of the collected data allows for it. These data should be hierarchically nested data of such a kind that an observation from one level is also reflected on the higher levels. The advantage of multilevel analysis in comparison to the classical approach can be seen by analyzing the changeability of the dependent variable from the first level, as predictors such as independent variables of the first level variables can be considered, as well as the variables from higher levels (Nezlek, 2012; Łaguna, 2018).

To establish whether there is any significant relation between the strategies applied by farm users and their place of residence (and agricultural production activities), territorial differentiation should be analyzed in regards to farm profitability, assuming that farm incomes are the final effect of applied development strategies. In all of our analyses, we view territorial differentiation as divided by counties and we present the income in the form of a normalizing variable—income logarithm, similar to the methods of Domański and Pokropek (2011, p. 32). It should also be added that, in transformation of incomes to the form of a natural logarithm, the minimal values were ascribed to farms with no incomes, which made up 3% of all surveyed farms.

Is there any interdependence between the profitability of the farm and the county where the farm user resides? To determine whether this interdependence is statistically significant and what percentage of total income changeability can be linked to the county division, we will first apply the traditional method of statistical inference, namely the analysis of variance (model ANOVA). Its results—estimated in SPSS—are presented in Table 10.2.1.

Table 10.2.1. The results of the analysis of variance ANOVA (without weights) for farm incomes according to division by counties. Number of counties N = 101. Number of farms (farmers) = 3543

	Sum of squares	df	Mean square	F	p
Explained variation	2001.651	100	20.017	12.558	0.000
Unexplained variation	5486.265	3442	1.594		
Total variation	7487.916	3542			
η^2	0.267				

Source: Own research.

Table 10.2.1 contains information on the components of the sum of squares of the differences: explained variation is the sum of squares of the differences between the mean values of the incomes in the samples and the means in the counties, unexplained variation is the sum of the squares of the differences between the farm incomes of particular counties and means for these counties, and the total variation of the model is the sum of the squares of the differences between the incomes of particular farms and the mean income value in the sample. In the subsequent columns, the following characteristics can be found: number of the degrees of freedom (df), mean square of differences, the value of the F test which, in this model, serves to verify the hypothesis indicating that mean farm incomes depend on the place of residence of the farm user and the probability (p) of rejection of null hypothesis (indicating lack of differences between the counties in regards to farm incomes), if it were true. As indicated by the minimal value of probability to reject such an hypothesis (0.000), the differences between the counties in regards to farm incomes are quite significant. The significance of interdependence between the county of residence and the farm income is described by the value of the correlation ratio (η^2). It was estimated to be at the level of 0.267, which points to a very strong interdependence—the county of residence explains almost 27% of total changeability in farm income. Therefore, it is justified to apply the multilevel model for the analysis of territorial differentiation due to farm profitability.

Multilevel modeling divides the analysis process into stages, which means constructing a simple model at first, and then moving onto more complex models. Application of the most simple model, known as a null model (or ANOVA with random effects), serves to determine the variation of dependent variables at all levels. It does not consider any independent variables, whether they are individual variables from the first level or group variables from other levels. These variables are introduced to the more complex models in the subsequent stages of analysis.

As was mentioned before, due to the sample size and the data structure we will focus on the two-level analysis, where level-1 will be represented by farms (farmers) and level-2 will be represented by counties. All of the analyses will be conducted by use of statistical program HLM 7.03, which specifically serves to process multilevel modelling and is described in detail in the introductory part of this publication. Let us consider: how meaningful is the income at the level of farms and at the level of counties? Estimation of income variation for both levels relies on obtaining certain parameters for the null model by introduction of commands and starting the calculation procedure. From the list of variables an dependent variable is selected. In our situation it is the D_LN income logarithm, but in the more complex models independent variables would also be selected. It should be noted that the variables used in the analyses must be previously prepared in separate files for each level. In our two-level model there are two files: the file with data related to farms (level-1) and the file with data related to counties (level-2). After selection of suitable variables the program creates the model pattern, in the system of equations form. Before commencing the calculation, weights are introduced to either one level or both levels. The estimation results of the null model for mean farm incomes are presented in Frame E2.1.

Frame E2.1. Estimation results of the two-level model in HLM 7.03

Summary of the model specified

Level-1 Model

$D\_LN_{ij} = \beta_{0j} + r_{ij}$

Level-2 Model

$\beta_{0j} = \gamma_{00} + u_{0j}$

Mixed Model

$D\_LN_{ij} = \gamma_{00} + u_{0j} + r_{ij}$

Final results: Iteration 5

Iterations stopped due to small change in likelihood function

$\sigma^2 = 1.52798$

Standard error of $\sigma^2 = 0.03683$

τ

INTRCPT1, β_0	0.56222

Standard error of τ

INTRCPT1, β_0	0.08578

Random level-1 coefficient	Reliability estimate
INTRCPT1, β_0	0.922

The value of the log-likelihood function at iteration 5 = $-5.907013E + 003$

Final estimation of fixed effects (with robust standard errors)

Fixed Effect	Coefficient	Standard error	t-ratio	Approx. $d.f.$	p-value
For INTRCPT1, β_0					
INTRCPT2, γ_{00}	2.403165	0.086361	27.827	100	< 0.001

Final estimation of variance components

Random Effect	Standard Deviation	Variance Component	$d.f.$	χ^2	p-value
INTRCPT1, u_0	0.74981	0.56222	100	1426.42224	< 0.001
level-1, r	1.23611	1.52798			

Statistics for the current model

Deviance = 11814.025576

Number of estimated parameters = 3

In the upper part of the frame there is a specification of the 2-level model for average farm incomes, with integration of that part which identifies the individual level and the county level, wherein:

D_LN_{ij} is the dependent variable, namely incomes (income logarithm) for the I unit (farm) from the J unit (county),

γ_{00} is the mean value of averages of farm incomes in the population of counties

u_{0j} is the deviation of average incomes of the J unit (county) from the mean value in the population, which means "the remainder" for the level-2 (counties),

r_{ij} is the deviation of incomes of the I unit (farm) from J unit (county) from the average income f this county, which means "the remainder" for the level-1 (farms),

β_{0j} is the deviation of the mean value of incomes for the J unit (county) from the average value for the first selected county.

In the next part, there is information that the estimation of parameters was completed in its fifth iteration, as subsequent iterations did not improve the fitting of the model or information on random effects and reliability for group effects. The value of reliability indicator notes the accuracy of the estimation of average farm income in particular counties. It can assume values starting from 0, the lowest reliability value, and go up to 1, which is the highest. Therefore, the value of 0.922 presents the highest precision level of estimated average incomes for the counties.

The last part contains the most important information regarding values of estimated parameters. First, parameters of fixed effects are considered; in our case, they are the mean value of average farm incomes in the surveyed counties ($\gamma_{00} = 2.403165$), with the standard error of the average (0.086361), as well as the test value T (27.827), with the number of degrees of freedom (100) and the relevance (< 0.001) for the null hypothesis assuming that the permanent parameter is 0. The test results allow us to accept the competing hypothesis, stating that the field parameter is not 0. Next, the parameters of random effects for level-1 and level-2 are presented in the form of standard deviations and variation. Standard deviation for random effects on level-2 (counties) is 0.74981 and variance is 0.56222. Standard deviation for random effects from level-1 (farms) equals 1.23611 and variance equals 1.52798. Furthermore, the value of the test Chi-square is 1426.42, with the number of degrees of freedom (100) and relevance (< 0.001) for null hypothesis, which states the random effect for level-2, equaling 0. Considering the test results, it seems rational to accept the competing hypothesis, which states that the random parameter for level-2 does not equal 0. This relates to fitting of the model (deviance = 11814.025576) and the number of estimated parameters (3). This is important for verification if various models fit to the data.

Referring to the data above, let us try to answer the question as to whether there is any significant differentiation between the incomes at the farm level and the incomes at the county level. Once again, we present the most important parameters in Table 10.2.2, which helps to answer this question.

Table 10.2.2. Estimation of variation of farm incomes for the level-1 (farms) and for the level-2 (counties). Number of counties N = 101. Number of farms = 3543. Program HLM 7.03—null model

Parameter	Value
γ_{00}	2.40 (0.09)
$Var(u_{0j}) = \tau_{00}$	0.56 (0.09)
$Var(r_{ij}) = \sigma^2$	1.53 (0.04)
$ICC = \dfrac{\tau_{00}}{\tau_{00} + \sigma^2}$	0.269
Deviance	11814.030

Source: Own work.

In Table 10.2.2, in addition to the estimated average of income (γ_{00}), the values of variance components for the level-2 (counties: τ_{00}) are presented, as well as those for the level-1 (farms: σ^2). These are the inter- and intraclass variations. The table also contains values of standard deviations, provided in parentheses. Furthermore, the value of interclass correlation coefficient—ICC is calculated as the ratio of interclass variation to total variation, that being the sum of interclass and intraclass variations. This is analogous to the correlation ratio discussed with the ANOVA model (Hox, 2010; Radkiewicz & Zieliński, 2010; Domański & Pokropek, 2011). This is the indicator of the homogeneity of counties in terms of farm incomes. In other words, one can estimate to what extent the farms within a county are more similar to each other in terms of acquired incomes than farms located across various counties. Greater homogeneity of farms from within one county also means greater income differences between the farms of various counties. Considering the values of intergroup variation (for the counties: 0.56) and intraclass variations (for farms: 1.53), as well as the value of the indicator of the intraclass correlation (0.269) we can state that the differentiation of incomes at the farm level and at the county level is significant. To be precise, the differentiation is significantly greater at the farm level than at the county level, nevertheless, the acquired results indicate quite clearly that the differences in farm profitability have a multilevel character. With this being the case, let's determine whether the characteristics of farms or counties have any influence on this phenomenon. That is, we will apply the analysis of a more complex two-level model and take into consideration independent variables. Let's start from group variables, such as characteristics of the counties.

10.3. Two-level Model with Independent Group Variables

What characteristics of the communities at the county level could impact the incomes of farms functioning in these communities? Initially, it could be stated that there are two categories of these characteristics. As the great majority of family farms sold their products on local markets it seemed that farm profitability might impact the purchasing power of local consumers, which depends on the socio-economic level of local community development. On one hand, the indicators of development might be as general as GDP or the county expenditures per person, residents' incomes according to tax forms, entrepreneurship level measured by the number of businesses functioning in the county, percentage of people employed outside of agriculture, or the level of urbanization measured by the percentage of urban population yet, on the other, these indicators might illustrate the level of unemployment in the county or the percentage of households that use the social welfare system. The second category was made by the characteristics of the counties, which directly influenced the developmental abilities of farms. In this case, the level of land resources in the county appeared to be the most important factor. Farm area—as was often highlighted—was a basic determinant of the volume of agricultural production, as well as income level. Modest farm land resources in a particular county narrowed the development abilities of farms functioning within its territory, in terms of the operated utilized land area and, at the same time, limited the profitability of these farms.

Were any of these characteristics significantly connected with farm incomes? To answer this question using the HLM program, one needs to expand the null model (discussed in the previous subchapter) to include independent group variables, such as characteristics of counties. This is done by choosing the appropriate independent variable (or variables) from the variables for level-2 (counties). After the selection of variables, the program creates a model equation—compared to the null model—and the independent variable (x_j) of permanent character for J unit (county) appears as well as the ratio (γ_{01}) describing the strength of the relationship between the independent variable and the dependent variable from level-2. Moreover, the remainder from the level-2 (u_{0j}) and its variance (τ_{00}), which are conditioned on variable (x_j), take on new meaning. As a result, intergroup correlation also assumes a conditional character, and the intraclass correlation coefficient (ICC) opens itself to a different interpretation. It indicates the percentage of explained variation of the independent variable and explains the division into categories of higher level (counties), on the assumption that, for all of these categories (counties), the value of the variable (x_j) is the same. The explanatory force meter of the model R^2 is the new indicator in the model with independent variable—in comparison to the null model. In this model, with the independent variables only from the level-2 (county characteristics), the explanatory force meter $R^2(p2)$ is a difference ratio between intraclass variance for the group model and intergroup variance for the null model. It provides the percentage for which unexplained intergroup variation from the null model is reduced, when the independent variable (or variables) from level-2 is/are considered in the model (Domański & Pokropek, 2011, pp. 80–83).

Let us analyze the results of the estimation of the farm incomes variance by inputting one of the categories of county characteristics at the level of farm land resources as an independent variable. The indicator of farm resources was created on the basis of the variable from an individual level, that being a farm area operated by surveyed farmers. It presents the mean farm area in the county, calculated as the mean farm area in the farm sample of a particular county (*SGR*). It can therefore be assumed that the large resources of farm land in the county go hand in hand with farmers' tendency to have large farms and that, in turn, gives them enhanced abilities to increase profitability. The estimation results of the model for average farm incomes while taking into account the average farm area in a county is presented in Frame E2.3.1.

Frame E2.3.1. Estimation results of the two-level model with independent group variable in HLM 7.03

Summary of the model specified

Level-1 Model

$D\_LN_{ij} = \beta_{0j} + r_{ij}$

Level-2 Model

$$\beta_{0j} = \gamma_{00} + \gamma_{01} * (SGR_j) + u_{0j}$$

Mixed Model

$$D\_LN_{ij} = \gamma_{00} + \gamma_{01} * SGR_j + u_{0j} + r_{ij}$$

Final results: Iteration 6

Iterations stopped due to small change in likelihood function

$\sigma^2 = 1.52776$

Standard error of $\sigma^2 = 0.03683$

τ

INTRCPT1, β_0	0.36215

Standard error of τ

INTRCPT1, β_0	0.05762

Random level-1 coefficient	Reliability estimate
INTRCPT1, β_0	0.884

The value of the log-likelihood function at iteration 6 = −5.886679E + 003

Final estimation of fixed effects (with robust standard errors)

Fixed Effect	Coefficient	Standarderror	t-ratio	Approx. d.f.	p-value
For INTRCPT1, β_0					
INTRCPT2, γ_{00}	1.631552	0.175730	9.284	99	< 0.001
SGR, γ_{01}	0.078991	0.015812	4.996	99	< 0.001

Final estimation of variance components

Random Effect	Standard Deviation	Variance Component	d.f.	χ^2	p-value
INTRCPT1, u_0	0.60179	0.36215	99	973.36902	< 0.001
level-1, r	1.23603	1.52776			

Statistics for the current model

Deviance = 11773.357394

Number of estimated parameters = 4

Estimated model is also presented as an equation: $D\_LN_{ij} = \gamma_{00} + \gamma_{01} * SGR_j + u_{0j} + r_{ij}$, wherein:

D_LN_{ij} is the dependent variable, namely incomes (logarithm of incomes) for I unit (farm), from J unit (county),

γ_{00} is the mean value of averages of farm incomes for the population of counties,

γ_{01} is the relationship strength factor between independent variable and dependent variable,

SGR_j is the independent variable, namely average farm area for J unit (county),

u_{0j} is the deviation of average incomes of the J unit (county) from the mean value in the population, which means "the remainder" for the level-2 (counties),

r_{ij} is the deviation of incomes of the I unit (farm) from the J unit (county) from the average income of the given county, which means "the remainder" for the level-1 (farms).

The most important information here relates to parameters of fixed effect such as: γ_{00} (mean value of farm incomes for the counties) and $\gamma_{01}(SGR)$ regression coefficient for independent variable SGR and random effects for level-1 (farms) and level-2 (counties), in the form of standard deviations and variances. HLM does not calculate intraclass correlation coefficient (ICC) and R^2. All the data and information are presented in Table 10.3.1.

As indicated by the data, when the estimated mean of incomes (γ_{00}) equals 1.63, and differentiation connected with land resources is not considered, then the regression coefficient for the independent variable SGR ($\gamma_{01}(SGR)$) equals 0.08. This indicates that the value of the growth of county incomes will increase by one standard deviation, along with the growth of average farm area. The parameter $Var(r_{ij})$, which serves as a variance of incomes at the farm level, has a value of 1.53, while the parameter $Var(u_{0j})$, which is a characteristic of income variance at the county level, equals 0.36. This confirms that the differentiation of incomes is much greater at the farm level than at the county level. The value of interclass correlation coefficient (ICC) indicates, however, that division into counties explains up to 19% of farm income changeability, assuming that average farm area in all the counties is the same. Furthermore, as indicated by the value $R^2(p2)$, consideration of differentiation between the counties, in terms of farm area resources, reduced the percentage of unexplained interclass income variation by 36%. It is also worth noting that the deviance value is 11773.36, which means that the fitting of this model to data is greater than the fitting of the null model, for which the deviance is 11814.03. Would such fitting increase with consideration of other dependent group variables in the model—thus characterizing the purchasing power of local consumers?

Table 10.3.1. Parameters for the two-level model of farm income regression given in farm areas in the counties. Number of counties N = 101. Number of farms = 3543. Program HLM 7.03—model with independent variable (M1)

Parameter	Value
γ_{00}	1.63 (0.18)
$\gamma_{01}(SGR)$	0.08 (0.01)
$Var(u_{0j}) = \tau_{00}$	0.36 (0.06)
$Var(r_{ij}) = \sigma^2$	1.53 (0.04)
$ICC = \dfrac{\tau_{00}}{\tau_{00} + \sigma^2}$	0.19
$R^2(p2) = \dfrac{\tau_{00}(M0) - \tau_{00}(M1)}{\tau_{00}(M0)}$	0.36
Deviance	11773.36

Source: Own work.

The number of conducted analyses confirmed the lack of significant connection between the changeability of farm incomes and the majority of variables describing the level of socio-economic development in the community on the county level and the purchasing power of its residents, considering a variety of characteristics. They are: GDP, county expenditures, residents' incomes as stated in tax forms, level of entrepreneurship as measured by the number of businesses operating in the county, percentage of people employed outside of agriculture, level of urbanization as measured by the percentage of urban dwellers, and percentage of households utilizing social welfare services. The only characteristic that appears to be meaningful here is county unemployment level. It should be assumed that—unlike farm land resources—a county's unemployment level, which should be negatively correlated with the purchasing power of consumers, will also have a negative influence on farm incomes. The results of the model's estimation of average farm income, considering average farm area (*SGR*) and county unemployment level (*BEZROBOC*) are presented in Frame E2.3.2.

Frame E2.3.2. Results of the estimations of the two-level model with independent group variables in HLM 7.03

Summary of the model specified

Level-1 Model

$$D\_LN_{ij} = \beta_{0j} + r_{ij}$$

Level-2 Model

$$\beta_{0j} = \gamma_{00} + \gamma_{01} * (SGR_j) + \gamma_{02} * (BEZROBOC_j) + u_{0j}$$

Mixed Model

$$D\_LN_{ij} = \gamma_{00} + \gamma_{01} * SGR_j + \gamma_{02} * BEZROBOC_j + u_{0j} + r_{ij}$$

Final results: Iteration 6

Iterations stopped due to small change in likelihood function

$\sigma^2 = 1.52779$

Standard error of $\sigma^2 = 0.03683$

τ

INTRCPT1, β_0	0.33358

Standard error of τ

INTRCPT1, β_0	0.05360

Random level-1 coefficient	Reliability estimate
INTRCPT1, β_0	0.876

The value of the log-likelihood function at iteration 6 = $-5.883054E + 003$

Final estimation of fixed effects (with robust standard errors)

Fixed Effect	Coefficient	Standard error	t-ratio	Approx. d.f.	p-value
For INTRCPT1, β_0					
INTRCPT2, γ_{00}	2.059611	0.238187	8.647	98	< 0.001
SGR, γ_{01}	0.085239	0.015351	5.552	98	< 0.001
BEZROBOC, γ_{02}	−0.041474	0.012785	−3.244	98	0.002

Final estimation of variance components

Random Effect	Standard Deviation	Variance Component	d.f.	χ^2	p-value
INTRCPT1, u_0	0.57756	0.33358	98	894.93314	< 0.001
level-1, r	1.23604	1.52779			

Statistics for the current model

Deviance = 11766.108054

Number of estimated parameters = 5

In the above case, the estimated model is presented by the following equation:

$$D\_LN_{ij} = \gamma_{00} + \gamma_{01} * SGR_j + \gamma_{02} * BEZROBOC_j + u_{0j} + r_{ij}$$

In comparison to the equation with just one independent group variable (SGR), a new component of the equation here is $\gamma_{02} * BEZROBOC_j$, wherein:

γ_{02} is the relationship strength factor between the independent variable and the dependent variable (BEZROBOC),

$BEZROBOC_j$ is the independent variable, namely employment level for J unit (county)

The most essential information is related to parameters of fixed effects: γ_{00} (average incomes for counties), $\gamma_{01}(SGR)$ and $\gamma_{02}(BEZROBOC)$ (regression coefficient for independent variables SGR and BEZROBOC) and random effects for level-1 (farms) and level-2 (counties), in the form of standard deviations and variances, as well as the intraclass correlation coefficient (ICC) and R^2 presented in Table 10.3.2.

When controlling for the differentiation related to farm land resources and the employment level, the estimated mean of farm incomes (γ_{00}) equals 2.06, the regression coefficient for independent variable SGR ($\gamma_{01}(SGR)$) assumes the value 0.09, and the variable BEZROBOC ($\gamma_{02}(BEZROBOC)$) equals 0.04. This indicates that, with controlled unemployment level, farm incomes in the counties will grow by 0.09, in sync with the growth of average farm land area by one standard deviation, and farm incomes will drop by 4% with the growth of unemployment level by one standard deviation and with controlled level of farm land area resources. These data confirm that the differentiation of incomes is much greater at the farm level than at the county level: parameter $Var(r_{ij})$, describing the variation of incomes on the farm level, reaches a value of 1.53, while the parameter $Var(u_{oj})$, characterizing the variation of incomes on the county level, comes to 0.33. The value of intraclass correlation coefficient (ICC) indicates that the division into counties explains up to 18% of farm income changeability, assuming that the level of unemployment and the farm land resources are the same in all surveyed counties. The value of $R^2(p2)$ states that consideration of differentiation between the counties in terms of farm land resources (here meaning farm area) and unemployment level reduced the percentage of unexplained variance of group incomes by 36%, and deviance reached a value of 11776.11. This shows that the fitting of the model into the data does not differ significantly from the fitting of the model with one independent group variable (average farm area in the county). To what extent could this fitting change if we consider individual independent variables (variables from Level-1) in the model?

Table 10.3.2. Parameters for the two-level model of farm income regression considering unemployment level of farm areas in the counties. Number of counties N = 101. Number of farms = 3543. Program HLM 7.03—model with independent variable (M2)

Parameter	Value
γ_{00}	2.06 (0.24)
$\gamma_{01}(SGR)$	0.09 (0.02)
$\gamma_{02}(BEZROBOC)$	−0.04 (0.01)
$Var(u_{oj}) = \tau_{00}$	0.33 (0.05)
$Var(r_{ij}) = \sigma^2$	1.53 (0.04)
$ICC = \dfrac{\tau_{00}}{\tau_{00} + \sigma^2}$	0.18
$R^2(p2) = \dfrac{\tau_{00}(M0) - \tau_{00}(M2)}{\tau_{00}(M0)}$	0.36
Deviance	11766.11

Source: Own work.

10.4. Two-level Model with Individual Independent Variables

What are the characteristics of farms and their users that could be connected with their acquired incomes? What seems most essential here is the set of activities undertaken by farmers and related to the functioning of the farm. In more general terms, this can be understood as the type of development strategy adopted by a farm, which is reflected in the analyses of profitability of small and large farms. The elements of such strategies include: activities related to the size of farm area connected with the use of rental mechanisms, level of farm mechanization, employment policy (including the use of family labor), policy for financing the agricultural production, type of production (cultivation of plants, breeding of animals, combination thereof), and use of non-local markets for farm supply and sales. Obviously, the directions of farm strategies adopted by farmers/farm users were influenced significantly by the demographic characteristics of farm managers (sex, age, education) but these aspects will not be analyzed here. What appears more interesting than the individual characteristics of farmers is how their choice of particular farm strategies impacts the farm profitability level.

To designate farm strategies implemented on the surveyed farms, as many as 12 individual variables from level-1 were considered and then, based on the analysis of the main components for polychoric correlation matrix between these variables (see Domański & Pokropek, 2011, p. 64), their two main types were distinguished. The list of considered variables with their values can be found in Table 10.4.1.

Table 10.4.1. Component variables of the indicator of development strategy

Variable/question	Value of the variable/answer
Have there been any changes to the farm area that you operate? Has it increased, decreased, or remained the same?	0 = decreased 1 = no changes 2 = increased
Are you the owner of all used/operated land? Do you rent out part of your land to others or rent from others to operate more land?	0 = I rent part of my land to others 1 = I am the owner of all operated land 2 = part of operated land is rented from others
What is the value of your machines and equipment?	0 = no machines and equipment owned 1 = up to 50,000 PLN 2 = above 50,000 PLN
Have you taken any lines of credit or loans for agricultural activity in the last 3 years?	0 = yes 1 = no
What kind of production do you engage in?	0 = plant production only 1 = animal and plant production
What percentage of plant production is devoted to sustaining animal production?	0 = 0% 1 = up to 50% 2 = over 50%
Where do you buy material and equipment needed for production?	0 = on the local market only 1 = on the local market and beyond
Where do you sell your products?	0 = only on the local market 1 = on the local market and beyond
How many household members help you with farm work?	0 = none 1 = one person 2 = two or more persons
Do you engage in reciprocal work exchange with other farmers?	0 = yes 1 = no
Do you employ anyone on your farm to do paid work?	0 = no 1 = yes
Do you use services of external businesses/companies?	0 = no 1 = yes

Source: Own work.

The first strategy, which is described here as sustainable development, can be characterized as having mixed agricultural production including plant cultivation and animal breeding and devoting a significant portion of plant production for the use of farm animals. The second strategy, presented here as commodity strategy, involves the following characteristics: engagement in nonlocal markets for sale of products and for supplying farms with materials and means of production, increasing the land area of farm operations in the last three years, use of land rental mechanisms, as well as bank and farm loans. For both strategies, a high level of mechanization, as well as use of family labor, were quite significant, with mechanization playing a more important role in the commodity strategy and use of family labor in the sustainable development strategy.

To analyze how the adoption of a particular strategy influences the changeability of farm incomes, the expansion of the null model to include independent variables of an individual character is needed. In our case, these variables are the indicators of particular strategies, measured by the values of factor loadings obtained in the analysis of main components. The whole procedure for introduction of such variables to the model is similar to the situation of introduction of independent group variables. In the HLM program it is done by selecting the appropriate independent variable from the variables for level-1 and the program automatically creates the model equation, in which the independent variable (x_{ij}) appears for I unit (farm) from J category (county) as well as the regression coefficient (γ_{10}) describing the mean dependence between the dependent and independent variables in J unit categories of the level-2 (counties). It should be noted here that this is a random intercept model, where it is assumed that the regression constant for the J unit category of level 2 (county) has the character of a random variable ($\beta_{0j} = \gamma_{00} + u_{0j}$), and that the regression coefficient between the dependent variable and independent variable in all the categories of level-2 (counties) has the same value (Domański & Pokropek 2011, p. 100).

What is the relationship between the farm incomes and the extent of implementation of the sustainable development strategy by farmers (ST_ZR)? The results of the estimation model which takes that indicator into account are presented in Frame E2.4.1.

Frame E2.4.1. Results of the estimation of the two-level model with individual independent variables (ST_ZR) in HLM 7.03

Summary of the model specified

Level-1 Model

$D\_LN_{ij} = \beta_{0j} + \beta_{1j} * (ST\_ZR_{ij}) + r_{ij}$

Level-2 Model

$$\beta_{0j} = \gamma_{00} + u_{0j}$$

$$\beta_{1j} = \gamma_{10}$$

Mixed Model

$$D\_LN_{ij} = \gamma_{00} + \gamma_{10} * ST\_ZR_{ij} + u_{0j} + r_{ij}$$

Final results: Iteration 4

Iterations stopped due to small change in likelihood function

$\sigma^2 = 1.43212$

Standard error of $\sigma^2 = 0.03452$

τ

INTRCPT1, β_0	0.60850

Standard error of τ

INTRCPT1, β_0	0.09188

Random level-1 coefficient	Reliability estimate
INTRCPT1, β_0	0.932

The value of the log-likelihood function at iteration 4 = $-5.798979E + 003$

Final estimation of fixed effects (with robust standard errors)

Fixed Effect	Coefficient	Standard error	t-ratio	Approx. d.f.	p-value
For INTRCPT1, β_0					
INTRCPT2, γ_{00}	2.389004	0.089748	26.619	100	< 0.001
For ST_ZR slope, β_1					
INTRCPT2, γ_{10}	0.354746	0.043785	8.102	3441	< 0.001

Final estimation of variance components

Random Effect	Standard Deviation	Variance Component	d.f.	χ^2	p-value
INTRCPT1, u_0	0.78007	0.60850	100	1658.67865	< 0.001
level-1, r	1.19671	1.43212			

Statistics for the current model

Deviance = 11597.958408

Number of estimated parameters = 4

Estimated model can be presented by the following equation:

Level-1 (farms): $D\_LN_{ij} = \beta_{0j} + \beta_{1j} * (ST\_ZR_{ij}) + r_{ij}$,

Level-2 (counties): $\beta_{0j} = \gamma_{00} + u_{0j}$

$\beta_{1j} = \gamma_{10}$

Mixed model: $D\_LN_{ij} = \gamma_{00} + \gamma_{10} * ST\_ZR_{ij} + u_{0j} + r_{ij}$,

where:

D_LN_{ij} is the dependent variable, namely incomes (logarithm of incomes) for I unit (farm) from J unit (county),

β_{0j} is the regression constant for J unit (county),

β_{1j} is the regression coefficient for the independent variable (strategy of sustainable development) in J unit (county),

γ_{00} is the regression constant, namely mean of incomes, as specified during control for the independent variable,

γ_{10} is the regression coefficient describing the mean dependency between dependent (incomes) and independent variables (indicator of the strategy of sustainable development) in J unit (county),

ST_ZR_{ij} is the independent variable, namely the indicator of implementation for the sustainable development strategy for I unity (farm) from J unit (county),

u_{0j} is the random effect from level-2 (deviation of mean incomes for J unit (county) from the mean in the population), namely "the remainder" for level-2 (counties),

r_{ij} is the random effect from level-1 (deviation of mean incomes for I unit (farm) from J unit (county) from the mean of that county), namely "the remainder" for level-1 (farms).

Information related to parameters of fixed effects: γ_{00} (mean incomes for counties) and $\gamma_{10}(ST\_ZR)$ (regression coefficient for the independent variable ST_ZR), random effects for level-1 (farms) and level-2 (counties) in the form of variance, as well as calculated intraclass correlation coefficient (ICC) and $R^2(pl)$ are presented in Table 10.4.2.

Table 10.4.2. Parameters of the two-level model for farm income regression due to implementation of the sustainable development strategy by farmers. Number of counties N = 101. Number of farms = 3543. Program HLM 7.03—model with independent variable (M3A)

Parameter	Value
γ_{00}	2.39 (0.09)
$\gamma_{10}(ST\_ZR)$	0.35 (0.04)
$Var(u_{0j}) = \tau_{00}$	0.61 (0.09)
$Var(r_{ij}) = \sigma^2$	1.43 (0.03)
$ICC = \dfrac{\tau_{00}}{\tau_{00} + \sigma^2}$	0.30
$R^2(pl) = \dfrac{\sigma^2(M0) - \sigma^2(M3A)}{\sigma^2(M0)}$	0.07
Deviance	11597.96

Source: Own work.

When controlling for the differentiation related to implementation of the sustainable development strategy, the estimated mean of incomes (γ_{00}) equals 2.39 and the regression coefficient for independent variable ST_ZR ($\gamma_{10}(ST\_ZR)$) equals 0.35. This shows that for every standard deviation of the indicator of sustainable development strategy there is a 35% increase in mean farm incomes, expressed in a logarithmic scale. As indicated by the data, the value of the parameter $Var(r_{ij})$, which characterized the variance of incomes at the farm level, equals 1.43, and the value of the parameter $Var(u_{0j})$, which characterized the variance of incomes at the county level, equals 0.61. Differentiation of incomes is much greater on the farm level than on the county level, but it is significant on both levels. With the assumption that average farms in all counties adopt the strategy of sustainable development to the same extent, the value of interclass correlation coefficient (ICC) indicates that county divisions explain up to 30% of farm income changeability. The indicator of the strategy of sustainable development explains about 7% of income differentiation, which is stated by the value of R^2(p1). The results indicate quite clearly that the differentiation of farm profitability has a multilevel character.

And what would the above parameters look like if the model took into account the commodity strategy (ST_TW)? The estimation results for this model considering that indicator is presented in Frame E2.4.2.

Frame E2.4.2. Results of the estimation of two-level model with individual independent variables (ST_TW) in HLM 7.03

Summary of the model specified

Level-1 Model

$D\_LN_{ij} = \beta_{0j} + \beta_{1j} * (ST\_TW_{ij}) + r_{ij}$

Level-2 Model

$\beta_{0j} = \gamma_{00} + u_{0j}$

$\beta_{1j} = \gamma_{10}$

Mixed Model

$D\_LN_{ij} = \gamma_{00} + \gamma_{10} * ST\_TW_{ij} + u_{0j} + r_{ij}$

Final results: Iteration 5

Iterations stopped due to small change in likelihood function

$\sigma^2 = 1.09353$

Standard error of $\sigma^2 = 0.02636$

τ

INTRCPT1, β_0	0.49729
Standard error of τ	
INTRCPT1, β_0	0.07475

Random level-1 coefficient	Reliability estimate
INTRCPT1, β_0	0.936

The value of the log-likelihood function at iteration 5 = $-5.324324E + 003$

Final estimation of fixed effects (with robust standard errors)

Fixed Effect	Coefficient	Standard error	t-ratio	Approx. d.f.	p-value
For INTRCPT1, β_0					
INTRCPT2, γ_{00}	2.423676	0.079525	30.477	100	< 0.001
For ST_TW slope, β_1					
INTRCPT2, γ_{10}	0.866032	0.043062	20.111	3441	< 0.001

Final estimation of variance components

Random Effect	Standard Deviation	Variance Component	d.f.	χ^2	p-value
INTRCPT1, u_0	0.70519	0.49729	100	1717.87182	< 0.001
level-1, r	1.04572	1.09353			

Statistics for the current model

Deviance = 10648.648506

Number of estimated parameters = 4

The estimated model can be presented with the equation: $D\_LN_{ij} = \gamma_{00} + \gamma_{10}$ * * $ST\_TW_{ij} + u_{0j} + r_{ij}$, where ST_TW_{ij} is an independent variable, namely, the indicator of implementation of the commodity strategy for I unit (farm), from J unit (county).

Information on parameters of fixed effects $-\gamma_{00}$ (mean income for farms), γ_{10}(ST_TW) (regression coefficient for independent variable ST_TW), as well as random effects for level-1 (farms) and level-2 (counties) in the form of variance and intraclass correlation coefficient (ICC), and R^2(p1) are presented in Table 10.4.3.

When controlling for the differentiation related to implementation of the commodity strategy, the estimated mean of incomes (γ_{00}) equals 2.42, and the regression coefficient for independent variable equals ST_TW (γ_{10}(ST_TW)) 0.87. This means that, for every standard deviation of the commodity strategy indicator, there is an 87% increase in mean farm incomes expressed in a logarithmic scale. The value of the parameter $Var(r_{ij})$, which characterized the variance of incomes on the farm level, equals 1.09, and the value of parameter $Var(u_{0j})$, which characterized the variance of incomes at the county level, equals 0.50. Therefore, the differentiation of incomes on both levels is significant but is more than double at the farm level what is found at the county level. With the assumption that all farms implement the commodity strategy to the same extent, the value of intraclass correlation coefficient (ICC) indicates that division into counties explains as much as 31% of the changeability in farm incomes. The value R^2(p1) states that the indicator of commodity strategy explains 29% of income differentiations. These results show quite clearly that differentiation in farm profitability has a multi-level character and the indicator of commodity strategy for farms is a significant predictor of their profitability. The value of deviance is 10648.65, which means that the fitting of that model to the data is greater than the fitting of the model with the indicator of sustainable development strategy as an individual independent variable, for which the statistical deviance equals 11597.96. It should be noted that the level of significance of the Chi- square (1717.87182) with 100 degrees of freedom for random effect is < 0.001 (see Frame E2.4.2). This means that when controlling the indicator of commodity strategy, average incomes in counties still differ in a statistically significant manner.

Table 10.4.3. Parameters for the two-level model of farm income regression due to implementation of the commodity strategy by farmers. Number of counties N = 101. Number of farms = 3543. Program HLM 7.03—model with independent variable (M3B)

Parameter	Value
γ_{00}	2.42 (0.08)
$\gamma_{10}(ST\_TW)$	0.87 (0.04)
$Var(u_{0j}) = \tau_{00}$	0.50 (0.08)
$Var(r_{ij}) = \sigma^2$	1.09 (0.03)
$ICC = \dfrac{\tau_{00}}{\tau_{00} + \sigma^2}$	0.31
$R^2(p1) = \dfrac{\sigma^2(M0) - \sigma^2(M3B)}{\sigma^2(M0)}$	0.29
Deviance	10648.65

Source: Own work.

And what results can be obtained by estimating the model with independent variables of an individual and group character?

10.5. Two-level Model with Independent Individual and Group Variables

Let's introduce to the model those variables which have the strongest relationship to farm profitability. The indicator of useable farm land in the county (*SGR*) will be presented as an independent group variable and the indicator of the implementation of the commodity strategy (ST_TW) as an independent individual variable. After introduction of these variables, they can be presented in the following equation:

$$D\_LN_{ij} = \gamma_{00} + \gamma_{01} * SGR_j + \gamma_{10} * ST\_TW_{ij} + u_{0j} + r_{ij}$$

The results of its estimation are presented in Frame E2.5.1.

Frame E2.5.1. The results of the estimation of the two-level model with group and individual independent variables (SGR and ST_TW) in HLM 7.03

Summary of the model specified

Level-1 Model

$$D\_LN_{ij} = \beta_{0j} + \beta_{1j} * (ST\_TW_{ij}) + r_{ij}$$

Level-2 Model

$$\beta_{0j} = \gamma_{00} + \gamma_{01} * (SGR_j) + u_{0j}$$

$$\beta_{1j} = \gamma_{10}$$

Mixed Model

$$D\_LN_{ij} = \gamma_{00} + \gamma_{01} * SGR_j + \gamma_{10} * ST\_TW_{ij} + u_{0j} + r_{ij}$$

Final results: Iteration 6

Iterations stopped due to small change in likelihood function

$\sigma^2 = 1.09354$

Standard error of $\sigma^2 = 0.02636$

τ

INTRCPT1, β_0	0.37131
Standard error of τ	
INTRCPT1, β_0	0.05702

Random level-1 coefficient	Reliability estimate
INTRCPT1, β_0	0.916

The value of the log-likelihood function at iteration 6 = $-5.310677E + 003$

Final estimation of fixed effects (with robust standard errors)

Fixed Effect	Coefficient	Standard error	*t*-ratio	Approx. *d.f.*	*p*-value
For INTRCPT1, β_0					
INTRCPT2, γ_{00}	1.814001	0.175569	10.332	99	< 0.001
SGR, γ_{01}	0.062335	0.017030	3.660	99	< 0.001
For ST_TW slope, β_1					
INTRCPT2, γ_{10}	0.860823	0.042298	20.351	3441	< 0.001

Final estimation of variance components

Random Effect	Standard Deviation	Variance Component	*d.f.*	χ^2	*p*-value
INTRCPT1, u_0	0.60935	0.37131	99	1289.42466	< 0.001
level-1, r	1.04573	1.09354			

Statistics for the current model

Deviance = 10621.353006

Number of estimated parameters = 5

The most important data related to parameters of fixed effects: γ_{00} (mean incomes for counties), $\gamma_{01}(SGR)$ (regression coefficient for independent variable *SGR*), $\gamma_{10}(ST\_TW)$ (regression coefficient for independent variable *ST_TW*), and random effect for level-1 (farms) and level-2 (counties) in the form of variance, as well as calculated value of interclass correlation coefficient (ICC), R^2(p2) and R^2(p1), are presented in Table 10.5.1.

When controlling for the differentiation of the farm land resources indicator and the commodity strategy indicator, the estimated mean of incomes (γ_{00}) has the value of 1.81. The regression coefficient for the independent variable *SGR* ($\gamma_{01}(SGR)$) equals 0.06, and for the independent variable *ST_TW* ($\gamma_{10}(ST\_TW)$) it is 0.86. This shows that farm incomes increase with the growth of utilized agricultural land in the county, as expressed by average farm area as well as by the level of commodity strategy implementation. In controlling for the level of implementation of the commodity strategy, an increase by one unit in the resources of agricultural utilized area raises mean farm incomes by 6%. In controlling for the indicator of utilized farm area resources, an increase of the commodity strategy implementation indicator by one unit translates to 86% growth of these incomes (on the logarithmic scale). The differentiation of incomes is significant on both levels but it is still greater on the farm level ($Var(r_{ij}) = 1.09$) than on the county level ($Var(u_{oj}) = 0.37$). The value of the intraclass correlation coefficient (ICC) indicates that division into counties explains 25% of farm income changeability, assuming that all farms implement commodity strategy to the same extent, and that average farm area in all counties is about the same. Both of these variables, namely, the level of commodity strategy implementation and utilized agricultural area resources in the county, explain up to 34% of income differentiation between the counties ($R^2(p2) = 0.34$), and the level of commodity strategy implementation explains 29% of income differentiation between the farms ($R^2(p1) = 0.29$). It should be noted that the deviance value for this model equals 10621.35, which shows that its fitting to the data is greater than its fitting with the models described earlier. Additionally, the level of significance of the Chi-square test for random effects (1289.42466 with 99 degrees of freedoms) is $p < 0.001$ (see Frame E2.5.1). This shows that by controlling the indicator of commodity strategy and the indicator of utilized agricultural area, average farm incomes still differ to a statistically significant degree.

Table 10.5.1. Parameters of the two-level model of farm income regression due to agricultural land resources in counties and the degree of implementation of commodity strategy by farmers. Number of counties N = 101. Number of farms = 3543. Program HLM 7.03—model with group and individual independent variables (M4)

Parameter		Value
γ_{00}		1.81 (0.18)
$\gamma_{01}(SGR)$		0.06 (0.02)
$\gamma_{10}(ST\_TW)$		0.86 (0.04)
$Var(u_{oj}) = \tau_{00}$		0.37 (0.06)
$Var(r_{ij}) = \sigma^2$		1.09 (0.03)
$ICC = \dfrac{\tau_{00}}{\tau_{00} + \sigma^2}$		0.25
$R^2(p2) = \dfrac{\tau_{00}(M0) - \tau_{00}(M4)}{\tau_{00}(M0)}$		0.34
$R^2(p1) = \dfrac{\sigma^2(M0) - \sigma^2(M4)}{\sigma^2(M0)}$		0.29
Deviance		10621.35

Source: Own work.

Let us introduce one more independent group variable, which has been shown to have a connection with farm incomes, namely, the level of unemployment on the county level (*BEZROBOC*). After its introduction, the following model of equation is presented:

$$D\_LN_{ij} = \gamma_{00} + \gamma_{01} * SGR_j + \gamma_{02} * BEZROBOC_j + \gamma_{10} * ST\_TW_{ij} + u_{0j} + r_{ij},$$

where $BEZROBOC_j$ is an independent variable, namely, the indicator of unemployment level for the J unit (county). The most important results of this model's estimation are presented in Frame E2.5.2. Table 10.5.2 contains information on the value of parameters of fixed effects and random effects, as well as calculated values of the interclass correlation coefficient (ICC), $R^2(p2)$ and $R^2(p1)$.

Frame E2.5.2. Results of estimation of the two-level model with group and individual independent variables (SGR, $BEZROBOC$ and ST_TW) in HLM 7.03

. . .

Final estimation of fixed effects (with robust standard errors)

Fixed Effect	Coefficient	Standard error	t-ratio	Approx. d.f.	p-value
For INTRCPT1, β_0					
INTRCPT2, γ_{00}	2.042617	0.229315	8.907	98	< 0.001
SGR, γ_{01}	0.065682	0.016966	3.871	98	< 0.001
BEZROBOC, γ_{02}	−0.022161	0.012305	−1.801	98	0.075
For ST_TW slope, β_1					
INTRCPT2, γ_{10}	0.859599	0.042240	20.350	3441	< 0.001

Final estimation of variance components

Random Effect	Standard Deviation	Variance Component	d.f.	χ^2	p-value
INTRCPT1, u_0	0.60251	0.36302	98	1259.26254	< 0.001
level-1, r	1.04573	1.09356			

. . .

The data show that when controlling for differentiation of counties in terms of the resources of utilized agricultural areas and the degree of implementation of commodity strategy, an increase in the unemployment level by one unit results in a 2% drop in farm incomes. However, with the similar meaning of the indicator for the commodity strategy implementation level ($\gamma_{10}(ST\_TW) = 0.86$) for income changeability that is similar to the previous model, without consideration for the unemployment level, there is a slight increase of the indicator for agricultural land resources

$(\gamma_{01}(SGR) = 0.07)$. It can be inferred that, with a lack of significant disparities in the purchasing power of the local community which could, presumably, be measured by the unemployment level, the relationship between the farm area utilized in the county and farm profitability becomes slightly more significant. Considering the indicator for the unemployment level causes an increase in the value $R^2(p2)$, which then equals 0.36. This shows that all three variables combined account for 36% of the differentiation between counties and, therefore, somewhat more than the level of commodity strategy implementation and the resources of utilized agricultural land without the unemployment level indicator, which points to 0.34). This slightly improves the fitting of the model to the data, with the value of deviance being 10619.31.

Table 10.5.2. Parameters of the two-level model of farm income regression due to the resources of utilized agricultural land and unemployment level, as well as the degree of implementation of commodity strategy on farms. Number of counties N = 101. Number of farms = 3543. Program HLM 7.03—model with group and individual independent variables (M5)

Parameter	Value
γ_{00}	2.04 (0.23)
$\gamma_{01}(SGR)$	0.07 (0.02)
$\gamma_{02}(BEZROBOC)$	−0.02 (0.01)
$\gamma_{10}(ST\_TW)$	0.86 (0.04)
$Var(u_{0j}) = \tau_{00}$	0.36 (0.06)
$Var(r_{ij}) = \sigma^2$	1.09 (0.03)
$ICC = \dfrac{\tau_{00}}{\tau_{00} + \sigma^2}$	0.25
$R^2(p2) = \dfrac{\tau_{00}(M0) - \tau_{00}(M5)}{\tau_{00}(M0)}$	0.36
$R^2(p1) = \dfrac{\sigma^2(M0) - \sigma^2(M5)}{\sigma^2(M0)}$	0.29
Deviance	10619.31

Source: Own work.

To finish this part of the analysis, one more model should be analyzed where, instead of commodity strategy, the strategy of sustainable development will be considered (ST_ZR), even though—as indicated by previous analyses—this is an inferior predictor of changeability in farm incomes than the indicator for commodity strategy. In this case, we have the following equation model:

$$D\_LN_{ij} = \gamma_{00} + \gamma_{01} * SGR_j + \gamma_{02} * BEZROBOC_j + \gamma_{10} * ST\_ZR_{ij} + u_{0j} + r_{ij}$$

The most important results of the estimation for this model are presented in Frame E2.5.3, and information related to the value of parameters of fixed effects and random effects, as well as calculations for the value of intraclass correlation coefficient (ICC), R^2(p2) and R^2(p1), are all presented in Table 10.5.3.

Frame E2.5.3. The results of the estimation of the two-level model with group and individual independent variables (*SGR, BEZROBOC* and *ST_ZR*) w HLM 7.03

. . .

Final estimation of fixed effects (with robust standard errors)

Fixed Effect	Coefficient	Standard error	*t*-ratio	Approx. *d.f.*	*p*-value
For INTRCPT1, β_0					
INTRCPT2, γ_{00}	2.082011	0.261000	7.977	98	< 0.001
SGR, γ_{01}	0.084730	0.014414	5.878	98	< 0.001
BEZROBOC, γ_{02}	−0.044161	0.014300	−3.088	98	0.003
For *ST_ZR* slope, β_1					
INTRCPT2, γ_{10}	0.350077	0.043388	8.069	3441	< 0.001

Final estimation of variance components

Random Effect	Standard Deviation	Variance Component	*d.f.*	χ^2	*p*-value
INTRCPT1, u_0	0.61643	0.37999	98	1087.58687	< 0.001
level-1, r	1.19667	1.43201			

. . .

It turns out that by controlling the differentiation between the counties in regards to the resources of utilized agricultural land and the level of sustainable development strategy implementation, the growth of unemployment level in the county by one unit causes a drop in farm incomes of 4%, which is double that which could be seen when controlling the level of commodity strategy implementation. Moreover, the significance of the resources of utilized farm areas also increases (γ_{01}(SGR) = 0.08). It can be assumed that incomes of farms implementing the strategy of sustainable development are more connected with the consumers' purchasing power and the farm area (implying the resources of utilized agricultural areas in the county) than the income of farms implementing other strategies.

Table 10.5.3. Parameters of the two-level model of farm income regression due to resources of utilized agricultural land and unemployment level in the counties, as well as degree of sustainable development strategy implementation on farms. Number of counties N = 101. Number of farms = 3543. Program HLM 7.03—model with group and individual independent variables (M6)

Parameter	Value
γ_{00}	2.08 (0.26)
$\gamma_{01}(SGR)$	0.08 (0.01)
$\gamma_{02}(BEZROBOC)$	−0.04 (0.01)
$\gamma_{10}(ST\_ZR)$	0.35 (0.04)
$Var(u_{0j}) = \tau_{00}$	0.38 (0.06)
$Var(r_{ij}) = \sigma^2$	1.43 (0.03)
$ICC = \dfrac{\tau_{00}}{\tau_{00} + \sigma^2}$	0.21
$R^2(p2) = \dfrac{\tau_{00}(M0) - \tau_{00}(M6)}{\tau_{00}(M0)}$	0.32
$R^2(p1) = \dfrac{\sigma^2(M0) - \sigma^2(M6)}{\sigma^2(M0)}$	0.07
Deviance	11554.17

Source: Own work.

Considering all the analyses presented thus far, it should be stated that the differentiation of farm incomes had a multilevel character. The characteristics of level-1 (farms), and level-2 (counties) are significant here but those from level-1 are more meaningful. In other words, farm incomes influence the decisions of farm users and operators but the effectiveness of these decisions, which are embodied by strategies, are determined by local context. This context includes economic aspects such as resources of utilized agricultural areas (which translates to the area of particular farms), as well as aspects of a social nature, such as the purchasing power of the consumers. What is important here is not necessarily the same level of social prosperity and well-being (measured by GDP or by county expenses per resident, residents' incomes according to tax forms, level of urbanization, or level of entrepreneurship) but the scale of the phenomena limiting the purchasing power of part of the community, for example, through the existing unemployment level.

10.6. Resources of Utilized Agricultural Land, Farm Profitability, and Farm Development Strategies in Territorial Cross-Section

The general conclusion from the analyses presented above is quite clear, indicating that the incomes of family farms are determined by certain characteristics of local communities where these farms operate. However, to a much greater extent they are connected to activities undertaken by farm users/operators. When it comes to local context, resources of utilized agricultural land available to local communities constitute the most important factor, as they determine the area sizes of particular farms. In other words, in communities (here, counties), where it is possible to buy or rent more agricultural land, there are more large-area farms which can engage in large-scale production and increase their profitability. However, the farm area is not the only factor connected with the income levels achieved by farms. As can be inferred from the above analyses, the characteristics of farms and their users/operators are also important. Even more important are the decisions of farmers/farm users related to the implementation of certain development strategies. It can be stated that farms in counties with similar resources of utilized agricultural land are also similar in terms of farm area but they can, nevertheless, implement various development strategies and attain different levels of profitability.

What are the relationships between profitability, farm area, and implemented strategies in territorial cross-section? This is illustrated by pictures 10.6.1–10.6.4, which present the distribution of remainders from level-2 for the null model that can be interpreted as means/averages, respectively, for: value of farm incomes (income logarithm), farm areas, indicator of commodity strategy implementation, as well as the indicator of sustainable development strategy implementation. The highest mean values are illustrated with the most intense colors and the lowest with the least intense colors.

Comparisons made between the first two maps (pictures 10.6.1 and 10.6.2), which present the distribution of average incomes and average farm areas, indicate quite clearly that these distributions do not coincide. In other words, the counties where farms bring in the highest incomes do not always have the largest resources of utilized agricultural land areas, as measured by the average areas of actual farms in the county. At the same time, it should be noted that counties where farms generate the lowest incomes are often those with the lowest resources of utilized agricultural areas. Moreover, the distribution of counties according to the level of acquired incomes is relatively normalized as the counties of various income levels

Picture 10.6.1. Distribution of average farm incomes in the counties. Number of surveyed counties N = 101

Source: Own elaboration.

Picture 10.6.2. Distribution of average farm areas in the counties. Number of counties N = 101

Source: Own elaboration.

Picture 10.6.3. Distribution of mean values for the indicator of commodity strategy implementation in the counties. Number of counties N = 101

Source: Own elaboration.

Picture 10.6.4. Distribution of mean values for the indicator of sustainable development strategy implementation in the counties. Number of counties N = 101

Source: Own elaboration.

are located in various parts of Poland, with the exception of those counties with the lowest incomes that are concentrated in southeastern Poland, where counties with the most limited resources of useable agricultural land prevail. However, such counties are also predominant in Central Poland, as indicated by Picture 10.6.2, but their incomes are still, relatively, higher. The counties where farm incomes are the highest can be found in Kuyavian-Pomerania Province (counties: Rypiński and Wąbrzeski), in Lublin Province (counties: Lubelski and Bialski), in Łódzkie Province (counties: Sieradzki and Poddębicki), in Holy Cross Province (Sandomierski County), in Greater Poland Province (counties: Gnieźnieński, Czarnkowsko--Trzcianecki, and Szamotulski), in Western Pomerania Province (Gryficki County), in Podlaskie Province (counties: Łomżyński, Kolneński, and Suwalski), and in Mazovian Province (counties: Sokołowski, Ostrowski, Ostrołęcki, and Grójecki). The highest resources of usable agricultural area can be found in the following counties of Northern and Western Poland: Wąbrzeski, Ostrołęcki, Łomżyński, Szamotulski, Gryficki, Sulęciński, and Żarski (Lubuskie Province), Nyski (Opolskie Province), Bartoszycki and Olecki (Warmian-Masurian Province), as well as Koszaliński (Western Pomerania).

How do the development strategies implemented by farms correspond with the above distribution of data? It turns out that an orientation towards commodity strategy, with its attendant characteristics such as increasing the farm area by buying or renting more land, a high level of mechanization, searching for contractors outside of local markets, and use of lines of credit and loans might not be determined by farm area, although this strategy is mostly implemented in counties where there are farms of at least a medium land area. Analyzing the territorial distribution of counties with a relatively high level of implementation of this strategy, it seems that having a relatively close proximity to large metropolitan areas is an important factor here. For example, the commodity strategy is quite often implemented on the farms of Oleśnicki County near Wrocław, Lubelski County surrounding the City of Lublin, in Poddębicki County (near the City of Łódź) and in Sieradzki County (surrounding the town of Sieradz, and relatively close to the City of Łódź), in the counties of Upper Silesia (Gliwicki County, near the town of Jastrzębie-Zdrój), near the City of Gdańsk, and in Koszaliński County, surrounding the town of Koszalin (Picture 10.6.3). Although the implementation of this strategy usually ensures medium level incomes, there are some cases of incomes at the highest level, as in the counties: Lubelski, Sieradzki, Szamotulski, or Ostrowski. Somewhat different is the distribution of counties where farms/farm operators prefer the strategy of sustainable development, as characterized by mixed production of plants and animals, a large part of plant production devoted to the needs of farm animals, and participation by a larger number of family members in agricultural activities. This type of strategy prevails markedly in counties that are the smallest in terms of area and report the lowest incomes, and they are mostly located in Southeastern Poland (Picture 10.6.4). It does not mean, however, that the farms following this strategy cannot have incomes at the highest level, as indicated by the situation of farms in the following counties: Wąbrzeski, Ostrołęcki, Ostrowski, Sokołowski, Kolneński, Łomżyński, or Suwalski. It seems that the useable farm area resources of the county play an important role here, especially when it comes to the area of land that a farmer can utilize for agricultural purposes. It can be therefore maintained that the profitability connected with the implementation of sustainable development strategy is greater with greater farm area.

To conclude, it should be emphasized once again that the differentiation of farm incomes between counties is of a multilevel character. The local context is particularly important here, in terms of resources of useable agricultural land, economic conditions such as unemployment level, geographic location, and distance from metropolitan areas, in particular. The farm characteristics or, even more, the characteristics of farm users/operators, who make decisions that have consequences in the given development strategy's implementation, are even more important here. In our analyses, we have focused on the strength of the relationship between three variables: a) farm profitability, b) resources of usable agricultural land in the counties, as well as c) strategies implemented on farms, trying to show how diverse the map of Poland is in these contexts. Such conclusions can only be made with the awareness that the economic situation of farms in a territorial cross-section is also determined by many other factors that have not been addressed in this chapter.

REFERENCES

Domański, H. & Pokropek, A. (2011). *Podziały terytorialne, globalizacja a nierówności społeczne. Wprowadzenie do modeli wielopoziomowych* [Territorial divisions, globalization and social inequalities. Introduction to multilevel models]. Wydawnictwo Instytutu Filozofii i Socjologii PAN.

Hox, J.J. (2010). *Multilevel Analysis Techniques and Applications*. Routledge.

Łaguna, M. (2018). Wprowadzenie do wielopoziomowej analizy danych [Introduction to multilevel data analysis]. *Polskie Forum Psychologiczne* 2(23), pp. 377–394.

Nezlek, J.B. (2012). Multilevel modeling for psychologists. In: H. Cooper, P.M. Camic, D.L. Long, A.T. Panter, D. Rindskopf, & K.J. Sher (eds.), *APA Handbook of Research Methods in Psychology*, Vol. 3: *Data Analysis and Research Publication*, pp. 219–241. American Psychological Association.

Radkiewicz, P. & Zieliński, M.W. (2010). Hierarchiczne modele liniowe. Co nam dają i kiedy warto je stosować [Hierarchical linear models: What do they give us and when is it worth using them?]. *Psychologia Społeczna* 2–3(14), pp. 217–233.

Raudenbush, S.W., Bryk, S.A., Cheong, Y.F., Congdon, R.T., Jr., & Toit, M. du (2011). *HLM 7: Hierarchical Linear and Nonlinear Modelling.* SSI Scientific Software International.

Think Locally, Act Globally: Polish farmers in the global era of sustainability and resilience, ed. by Krzysztof Gorlach and Zbigniew Drąg in collaboration with Anna Jastrzębiec-Witowska and David Ritter
Jagiellonian University Press, Kraków 2021, pp. 459–518
ISBN 978-83-233-4949-5
DOI: http://dxdoi.org/10.4467/K7195.199/20.20.12738

Chapter Eleven: Daily Life Strategies among Farming Families in Poland of 2017

Zbigniew Drąg ⓘ https://orcid.org/0000-0002-9106-7758/

11.1. Introductory Remarks

The analyses presented in the previous chapters dealt with various aspects of functioning of family farms, their economic and social condition, as well financial situation, all of which affected their development strategies. The farmers, as a social category, were also the subject of the analyses and attention was given to their socio-demographic characteristics, as well as the professional and social roles they fulfilled. Numerous analyses addressed various aspects of the functioning of the farms, including their material, social, and demographic situation. In this chapter, attention will be given to farmers and their families in a different context, namely, the context of their everyday life and its conditions in the contemporary globalized world. Both of these terms, quality of life and globalization, are central here and could be the subject of interest of representatives of numerous academic disciplines. Globalization, although it prompts great interest from economists, also attracts the attention of lawyers, political scientists, pedagogy experts, psychologists and, especially, sociologists (see, e.g., Beck, 1992, 2000; Bauman, 1998, 2000, 2005, 2007; Stiglitz, 2002; Domański&Pokropek, 2011; Chimiak & Fronia, 2012; Wasilewski, 2014; Walas--Trębacz, 2017). Globalization involves processes such as:

> . . . flow of people, capital, commodities and information, which as a consequence in-
> creases dependencies between states and economies and leads to the emergence of cul-
> ture that goes beyond national cultures (Domański, 2011, p. 10).

These processes have concrete social consequences for family farms, as well as the members of farm households, and the way they reach farms and affect their functioning is presented in an earlier chapter of this publication. As stated by Polish sociologist Jacek Wasilewski,

> . . . phenomena and social processes occur that we, as individuals and members of so-
> cial groups, have no control over them. They are imposed on us, happen outside of the
> realm of possibility of our influence but—whether we want it or not—impact our lives,
> force changes, get us out of the routine, put us in new situations and generally: demand
> our reactions, whether adapting to changing external conditions or fighting with them.
> Generally, we don't like when things go that way. We like stabilization, predictability,
> repetition and order (Wasilewski, 2014, p. 7).

It can therefore be stated that globalization processes reorganize all aspects of our everyday lives, affect the objective conditions of our lives, thus modifying our present actions, as well as psychological states, which determine the quality of our lives. The question worth asking is whether this means a crisis of values, convictions, habits, and ways of thinking and doing things that were highly regarded until now. If there is such a crisis, could it bring about a reduction in the quality of life?

The term "quality of life," similar to "globalization," is frequently used by economists, sociologists, psychologists, and educators, as well as representatives of the medical sciences, and they define it in a variety of ways (see Nussbaum & Sen, 1992; Felce & Perry, 1995; Czapiński & Panek 2011, 2013, 2015; Trzebiatowski, 2011; and Włodarczyk, 2015). Polish sociologist Jakub Trzebiatowski, reviewing the quality of life approaches in social sciences, distinguished four categories of definitions (2011, pp. 26–29). "Existential" definitions refer, to a certain degree, to the "to be or to have?" opposition. The quality of life is reflected in the "to be" motivation and not "to have" motivation. The desire to have, to possess is juxtaposed here to the cornucopia of personal experience, emotions, activities, acquiring knowledge, as well as participation in social life. The quality of life is identified here with "the quality of individual," as an active social subject, who is not indifferent and can be compassionate. In the definitions concentrating on "development" and "life," the quality of life is—generally speaking—linked with the ability to play the allotted and accepted social roles in the public and private realm. The third group of definitions refers to the category of individual needs (both objective and subjective) and accentuates the connection between the quality of life and the level of fulfillment of human needs. Finally, the fourth category of definitions is characterized by recognition of objective and subjective indicators of the quality of life. The objective aspects of the quality of life are seen in the level of fulfilling of people's needs, including material needs as well as the needs related to health, social activity, personal development, and leisure. The subjective dimension of the quality of life is determined by the psychological states that accompany the processes of the fulfillment of these needs. It should be noted however that multiple authors distinguish between the terms "standard of living" meaning life conditions and the term of "quality of life." As a consequence, the standard of living is understood as objective life situations (conditions of living) and the quality of life means its subjective individual evaluation. As stated by sociologists Janusz Czapiński and Tomasz Panek,

> . . . the division of social indicators into living conditions and the quality of life roughly corresponds to the distinction between the objective description of the living circumstances (conditions) and their psychological significance as expressed by the respondent's subjective assessment (the quality of life) (Czapiński & Panek, 2015, p. 14).

Taking note on the quality of life based on individual evaluations leads to subjective (perceived) concept of the quality of life with its main indicator being the level of satisfaction with various aspects of life (Trzebiatowski, 2011, p. 29; Włodarczyk, 2015, pp. 5–6). Such a concept of quality of life is also adopted in his chapter's reflection on farmers. The issue of farmers' quality of life and their psychological well-being will be discussed in the next subchapter. In addition to the universal indicators such as general satisfaction with life, references will be made to psychological aspects of well-being, such as one's sense of optimism, internal harmony, social support or sense of adaptation-related activism as well as perception of threats and challenges/requirements that stem from the already mentioned globalization processes. The following subchapter will attempt to analyse farmers' life strategies and lifestyles and will try to determine to what extent they affect farmers' quality of life. The central point of this reflection is the analysis of quality of life and farmers' life strategies with the consideration of territorial divisions and with the use of multilevel model. The crucial question is about the extent to which the quality of a farmers' life depends on their place of residence, and belonging to certain local community. It is particularly interesting if some characteristics of local communities might weaken or strengthen farmers' perceptions of threats and challenges coming from postmodern, globalized world. In other words, the question worth asking is whether local communities can be seen as some protective buffer zone for farmers, mitigating any upcoming (or perhaps already occurring?) confrontation with the new aspects of everyday life and, at the same time, determining the quality of life for individual farmers.

11.2. Indicators of Farmers' Quality of Life

As already mentioned, the level of satisfaction with life can be viewed as the most universal indicator of the quality of life, especially if a subjective category such as psychological well-being is being considered. While conducting the in-depth analysis of farmers' quality of life, other indicators measure their long-term psychological states, as those might also be related to their quality of life. Before presenting the related study results, it might be helpful to first take a look at the results of the study on the feelings of satisfaction with life in Polish society.

When, in 2014, CBOS (Centrum Badań Opinii Społecznych, Public Opinion Research Center) asked Poles the following question: "Are you generally content/satisfied with your whole life?" the responses revealed that 72% of Poles were content with their lives, 24% were more or less content and only 3% expressed discontentment (CBOS, 2015, p. 6). The same question was asked in 2019. Satisfaction/contentment was reported by 83% of Poles, with 24% being very content and 59% rather content. Only 15% of respondents were only somewhat content and 2% were discontent (CBOS, 2020, pp. 8–9). Successive increase in the quality of life of Poles in the recent years was confirmed in the results of other studies. The question "How do you perceive your entire life?", presented in the cyclical studies of "Social Diagnosis", in 2011 was met with 78.2% positive responses expressing contentment and satisfaction (delightful, pleasing, mostly satisfying). Responses "neither good nor bad" made up 16.0% of the total, and answers indicating discontentment/dissatisfaction from life (mostly dissatisfying, unhappy, terrible) constituted 5.8% (Czapiński & Panek, 2015, p. 115). In 2015 life satisfaction was reported by 81.5% of respondents and dissatisfaction by 4.8% of them. It should be noted that 13.7% of respondents expressed a neutral attitude with responses such as "neither good nor bad" given to the question on respondents' life perception (Czapiński & Panek, 2015, p. 193).

How to place farmers in the context of general level of social content/satisfaction in Poland? To measure farmers' level of satisfaction, the respondents were asked to evaluate the accuracy of the statement: "Generally, I am content/satisfied with my life" on the 1–7 scale, with 1 meaning complete rejection of the statement illustrated by "It does not apply to me" answer and 7 meaning the full approval expressed by "It fully applies to me" answer. There is also the important in-between opinion with the allotted number 4 indicating that a statement does not apply to the respondent. The distribution of answers to this question, which was evaluated on the 7-point scale, can be seen in Table 11.2.1. This question, as well as the others also evaluated on the 7-point scale, are presented here with the permission of Jacek Wasilewski, who is the author of the study (see Wasilewski, 2014).

Table 11.2.1. Farmers' level of life contentment, optimism and sense of internal harmony in 2017. Scale 1–7, where 1 is the lowest value, 4 is an average value, 7 is the highest value. Distribution in %. N = 3551

Statement (indicator)	Categories of answers		
	Approval (sum of answers 5–7)	Neither yes nor no (answer 4)	Disapproval (sum of answers 1–3)
Generally speaking I am content/satisfied with my life (life contentment, satisfaction from life)	79.7	14.2	6.1
I am looking towards future with optimism (optimism)	74.6	16.7	8.7
Presently, it is easier for me to live in accordance to my moral values (internal harmony)	60.3	26.5	13.2

Source: Own work.

The data in Table 11.2.1 indicate that when it comes to level of general satisfaction with life, farmers generally do not differ from general Polish population. If one was to accept the location of answers between the points 5 and 7 on the already mentioned 7-point scale as an indicator of satisfaction than as many 79.7% of farmers were satisfied with their lives and 14.2% were partially satisfied. The dissatisfied made 6.1% of all respondents. The sense of optimism was also relatively high among farmers and it appeared to be a good indicator of psychological well-being (Table 11.2.1). As many as 74.6% of farm users viewed their future with optimism, 16.7% displayed an ambivalent attitude, and 8.7% of respondents admitted having a pessimistic outlook on their future. This could be interpreted as farmers being more optimistic about their future than Polish society in general. As indicated by the CBOS studies conducted in 2019, 51% of Poles were satisfied with their future life perspectives, 33% were more or less satisfied, and 9% were dissatisfied (CBOS, 2020, p. 2).

The sense of psychological well-being among farmers sketched in this chapter so far as having its ground in the general level of satisfaction with life and level of optimism regarding the future might be a bit weakened by the level of internal harmony measured by the reaction to the statement: "Presently, it is easier for me to live in accordance with my moral values." Although the respondents agreeing with the above statement constituted the prevailing majority of 60.3%, the percentage of those with an ambivalent attitude was also quite significant, reaching 26.5%. Respondents who disagreed on how easy it was to live nowadays in accordance to moral values, which could be seen as indicative of internal disharmony, made up 13.2% of the sample (Table 11.2.1). Perhaps the above data should be interpreted as a sign of moral dilemmas experienced by a certain percentage of farmers but not to the extent to affect the core elements of their value systems and—at least for some farmers—to significantly weaken the level of their psychological well-being.

Conviction about one's ability to deal individually with every new situation might be considered one of the crucial factors determining the quality of life (level of psychological well-being). Such self-confidence, trust in one's own adaptation abilities, will be defined here as adaptation-related activism, which appears to be strongly linked to satisfaction with life, optimism, and internal harmony. The respondents were asked to give their attitude towards the following statement (on the already described 7-point scale): "I consider myself to be a person, who, when facing new situations can actively search for as much information as possible." The distribution of the collected answers is presented in Table 11.2.2.

Table 11.2.2. Level of adaptation-related activism and sense of social support for farmers in 2017. Scale 1–7, where 1 is the lowest value, 4 is an average value, 7 is the highest value. Distribution in %. N = 3551

Statement (indicator)	Category of answers		
	Approval (sum of answers 5–7)	Neither yes nor no (answer 4)	Disapproval (sum of answers 1–3)
I consider myself to be a person, who, when facing new situations, can actively search for as much information as possible (adaptation-related activism)	58.0	20.9	21.1
There are people who will help me when I am in need (social support)	88.7	7.4	3.9

Source: Own work.

In the study, the sense of adaptation-related activism among farmers reached similar values as their experience of internal harmony. As many as 58% of farmers thought of themselves as individuals who, in new situations could search for as much information as possible, but 21.1% of respondents did not share such self-perception and another 20.9% only partially identified themselves as information seekers when confronting new situations. It can be inferred that, for a significant number of farmers, a low level of adaptation abilities to new situations or even a lack thereof does not necessarily weaken their level of satisfaction with life or optimism and might not even have a strong link with their sense of satisfaction or optimism. Perhaps the importance of this factor for the sense of psychological well-being is limited due to a sense of social support which is very common among farmers. The statement: "There are people who will help me when I am in need" was met with a positive reaction from 88.7% of farm users. Only 7.4% of respondents had an ambivalent attitude to that statement, while another 3.9% displayed a negative attitude (Table 11.2.2).

Assuming that all characteristics presented above affect the quality of life (psychological well-being), they should correlate with each other in a significant way. Is that reflected in the research? Data presented in Table 11.2.3. have confirmed that. Similar results were also reported in the aforementioned study by Jacek Wasilewski, but the characteristics of psychological well-being analyzed in his study were constructed differently than those in the 2017 study on farm users (see Wasilewski, 2014, pp. 56–59). The values of correlation coefficients fell into the 0.187–0.559 interval, which meant they were rather high. As predicted, the correlation was the strongest between satisfaction with life and optimism (0.559) and then between the satisfaction with life and social support (0.448). Optimism and social support were also correlated to a significant degree and the value of correlation coefficient was 0.470. The weakest mutual relation could be noted between internal harmony and the adaptation-related activism (0.187).

Table 11.2.3. Pearson's correlation between the satisfaction with life, optimism, sense of internal harmony, adaptation-related activism and social support (all variables are measured in the 7-point scale)*

	Satisfaction with life	Optimism	Internal harmony	Adaptation-related activism	Social support
Satisfaction with life	1	0.559	0.283	0.300	0.448
Optimism		1	0.315	0.348	0.470
Internal harmony			1	0.187	0.235
Adaptation-related activism				1	0.272
Social support					1

* All the correlation coefficients are significant at the 0.01 level.

Source: Own work.

The five characteristics of psychological well-being analyzed above can be seen as universal indicators in the sense that, in essence, they are not specifically connected with current globalization processes. In other words, globalization does not constitute satisfaction with life, optimism, or social support. People in various times of human history, including the times before globalization, were either more or less satisfied with their lives than the people of the 21st century and could be equally optimistic or pessimistic about their future. Even before globalization, people had good or poor capabilities to deal with new situations and their sense of living according to their own moral values was sometimes strong and sometimes weak. Undoubtedly, the psychological well-being of contemporary people—as described by these universal characteristics—is shaped to a greater or lesser degree by new situations, or phenomena, which directly stem from globalization processes and may appear to individuals as threats or challenges. Their analysis will be presented in the following subchapter. Before that, however, these five universal indicators of quality of life should be examined again but this time within the context of territorial divisions. Importance of territorial divisions for the quality of life of farm users is naturally a central point of analyses for this entire chapter. Here, this issue is presented through the five discussed characteristics, such as level of satisfaction with life, optimism, internal harmony, as well as adaptation-related activism, and social support according to farmers' region of residence. Average values (with standard deviation) for these five characteristics in particular regions are presented in tables 11.2.4 and 11.2.5.

Table 11.2.4. Averages and standard deviations for the scales of satisfaction with life, optimism, and internal harmony according to farmers' region of residence. Scale 1–7, where 1 is the lowest value, 4 is average value, and 7 is the highest value

No.	Region	Life contentment scale*	Optimism scale**	Internal harmony scale*
1	Central	5.6 (1.2)	5.3 (1.2)	4.9 (1.4)
2	Southern	5.4 (1.1)	5.3 (1.2)	5.0 (1.3)
3	Eastern	5.4 (1.3)	5.3 (1.4)	5.0 (1.4)
4	Northwestern	5.6 (1.2)	5.4 (1.4)	4.8 (1.6)
5	Southwestern	5.4 (1.4)	5.2 (1.3)	4.7 (1.7)
6	Northern	5.2 (1.3)	5.1 (1.3)	4.6 (1.6)
Total		5.5 (1.3)	5.3 (1.3)	4.9 (1.5)

* $p < 0.01$, ** $p < 0.05$.

Source: Own research.

As indicated by the data, the differences between the average values for particular regions were statistically significant for all five scales, although the weakest significance was noted on the scale of optimism. This meant that the level of optimism about one's future was the characteristic that was the least differentiating among the regions. It is worth noting that the highest and the lowest average values on particular scales were noted in various regions. Northwestern Region reached the highest average values on all four scales: satisfaction with life, optimism, adaptation-related activism and social support and the region did not have aby lowest average value on any scale. Central Region was close to such situation as the average values on the three above scales were the highest there. The average value of the optimism scale was not as high as in the Northwestern Region but it did not fall to the lowest value. On two scales such as internal harmony and social support the highest values on the scales were reported for the Southern and Eastern Region. It should be added here that the Southern Region did not fall into the lowest average value on any scale and the Eastern Region was the one with the lowest value on the scale of adaptation-related activism. Southwestern Region and Northern Region did not reach the highest average value on any scales. The former fell into the lowest average value on the scale for social support and the latter had the lowest average values on as many as three scales: satisfaction with life, optimism and internal harmony. As illustrated by the data the psychological well-being of farmers in particular regions with consideration of five characteristics was quite diverse. This could justify the use of multilevel analysis in studying of farmers' quality of life. Such analysis will be conducted later in this chapter and it will feature divisions by counties.

Table 11.2.5. Averages and standard deviations for the scales of adaptation-related activism and social support according to farmers' region of residence. Scale 1–7, where 1 is the lowest value, 4 is average value, and 7 is the highest value

No.	Region	Scale of adaptation-related activism*	Scale of social support*
1	Central	4.9 (1.6)	5.9 (1.1)
2	Southern	4.6 (1.5)	5.9 (1.0)
3	Eastern	4.5 (1.7)	5.9 (1.2)
4	Northwestern	4.9 (1.6)	5.9 (1.2)
5	Southwestern	4.6 (1.8)	5.6 (1.4)
6	Northern	4.6 (1.5)	5.7 (1.1)
Total		4.6 (1.6)	5.9 (1.3)

* p < 0.01.

Source: Own research.

11.3. Quality of Life Determinant: Threats and global challenges

The analyses of Jacek Wasilewski on the awareness of globalization changes among non-metropolitan respondents of the young and middle age (16–43 years) indicated that farmers made the social category that perceived threats and challenges stemming from ongoing globalization processes in the lowest degree (Wasilewski, 2014). Could this also be reflected in the study featured specifically in this publication? Referring only to some indicators of these threats and challenges, these issues will be analysed for the whole category of farmers without the introduction of the age criterion. It should be added here that the awareness of globalization changes was a secondary issue in the 2017 study, so crucial for this publication and therefore not all indicators used in the Wasilewski study were used in the study and could be described here.

Table 11.3.1. Sense of global threats experienced by farmers surveyed in 2017 in the 5 years prior to the study. Scale 1–7, where 1 means "it does not apply to me at all," 4—"it neither applies to me, nor it does not apply," and 7—"it fully applies to me." Distribution in %. N = 3551

Statement (indicator)	Category of answers			Average and standard deviation
	Disapproval (sum of answers 1–3)	Neither yes nor no (answer 4)	Approval (sum of answers 5–7)	
"It is now harder to plan a professional career" (work)	29.0	21.8	49.2	4.3 (1.9)
"Now it is more likely for me to lose control over my finances due to modern ways of using credit cards, easy access to credits, loans and purchasing good on instalment plans" (existence: finances)	49.2	21.1	29.7	3.4 (2.0)
"My everyday life is determined by customs and traditions of other cultures to a greater degree" (existence: culture)	67.0	16.7	16.3	2.7 (1.7)
"Nowadays, there are fewer clues on what is right and what is wrong" (morality: moral relativism)	19.8	31.5	48.7	4.5 (1.5)
"I am more and more worried where we are going as a country and as a society" (morality: direction of development)	15.5	19.4	65.1	4.9 (1.6)

Source: Own work.

For the part of study that dealt with threats and challenges the respondents were asked to consider the last five years and evaluate 11 statements on the 7-point scale, where 1 meant "does not apply at all," and 7 meant "fully applies." Five statements referred to threats and six to challenges and requirements attributed to the globalization processes.

The threats were identified in the following areas: work (statement: "It is now harder to plan a professional career"), existential area related to finances ("Now it is more likely for me to lose control over my finances due to modern ways of using credit cards, easy access to credits, loans and purchasing good on installment plans"), cultural area ("My everyday life is determined by customs and traditions of other cultures to a greater degree") and moral area (statements: "Nowadays, there are fewer clues on what is right and what is wrong" and "I am more and more worried where we are going as a country and as a society"). The level of respondents' sense of threat is presented in Table 11.3.1.

The most intensely perceived threats were the ones related to the area of morality. Next were the threats related to work and the perception of existential threats was the weakest. The average value on the scale of threats related to the morality and direction of the country and social development was 4.9 and on the scale of moral relativism, in the absence of clues of what was right and what was wrong, it was 4.5. The work-related threats regarding difficulties on planning the career had the average value of 4.3 on the respective scale. The average value on the scales of existential threats related to finances was located well below the mean value of 4 and it was 3.4. Even lower was the average value on the scale of threats of cultural character, which was calculated as 2.7. It might be assumed that the sense of threat could be decreasing with growing sense of individual "control" over various aspects of social reality of everyday life. In other words, the financial problems related to excessive debit/overdraft on the payment card might be seen as an individual experience that a respondent could and should react to. The same could be said about the necessity to make adjustments to some customs from outside of local culture in everyday life as such situations might require direct reaction of the individual. In both presented cases, the individual would most likely be able to maintain a certain level of "control" over the situation. The threats in the existential area might be verified empirically to the extent to which they are part of the individual experience in everyday life. A low level of existential threats among surveyed farmers could to be connected to the lack of problems with the use of debit and payment cards or acquired loans. Very limited exposure to the influence of foreign culture requiring farmers to modify existing cultural habits might also serve as an explanation of the low level of threat. Respondents' life experience seemed to be reflected in how they perceived work-related threats measured here by difficulties in career planning and uncertainty about the future of the farm. The extent of farmers' "control" of the situation, where there was strong market competition and lots of regulations stemming from the implementation of agricultural policy, appeared to be limited. Therefore, the experience of farm situation being dependent on external factors in everyday functioning could potentially determine the perception of threats about the professional future of the farmer and the farm, as well. Growing dependence could bring about an increased sense of work-related threats. Assuming that one could ask, how to explain respondents' perceiving threats in the moral area so intensely? The reason for this might be the shared feeling that this area is totally beyond their direct "control." Observing the global trends in the realm of morality, farmers nowadays might lose their certainty about the stability of norms and values and at the same time they are aware of not having any meaningful influence on the process of redefining what is right and what is wrong. They could also feel uncertain and lacking any impact on the general direction of development followed by country and society. It might be helpful here to juxtapose the results of the 2017 study on farm users with the results of the study conducted by Jacek Wasilewski, which indicated that farmers were less concerned about the threats related to globalization processes than the general population of nonmetropolitan areas. Considering the average values on the scale of the threats discussed here, the analysis of the 2017 study results confirmed the findings from the Wasilewski study only in regards to existential and work-related threats. The situation was quite different with the threats related to morality. The farmers surveyed in the 2017 study scored higher on the scales of threats for moral relativism and morality issues compromised by direction of social development than the farmers or even general population of nonmetropolitan residents in the study conducted by Jacek Wasilewski (2014, pp. 39, 107–109). This could possibly be an effect of the eight year time gap, as Wasilewski conducted his study in 2009. In the following years, certain conflicts of a normative character were externalized in Poland, including those connected with the refugee crisis initiated in 2015, or the rights of sexual minorities. It appears that the age of respondents could be meaningful here. Attachment to well-established norms and values is more of a domain of older people rather than the young and therefore emergence of processes leading to changes to the existing normative order could be perceived as more of a threat by older respondents. The 2009 study was conducted by Jacek Wasilewski with a sample of relatively young people (ages 16–43), while the average age of farmers participating in the 2017 study was close to 51.

While the threats have negative connotations, the challenges (requirements) are not interpreted in such an unambiguous manner. For some people they might

awaken negative feelings and emotions due to perception of challenges and requirements forcing individuals to take actions, in which they would not like to engage
otherwise, in different situation and having the freedom of choice. For others, some
challenges might be positively motivating to take actions, especially if these individuals view the results of these actions as valuable. In this chapter the challenges
selected for analysis applied to the area of work and existential matters. Work-related
challenges considered here were connected with the requirement of independence
and autonomy and the need to improve professional qualifications. The former were
measured by the reactions to the statement: "My work now requires me to be more
independent and autonomous" and the latter by opinions on the statement: "Currently, I must spend more of my free time on continuous education and improving
my qualifications." In the existential area challenges of cultural character were considered ("Currently I am required to be more accepting of other people's lifestyles"),
as well as educational challenges ("Currently, I need to expand my horizons to
keep up with technical novelties of everyday life"), financial challenges ("I need
to make sure I have enough financial means to secure my needs in old age"), and
those related to physical appearance and fitness ("I must pay attention to my looks
and my figure/shape to be noticed by others"). Table 11.3.2 presents farmers' level of
identification of challenges.

Generally, increasing number of challenges and demands that contemporary
individuals might have to face in various areas of their lives got noted by farmers but
not all of them were given equal meaning or importance. So, in the realm of work,
respondents accentuated the requirement for more independence and autonomy,
which could probably be a consequence of the necessity for making independent production decisions based on the autonomous and individual analysis of the situation
on the agricultural market. Independence and autonomy appeared more important
than the demand to increase the qualifications, which could indicate that the respondents perceived their already acquired knowledge on farm management as adequate and sufficient in current socio-economic conditions. It should be mentioned
here the increasing demand for further education was the one that the respondents
paid the least attention to with a 3.8 average on the 7-point scale. There was only
one existential challenge/demand that was perceived as more meaningful than the
requirement for independence and autonomy, with an average of 4.9 on the appropriate scale and it was the necessity to take action ensuring the fulfillment of needs
and financial security in the old age (average 5.2). Could this indicate that more
and more farmers rejected the model of the retired farmer, leading a modest life
thanks to a low pension from the Agricultural Social Insurance Fund (KRUS, Kasa
Rolniczego Ubezpieczenia Społecznego)? This seems quite accurate. The material

and social aspirations of farmers might be indirectly inferred here as well as the refusal to accept the image of the traditional peasant and the preference for a vision of a farmer as a modern entrepreneur, representative of the middle class who, thanks to their own professional abilities, can secure for themselves a decent standard of living in old age. To some extent, this image of a farmer as a modern entrepreneur is also reflected in the reaction to the opinion on the necessity of expanding knowledge and horizons as a condition to keep up with technical novelties of everyday life with the average on the appropriate scale being 4.4. At the same time, awareness about the demands of modern times did not go hand in hand with attention being paid to looks and body shape. The average value on the appropriate scale was 3.8. Surveyed farmers were also quite moderate in their perception of the necessity to be more accepting of the lifestyles of other people, with the average on the scale being 4.0. These study results provide evidence that, in 2017, Polish rural areas still remained culturally homogenous and farmers rarely experienced direct contacts with representatives of different cultures and lifestyles in their everyday lives.

Table 11.3.2. Sense of global challenges perceived by farmers surveyed in 2017. Scale 1–7, where 1 means "it does not apply to me at all," 4—"it neither applies to me nor it does not apply," and 7—"it fully applies to me." Distribution in %. N = 3551

Statement (indicator)	Category of answers			Average and standard deviation
	Disapproval (sum of answers 1–3)	Neither yes nor no (answer 4)	Approval (sum of answers 5–7)	
My work now requires me to be more independent and autonomous (work: independence and autonomy)	17.8	17.9	64.3	4.9 (1.7)
Currently, I must spend more of my free time on continuous education and improving my qualifications (work: improving qualifications)	50.6	19.4	30.0	3.4 (1.8)
Currently I am required to be more accepting of other people's lifestyles (existential: other cultures)	33.4	27.2	39.4	4.0 (1.7)
Currently, I need to expand my horizons to keep up with technical novelties of everyday life (existential: expanding knowledge, educational)	28.1	20.4	51.5	4.4 (1.8)
I need to make sure I have enough financial means to secure my needs in the old age (existential: financial security)	13.7	12.6	73.7	5.2 (1.6)
I must pay attention to my looks and my figure/shape to be notices by others (existential: physical appearance)	41.0	24.3	34.7	3.8 (1.8)

Source: Own work.

In general, the global challenges described above, similar to the threats, were felt and experienced by farmers to a lesser degree than by the general population of Polish nonmetropolitan areas (see Wasilewski, 2014, pp. 45, 103). However, there was one exception. Farmers, more than the general nonmetropolitan population, felt compelled to take action to secure their financial situation in the old age. Not surprisingly, the age structure of both studied populations is of major importance. It should be remembered here that the surveyed farmers were significantly older than the nonmetropolitan residents in the study conducted by Wasilewski and, therefore, the challenge to secure financial means for old age was more crucial to them than to younger people. This is not to negate that farmers could also be influenced by growing aspirations to have a decent or even comfortable life.

The general observation about farmers' perception of threats and global challenges being less intense than the experience of other social categories might lead to the conclusion that globalization processes still reach the Polish countryside with a delay, in the form of a general social message that is open to interpretation and perceived in a rather fragmentary manner. Could these processes be having a gentler impact on the psychological well-being (quality of life) of farmers than on other social groups? The analyses of Jacek Wasilewski indicated a rather negative correlation between the feeling of growing threats and the characteristic of psychological well-being in the nonmetropolitan population (see Wasilewski, 2014, p. 59). Could a similar negative correlation apply to farmers as well? Table 11.3.3 presents the data reflecting the correlation between the analyzed type of threats and the quality of life characteristics.

The results of these analyses were quite surprising. It turned out that only one type of threat, namely the one connected with the cultural area, was negatively correlated with characteristics of psychological well-being. However, although the relation between the conviction about the increased influence of customs and traditions of foreign culture on everyday life was accompanied by a lower level of all life quality indicators, in only two cases were such relations statistically significant. For social support the correlation coefficient had a value of negative 0.127 and for satisfaction with life it was negative 0.120. Even for these characteristics, the correlations were not very strong. The same could be said about the other correlations for other types of threats, but they nevertheless were positive correlations. A correlation coefficient above 0.200 was noted only in four cases and three of them had to do with adaptation-related activism and one with internal harmony. The adaptation-related activism was positively correlated with:

1) work-related threats measured by the difficulty in planning a professional career (0.261),

2) threats in the area of finances, measured by fears of losing control over one's own finances (0.201), and

3) threats in the moral realm measured by concerns about the direction of the development of the country and society (0.221).

Table 11.3.3. Pearson's correlation between global threats and the factors affecting quality of life: satisfaction with life, optimism, sense of internal harmony, adaptation-related activism and social support (all variables were measured on the 7-point scale)

Global threats	Quality of life characteristics (psychological well-being)				
	Satisfaction with life	Optimism	Adaptation--related activism	Social support	Internal harmony
Work-related threats: difficulties in planning professional career	0.026	0.058*	0.261*	0.076*	0.083*
Existential threats in the area of finances: loss of control over finances	0.022	0.087*	0.201*	−0.009	0.045**
Existential threats in the cultural realm: influence of foreign culture	−0.120*	−0.018	−0.014	−0.127*	−0.004
Threats in the moral realm: moral relativism	0.123*	0.129*	0.167*	0.152*	0.163*
Threats in the moral realm: concerns about the direction of the country's development and social development	0.150*	0.151*	0.221*	0.173*	0.248*

* $p < 0.01$.

Source: Own work.

The correlation coefficient between internal harmony and threats in the moral realm connected with the direction of country and societal development was 0.248. Now, the interpretation of these data requires prior analysis of data on the correlations between the characteristics of psychological well-being and various types of global challenges, which Table 11.3.4 presents.

As seen in Table 11.3.4, all but one correlation were statistically significant and positive. For a large number of cases these correlations were stronger than that between the characteristics of psychological well-being and threats stemming from globalization. The highest value of the correlation coefficient was 0.539 and it presented the relation between the requirement to expand the knowledge on technical novelties and adaptation-related activism. Adaptation-related activism also had the strongest correlations with all other types of challenges. The correlation coefficients for this characteristics with various types of challenges were the following: 0.373 for work related challenges of independence and autonomy, 0.381 for the work-related requirement to improve one's qualifications. The correlation coefficient exceeded the value of 0.300 also for the requirement of acceptance of lifestyles of other people

(0.332). It exceeded the value of 0.250 for the challenge of securing financial means in old age (0.259) and the necessity to take care of one's physical appearance, including one's shape/figure (0.256). Relations between the analyzed challenges and the remaining characteristics of psychological well-being were relatively weaker but in some cases the correlation coefficient still exceeded the value of 0.200. Correlations at such a level could be noted between the requirement for independence and autonomy at work and three characteristics of psychological well-being, such as internal harmony (0.325), optimism (0.268) and satisfaction with life (0.250). Correlations well above 0.250 were also noted between the requirement to expand one's knowledge on technical novelties and optimism (0.315) and also between that same requirement and satisfaction with life (0.273).

Table 11.3.4. Pearson's correlation between global challenges and the factors affecting quality of life: satisfaction with life, optimism, sense of internal harmony, adaptation-related activism and social support (all variables were measured on the 7-point scale)

Global challenges	Quality of life characteristics (psychological well-being)				
	Satisfaction with life	Optimism	Adaptation--related activism	Social support	Internal harmony
Work-related challenges: greater independence and autonomy	0.250*	0.268*	0.373*	0.165*	0.325*
Work-related challenges: improving qualifications	0.119*	0.165*	0.381*	0.025	0.043**
Existential challenges: accepting of different lifestyles of other people	0.155*	0.186*	0.332*	0.113*	0.135*
Existential challenges: expanding the knowledge on technical novelties	0.273*	0.315*	0.539*	0.177*	0.191*
Existential challenges: securing financial means for old age	0.178*	0.179*	0.259*	0.191*	0.169*
Existential challenges: taking care of physical appearance (looks and body shape/figure)	0.130*	0.198*	0.256*	0.091*	0.152*

* $p < 0.01$, ** $p < 0.05$.

Source: Own work.

Here, it might be helpful to return to the analyses on the relations between the characteristics of farmers' psychological well-being and their perception of threats and challenges constituted by globalization processes, leaving an open question about the causes and results of these relations. In other words, one might ponder whether psychological well-being determines the perception of threats and challenges or vice versa. The interpretation of relation between the perceived increase of the influence of foreign cultures on everyday life and a reported lower level of psychological well-being seems quite explicit. The experience of everyday situations requires adjustment to new cultural customs or even customs previously negated. This might be perceived as an attack on the individual identity, potentially destroying the internal integrity or even constitute the threat to the identity of the community, where existing social order has been maintaining relatively smooth social relations so far. This makes the future uncertain and, as a consequence, could lower the level of satisfaction with life. But how should one interpret the situation where the increase of psychological well-being is accompanied by a growing sense of confrontation with challenges and threats? Acknowledging that not all types of threats and challenges were considered here it should also be recognized that the sense of their increase did not lower respondents' evaluation of their life quality. These threats and challenges were not the consequences of real everyday life experience with its real situations but they mostly likely remained in the area of imaginary notions, founded on rather general information coming from outside of local communities and therefore did not take a central place in their consciousness. Not having such an important position in farmers' minds, threats and challenges did not negatively affect farmers' psychological well-being. To some extent such an interpretation could be confirmed by the thesis of the backwardness of Polish rural areas in confrontation with globalization processes and their consequences. This should not be seen as a situation where psychological well-being was provided by "the world at the doorstep" (Gorlach, 2001), but perhaps the situation where this was granted by "the world at the doorstep of the local community." Or perhaps reverse interpretation would be more accurate. Farmers' knowledge on globalization processes could be seen as sufficient to enable them to recognize their actual consequences for them and their most immediate environment. It should be emphasized here that all but one challenge were most positively correlated with adaptation-related activism, which promoted the concept of the individual as someone who sought as much new information as possible in new situations. Perhaps better recognition of the essence of these phenomena and processes that are described here as threats or challenges brings about the conviction that at least some of them could be positive and might serve individual development (i.e., requirement for greater independence and autonomy at work, requirement to improve qualifications, or requirement to expand one's knowledge on technical novelties) and they do not necessarily have to turn the existing world upside down. Consequently, this could foster the feeling that "normal life" is still possible in new conditions and such a way of thinking might surely serve the preservation of psychological well-being. It appears that the character of the relationship between the quality of life and the perceived threats and global challenges could also be influenced by strategies of adaptation to the contemporary world implemented by the respondents. These issues will be presented in the next subchapter, while the current one will address the perception of threats and challenges according to territorial divisions. It is assumed that the region of respondents' residence might be a factor differentiating the level of perceived threats and challenges. Averages (with standard deviations) for the scales of threats in particular regions are presented in Table 11.3.5, and for the scale of challenges in Table 11.3.6.

Table 11.3.5. Average and standard deviation for scales of work-related, existential and morality--related threats. Scale 1–7, where 1 means "it does not apply to me at all," 4—"it neither applies to me nor does it not apply," and 7—"it fully applies to me." Distribution in %. N = 3551

| Region | Threats | | | | |
| | Work related: difficulties in planning professional career* | Existential | | Morality-related | |
		In cultural realm: influence of foreign cultures*	Financial: loss of control over own finances*	Moral relativism*	Concerns about the direction of country and societal development*
Central	4.2 (1.8)	2.4 (1.5)	3.4 (1.8)	4.4 (1.5)	5.1 (1.4)
Southern	4.7 (1.5)	3.2 (1.8)	4.0 (2.0)	4.9 (1.2)	4.7 (1.5)
Eastern	4.3 (2.0)	2.7 (1.7)	3.1 (2.0)	4.5 (1.6)	4.9 (1.6)
Northwestern	3.7 (1.9)	2.3 (1.5)	3.0 (1.9)	4.1 (1.6)	4.8 (1.7)
Southwestern	4.0 (1.9)	2.8 (1.8)	3.8 (1.9)	4.4 (1.8)	4.7 (1.7)
Northern	4.3 (1.7)	2.9 (1.6)	3.7 (1.9)	4.5 (1.6)	5.2 (1.4)
Total	**4.3 (1.9)**	**2.7 (1.7)**	**3.4 (2.0)**	**4.5 (1.5)**	**4.9 (1.6)**

* $p < 0.01$.

Source: Own work.

Table 11.3.6. Average and standard deviation for scales of work-related and existential challenges. Scale 1–7, where 1 means "it does not apply to me at all," 4—"it neither applies to me nor does it not apply," and 7—"it fully applies to me." Distribution in %. N = 3551

| Region | Challenges | | | | | |
| | Work-related | | Existential | | | |
	Greater independence and autonomy*	Improving qualifications*	Accepting different lifestyles of other people*	Taking care of one's looks and shape/ figure*	Expanding knowledge on technical novelties*	Financial security in the old age*
Central	5.0 (1.7)	3.7 (1.8)	4.0 (1.7)	3.9 (1.7)	4.7 (1.7)	5.3 (1.6)
Southern	4.9 (1.5)	3.4 (1.7)	4.2 (1.4)	3.9 (1.7)	4.3 (1.6)	5.4 (1.4)
Eastern	4.8 (1.8)	3.1 (1.9)	3.9 (1.7)	3.7 (1.8)	4.1 (1.8)	5.0 (1.7)
Northwestern	5.1 (1.8)	3.4 (1.9)	3.8 (1.8)	3.5 (1.9)	4.6 (1.8)	5.3 (1.7)
Southwestern	5.0 (1.7)	3.3 (1.9)	3.9 (1.8)	3.3 (1.9)	4.4 (1.9)	5.1 (1.9)
Northern	5.1 (1.5)	3.8 (1.7)	3.8 (1.5)	3.8 (1.7)	4.4 (1.7)	5.3 (1.7)
Total	**4.9 (1.7)**	**3.4 (1.8)**	**4.0 (1.7)**	**3.8 (1.8)**	**4.4 (1.8)**	**5.2 (1.6)**

* $p < 0.01$.

Source: Own work.

As indicated by the above data, the differences between the averages for particular regions were statistically significant on all scales of threats and all scales of challenges. The highest and the lowest average values were noted in various regions. However, five out of eleven highest average values were calculated for the respondents in the Southern Region. They were the ones who had the most intense awareness of threats related to difficulties in planning professional career and growing moral relativism as well as the challenges, including the necessity to accept the lifestyles of other people, securing financial means for the old age and taking good care of physical appearance with the attention being paid to shape/figure. Farmers of this region were the least concerned about the direction of the country and societal development. Farmers in Central and Northern Region had the highest average values on three scales. The former scored highest averages on the scale of threats related to the development of the country and a society and on the scales for challenges reflecting necessity to expand knowledge on technical novelties and the need to take care of physical appearance including shape/figure. The latter had the highest average values for the scale of threats related to the influence of other cultures on everyday life and the scales of work-related challenges involving increase of independence and autonomy, as well as improving professional qualifications. Furthermore, the respondents of the Northern Regions had the lowest average value on the scale of challenged related to accepting the lifestyle of other people and the studied farmers of Central Region had the lowest average value on the scale of threat related to the influence of foreign cultures on everyday life. Farmers of the three regions, namely Southern, Central and Northern Region appeared to be the most perceptive of global threats and challenges than their counterparts in other regions of Poland. Farmers of the Southwestern Region had the highest average value on just one scale and it was the scale of financial threat related to losing control over finances. The same farmers had the lowest values on two scales: the scale of threat about the direction of the country and societal development and the scale of challenge involving the requirement to look good and be in shape. The respondents residing in the Northwestern Region had the highest average on the scale of work-related challenges pertinent to independence and autonomy and the lowest averages on three scales; scale of threat of moral relativism, scale of work-related threat addressing the difficulties with planning professional career and the scale of challenge of accepting different lifestyles of other people. The table shows the respondents of the Eastern Region as the least concerned with threats and challenges brought by globalization. They did not reach the highest average value on any of the scales but had lowest average values on five scales. One of them was the scale of threat related to losing control over finances and the other four included the scales of work-related challenges pertinent to having more independence and autonomy, necessity to improve qualifications, as well as existential challenges of expanding the knowledge on technical novelties and securing financial means for old age. As seen in the last two tables the perception of global threats and challenges differed significantly in various regions just like the sense of well-being, which was addressed earlier in this chapter differed in various regions. This provides yet another premise to apply multilevel analysis to studies on farmers' quality of life.

11.4. Strategies for Control of Globalization Changes in the Postmodern World

All social changes usually have a negative impact on the psychological well-being of individuals and even social groups. To maintain a sense of well-being, individuals might engage in various activities to adapt to a new social situation but the success of these efforts depends on their adequacy. To examine the effectiveness of adaptation strategies it is usually enough to study concrete actions by concrete individuals. However, in the contemporary world, when

> . . . global and postmodern changes are fast, multidimensional and all-encompassing it would be difficult to examine how people react to them and how they deal with them simply by asking them about concrete remedial activities. Today, this would not be a right path, considering the extent of changes and the variety of potential reactions to these changes (Wasilewski, 2014, p. 134).

What should be examined instead of concrete individual activities? One of the proposals refers in general terms to psychological predispositions that prompt individuals to achieve a certain goal or to abandon it depending on the type of situation. These general predispositions designate various strategies of control and adaptation to global changes taken by individuals. They also set the ways of dealing with postmodern threats and challenges. There are three types of active strategies oriented at goal achievement and two types of avoidance strategies, used when the goal needs to be abandoned. Psychological theory of control, conceptualized as a motivational theory of development in the cycle of life provides a theoretical ground for such approach. This was also the perspective that Jacek Wasilewski used in his study on globalization changes and strategies of their control (see more on this subject: Wasilewski, 2014, pp. 133–174). This approach will be also applied here in the attempts made in this chapter to analyze the adaptation reactions of surveyed farmers to identified global threats and challenges. The references will be made to active strategies of controlling the change (orientation at goal achievement) and avoidance strategies (orientation at abandoning the goal). However, due to the character of the 2017 study, in which the issues related to awareness about global changes had secondary importance, only general models of active strategies and avoidance strategies will be analyzed here and their variations will not be addressed. Furthermore, unlike in other standard studies these models will be operationalized on the basis of one single indicator, instead of the typical set of indicators. Instead of analyzing particular strategies for various types of threats and challenges, there will be one general scale for threats and one for challenges. In other words, detailed analyses typical for standard studies will be omitted here and the main goal of this subchapter will be a general presentation of the issues pertinent to the ways in which Polish farmers try to execute control over changes related to globalization.

The question characterizing the active strategy was really the statement about adaptation-related activism. It was already mentioned in this chapter and read: "I consider myself to be a person, who, when facing new situations can actively search for as much information as possible." The question presenting the strategy of avoidance was actually the statement about emotional religiosity: "My religious faith helps me to endure the hardest moments." The respondents were asked to evaluate how these statements applied to them, taking into consideration the last 5 years and using the 7-point scale, where 1 meant "It does not apply to me at all," and 7 meant "it fully applies to me." Recognition of adaptation-related activism as an indicator of active strategy is based on the premise that knowledge about global threats and challenges is a crucial way of controlling them. A different situation is with religiosity experienced as a strategy of avoidance as it is based on the assumption that the changes cannot be controlled on the basis of rational factors and the awareness about the futility of attempts to gather adequate information about the changes. This can turn one to the transcendent factor with fortitude and with hope to receive protection against something that is new and hard to understand. It should also be noted that these two types of reactions are not necessarily contradictory and ones does not have to exclude the other. Individuals might even use them at the same time and not only in confrontation with different threats and challenges but even with particular ones. In the study the correlation between both these strategies was positive (0.150) and statistically significant (0.01). Average and standard deviation for adaptation-related activism were 4.6 and 1.6, respectively, and for emotional religiosity 5.1 and 1.6, respectively.

As mentioned before, in order to determine the extent to which the above strategies could be used by farmers in response to emerging threats and challenges, two scales were constructed. The first one was a general scale based on five types of threats already analyzed earlier in this chapter and the other was a scale of challenges based on 6 types of challenges, described earlier in this chapter, as well. Average and standard deviation on the already mentioned 7-point scale constructed for challenges were 4.35 and 1.14, respectively, and on the scale on threats: 4.05 and 0.97, respectively. Obviously, one should be aware of the limitations of these scales as they lumped together various types of threats and challenges, which—as presented in the earlier analyses in this chapter—were experienced by farmers with various degrees of intensity. However, the purpose of the analysis is not a detailed examination of strategies that farmers could possibly adopt in order to deal with particular threats and challenges, but only to provide a general understanding of the intensity of perceiving threats and challenges and the possible dominant adaptation reactions.

Assuming that for farmers adopting certain strategies of controlling the changes, certain types of activities connected with social communication and ways of acquiring information might be meaningful, especially in regards to the functioning of local communities, where global threats and challenges affected the respondents, the survey was conducted on the frequency of acquiring information from various sources. The respondents were given a set of 14 questions and asked to mark their answers on the 7-point scale, where 1 meant "never," 7 meant "at least several times a week" and a value of 4 indicated "more or less once a month" answer. Three dimensions of social communication and sources for the acquisition of information were identified as a result of applied factor loads analysis. They were:

1) communication oriented towards the Internet community, with acquiring information on the Internet,
2) communication oriented toward the local community, with acquiring information from various categories of residents and traditional sources such as local press, posters, and announcements,
3) communication with a close-knit circle of people, which involved gathering information from family, acquaintances and neighbors.

The questions, as well as calculated factor loads, are presented in Table 11.4.1. Although the selected types of communication and ways of acquiring communication were independent from each other, they had some elements in common. Communication oriented towards the Internet community and towards local community both acquired information from municipal websites and other portals, as well as local press. Communication oriented towards the local community and towards a close-knit community usually involved acquiring information from family members, acquaintances, and other residents of the area.

Table 11.4.1. Factor loads of the main components of farmers' social communication strategies (Varimax rotations, loads below 0.20 were excluded)*

Variable (question): How often do you...?	Components		
	I. Internet community	II. Local community	III. Close-knit community
Peruse local news on the Internet	0.814		
Acquire information from other local Internet portals	0.807	0.223	
Acquire information from municipal websites	0.800	0.246	
Use electronic mail	0.799		
Acquire information from Facebook or other social networks	0.693		
Acquire information from leaflets, posters, and announcements		0.708	
Acquire information from the head of the village		0.674	
Acquire information from priests		0.667	
Acquire information from other residents		0.621	0.343
Acquire information from local press	0.280	0.560	
Acquire information from family members and acquaintances		0.552	0.407
Host friends and acquaintances at home			0.748
Talk to neighbors in front of the house			0.676
Visit with family members			0.676
Percentage of explained variation	23.1	18.4	12.9
Total percentage of explained variation	54.4		

* All variables were measured on the 7-point scale, where 1 meant "never," 4 meant "less than once a month," and 7—"at least several times a week."

Source: Own research.

Analysis of adaptation strategy will start with the active strategy oriented towards goal achievement such as exercising control over the ongoing changes. This strategy will be operationalized by references to adaptation-related activism understood as individual disposition to actively seek information on new situations. What would be the determinants of such activities? Could these activities be prompted by the sense of global threats and challenges? To answer this question regression analysis was used with Table 11.4.2 presenting regression coefficients for three models. Model 1 only addressed the basic socio-demographic characteristics (farm area, professional activity outside of the farm, sex/gender, age and education) and two subjective variables related to the perception of changes and globalization challenges. Model 2 meant expanding of Model 1 with as many as four additional models of social communication and acquiring of information. Three of these models were designed on the basis of factor analysis and the fourth one could be described as the model of "communication oriented towards a spiritual community," operationalized as participation in church services (measured on 7-point scale, from 1—"never," to 7—"at least several times a week"). Model 3 involved adding more subjective variables, related to psychological well-being: internal harmony, social support, optimism, and satisfaction with life.

Table 11.4.2. Regression analysis for the scale of adaptation-related activism—active strategy. Standardized regression coefficients (β)

Variable	Value of the variable	Model 1 β	Model 2 β	Model 3 β
Farm size	number of ha	0.051***	0.039**	0.032*
Work outside of farm	0. No 1. Yes	0.034*	0.024	0.017
Sex/gender	0. Male 1. Female	−0.007	−0.011	−0.018
Age	in years	−0.081***	−0.034*	−0.033*
Education	number of completed classes	0.144***	0.102***	0.094***
Perception of threats	scale 1–7	0.090***	0.092***	0.096***
Perception of challenges	scale 1–7	0.432***	0.401***	0.337***
Communication: spiritual community	scale 1–7		0.002	−0.008
Communication: Internet community	scale 1–7		0.157***	0.156***
Communication: local community	scale 1–7		0.041**	0.031*
Communication: close-knit community	scale 1–7		0.041**	−0.002
Internal harmony	scale 1–7			0.013
Social support	scale 1–7			0.108***
Optimism	scale 1–7			0.094***
Satisfaction with life	scale 1–7			0.056**
Corrected R² and standard calculation error		0.331 (1.343)	0.350 (1.324)	0.388 (1.285)

* p < 0.05, ** p < 0.01, *** p < 0.001.

Source: Own research.

It turns out that Model 1 explained as much as 33% of the variation of dependent variable and the introduction of subsequent variables resulted in further improvement of this proportion. In Model 2 up to 35% of the variation was explained and in Model 3 it was almost 39%. This was a very high result for sociological analyses. In regards to socio-demographic characteristics sex/gender did not have any significant net effect on the choice of active strategy with applied control over any other variables. A very weak effect on the choice of active strategy was noted for work outside of farm sector. It was only noticeable in the Model 1 but taking into account more variables in Model 2 and Model 3 made working outside of agriculture statistically insignificant. The net effect of the remaining characteristics was also weakened with introduction of more variables but still stayed at statistically significant level. Preference for active strategy increased with farm area and also as expected with higher level of education (measured in number of years/classes competed in school), and decreased with the respondents' age. It should be noted that the net effect of education on the choice of active strategy was decidedly greater than the net influence of farm area and age. Even with controlling all the other analyzed variables level of education maintained the highest level of significance (0.001). Perception of threats was also a determinant of strong influence. It was observed that the net effect increased gradually with consideration of subsequent variables, which was an exceptional case. Even stronger was the influence of perception of challenges on taking up active strategy to deal with globalization-related changes. To be precise, it decreased with addition of more variables but in all models it stayed at decidedly higher level than the influence of other variables. Introduction of various patterns of communication and ways to acquire information to Model 2 revealed that participation in religious services (communication oriented at spiritual community) did not have any effect on the preference of active strategy of dealing with changes. The visible effect of communication oriented towards close-knit community (Model 2) on the preference of active strategy lost its statistical significance when the control of personality variables related to psychological well-behind was applied (Model 3), and the impact of communication oriented towards local community on strategy preference was weakened. Among presented patterns of communication the most significant net effect was noted with the communication oriented towards Internet community (Model 2), and the level of its net effect was also maintained even if the analysis considered personality variables (Model 3). Considering the personality variables in Model 3 it should be stated that only the level of internal harmony did not have any effect on the active strategy preference. Optimism and social support had the relatively highest net effect on the preference of the active strategy. How to conclude the results of regression analysis of the adaptation-related activism scale? What determined the preference of active strategy? The perception of challenges of globalization and postmodern world was the strongest determinant. The perception of threats was also very important but not to such an extent. Use of Internet as a source of information about the social environment played a rather important role here and the scale of direct contacts within local community also did so as well but to a lesser extent. The active strategy was also fostered by the high level of psychological well-being (particularly the sense of social support, optimism and satisfaction with life), as well as higher level of education and a younger age. One might ask the question about the respondents preferring avoidance strategies. Table 11.4.3 presents the results of the regression analysis for the scale of experienced religiosity, which was conducted similarly to the regression analysis for the scale of adaptation-related activism by constructing three regression models.

Table 11.4.3. Regression analysis for the scale of lived/experienced religiosity—avoidance strategy. Standardized regression coefficients (β)

Variable	Values of variable	Model 1 β	Model 2 β	Model 3 β
Farm size	number of ha	−0.039*	−0.031*	−0.039**
Work outside of farm	0. No 1. Yes	−0.035*	−0.006	−0.011
Sex/gender	0. Male 1. Female	0.133***	0.101***	0.088***
Age	in years	0.143***	0.107***	0.079***
Education	number of	−0.054**	−0.027	−0.032*
Perception of threats	scale 1–7	0.146***	0.106***	0.101***
Perception of challenges	scale 1–7	0.177***	0.173***	0.042**
Strategy: spiritual community	scale 1–7		0.320***	0.290***
Strategy: Internet community	scale 1–7		−0.075***	−0.082***
Strategy: local community	scale 1–7		0.082***	0.067***
Strategy: close-knit community	scale 1–7		0.079***	0.017
Internal harmony	scale 1–7			0.212***
Social support	scale 1–7			0.059***
Optimism	scale 1–7			0.119***
Satisfaction with life	scale 1–7			0.114***
Corrected R² and standard calculation error		0.101 (1.541)	0.231 (1.426)	0.344 (1.316)

* p < 0.05, ** p < 0.01, *** p < 0.001.

Source: Own research.

It turned out that Model 1 only explained 10% of the variation in the dependent variable. However, considering these variables in conjunction with some patterns of communication (Model 2) increased the proportions of explained variation to 23%, and introduction of psychological well-being variables led to a further increase in the proportions of explained variation, up to 34%. Such results should be considered satisfying. In Model 1, all socio-demographic characteristics turned out to be statistically significant but the strongest net effects were noted with age and sex/gender. Preferences for the strategy of avoidance increased with age and were decidedly stronger among women, as well as respondents, who completed fewer years of education, had smaller farms, and did not work outside their own farms. One could say that such strategies were much more preferred among traditional peasants, and to be more precise in pointing out sex/gender, which played a role here—peasant women. The adoption of such strategies was determined to the greatest extent by perception of globalization challenges and to a lesser degree by perception of threats (Model 1). However, it should be added here that consideration of the subsequent variables greatly weakened the net effects of the majority of the above variables. Variables such as work outside of farm turned out to be statistically insignificant. Net effects of education, sex/gender and age also declined but for sex/gender and age they still maintained the highest level of significance (0.001). The net effect of perceived threats also diminished but the most notable drop of net effect could be seen with the perceived challenges, after the introduction of personality variables connected psychological well-being (Model 3). Consequently, the perception of threats turned out to be a more meaningful determinant for the preference of avoidance strategy than the perception of challenges. As far as patterns of communication were concerned, the strongest net effect was noted with the participation in religious services, while the communication oriented toward the local community and close-knit circles of family and friends had a much lower net effect. A similar effect, although of a negative character was revealed for use of the Internet as a source of information decreasing the level of emotional religiosity responsible for fostering the strategy of avoidance (Model 2). Introduction of personality variables related to well-being modified the net effects of particular types of communication. The net effect of the communication oriented at close-knit circle of people turned out to be statistically insignificant. Also somewhat weaker were net effects of participation in religious services and contacts within local community, while the negative effect of Internet communication (Model 3) increased. With personality variables the strong net effect of internal harmony and quite noticeable effects of other characteristics describing psychological well-being, namely: optimism, satisfaction with life, and social support (Model 3) should be emphasized here. What could be the strongest determinant for the preference of the avoidance strategy? First of all, it could be the participation in religious services but also active communication within local community, maintaining the high level of psychological well-being with particular attention given to internal harmony and at the same time the perception of increasing globalization threats. This was also a strategy preferred by women, who were rather older, not well educated, operating smaller farms, and not using Internet as a source of information on social reality.

It might be concluded that regression analysis revealed that different factors played a decisive role in respondents' preference for active strategy and different for their preference for avoidance strategy. However, both strategies seemed to serve to maintain psychological well-being. The active strategy was foremost a strategy of younger farmers, better educated, able to use the Internet and search for information, not only in their immediate surroundings but also about the postmodern world there. This strategy was the answer to the globalization challenges reaching farmers' consciousness but also served, albeit to a lesser degree, to identify the emerging threats and perhaps also to exercise the control over them in regards to everyday life. The strategy of avoidance, however, preferred by women and farmers defined as belonging to the traditional category of peasants appeared to be a reaction to perceived threats and, to a lesser degree, to challenges. Perhaps a lack of discernment of the essence of these threats and a lack in ability to apply the mechanism of control over these threats caused these respondents to follow this path, so they could feel safe and maintain internal harmony.

The question remains: Are the strategies analysed here independent from the characteristics of the local community or could territorial divisions perhaps be their important determinants? Could the strategies of dealing with global threats and challenges differentiate farmers from various regions and various counties in Poland? The answers to these questions will be searched for in the next subchapter.

11.5. Quality of Life, Global Threats, Challenges, and Strategies of Control in the Context of Territorial Divisions

The contents of this chapter so far have provided analysis of differences related to psychological well-being, a sense of global threats and challenges, and strategy for their control in the individual dimension focusing on the characteristics of respondents. The results of these analyses confirmed that the intensity of these factors varied significantly depending on farmers' place of residence and could be viewed within a territorial dimension. The question worth asking is this: To what extent did territorial divisions determine the quality of life (psychological well-being), the sense of threat and challenges, or the propensity for adopting a certain type of coping strategy to deal with these threats and challenges? Was it to a lesser or greater degree than individual characteristics that were analyzed earlier? Further questions might be asked about the links between the place of residence and the characteristics of the community in which the farmers functioned, and if these characteristics could play any significant role. To answer these questions, the multilevel model was applied and its most important premises were presented in the previous chapter. As was done previously, the two level analysis was applied here, with level one represented by farmers, and being "rooted" in the counties (local communities) constituting the second (higher) level of analysis. All the analyses were conducted with the use of statistical program HLM 7.03. Dependent variables were the variables that were analyzed in reference to individuals such as: perception of challenges (*SK_WZW*), perception of threats (*SK_ZAG*), use of active strategy (*SK_A*), and use of avoidance strategy (*SK_U*). The fifth dependent variable was a general sense of psychological well-being (*SK_DP*), which was constructed using a 7-point scale based on four previously analyzed variables: satisfaction with life, optimism, social support, and internal harmony. The value of general sense of psychological well-being was calculated as an average of these variables, with the mean value of that scales being 5.39, with Cronbach's alpha value of 0.706.

How meaningful could be the differentiation between psychological well-being, sense of threats and challenges, predisposition to apply the strategy of avoidance or the active strategy of controlling changes on individual level (farm users) and on the level of counties? Answers to these questions are delivered through parameters acquired through estimation of variations of these five variables for two level and based on the null model. Detailed description of the procedure of data preparation, introduction of commands and calculations for HLM 7.03 program was already presented in the previous chapter. The results of the estimation of an empty model for the averages of the variables such as: psychological well-being, sense of threats, sense of challenges, preferences for active strategy ad preferences for avoidance strategy are presented in frames E2.11.1–E2.11.5. In these four cases only the most crucial fragments of the frames are considered.

Frame E2.11.1. Results of estimation of the two level null model for the scale of psychological well-being in HLM 7.03

Summary of the model specified

Level-1 Model

$SK\_DP_{ij} = \beta_{0j} + r_{ij}$

Level-2 Model

$\beta_{0j} = \gamma_{00} + u_{0j}$

Mixed Model

$SK\_DP_{ij} = \gamma_{00} + u_{0j} + r_{ij}$

Final results: Iteration 6

Iterations stopped due to small change in likelihood function

$\sigma^2 = 0.71577$

Standard error of $\sigma^2 = 0.01725$

τ

INTRCPT1, β_0	0.18596

Standard error of τ

INTRCPT1, β_0	0.02929

Random level-1 coefficient	Reliability estimate
INTRCPT1, β_0	0.893

The value of the log-likelihood function at iteration 6 = $-4.547640E + 003$

Final estimation of fixed effects (with robust standard errors)

Fixed Effect	Coefficient	Standard error	t-ratio	Approx. d.f.	p-value
For INTRCPT1, β_0					
INTRCPT2, γ_{00}	5.398100	0.046836	115.255	100	< 0.001

Final estimation of variance components

Random Effect	Standard Deviation	Variance Component	d.f.	χ^2	p-value
INTRCPT1, u_0	0.43123	0.18596	100	1020.31528	< 0.001
level-1, r	0.84603	0.71577			

Statistics for the current model

Deviance = 9095.280198

Number of estimated parameters = 3

Frame E2.11.2. Results of estimation of the two level null model for the scale of challenges in HLM 7.03

. . .

Mixed Model

$SK\_WZW_{ij} = \gamma_{00} + u_{0j} + r_{ij}$

. . .

Final estimation of fixed effects (with robust standard errors)

Fixed Effect	Coefficient	Standard error	t-ratio	Approx. $d.f.$	p-value
For INTRCPT1, β_0					
INTRCPT2, γ_{00}	4.322165	0.057676	74.939	100	< 0.001

Final estimation of variance components

Random Effect	Standard Deviation	Variance Component	$d.f.$	χ^2	p-value
INTRCPT1, u_0	0.50850	0.25857	100	979.30636	< 0.001
level-1, r	1.02121	1.04288			

. . .

Frame E2.11.3. Results of estimation of the two level null model for the scale of threats in HLM 7.03

. . .

Mixed Model

$$SK\_ZAG_{ij} = \gamma_{00} + u_{0j} + r_{ij}$$

. . .

Final estimation of fixed effects (with robust standard errors)

Fixed Effect	Coefficient	Standard error	t-ratio	Approx. $d.f.$	p-value
For INTRCPT1, β_0					
INTRCPT2, γ_{00}	3.974238	0.057224	69.450	100	< 0.001

Final estimation of variance components

Random Effect	Standard Deviation	Variance Component	$d.f.$	χ^2	p-value
INTRCPT1, u_0	0.51311	0.26328	100	1429.67239	< 0.001
level-1, r	0.82696	0.68387			

. . .

Frame E2.11.4. Results of estimation of the two level null model for the scale of active strategy in HLM 7.03

. . .

Mixed Model

$ST\_A_{ij} = \gamma_{00} + u_{0j} + r_{ij}$

. . .

Final estimation of fixed effects (with robust standard errors)

Fixed Effect	Coefficient	Standard error	t-ratio	Approx.d.f.	p-value
For INTRCPT1, β_0					
INTRCPT2, γ_{00}	4.681427	0.062828	74.512	100	< 0.001

Final estimation of variance components

Random Effect	Standard Deviation	Variance Component	d.f.	χ^2	p-value
INTRCPT1, u_0	0.53997	0.29156	100	531.62314	< 0.001
level-1, r	1.55327	2.41263			

. . .

Frame E2.11.5. Results of estimation of the two level null model for the scale of avoidance strategy in HLM 7.03

. . .

Mixed Model

$ST\_U_{ij} = \gamma_{00} + u_{0j} + r_{ij}$

. . .

Final estimation of fixed effects (with robust standard errors)

Fixed Effect	Coefficient	Standard error	t-ratio	Approx. d.f.	p-value
For INTRCPT1, β_0					
INTRCPT2, γ_{00}	5.122329	0.078470	65.278	100	< 0.001

Final estimation of variance components

Random Effect	Standard Deviation	Variance Component	d.f.	χ^2	p-value
INTRCPT1, u_0	0.71394	0.50971	100	964.25590	< 0.001
level-1, r	1.43624	2.06279			

. . .

The most crucial parts of the frames are at the top and at the bottom. The upper parts of the frames present the specification of the two level model for average values of particular dependent variables (respectively for: SK_DP, SK_WZW, SK_ZAG, SK_A, SK_U) with consideration of the part identifying the level of individual and the county level, where:

SK_DP_{ij} (and respectively: SK_WZW_{ij}, SK_ZAG_{ij}, SK_A_{ij}, SK_U_{ij}) is a dependent variable, namely psychological well-being (and respectively: sense of threats, sense of challenges, preferences for active strategy, preferences for avoidance strategy) of I farmer, from J county,

γ_{00} is the mean value of averages for psychological well-being (and respectively: for sense of threats. sense of challenges, preferences for active strategy, preferences for avoidance strategy) in the population of counties,

u_{0j} is the deviation of average psychological well-being (and respectively: sense of threats. sense of challenges, preferences for active strategy, preferences for avoidance strategy) for J county from the mean value in the population, which means "the remainder" for the level-2 (counties),

r_{ij} is the deviation of psychological well-being (and respectively: sense of threats. sense of challenges, preferences for active strategy, preferences for avoidance strategy) of I farmer, from J county from the average of this county, which means "the remainder" for the level-1 (farmers),

β_{0j} is the deviation of the mean value (and respectively: sense of threats. sense of challenges, preferences for active strategy, preferences for avoidance strategy) for J county from the average of the first selected county (presented only in the frame E2.11.1).

The lower parts of the frames contain the most crucial information on the values of estimated parameters. First, presented are the parameters of fixed effects, mean (mean value of averages) of dependent variable (y_{00}), with a standard error for the mean, as well as the value of t-ration with the number of degrees of freedom and level of significance for the null hypothesis assuming that fixed parameter has a value of 0. (The results of the test determine whether the competing hypothesis can be accepted, stating that fixed parameter is not 0. If $p < 0.001$, the competing hypothesis is assumed). Next presented are the parameters of random effect for level-1—individuals/farmers (r) and for level-2—counties (u_0), in the form of standard deviation and variation. The frames also present the values of the Chi-square test with the number of degrees of freedom and level of significance for the null hypothesis, stating that the random effect for level-2 equals 0. (In this case the results of the test are also meant to determine whether to accept the competing hypothesis stating that the parameter of random effect is not 0. In case of $p < 0.001$ a competing hypothesis is assumed.) The most important results of the conducted estimations are presented in Table 11.5.1.

Table 11.5.1. The results of estimation of variation in psychological well-being, global threats and challenges and strategies for controlling the changes, namely active strategy and avoidance strategy for level-1 (farmers) and level-2 (counties). Number of counties = 101. Number of farmers = 3543. Program HLM 7.03—null model (M0)

Parameter	Value of SK_DP	Value of SK_WZW	Value of SK_ZAG	Value of SK_A	Value of SK_U
y_{00}	5.40 (0.05)	4.32 (0.06)	3.97 (0.06)	4.68 (0.06)	5.12 (0.08)
$Var(u_{0j}) = \tau_{00}$	0.19 (0.03)	0.26 (0.04)	0.26 (0.04)	0.29 (0.05)	0.51 (0.08)
$Var(r_{ij}) = \sigma^2$	0.72 (0.02)	1.04 (0.03)	0.68 (0.02)	2.41 (0.06)	2.06 (0.05)
$ICC = \dfrac{\tau_{00}}{\tau_{00} + \sigma^2}$	0.206	0.199	0.278	0.108	0.198
Deviance	9095.28	10424.59	8969.91	13334.79	12840.87

Source: Own work.

Besides the estimated average of analyzed characteristics (y_{00}) and its standard error (in bracket) Table 11.4 also presents the values of variation components for level-2—counties (τ_{00}) and for level-1—farmers (σ^2), namely interclass and intraclass variation, and in brackets values of standard deviations. The value of intraclass correlation coefficient (ICC) is also calculated as a ratio of interclass variation to total variation, which is the sum of interclass variation and intraclass variation. It allows us to estimate to what extent farmers from the same county are similar to each other in regards to certain characteristics than farmers residing in various other counties. The greater the value of the intraclass correlation coefficient, the greater the homogeneity of farmers from a particular county also, as well as the differences between farmers from various counties.

As expected, the differentiation of characteristics was decidedly greater at the level of individuals than at the level of the counties. In all cases, the value of intraclass variation was much greater than the value of interclass variation. Nevertheless the calculated values of parameters indicated that each of these differentiations had a multilevel character: value of the intraclass correlation coefficient reached 0.108 for active strategy and 0.278 for the perception of globalization threats. In other words, territorial divisions (namely counties) explained from 11% to 28% of total variation of analyzed respondents' characteristics. Pictures 11.1–11.5 illustrate the differentiation of these characteristics according to territorial divisions. They present distributions of the remainders from level-2 (counties) for null models, which can be interpreted appropriately and respectively for the following scales: psychological well-being, sense of threats, sense of challenges, as well as propensity to apply active strategies or avoidance strategies. On the map, the highest mean values are depicted with the most intense colors and the lowest mean values as presented with the least intense colors.

When analyzing Picture 1, it would be hard to say that sharp territorial differentiation could be seen between the farmers of various regions due to the quality of life (psychological well-being).

In this context, it would be hard to present Poland as a country of regions representing various tiers of quality of life, traditionally perceived as Poland A (with the highest quality of life), Poland B, and/or Poland C (with the lowest quality of life). The counties with the highest level of psychological well-being, indicating the most desirable quality of life, as well as the counties with lowest level of the quality of life, were scattered, located in various parts of Poland. The former could be found in eastern Poland (Jarosławski, Przemyski, Janowski, Garwoliński, Węgrowski, Łomżyński, and Sokólski counties), in western Poland (Oleśnicki, Gnieźnieński, Czarnkowsko-Trzcianecki, Koszaliński, and Wąbrzeski counties), as well as in central and southern parts of Poland (Radomszczański, Wieluński, Olkuski, Miechowski, and Cieszyński counties). The latter could be found in the south and west of Poland (Dąbrowski, Tarnowski, Sanocki, Sandomierski, Biłgorajski, Tomaszowski counties), in Mazovia and in northwestern Poland (Pułtuski, Mławski, Ostrołęcki, Bartoszycki, Olecki, Iławski counties), as well as in western and central Poland (Polkowicki, Wolsztyński, Nyski, Kłobucki, Poddębicki, and Słupski counties). The maps of perception of global challenges (Picture 11.2) and global threats

Picture 11.1. Distribution of averages for the scale of psychological well-being in the counties. Number of surveyed counties N = 101

Source: Own work.

Picture 11.2. Distribution of averages for the scale of challenges in the counties. Number of surveyed counties N = 101

Source: Own work.

Picture 11.3. Distribution of averages for the scale of threats in the counties. Number of surveyed counties N = 101

Source: Own work.

Picture 11.4. Distribution of averages for the scale of active strategy in the counties. Number of surveyed counties N = 101

Source: Own work.

Picture 11.5. Distribution of averages for the scale of avoidance in the counties. Number of surveyed counties N = 101

Source: Own work.

(Picture 11.3) differed somewhat from the map of psychological well-being. Farmers from eastern parts of Poland perceived the challenges less intensely than farmers from western Poland, but counties with a low sense of global challenges could also be found in the west, for example, Polkowicki, Wolsztyński, and Gnieźnieński counties. However, the global challenges were perceived most intensely by farmers from the central belt, stretching from Lesser Poland and Silesia to Łódzkie Province with its surrounding counties. Within this belt, five (Radomszczański, Wieluński, Limanowski, Ostrowski, and Kutnowski counties) out of six counties with the highest perception of challenges were located. Generally speaking, no single dominant model of relations between the well-being and sense of threats could be identified in the counties. There were counties where the correlation coefficients for psychological well-being and the perception of threats were among the highest (e.g., Radomszczański and Wieluński counties) but also one could easily identify counties, where these coefficients were the lowest, such as in Pułtuski and Polkowicki counties. The study also easily found counties where the psychological well-being coefficient was the highest, yet the sense of global challenges the lowest, such as in Gnieźnieński and Jarosławski counties. There was also the case of Sandomierski County, where there was a low coefficient of psychological well-being and a high coefficient of the perception of challenges. Generally, the thesis regarding psychological well-being and a high sense of challenges was positively correlated. To some degree, the maps of challenges was similar to the map of threats. Counties with the lowest sense of threat could mostly be found in eastern part of Poland and included Dąbrowski, Sanocki, Niżański, Chełmski, Janowski, Puławski, Ostrowiecki, Kozienicki, Biało-brzeski, Mławski, Łomżyński counties. Elsewhere, they were quite rare (Poddębicki and Polkowicki counties). Similarly, counties with the highest sense of threats were mostly concentrated in the central belt, which extended to the south to Lesser Po-land Province (with the dominant position of Limanowski, Nowotarski, Bocheński, Myślenicki, Miechowski counties) and to the east to include the counties of Subcar-pathian province (Dębicki and Kolbuszowski counties), Holy Cross Province (Buski and Sandomierski counties), and Lublin Province (Tomaszowski and Krasnostawski counties). In the case of threats perception—just like it was with the perception of challenges—there were counties that had, simultaneously, relatively high indicators of well-being and high indicators of threats (Miechowski and Kutnowski counties), as well as counties where the indicators of these two characteristics were low (Mławski and Wolsztyński counties). Other combinations of coefficients, such as a high level of psychological well-being and a low level of the sense of threats were also identified, for example in Łomżyński and Janowski counties. The reverse com-bination of coefficients, with a low level of well-being and a high sense of global threats, was noted in Sandomierski and Tomaszowski counties. Although a positive correlation between psychological well-being and a sense of threat was confirmed in the study, it was not as strong as the positive correlation between psychological well-being and the perception of challenges. The map of threats, and to a greater extent the map of challenges, shared some similarities with the map of preferences of active strategies (Picture 11.4), albeit not without important modifications. The coefficient of the preferences for active strategies was relatively higher in the western counties (with the exceptions of Polkowicki and Wolsztyński counties) than in east-ern counties but it was the highest in the counties of the central belt. It should be added here that the counties with the lowest indicator of the propensity to use active strategy spread from the eastern part of Lesser Poland Province (Dąbrowski, Tar-nowski and Gorlicki counties) and Subcarpathian (Sanocki county), through Holy Cross Province (Włoszczowski, Ostrowiecki counties) and Lublin province (Biłgora-jski and Lubelski counties), up to Podlasie and northern Masovia (Łomżyński, Olecki, Pułtuski, and Mławski counties). The counties with a low level of engagement in active strategy of dealing with changes did not include the areas at the country's border, such as Sokólski, Bialski, Chełmski, and Przemyski counties. This perhaps might be connected with greater engagement of farmers in those counties with partners from Ukraine and Belarus. The area covering counties with the highest coefficient of active strategy included central parts of Lesser Poland (Limanowski, Okulski, and Miechowski counties), Łódzkie Province (Wieluński, Radomszczański, and Kutnowski counties), part of Kuyavia (Rypiński County) but also extended to the counties of western parts of Masovia (Sierpecki, Sochaczewski, and Grójecki counties), as well as southern parts of Greater Poland (Kępiński and Ostrowski counties) or even northern parts of Lower Silesia (Oleśnicki County). Even the most cursory look at the map of the strategy of avoidance indicated a much different situation than what was noted with the active strategy. The counties with the highest coefficient of avoidance strategy were mostly located in eastern parts of Poland, including the areas near borders (Kolneński, Łomżyński, Sokólski, Bialski, Łęczyński, Chełmski, Janowski, Niżański, Jarosławski, Przemyski, Krośnieński, Gorlicki, Szydłowiecki, Garwoliński, and Węgrowski counties). The counties with the lowest coefficient of this strategy were prevalent in western Poland (Picture 11.5). High coefficient of active strategy was also seen in some of the counties of the central belt (Cieszyński, Nowotarski, Limanowski, Olkuski, Miechowski, Radomszczański, Wieluński, Kutnowski, Rypiński, as well as Oleśnicki and Gnieźnieński counties). At the same time, the strongest concentration of counties with the lowest level of this coefficient could be found in Greater Poland (Turecki, Kępiński, Szamotulski, Ostrowski, and Wolsztyński counties) and in the northern part of the country (Słupski, Wąbrzeski, Iławski, Bartoszycki, Olecki, Mławski, Pułtuski, and Suwalski counties). Such counties could also be found in eastern and central parts of Poland and they were usually located in the proximity to large cities or industrial centers (Tarnowski, Sandomierski, Biłgorajski, Lubelski, Kozienicki, Piotrkowski, Poddębicki, and Sieradzki counties). The analysis of the maps illustrating the preferences for active and avoidance strategies confirmed the thesis that they were not mutually exclusive but rather complementing each other and used in response to certain situations. They could be preferred to an equal degree (large or small) but there were also situations when one strategy was clearly preferred. High preference for both strategies applied to farmers from counties such as: Limanowski, Radomszczański, Wieluński, and Kutnowski, while low levels of preferences for both strategies were noted among farmers from Polkowicki, Wolsztyński, Tarnowski, and Mławski counties. The examples of counties with highly preferred active strategies were Sochaczewski and Grójecki counties and the counties with notable preference for strategy of avoidance included Nowotarski and Cieszyński.

To conclude the analysis on the quality of life of Polish farmers according to territorial divisions one should ask how farmers from various counties were located on the analyzed scales of five characteristics. There were lots of configurations: from highest locations on the scales in all 5 dimensions (farmers of Kutnowski County) to the lowest positions in all dimensions (farmers of Mławski county). Analysis of the estimation results of null models for all scales revealed that individual characteristics of farmers were of primary importance but the characteristics of the counties we also quite meaningful. Here, the multilevel model of analysis might be helpful with identifying of these characteristics that could determine location of farmers from the surveyed counties on the analyzed scales.

Some remarks should be made on the subjectivity of variables before a more in-depth analysis of them is undertaken. Variables such as sense of psychological well-being and sense of globalization-related threats and challenges could possibly impact one's readiness to engage in active (searching for information) or passive (relying on faith and religious beliefs) strategies for dealing with globalization-related challenges. At the same time, the choice of strategy taken in response to emerging threats and challenges might be crucial to maintaining a certain level of psychological well-being. What individual characteristics of farmers and context characteristics of counties could determine the above psychological states, which—as indicated by previous analyses—were correlated with each other, to a greater or lesser degree? Should emphasis be placed on finding objective characteristics or should analysis go in the direction of subjective characteristics, generally alluding to the social climate that farmers live in, day in and day out? The analysis presented here was meant to consider both types of characteristics. When analysing the situation of farmers, a multitude of variables, such as sex/gender, age, education, work outside the farm, farm area and profitability were taken into account, as well as variables related to patterns of communication and acquiring information. These were described in detail in the previous analyses.

The variables characterizing counties subjected to the analysis were those describing their socio-economic situation. The same variables were used in the analysis of farm income changeability in the previous chapter. Additionally, two other variables were established on the basis of the data from the sample and they were recognized as indicators of the social climate in the country: level of social trust and level of readiness to cooperate. To measure social trust the respondents from particular counties were asked the following question: *Which viewpoint do you identify with more: "Generally, the majority of people can be trusted" or "One should be very careful in contacts with other people"?* The choice of the first viewpoint option indicated a level of social trust and, in particular counties, the percentage who chose it varied widely, from several to 81%. The second variable was measured by the percentage of respondents that chose the first answer to the question: *Which viewpoint do you identify with more: "It is possible to achieve more by working together with other people than by working alone" or "Cooperation with other people rarely brings the desired effects"?* In particular counties the percentages differed from just several percent to 97%.

The analyses of the influence of the characteristics presented above on the psychological situation of farmers according to territorial divisions was done with the use of modeling, which was conducted through the null model equation. It should be remembered here that while analysis of the null model allows one to determine whether the dependent variable is significantly differentiated in the territorial aspect, the model analysis with independent variables points out whether such differentiation is the case when these variables are controlled. Earlier analyses in this chapter considered psychological well-being to be understood as the quality of life for farmers. As already known, this variable was significantly differentiated according to the division by counties and the analysis of its null model explained 21% of the total variation in psychological well-being. Could the characteristics presented above (of counties and individuals) provide some significant explanatory power, if they were considered in the analysis of reduced to significant unexplained variation of psychological well-being at the level of counties? First, the analysis considered all the characteristics of the counties, separately introducing each variable to the equation model. It turned out that, of the 16 variables, only 3 had impact—although with limited significance—on the differentiation of farmers' psychological well-being according to the division by counties: one objective variable, being the level of county expenditure per capita, and two other variables established on the basis of farmers' opinions, namely the level of trust in the county and level of readiness for cooperation. It could be assumed that the psychological well-being of individuals functioning in some community was conditioned (at least to some extent) by its general social atmosphere. It is generally assumed that in communities where people have a significant level of trust in others, have a readiness to cooperate, they feel better and are more satisfied with life. They know they can count on others, look into the future with greater optimism, feel internal peace. Here, the thesis of Piotr Sztompka relating to the importance of social climate for a sense of safety in everyday life (Sztompka, 2007) should find its confirmation. However, it is not just relations with other residents of a particular area that builds the quality of life. The actions of local authorities should also be seen as meaningful as they impact the volume and the structure of budget expenditure directly influencing the community life as a whole. Would the conducted analyses confirm these hypotheses? The most important fragments of the results of estimation models of psychological well-being (SK_DP) were considered in consecutive order: level of county expenditure per capita (WYDM), level of farmers' readiness to cooperate with others (WSPOLNE), and level of trust (ZAUFA) are presented in Frames E2.11.6–E2.11.8.

Frame E2.11.6. Results of estimation of two level model with group independent variable—coefficient of the county expenditure per capita in HLM 7.03

. . .

Mixed Model

$$SK\_DP_{ij} = \gamma_{00} + \gamma_{01} {}^* WYDM_j + u_{0j} + r_{ij}$$

. . .

Final estimation of fixed effects (with robust standard errors)

Fixed Effect	Coefficient	Standard error	t-ratio	Approx. $d.f.$	p-value
For INTRCPT1, β_0					
INTRCPT2, γ_{00}	6.448867	0.515127	12.519	99	< 0.001
WYDM, γ_{01}	−0.246909	0.120218	−2.054	99	0.043

Final estimation of variance components

Random Effect	Standard Deviation	Variance Component	$d.f.$	χ^2	p-value
INTRCPT1, u_0	0.42049	0.17681	99	967.78094	< 0.001
Level-1, r	0.84604	0.71578			

. . .

Frame E2.11.7. Results of estimation of two level model with group independent variable—coefficient of readiness for cooperation in the county, in HLM 7.03

. . .

Mixed Model

$$SK\_DP_{ij} = \gamma_{00} + \gamma_{01} {}^* WSPOLNIE_j + u_{0j} + r_{ij}$$

. . .

Final estimation of fixed effects (with robust standard errors)

Fixed Effect	Coefficient	Standard error	t-ratio	Approx. d.f.	p-value
For INTRCPT1, β_0					
INTRCPT2, γ_{00}	5.021976	0.194732	25.789	99	< 0.001
WSPOLNIE, γ_{01}	0.597245	0.303588	1.967	99	0.052

Final estimation of variance components

Random Effect	Standard Deviation	Variance Component	d.f.	χ^2	p-value
INTRCPT1, u_0	0.42092	0.17717	99	983.92384	< 0.001
level-1, r	0.84601	0.71574			

. . .

Frame E2.11.8. Results of estimation of two level model with group independent variable—coefficient of the level of trust in the county, in HLM 7.03

. . .

$$SK\_DP_{ij} = \gamma_{00} + \gamma_{01} * ZAUFA_j + u_{0j} + r_{ij}$$

. . .

Final estimation of fixed effects (with robust standard errors)

Fixed Effect	Coefficient	Standard error	t-ratio	Approx. d.f.	p-value
For INTRCPT1, β_0					
INTRCPT2, γ_{00}	5.497358	0.071768	76.599	99	< 0.001
ZAUFA, γ_{01}	−0.578542	0.245631	−2.355	99	0.020

Final estimation of variance components

Random Effect	Standard Deviation	Variance Component	d.f.	χ^2	p-value
INTRCPT1, u_0	0.42172	0.17785	99	976.69915	< 0.001
level-1, r	0.84603	0.71576			

. . .

In the above models the dependent variable is SK_DP_{ij}, which is the coefficient of psychological well-being of I-farmer from J-county, the following symbols should be explained: γ_{00} is the mean (mean value of averages in the counties, to be precise) of the coefficient of psychological well-being in the population, γ_{01} is a coefficient describing the relationship strength between the dependent variable (SK_DP) and independent variable from the level-2 x_j, which is the mean of the coefficient (respectively: *WYDM, WSPOLNIE, ZAUFA*) of constant character for J country, u_{0j} is the deviation of the mean value of the psychological well-being coefficient between J county and the average in the population, which is the "remainder" from level-2 (counties), and r_{ij} is a deviation of the psychological well-being coefficient between I farmer from J county and the mean value of that county, which is the "remainder" from level-1 (farmers). The most crucial information on the fixed effects parameters, γ_{00} and $\gamma_{01}(x_j)$, random effects for level-1 (farmers) and level-2 (counties) in the forms of standard deviations and variations of intraclass correlation coefficient (ICC) and R^2 were gathered and presented in Table 11.5.2. As already pointed out, when models with independent variables were compared to the null model, the remainder from level-2 (u_{0j}), as well as its variation, acquired new meanings (τ_{00}), which were conditioned by independent variable (x_j). Intraclass correlation coefficient (ICC) indicated what percentage of variation of dependent variable corresponded with division into counties with controlling independent variable also assumed conditional character when there was a premise that the value of dependent variable was the same in all counties. The explanatory force meter of model R^2, in the model with independent variables only from level-2 (characteristics of the county), or $R^2(p2)$ which is the ratio between interclass variation for the null model and interclass variation for the model with independent variables to interclass variation for null model should indicate by what percentage of unexplained interclass variation from the null model could be reduced, if the model considered independent variables from the level-2.

The results from estimation were quite surprising. First of all, the results did not confirm the initial hypothesis. Although, the level of psychological well-being increased with a higher level of readiness to cooperate, which was indicated by the value of the coefficient $\gamma_{01}(WSPOLNIE) = 0.60$), it also declined with an increase in the expenses coefficient ($\gamma_{01}(WYDM) = -0.25$) and the level of trust ($\gamma_{01}(ZAUFA) = -0.58$). The thesis of Piotr Sztompka (2007) on the importance of an atmosphere of trust for a sense of safety in everyday life might be quite problematic in the farmers' social environment. How to interpret the study results? What about the specifics of the farmers' mentality? Could the experience of everyday life be meaningful here? Positive impact of the readiness for cooperation on psychological well-being could be connected with factual experience of such cooperation in

Table 11.5.2. Parameters of the two level model of regression of psychological well-being, according to county characteristics (x_j). Number of counties N = 101. Number of farmers = 3543. Program HLM 7.03—model with group independent variable (M1)

Parameter	Expenses $x_j = WYDM$	Cooperation $x_j = WSPOLNIE$	Trust $x_j = ZAUFA$
γ_{00}	6.45 (0.51)	5.02 (0.19)	5.50 (0.07)
$\gamma_{01}(x_j)$	−0.25 (0.12)	0.60 (0.30)	−0.58 (0.25)
$Var(u_{0j}) = \tau_{00}$	0.18 (0.03)	0.18 (0.03)	0.18 (0.03)
$Var(r_{ij}) = \sigma^2$	0.72 (0.02)	0.72 (0.02)	0.72 (0.02)
$ICC = \dfrac{\tau_{00}}{\tau_{00} + \sigma^2}$	0.198	0.198	0.199
$R^2(p2) = \dfrac{\tau_{00}(M0) - \tau_{00}(M1)}{\tau_{00}(M0)}$	0.05	0.05	0.04

Source: Own work.

everyday life. Instances of mutual help and cooperation among neighbors, whether on the farm or with the building of rural infrastructure, are, in many social circles of famers, quite frequent and rather positively perceived. Still, they might not foster an atmosphere of social trust and this might be inferred from the mean values of coefficients for cooperation and trust. On average, 64% of farmers in the surveyed counties declared a readiness for cooperation and only slightly over 18% of them were convinced that the majority of people could be trusted. This might confirm the continuous presence of the "wary, mistrustful peasant" syndrome. Why would an increase in trust be negatively affecting the psychological well-being? Could this possibly mean that in the counties with a higher level of trust, farmers were more likely to engage in interpersonal contacts and, because of that, often experienced disappointments in others, which negatively affected their psychological well-being? And consequently, could the selection of social contacts be greater in places where people did not trust each other, consequently limiting these contacts to the close-knit circles, and significantly reducing their sense of disappointment and stressful situations? Then the question arises as to how to interpret the negative impact of expenditure at the county and municipal level on psychological well-being? Could this again be explained by the "wary, mistrustful peasant," but this time aimed not towards other rural inhabitants but towards the authorities? Perhaps in richer communities there is a growing conviction, which might not be accurate, but could be based on lack of trust that local authorities allot financial means to social causes in an unjustified manner. It cannot be precluded, however, that such opinions might be based on facts and the higher per capita expenses of those counties do not translate into better fulfilment of farmers' needs and serve to fulfill the needs of other social groups. Both situations could possibly foster the sense of an increase in relative deprivation among farmers. It is worth noting that only 11% of respondents agreed with the statement: *Generally speaking, our society is organized to ensure justice* and 15% of surveyed farmers agreed with another statement *In Poland, everybody now has almost the same chances to be rich and happy.* Which of these interpretations could be justified? It is hard to say with certainty especially that the relations described here oscillate around statistical significance (0.05) and after the introduction of subsequent dependent variables while controlling for all the other variables, these relations might prove false. Before verifying that, it should be noted that, just as expected, the differentiation of psychological well-being in every model was much greater at the individual level ($Var(r_{ij}) = 0.72$) than at the level of counties ($Var(u_{0j}) = 0.18$). Division into counties while controlling for the adequate dependent variable in a particular model explained not quite 20% of the well-being changeability as the intraclass correlation coefficient (ICC) reached 0.198–0.199, which was not much less than in the empty model with the value of 0.206. Taking into consideration the differentiation between the counties due to the level of expenditure, readiness for cooperation and the level of trust (for every variable being analyzed separately) did not significantly reduce the percentage of unexplained interclass variation of psychological well-being. The explanatory force meter for the analyzed models $R^2(p2)$ reached the value of 0.04–0.05, which indicated that the reduction in unexplained variation of psychological well-being was around 4–5%. These data confirmed the co-efficiency of the considered variables in the context of psychological well-being and were situated at the border of significance. Unfortunately, these were still the most indicative variables in the study.

A procedure involving introduction of independent variables to the regression model of psychological well-being, similar to the one applied in the analysis of characteristics of the counties, was used in the consideration of individual qualities of farmers. As expected, and while taking into account results of earlier analyses, the explanatory powers of all these characteristics (analyzed separately) were statistically significant but introducing them to the model did not fully explain the differentiation in farmers' quality of life (understood as psychological well-being) between the counties. Here, two models were chosen for the presentation: model of active strategy and model of avoidance strategy. In both models, one of the group characteristics was considered—level of trust in the county. The most important fragments of the results of estimation of these models are presented in frames E2.11.9–E2.11.10.

Frame E2.11.9. Results of estimation of two level model with group and individual independent variable (level of trust in the county: ZAUFA and active strategy: ST_A), in HLM 7.03

. . .

Mixed Model

$$SK\_DP_{ij} = \gamma_{00} + \gamma_{01} * ZAUFA_j + \gamma_{10} * ST\_A_{ij} + u_{0j} + r_{ij}$$

. . .

Final estimation of fixed effects (with robust standard errors)

Fixed Effect	Coefficient	Standard error	t-ratio	Approx. $d.f.$	p-value
For INTRCPT1, β_0					
INTRCPT2, γ_{00}	4.593645	0.121825	37.707	99	< 0.001
ZAUFA, γ_{01}	−0.483868	0.230244	−2.102	99	0.038
For ST_A slope, β_1					
INTRCPT2, γ_{10}	0.189515	0.015431	12.281	3441	< 0.001

Final estimation of variance components

Random Effect	Standard Deviation	Variance Component	$d.f.$	χ^2	p-value
INTRCPT1, u_0	0.37505	0.14066	99	885.02050	< 0.001
level-1, r	0.79512	0.63221			

. . .

Frame E2.11.10. Results of estimation of two level model with group and individual independent variable (level of trust in the county: *ZAUFA* and strategy of avoidance: *ST_U*), in HLM 7.03

. . .

Mixed Model

$$SK\_DP_{ij} = \gamma_{00} + \gamma_{01} * ZAUFA_j + \gamma_{10} * ST\_U_{ij} + u_{0j} + r_{ij}$$

. . .

Final estimation of fixed effects (with robust standard errors)

Fixed Effect	Coefficient	Standard error	t-ratio	Approx. $d.f.$	p-value
For INTRCPT1, β_0					
INTRCPT2, γ_{00}	4.358251	0.123429	35.310	99	< 0.001
ZAUFA, γ_{01}	−0.446539	0.191080	−2.337	99	0.021
For ST_U slope, β_1					
INTRCPT2, γ_{10}	0.217891	0.019234	11.328	3441	< 0.001

Final estimation of variance components

Random Effect	Standard Deviation	Variance Component	$d.f.$	χ^2	p-value
INTRCPT1, u_0	0.32355	0.10468	99	681.50239	< 0.001
level-1, r	0.79064	0.62512			

. . .

In these models, in comparison to models with an independent group variable from level-2, the new elements are as follows: x_{ij}—independent variable from level-1 (ST_A, ST_U) for I farmer from J county and y_{10}—regression coefficient describing the average relation between dependent and independent variable from level-1, which is the same (constant) for J counties. As recalled from earlier, the model considering an independent variable from level-1 was a model with random regression constant. The most important data on parameters of fixed effects, such as γ_{00}, $\gamma_{01}(ZAUFA)$ and $\gamma_{10}(x_{ij})$ and random effects— variation for level-1 (farmers) and level-2 (counties), as well as the value of the intraclass correlation coefficient (ICC) and explanatory force meter for models $R^2(p2)$ and $R^2(p1)$ were all presented in Table 11.5.3. Model $R^2(p2)$ should explain the differentiation decrease in well-being between the counties in comparison to the null model, with consideration made for both independent variables. The value of $R^2(p1)$ should indicate the differentiation decrease of psychological well-being between farmers in comparison to the empty model, with consideration for the variable from level-1 (coefficient of strategy).

Table 11.5.3. Parameters of the two level model of regression of farmers' psychological well-being according to the level of trust in the counties and strategy of controlling changes preferred by farmers (x_{ij}). Number of counties N = 101. Number of farmers = 3543. Program HLM 7.03—model with group and individual independent variables (M2)

Parameter	$X_{ij} = ST\_A$ (active strategy)	$X_{ij} = ST\_U$ (avoidance strategy)
γ_{00}	4.59 (0.08)	4.36 (0.12)
$\gamma_{01}(ZAUFA)$	−0.48 (0.27)	−0.45 (0.19)
$\gamma_{10}(x_{ij})$	0.19 (0.02)	0.22 (0.02)
$Var(u_{0j}) = \tau_{00}$	0.14 (0.02)	0.10 (0.02)
$Var(r_{ij}) = \sigma^2$	0.63 (0.02)	0.63 (0.02)
$ICC = \dfrac{\tau_{00}}{\tau_{00} + \sigma^2}$	0.18	0.14
$R^2(p2) = \dfrac{\tau_{00}(M0) - \tau_{00}(M2)}{\tau_{00}(M0)}$	0.24	0.44
$R^2(p1) = \dfrac{\sigma^2(M0) - \sigma^2(M2)}{\sigma^2(M0)}$	0.12	0.13

Source: Own work.

In both models individual independent variables had a positive influence on the level of psychological well-being. This meant that its level increased with the growing level of readiness to take up active strategy $(y_{10}(ST\_A) = 0.19)$, as well as avoidance strategy $(y_{10}(ST\_U) = 0.22)$, while the differentiation of the level of trust coefficient between the counties was controlled. The results of the analyses confirmed positive correlation between psychological well-being and both strategies. Consideration of other results and analyses revealed that these strategies were not in opposition with each other but were, in fact, complementary to each other and each of them could have a positive impact on the well-being. However, introduction of these variables to the model did not change the character of interdependencies between the well-being and the level of trust in the county. In both models, with controlling the strategy coefficient, an increase in the level of trust in the county caused a decline in the level of well-being and $y_{01}(ZAUFA)$ assumed the values −0.48 and −0.45 respectively. Although these relations had limited statistical significance (0.05) it should be examined whether the "mistrustful peasant" syndrome, as an element of peasant mentality cultivated to this day, might have been an unconscious mechanism of longing for a "world at the doorstep" that would protect peasants— albeit to the limited degree—from the influx of what is unknown and external, and could potentially destroy psychological well-being. As expected, in both models, the differentiations of well-being on both levels were significant but in comparison to null model the differentiation was decidedly greater at the individual level ($Var(r_{ij})$) taking the respective values of 0.63 and 0.63 than at the county level ($Var(u_{oj})$), where the respective values were 0.14 and 0.10. With controlling the level of trust in the counties and the strategy of controlling the changes, the correlation coefficient fell respectively to 0.18 and 0.14. This meant that in the model with active strategy the differentiation into counties could have accounted for about 18% of the psychological well-being of farmers and, in the model with avoidance strategy, for 14%. Both variables in the first model explained a total of around 24% of differentiation of well-being between the counties and in the second model around 44%, as indicated by the values of $R^2(p2)$—respectively: 0.24 and 0.44. To be more precise, this meant a decrease in the percentage of differentiation of the quality of life between the counties in comparison to the null model after consideration of dependent variables. It should also be noted that both strategies reduced psychological well-being differentiation among farmers by more or less the same percentage. Readiness to apply (Model 1 explained about 12% of the changeability in farmers' well-being, and the strategy of avoidance 13%. The values of ($R^2(p1)$) measuring the differentiation of well-being for level-1 were, respectively, 0.12 and 0.13. In both models, average levels of the quality of life also understood as psychological well-being in the counties differed in a statistically significant manner. This was confirmed by the level of significance of the Chi-square test for random effect which was $p < 0.001$, as seen in the lower part of the frames E2.11.9 and E2.11.10).

Multilevel analysis was conducted not just to recognize whether a certain characteristic (dependent variable) was differentiating in the territorial aspect, such as the division into counties with application of the estimation of the null model, or to state whether differentiation occurred while controlling for other characteristics, independent variables thanks to the estimation of the model with random constant. It was also meant to determine whether the relations between dependent variable and independent variable were similar in all counties or if significant differences were present between the counties. For that purpose, a model with random slope was applied, which in fact was the developed model with random constant. In this model there was an assumption about the presence of a random constant and random regression coefficient, serving as a measure of differentiation of the relation between dependent and independent variable in J categories, which are counties. The description of this model and the procedure of specification of appropriate equation in the HLM program was well presented in the work of Henryk Domański and Artur Pokropek (2011, pp. 115–136). Here the analysis should only be concentrated on explaining the results of the estimation of two cases of such model, in which education understood as number of school years completed was an independent variable. The dependent variable was either coefficient of active strategy or coefficient of avoidance strategy. It should be known from earlier analyses that division into counties differentiated both coefficients and that both of them were significantly affected by education; positively correlated with active strategy and negatively correlated with the avoidance strategy. Were these interdependencies similar in all counties or were there significant differences between them? The most important parts of the results of estimation of adequate models are presented in frames E2.11.11 and E2.11.12. In both models education WYK was centered according to the average in the population. More about centering can be found in the work of Domański and Pokropek (2011, pp. 120–121).

Frame E2.11.11. Results of estimation of the two level model with random slope (dependent variable—active strategy ST_A, independent variable—education in years completed WYK), in HLM 7.03

. . .

WYK has been centered around the grand mean.

Mixed Model

$$ST\_A_{ij} = \gamma_{00} + \gamma_{10} * WYK_{ij} + u_{0j} + u_{1j} * WYK_{ij} + r_{ij}$$

. . .

τ

INTRCPT1, β_0	0.26577	−0.01230
WYK, β_1	−0.01230	0.00161

Standard errors of τ

INTRCPT1, β_0	0.04722	0.00713
WYK, β_1	0.00713	0.00204

τ (as correlations)

INTRCPT1, β_0	1.000	−0.594
WYK, β_1	−0.594	1.000

. . .

Final estimation of fixed effects (with robust standard errors)

Fixed Effect	Coefficient	Standard error	t-ratio	Approx. $d.f.$	p-value
For INTRCPT1, β_0					
INTRCPT2, γ_{00}	4.692389	0.060354	77.748	100	< 0.001
For WYK slope, β_1					
INTRCPT2, γ_{10}	0.175774	0.012990	13.532	100	< 0.001

Final estimation of variance components

Random Effect	Standard Deviation	Variance Component	$d.f.$	χ^2	p-value
INTRCPT1, u_0	0.51553	0.26577	100	518.03794	< 0.001
WYK slope, u_1	0.04015	0.00161	100	109.01048	0.253
level-1, r	1.49927	2.24782			

. . .

Frame E2.11.12. Results of estimation of the two level model with random slope (dependent variable—avoidance strategy ST_U, independent variable—education in years WYK), in HLM 7.03

. . .

WYK has been centered around the grand mean.

Mixed Model

$$ST\_U_{ij} = \gamma_{00} + \gamma_{10} * WYK_{ij} + u_{0j} + u_{1j} * WYK_{ij} + r_{ij}$$

. . .

τ

INTRCPT1, β_0	0.50658	−0.02323
WYK, β_1	−0.02323	0.00303

Standard errors of τ

INTRCPT1, β_0	0.08023	0.00956
WYK, β_1	0.00956	0.00210

τ (as correlations)

INTRCPT1, β_0	1.000	-0.593
WYK, β_1	-0.593	1.000

. . .

Final estimation of fixed effects (with robust standard errors)

Fixed Effect	Coefficient	Standard error	t-ratio	Approx. $d.f.$	p-value
For INTRCPT1, β_0					
INTRCPT2, γ_{00}	5.120339	0.078065	65.591	100	< 0.001
For WYK slope, β_1					
INTRCPT2, γ_{10}	-0.047947	0.012960	-3.700	100	< 0.001

Final estimation of variance components

Random Effect	Standard Deviation	Variance Component	$d.f.$	χ^2	p-value
INTRCPT1, u_0	0.71175	0.50658	100	920.66959	< 0.001
WYK slope, u_1	0.05506	0.00303	100	115.22955	0.142
level-1, r	1.42684	2.03588			

. . .

In these models, as compared to the models with random constant, $u_{1j} * WYK_{ij}$, which was a random effect of the slope, understood as the regression coefficient for independent variable WYK, constant for the J county constituted the new element. The effect of the estimation model was to determine the value of regression constant γ_{00}, describing the averages of the strategy coefficient for farmers with average number of years completed in school (due to centering education according to the population average) and value of the regression coefficient $\gamma_{10}(WYK)$, describing the relationship force between the strategy coefficient and farmers' education, as well as to determine random effect for the constant (u_{0j}) and for regression coefficient (u_{1j}). The first situation presented the value of differentiation of the strategy between the counties and the other the value of differentiation of the relation between the strategy and the education according to division by counties. The most important data related to the values of fixed effect and random effect, as well as the value of co-variance and correlation between random effects of the constant and regression coefficient—$Cov(u_{0j}, u_{1j})$ and $Cor(u_{0j}, u_{1j})$—were presented in Table 11.5.4. The value of co-variation allowed us to state whether the differentiation of the counties due to dependency of the strategy coefficient on education was connected with the average level of strategy coefficient (value of regression constant) for the counties. The value of these correlation indicated the variation and the strength of this relationship.

Table 11.5.4. Parameters in the two level model for interdependencies between the strategies of controlling changes (ST_A and ST_U) and education. Number of counties N = 101. Number of farmers = 3543. Program HLM 7.03—model with random slope

Parameter	ST_A (active strategy)	ST_U (avoidance strategy)
γ_{00}	4.69 (0.06)	5.12 (0.08)
$\gamma_{10}(WYK)$	0.18 (0.01)	−0.05 (0.01)
$Var(u_{0j}) = \tau_{00}$	0.27	0.51
$Var(u_{1j}) = \tau_{10}$	0.002	0.003
$Var(r_{ij}) = \sigma^2$	2.25	2.04
$Cov(u_{0j}, u_{1j}) = \tau_{01}$	−0.01	−0.02
$Cor(u_{0j}, u_{1j})$	−0.59	−0.59

Source: Own work.

The value of regression constant in both models (γ_{00}) indicated the level of readiness to apply a particular strategy among farmers who completed the average number of years in school. Parameter $\gamma_{10}(WYK)$ described the strength of the relationship between the readiness to apply a particular strategy with a farmers' education. In both cases these relationships were statistically significant (see frames E2.11.11 and E2.11.12). As expected, the readiness to take up active strategy increased with a higher level of education and each additional year of schooling caused an increase in this strategy coefficient by 0.18. Consequently, the readiness to take up avoidance strategy decreased with the level of education and each additional year of school caused the decline of the coefficient of this strategy by 0.05. The random effect for the constant ($Var(u_{0j}) = \tau_{00}$) (see frames E2.11.11 and E2.11.12) in both models was also statistically significant, which meant that the level of readiness to take up active strategy and/or avoidance strategy was determined territorially and connected with the division into counties. It should also be stated that such a relation was stronger in the avoidance strategy ($Var(u_{0j}) = 0.51$) than in the active strategy ($Var(u_{0j}) = 0.27$). The values of random effects ($Var(u_{1j}) = \tau_{10}$) indicated that the relation between the tendency to take any strategy and education was not the same in the all counties, although these differences were very small (respectively: 0.002 and 0.003) and statistically insignificant (see frames E2.11.11 and E2.11.12). It might be worth noting that in the counties with the higher level of readiness to take up any of the two strategies its relation with the level of education was weaker in the counties with a lower level of coefficient of these strategies. This was indicated by the negative values of $(Cov(u_{0j}, u_{1j}) = \tau_{01})$ and correlation $(Cor(u_{0j}, u_{1j}))$. If the differences between the counties in regards to the described strategies and education were statistically significant, could one ask about their causes and the county characteristics that were decisive? The multilevel analysis, estimating the hierarchical model, meaning the developed model with a random slope might provide the answer to this question. This will not be done here mainly because of the noted statistical insignificance of differences between the counties in regards to the relation between the strategies presented above and educational level. More information on these very issues can be found in the works Henryk Domański and Artur Pokropek (2011) or Stephen W. Raudenbush et al. (2011). It should also be added here that—as initially expected—in both models differentiation of the strategy coefficient on both levels was significant, but decidedly greater at the individual level $(Var(r_{ij}))$ taking the respective values of 2.25 and 2.05.

11.6. Concluding Remarks

The reflection of this chapter concentrated on matters of the quality of the everyday lives of farmers in the contemporary, postmodern world. Generally, it might be stated that the consequences of globalization-related changes, in the form of perceived threats and challenges, have already reached the Polish countryside and farmers' circles. However they did not have a negative impact on farmers' quality of life, also defined as psychological well-being. It might appear that farmers already elaborated a set of strategies for controlling globalization-related changes and dealing with their consequences. They varied, including active strategies that were rational and involved searching for information about what was new and unknown, as well as strategies of avoidance, with a strong emotional component. The avoidance strategies concentrated on searching for answers to new situations through references to what might appear as irrational factors, and often meant turning to faith. It should be noted that these strategies were not in opposition, but were rather complementary to each other and were applied depending on the particular given situation. The complementary character of the strategies taken by farmers allowed them to maintain a relatively high level of psychological well-being. Obviously, globalization-related challenges did not interfere with the everyday lives of all farmers and all farming circles—defined here as county-based communities—to the same degree. The level of threats and challenges was as diverse as the farmers' level of readiness to take up particular strategies for controlling changes and the perceived level of psychological well-being. The application of multi-level analysis allowed us to determine that mostly individual characteristics impacted the differentiation of the level of quality of life, as well as the readiness to take up adaptation strategies and the sense of perceived threats and challenges. However, the meaning of the place of residence, belonging to certain local community defined through division into counties was also meaningful. Some of these individual characteristics were identified. It was more difficult to identify these characteristics for local communities. It turned out that various factors and indicators describing the economic and material situation of the community either had no impact or had a limited, but negative, influence. Factors such as the intellectual potential of the community, level of aging, level of economic development (measured by unemployment rate, extent of social assistance allotted to households, or GDP per capita), level of entrepreneurship, and/or crime rate, did not affect the quality of life in particular counties in any meaningful way. This indicated that quality of life in various communities could be similar regardless of their material wealth status, age structure, intellectual potential, proximity to metropolitan areas, or peripheral character. At the same time the influence—albeit limited—of other characteristics attributed to communities at the county level could be noted. These characteristics constituted general social climate and determined character of interpersonal relations. They were: level of readiness for cooperation and level of trust in the county. As expected, a higher level of readiness to cooperate in a particular county fostered an increase in farmers' psychological well-being. It was surprising, however, that the level of psychological well-being was declining in those counties with a higher level of social trust. As already stated earlier in this chapter, this could be interpreted within the category of "mistrustful peasant" syndrome, indicating the persistence of the peasant mentality and perhaps a longing for "the world at their doorstep."

The question about the agricultural picture of Poland from a territorial perspective might be answered by summarizing this chapter. Is it just one homogenous entity, in which farmers from the counties of the Subcarpathian Region do not differ from farmers of Western Pomerania? Or perhaps the division into the traditional, peasant countryside of eastern Poland and the more modern, enterprise-oriented countryside of western Poland, so present in the public imagination, is still valid in real life? The second picture seems more adequate on a higher level of generality but it might be even better to select another type of Polish countryside. This would be the countryside of the central belt extending from Lesser Poland) and Upper Silesia to Łódzkie Province and surrounding areas. Referring to the legacy of Poland's partition, which is common even today, one might say with some simplification that Polish agricultural areas were strongly impacted by the agricultural styles of

Picture 11.6. Level of social trust in the counties. Number of surveyed counties N = 101

Source: Own work.

occupying states, namely Russia, Prussia, and Austro-Hungary. The countryside of the central belt is located more or less in the area where the partition borders used to be. Considering the context, quality and course of farmers' everyday lives, it is quite noticeable that counties with the highest level of social trust are located in that central belt (Picture 11.6). And perhaps this is a reason why these counties do not tower over others in terms of the quality of life, as they do in regards to the level of perceived threats and globalization-related challenges or readiness to take up strategies within their control, especially active strategy.

REFERENCES

Bauman, Z. (1998). *Globalization: The human consequences*. Polity Press.

Bauman, Z. (2000). *Liquid Modernity*. Polity Press.

Bauman, Z. (2005). *Liquid Life*. Polity Press.

Bauman, Z. (2007). *Liquid Times: Living in an age of uncertainty*. Polity Press.

Beck, U. (1992). *Risk Society: Towards a new modernity*. SAGE Publications.

Beck, U. (2000). *What Is Globalization?* Polity Press.

CBOS (2015). *Zadowolenie z życia* [Satisfaction with life]. Komunikat z badań No. 3.

CBOS (2020). *Zadowolenie z życia*. Komunikat z badań No. 2.

Chimiak, G. & Fronia, M. (eds.) (2012). *Globalizacja a rozwój. Szanse i wyzwania dla Polski* [Globalization and development: Opportunities and challenges for Poland]. Wydawnictwo Naukowe Scholar.

Czapiński, J. & Panek, T. (eds.) (2011). Social diagnosis 2011. Objective and subjective quality of life in Poland. Report. *Contemporary Economics* 5(3), pp. 1–366.

Czapiński, J. & Panek, T. (eds.) (2013). Social diagnosis 2013. Objective and subjective quality of life in Poland. Report. *Contemporary Economics: Special issue* 7, pp. 1–490.

Czapiński, J. & Panek, T. (eds.) (2015). Social diagnosis 2015. Objective and subjective quality of life in Poland. Report. *Contemporary Economics* 4(9), pp. 1–538.

Domański, H. & Pokropek, A. (2011). *Podziały terytorialne, globalizacja a nierówności społeczne. Wprowadzenie do modeli wielopoziomowych* [Territorial divisions, globalization and social inequalities: Introduction to multi-level models]. Wydawnictwo Instytutu Filozofii i Socjologii PAN.

Felce, D. & Perry, J. (1995). Quality of life: Its definition and measurement. *Research in Developmental Disabilities* 16(1), pp. 51–74.

Gorlach, K. (2001). *Świat na progu domu: Rodzinne gospodarstwo rolne w Polsce w obliczu globalizacji* [The world at the doorstep: A family farm in Poland in the face of globalization]. Wydawnictwo Uniwersytetu Jagiellońskiego.

Nussbaum, M. & Sen, A. (1992). *Quality of Life*. Oxford University Press.

Raudenbush, S.W., Bryk, A.S., Cheong, Y.F., Congdon, R.T., Jr., & Toit, M. du (2011). *HLM 7. Hierarchical linear and nonlinear modeling*. SSI Scientific Software International.

Stiglitz, J.E. (2002). *Globalization and Its Discontents*. W.W. Norton & Company.

Sztompka, P. (2007). *Zaufanie. Fundament społeczeństwa* [Trust: The foundation of society]. Wydawnictwo Znak.

Trzebiatowski, J. (2011). Jakość życia w perspektywie nauk społecznych i medycznych – systematyzacja ujęć definicyjnych [Quality of life in the perspective of social and medical sciences: Classification of definitions]. *Hygeia Public Health* 46(1), pp. 25–31.

Walas-Trębacz, J. (2017). Globalizacja – przesłanki i wyzwania dla przedsiębiorstw funkcjonujących na rynkach międzynarodowych. In: J. Wiktor (ed.), *Zarządzanie przedsiębiorstwem międzynarodowym: Integracja różnorodności* [Managing an international company: Integrating diversity], pp. 19–62. Wydawnictwo C.H. Beck.

Wasilewski, J. (2014). *Świadomość zmian globalizacyjnych na polskiej prowincji* [Awareness of globalization changes in the Polish provinces]. Wydawnictwo Naukowe Scholar.

Włodarczyk, K. (2015). Jakość życia postrzegana przez Polaków w XXI wieku [The quality of life perceived by Poles in the 21st century]. *Konsumpcja i Rozwój* 1(10), pp. 3–16.

PART FOUR

SOME INDEPENDENT STUDIES

Think Locally, Act Globally: Polish farmers in the global era of sustainability and resilience, ed. by Krzysztof Gorlach and Zbigniew Drąg in collaboration with Anna Jastrzębiec-Witowska and David Ritter
Jagiellonian University Press, Kraków 2021, pp. 521–522
ISBN 978-83-233-4949-5
DOI: http://dxdoi.org/10.4467/K7195.199/20.20.12739

Some Introductory Remarks by the First Editor to Part Four

Krzysztof Gorlach   https://orcid.org/0000-0003-1578-7400/

The last part of this work, containing chapters 12 through 15, has a rather specific character. It consists of relatively independent studies, conducted at the proverbial outskirts of the mainstream research, by authors that include some members of the project team, employed as post-doctoral researchers or as scholarship recipients. This part of the publication also contains writings by authors who are loosely connected with the research team, as either volunteers or informal collaborators.

The first chapter of the discussed part comprises Chapter 12 of the entire publication. Its author, Adam Mielczarek, PhD, focuses on qualitative analysis of the functioning of a small group of dairy farms located in central Poland. The main goal of Mielczarek's analysis is to elucidate the various lifestyles of several generations of farmers engaged in the dairy sector.

The second chapter of this part, Chapter 13, addresses various aspects of food safety and food security. Its three authors are: Zbigniew Drąg, PhD (second editor); Professor Piotr Nowak, PhD (informal collaborator in the project); and Martyna Wierzba-Kubat (PhD candidate and a scholarship recipient in this project). This particular chapter presents opinions and attitudes of the surveyed farmers on matters pertinent to food safety and food security.

The third chapter of this part, comprising Chapter 14 of the entire publication, concentrates on the role, and increasing importance, of information technologies in regard to various aspects of farm management. The content of this chapter is provided by Adam Dąbrowski (PhD candidate and a volunteer in the project), Maria Kotkiewicz (PhD candidate and a scholarship recipient), and Professor Piotr Nowak (informal collaborator on the project). The authors examine and interpret the results of the study done specifically for this project, as well as earlier studies that they conducted.

Finally, Chapter 15, which is the last chapter of this part, and written by Professor Grzegorz Foryś (informal collaborator on the project), deals with problems of farmers' political mobilization. The reflection's focus for this chapter is based on the results of the study customized for this project and other sources considered by the author.

Think Locally, Act Globally: Polish farmers in the global era of sustainability and resilience, ed. by Krzysztof Gorlach and Zbigniew Drąg in collaboration with Anna Jastrzębiec-Witowska and David Ritter
Jagiellonian University Press, Kraków 2021, pp. 523–559
ISBN 978-83-233-4949-5
DOI: http://dxdoi.org/10.4467/K7195.199/20.20.12740

Chapter Twelve: The Modern Barn in the Local Cultural Field: Habitus of strong family farms in the face of changes in the contemporary Polish countryside

Adam Mielczarek   https://orcid.org/0000-0002-8932-2932/

12.1. Introductory Remarks

The period of time from the decline of Communism to the present day was a time of deep and radical changes in the functioning of the Polish countryside. After 1956, when Polish authorities ceased their push for collectivization in agriculture, the slow process of agricultural modernization began. It meant a relative improvement in access to agricultural machinery and the industrial means of production, but it did not lead to farm enlargement or specialization in agricultural production. In the second half of the 1980s, Polish agriculture appeared to function symbiotically with the economy of deprivation which characterized that time. It generated a rather specific rationality for farming, in which limited access to the means of production was compensated by the comforting certainty that the agricultural products will be sold, even if they were not produced in an efficient way. This preserved farm fragmentation and thus, the agricultural sector at the time was not really able to maximize production (Halamska, 2015, pp. 113–114).

This time period was interrupted by changes related to the transformation of the political system. These were all-encompassing transformations of the economic system and various forms of support for the agricultural sector ceased to exist. The certainty of sales of agricultural production ended rapidly and there was a significant drop in profitability, while competition increased substantially. The countryside became a buffer zone absorbing the workforce surpluses that resulted from rapid economic changes occurring in other sectors. The difficult conditions in which agriculture existed at the time contributed to the unfolding of a gap between small subsistence farms, that produced for their own needs and only aimed to maintain their existence, and larger, specialized farms that were on the rise and actively adjusted themselves to the new conditions of farming (Gorlach, 2001, pp. 173–187; 2009, pp. 73–80).

The accession to the European Union, as well as the prior emergence of pre--accession instruments of support for Polish agriculture and rural areas, led to another change in that realm. This allowed for access to numerous programs which served to bolster the agricultural modernization process (Fedyszak-Radziejowska, 2005, 2008), as well as the acceleration of market elimination of some farms that could not adjust well to the preferred production modes (Bukraba-Rylska, 2014). For farmers and rural residents, the direct agricultural subsidies that owners of agriculturally utilized land began to receive became an important factor in their economic calculations.

It was a time of growth for specialized commodity farms and small farms wherein production was, for the most part, merely a modest supplement to one's income, which was mostly acquired otherwise. The earlier observed trend, wherein farms of average potential fell out of operation, changed to a more "survival of the fittest" scenario, in which the weakest farms ceased to exist (Gorlach, 2009, pp. 80–95). This resulted in a decreasing percentage of farmers among the total rural population and a growing percentage of rural residents who worked in urban and suburban areas in professions unrelated to agriculture or rural areas.

Generally, in the last few decades, agriculture and rural areas have experienced several serious and consecutive tremors affecting their functioning, the results of which could be seen in an overall shrinkage of the rural population and a decline in the percentage of the agriculturally-involved population. The number of residents earning income outside of agriculture increased, and that included people commuting to work in urban areas where they held professions more typical of urban areas (Halamska, 2016). The farm structure changed but, for the most part, the farms preserved their family character (Gorlach, 2009, pp. 97–107). Serious civilization changes were parallel, having an effect on the technologies of farm work and shortening the previously significant cultural and societal between urban and rural areas.

The aim of the research described herein was to answer the question of how the farmers themselves dealt with the cumulative changes. Keeping in mind the "end of peasantry" hypothesis and the notable lifestyle and customs-related changes occurring in Polish rural areas, the research addressed the question of the persistence of traditional cultural patterns shared by Polish farmers. In addition to the large comprehensive, quantitative research described in the other parts of the following publication, a relatively small qualitative research project was conducted which tracked the processes taking place on prosperous, specialized family farms. It was assumed that these strong, well-adjusted farms and not weak, subsistence farms, would be setting trends for the definitive evolutionary model of a strong farm. As for the theoretical language used to describe the subjective part played by farmers facing the changes affecting their farms, the concept of "habitus," as popularized by Pierre Bourdieu, was applied. This concept served to explain the phenomena of changes in, as well as the permanence of, certain "habituses" which were able to determine the existence or decline of social categories such as the "peasant class" in Poland.

12.2. Continuity and Change from the Theoretical Perspective of Pierre Bourdieu

According to the French sociologist Bourdieu, a human person in his or her life acts within many parallel social realms, which he calls "fields." Individuals conducting their activities in various fields follow a different set of rules in each field, playing the game accepted in each particular field. Domination is the objective of the game. Individuals active in a certain field strive to have the best possible, as well as most influential, position in that field (Bourdieu, 2005, 2007, 2009).

One's chances in any given competition and ambitions might not be equal. Generally, the entire field and its condition, shape, and rules, are historically determined. The characteristics of the field depend on what happens both inside and outside of the field. Some fields have an impact on other fields and their combined influences are iteratively complex, with variables not always resulting in a well-rounded entity which could be described as a uniformly structured system. The situation in the field is always dynamic: in some way it is a self-regulating balance of power which, through the actions of participants, constantly changes (Crossley, 2003; Thomson, 2008).

To characterize acting individuals from the perspective proposed by Bourdieu, one should mention two main features: they have a certain "habitus" and possess some kind of capital. Habitus is a set of dispositions shaped by socialization and experience. It derives from the starting position of the individual and the environment that has shaped that individual, but also changes with his or her new, accumulated experiences. This capital can be seen as (material and nonmaterial) resources that are at an individual's disposal and can be acquired, gained, lost, or socially valued in various ways. The capital resources that are in an individual's possession also have a reciprocal influence on the habitus. Consequently, habitus plays a crucial role in the use of capital resources and their abilities to reproduce (see Bourdieu, 2000).

Habitus can be seen as a factor contributing to the continuity and permanence of social life. It is through habitus that the rules of social games, taking place in various fields, are processed and reproduced, as well as the behaviors of representatives of various groups. Habitus also ties in with a certain "feel of the game." It is a factor that enables individual players to understand which rules to follow, how to use them, and, sometimes, how to stretch, or even change them. Habitus belongs to individuals, as well as to groups, that occupy similar positions in a field and are in possession of similar types of capital. These groups reproduce it, handing down those predispositions and inclinations that characterize them to the individuals aspiring to gain access to them. These features, however, remained unchanged. The acts of individuals are based on their habitus, which was already shaped in the past, but individuals still remain self-reliant entities, able to adjust to circumstances and, therefore, are subjectively shaping the rules and solutions within the habitus (Maton, 2008).

This dynamic has an influence on the field and on habitus, which can undergo change through mutual adjustment. External circumstances, including changes taking place in other fields, can more or less rapidly "downgrade" the reproduced habitus, forcing individuals to accelerate their adaptation of this habitus to the new reality. The adaptation of habitus might sometimes be outpaced by the dynamics of the change. This is known as the hysteresis effect (Hardy, 2008) and it describes the situation in which patterns of conduct that were shaped over several decades maintain a certain permanence, even if they are not fully compatible with the new rules.

This conceptual grid elaborated by Bourdieu allows for an examination of the reality of the Polish countryside as a specific, distinct field, wherein the conditions of quite rapid change exist, and the struggle for domination takes place. Heretofore, the field was relatively homogenous, with the dominant habitus being family farms. Due to civilizational and sociopolitical changes, the bearers of the new habitus appeared in the field and began to question this domination. As such, despite the preservation of the traditional habitus connected with the family-based economy, the rural field as a socio-cultural space is experiencing disintegration. The farmers who used to dominate are currently not numerous enough to impose their own model of the field.

Other social groups are not adequately rooted in that field to propose new cultural patterns which could be suitable for adaptation and acceptable to all involved parties, and thus replace the old community.

The question regarding preservation of the family farm cultural model is really a question about the future of the habitus that is tied to it. Considering the far-reaching transformations taking place in agriculture and rural areas, it is hard to imagine that this habitus could remain unchanged. Following the logic of the above approach, there are three possible scenarios for farms' destinies.

In the first farm destiny scenario, the extant habitus will adapt to the undergoing changes and will reproduce through the absorption and use of emerging innovations, which could effectively contribute to the creation of vital capital types. Livestock farmers will modernize their barns, reorganize their work, and learn how to use the inventions and achievements of contemporary civilization in everyday life. Their lifestyle, however, will still be based on old cultural patterns and they will determine what is valuable and meaningful for people involved in farming.

In the second scenario, the attachment to the old rules of the game will become an obstacle to the use of novelties and innovations that allow for the possibility of building one's own position. This would result in the hysteresis effect, described as a certain inertia of the habitus. In such a situation the old patterns are reproduced and would not guarantee the existing capital resources to build upon the current position in the changing world. This could mean that farms trying to maintain their position will invest in outdated solutions that do not strengthen their position in relation to other actors that are present in this particular field. The farms will maintain their traditional form but this will happen at the expense of the farmers' social position.

The third scenario deals with the situation of the cultural reorientation of traditional, established farmers, who find themselves within the orbit of heretofore foreign cultural influences, which they would make the object of their aspirations. By using the new possibilities brought forward by civilization, they will conclude that the habitus, which until now has formed their attitudes, has become a burden. Consequently, the principles and values they have adhered to will be replaced with principles and values generated by other, stronger groups, which would then become the objects of their aspirations. In contemporary Poland this could probably lead to the watering down of the rural ethos and customs in favor of developing "bourgeoisie" tendencies in previously traditional farmers, who might be currently aspiring to become entrepreneurs and adopt some imagined patterns of urban life.

12.3. Research Concept and Survey Sample

The aim of the research was to acquire knowledge of changes in habitus taking place within Polish family farms. It was focused on the characteristics of this category of farms which, according to researchers, had a chance not only for surviving the time of changes, but also were thought capable of establishing cultural patterns for farming in the near future. Therefore, farms that appeared to be strong and thriving in the new reality were chosen as subjects for the research.

The idea of the research was to reach representatives of two generations of farmers continuously operating the same farms. It seemed worthwhile to compare how the changes occurring in family farms were perceived by those farmers who were affected by the farming experience in the 1980s and 1990s with the way of thinking of young farmers who were currently in the process of taking over farms. Before the research, it was presumed that through an examination of these often contradictory views from these two different categories of farmers, who derived different life experiences from operation of the same farms, a clear picture could possibly emerge of both the changes in the characteristics of family farm life, as well as the family farm styles of yesterday and today.

In the research described here, a significant amount of attention was given to the phenomena of habitus preservation, as well as the direction of changes in which the traditional habitus of the family farm is moving. The researchers examined the changes affecting the farm and farming methods, as well as the metamorphoses rural areas were undergoing and the social relations that existed there. There were questions about the extent to which the actors of social processes occurring in rural areas were able to define these changes and what tools they used to deal with them. In the reality of observed far-reaching transformations affecting the contemporary rural culture, the researchers were looking for the right language to answer this question: What has changed and what has remained the same?

The choice of the group of farmers which could serve as the population sample was not obvious. There was awareness about farmers and rural residents having various strategies of adjustment to social and economic changes. A natural point of reference for the planned research was the study of Amanda Krzyworzeka, conducted in the first decade of the 21st century and devoted to the paths of adaptation followed by farmers (Krzyworzeka, 2014). In her empirical research the author conducted a thorough study of residents of just one village and analyzed the farms that dealt with the economic changes in a variety of ways. Although Krzyworzeka stated that the most frequently applied solutions were the mixed ones, she also elaborated their systematization. She distinguished three types of adaptation strategies characteristic of farms. The most beneficial strategy (although only applied by a rather small number of farmers, and generally considered risky) was that of specialization. It meant a total commitment to agricultural production and significant investments in modern, high-efficiency, technical solutions. The second strategy, more frequently observed, had to do with diversification. It involved combining mercenary work of an "urban" character with maintaining some forms of farms or with multitasking agricultural production. The third strategy entailed having direct subsidies from the state as the main source of income and simultaneous partial fulfillment of one's own needs through agricultural production.

Aware of the scope breadth of the planned qualitative research, the author of the following text decided to analyze only one strategy and, thus, only one category of farmers. As the destinies of family farms are the center of interest of this entire publication, the decision was made to concentrate on specialized farmers. They were the group best able to keep the family profile of the farm (at least in the studied region) and, at the same time, remain oriented towards agricultural (livestock) production and acquiring satisfactory profits from same.

There was no expectation to obtain cross-sectional data or material representative of the regional farming population. It was decided that the study should be devoted to a somewhat homogenous group of farms operating in one region and not considered to be particularly diverse. The dairy cattle-breeding sector, already well-studied by experts in Western Europe (Ploeg, 2003), was specifically chosen for this research. The study was conducted in the Mazovian, where this category of agricultural production was strongly represented and already explored by Amanda Krzyworzeka, quoted earlier in this text.

The research was conducted from January 2, 2018 to March 9, 2018 in the northern and north-eastern parts of the Mazovian in 5 counties (Płoński, Ciechanowski, Makowski, Sokołowski, and Miński). On average, the farms were located about 100 km from Warsaw. Visits were paid to 10 farms that had barns (cowsheds) at various levels of modernization, containing from 16 to 70 dairy cows. As the study purposely focused on the farms that were thriving, the criterion for study participation was based on successful application for funds from the Rural Development Programme (measures: M4.3: support for investments in infrastructure related to development, modernization, or adaptation of agriculture and forestry, and M6.1 business start-up aid for young farmers with a strong orientation for family farms). The farm visits, as well as the interviews, were organized and carried out by the employees of the Agricultural Advisory Centres. For each of the studied farms, 3 interviews were conducted, in the following manner:

- one with the current operator of the farm,
- one with the previous operator (father or mother of current operator) and, where possible, with the representative of an older generation, as well
- one group interview, to which all people present at the time on the farm were invited. (This meant that the two people who were interviewed earlier on an individual basis were joined by other household members, i.e. wives, siblings, grandparents, for the group interview, if they agreed to do so.)

Each type of interview was handled according to a separate and previously prepared research scenario. The interview with the present (young) farm owner/operator mostly dealt with contemporary manners of farming and was designed to reveal to what degree the respondent perceived their style of farming as a continuation of the model followed by their parents. The interview with the previous farm operator/owner centered on past conditions and manners of farming. The goal was to obtain opinions on the differences between the current farm management of the young farmers and the previous management of the older respondents, as seen from the perspective of the older generation. The group interview concentrated on the respondents' perceptions of changes taking place in the countryside, understood here as a sociocultural space. It was our intention during this part of the interview that differences in the opinions of the respondents that were detected during individual interviews would be re-addressed and followed up with further inquiries.

In practice, the interviewers were able to conduct 10 individual interviews with current farm operators, 9 with previous farm operators (father or mother of the farm operator), 2 interviews with the grandfathers of current owners and 9 short group interviews with at least 2 respondents. It should be noted here that interviews were carried out in the kitchens and living rooms of farmhouses, while the everyday life of the farm was going on. Therefore, the line between the individual and group character of the interviews was rather fluid. The household residents and family members who were at home during the interviews often listened to the answers of the other respondents and did not hesitate to offer their own input. Such situations were not discouraged, as they were considered somewhat beneficial for research purposes.

All current farm operators that participated in the research were men, aged 24–48, with a significant majority of them under 30. They took over the farm relatively recently but their experience with managing the farm was not too short for the purposes of the research. The older generation was represented by individuals aged 54–80, and included two women. Women actively participated in 7 group interviews and their input was significant.

Despite the differences in age, affluence, and farm location in various parts of the Mazovian, the research results turned out to be quite consistent. There were a lot of similarities between the studied farms in how the respondents described the reality of farming and dealing with it. Even the differences observed between the generation of 20–30 year olds and that of the older farm operators did not vary a great deal from farm to farm. It should be stressed that the interviewed farmers defined themselves as rural residents, farmers, and members of family farms, and that was important for the purposes of the research questions. All respondents described the character of observed changes in similar ways and their thoughts on farming did not differ much. Comparing the experience of previous farming generations that passed the farm to their children with the experience of young farmers, respondents acknowledged that the contemporary conditions of farming and rural life were now much different. They also expressed the notion that the most basic conditions for agricultural family economy had changed. Nevertheless, despite the ongoing changes, family farms were still described by respondents as a phenomenon that remained unchanged in its principles. At the same time, all the respondents observed the transformations affecting the Polish countryside with some distress. From their perspective, social changes occurring in rural areas were changing rural customs in a significant way, as well as the foundation(s) for the functioning of a traditional rural community.

12.4. Farm History in the Stormy Time of Changes

As can be inferred from the interviews, the respondents' perceptions of the social world, where the lives of members of agricultural households and farms were happening, could be seen as a discrete universe, different than other fields of social life. The history of the farm and the family living on that farm is construed by household members in reference to the events occurring within the family and on the farm. These events could relate to people (births, marriages, deaths) or significant changes in ownership of buildings, machines, livestock, and equipment. This history is perceived as separate from the external events, which can be identified objectively as causes and factors of significant changes taking place in the Polish countryside over the last few decades. Such perceptions can be explained by the limited participation of farmers in other fields of social life (at least this is the case with the group of farmers that participated in our research). This is particularly true of those fields which, over the recent decades, were the sites of struggles that had determined the directions and extent of political and economic transformation in the region and the entire country. Quite different was the case of those few farmers who had displayed some political engagement and used the historical caesuras typical of urban public space and discourse. They saw the history of the same farm as a monotonous sequence of events on the fringe of world history, somewhat influenced by it but, for the most part, being governed by its own rules and characterized by significant inertia.

One of the most striking characteristics of how the respondents described the changes in the countryside was the almost complete absence of any political component. Although the memory of respondents usually went back to at least the end of 1980, and therefore covered the times of great, historical changes within the political and economic systems, their recollections were rarely rooted in political caesuras. Although the interviewers pressed the respondents to estimate when the events and situations they described took place, it was rather difficult to establish the historical context or references based on the answers they provided. When asked directly, they accepted the time frames proposed by the interviewer: "during the Communist times," "in the 1990s," "after joining the European Union" and were able to follow them. Nevertheless, they quickly abandoned them and referred to private histories which were rather obscure (at least from the interviewer's perspective), marked by events such as births, marriages, deaths of family members, moving to new homes and suchlike. They pointed to the dates when the farms were taken over (formally or factually) by consecutive generations of farmers and considered those to be the crucial moments in their private history. When reporting on the farm history, the respondents often mentioned the acquisition of new machines. It was rare for the respondents to talk about these events as turning points that would significantly modify their way of life and farming.

Generally, the narrations presented by the interviewed farmers were lacking important caesuras and breakthroughs. They perceived the history of their farms in categories of continuity and viewed the numerous changes occurring within them as a result of slow processes of gradual adjustment to certain trends observed in the world that remained external to the farm. Although they appeared to hold the conviction that over the course of their lives some deep changes had taken place, both on the farm and in rural areas generally, they still did not note the changes in the overall conditions for farmers. It was obvious to them that the root of farm management was to adjust to circumstances, meaning, for example, the reorientation of production, while using the right tools for it or financial instruments.

In their stories about farm history, farmers sometimes identified periods of time of better or worse economic conditions for the state or agricultural policies. However, their placement of events in time was, for the most part, just estimated.

In the 1980s and at the end of the 1970s we had the biggest acquisition of machinery for our farm. Whoever wanted to buy machines, bought them. (village of Torby, young farmer)

I remember such times… I am not sure whether it was in the 1990s or 1980s, but probably the 1980s. There was money, but you could not buy any merchandise. (village of Zastały, mother)

The year 1989 as the moment of Communism's collapse in Eastern Europe was almost completely absent from farmers' perceptions of the farm past. The connection between the rapid changes in the economic system and this particular moment in the history of the farm was only reported by one farmer, who was quite politically active. He was an activist in "Rural Solidarity" ("Solidarność Wiejska"), at the regional level, and a member of the municipal board after the 1990 election. His recollection of the time of economic transformation was as follows:

I produce milk, I have 8 cows. But nobody wants that milk . . . So, I go to the apartment complexes to sell the milk there . . . Capitalism is built like that. These are the things I used to do in order to survive somehow . . .
I stood with my milk by the block of flats, I had my [regular] clients. (village of Główki Wielkie, father)

His testimony was the exception. Other respondents took almost no notice of the differences between the decline of The People's Republic of Poland (the country's former name) and the harsh beginnings of the market economy in the Republic of Poland (the present name, after the official renaming in 1989). Their narratives did not include terms such as capitalism or socialism. The interviewed farmers were good managers, able to cushion the tremors stemming from the early transformations. As the time of these transformations did not mark any important events for the farm, the respondents could easily forget them.

[Q: Were the 1990s similar to the 1980s?]
—Yes, in some ways. This was going on and on. So, the production was increased from year to year. (village of Gadziny, the father)

According to the respondents, the years following the collapse of Communism in Poland were not more memorable than the previous ones. The interviewed farmers recollected such events as the change in the organization of milk purchases, which forced them to invest in cooling equipment. The respondents associated the moment of Poland's accession to the European Union with the accessibility of agricultural aid programs and direct subsidies.

> I think that the aid from the EU helped us a lot. Things sure have changed in the countryside after the accession to the European Union . . . As can be seen in the countryside, the people have done a lot since that time. It is not just the purchasing of machinery but also expanding the farms and... other stuff, too. (village of Olinka, young farmer)

Another issue was the introduction of phytosanitary regulations, which forced livestock farmers to separate the barn for the cows from other types of animal production. In some cases, farmers resigned from having additional animals on their farms, finding that combining of other types with their main livestock (cattle) to be too challenging.

12.5. Changes in the Manner of Farming

12.5.1. Changes on Farms

> Today's economy is lighter than it was before . . . [In the past] you took your horses to work the field. You plowed. Birds were singing, a little sparrow was singing. You came home, you ate your meal and your hands were clean. Now, you leave for work clean and you come back dirty. You need to add some gasoline and some oil may drip on you. The farm work is lighter but it makes you more nervous. Back in the day you lived comfortably like a butter donut. You kept calm and you were doing things slowly. And now everything is speeding... Faster and faster... (village of Przeszkody Dworskie, grandfather)

In our study, the area of changes perceived by farmers as important to their lives related most strongly to the mechanization of agriculture. Mechanization greatly affected farm work in terms of physical effort and potential production volume from family farms. Even though not all respondents of the older generation personally worked with the help of horses, they all remembered the time when horses were not rare in agriculture. All of the interviewed farmers also remembered, however, how much physical labor they had to put into farm work, when they only had simple tools and basic technologies of production at their disposal. They recalled how much their worked changed after their farms acquired machinery and applied modern farming solutions.

> Back in the day, it was such a joy to buy a manure spreader or a harvest-binder! (village of Zastały, mother)

The earliest changes were related to outdoor agricultural works. Mechanization made them faster and less labor-intensive. The need to engage a large number of people all at once in farm work became quite rare. Also, farm work became a bit more independent from the weather. It became possible to operate larger farm areas with the labor of family members only. Consequently, these changes also brought an increase in production.

There were changes in the organization of work and the division of labor on the family farm. The phasing-out of lighter, auxiliary farm work, which used to be the women's domain, could serve as an example here. One of the respondents described it in detail:

> When I was farming together with my wife, we also did all the work in farm buildings, the two of us. The field work we did together, as well. My wife was my helper... Now [my wife] does not drive to the field. The machines are doing the work. Collecting of hay is no longer done the way it used to be. Back in the day there was collecting from the field and drying. There was also baling, making the bales of hay. There was a lot of manual work with pitchforks and the rake. Now, this is done by machines. Grass faded and dried. Then it was packed and all wrapped. There were other types of women's work as well, such as collecting the potatoes. In the old times—and I remember my parents' farm work—women even helped with spreading fertilizer on the fields. Later, during our farming days, this was less common. There was a harvester and it was easier . . . The collected hay was taken from the fields with a self-binder and then threshing was done in the winter. The women did that, too. They maybe did some lighter work, but also this type of work.
> [Now m]y wife . . . does not go to the fields . . . When my parents were farming, such changes were unheard of. (village of Gadziny, father)

Agricultural activities are no longer conducted with the same diligence as they were in the past. Crop remains left in the field are considered waste and have no value, while in the past farmers gleaned as much of the crops as possible, even imperfect remainders. Sometimes this is considered a generational difference: representatives of the older generation still have the propensity for conservation so characteristic of the times before mechanization. They are (almost instinctively) opposed to wasting the crops left behind in the field by machines. This reaction, however, is not the basis for any actual conflict, as the older farmers (ultimately) accept the explanation that collecting these remains would go against the current rationality of farming and there cannot be any profit from it.

These changes do not just affect the field work. In the case of the studied group of farmers, there is a rather important change related to the factors of production profitability. The decision to specialize in livestock farming makes it impossible to diversify livestock on a farm. All respondents reported that, in the past, they raised several species of farm animals but now that situation is rare. The fact that there were no hens or other small animals in the farmyards of the studied households could also be seen as part of this change. This may not have much significance in how the farm functions, but can still be seen as an element of change in the agricultural landscape that is easily observed and generally acknowledged.

> Back in the day, there were cows, there were pigs and hens running around. Now, it is not allowed. (village of Torby, old farmer)

The changes that were particularly important to the studied group of farmers had to do with innovations in a) milking and the transport and storage of milk, b) providing feed for the livestock, and dealing with the ground surface bedding on the barn premises. During the study, most of the respondents were in the process of rebuilding and/or modernizing their barns or had recently undertaken such improvements. For that reason, these issues were extensively discussed by research participants. On several farms, modernization dealt with some elements of livestock care and included the introduction of solutions at various levels of technological advancement. All of these innovations were perceived by respondents as facilitating their work and improving their milk yields. They could not, however, change one crucial factor: farmers' everyday duty of tending to the cows in the mornings and evenings.

> Despite the different machines and different work conditions this duty still remains. Constant duty. Especially if you are raising dairy cattle. (village of Gadziny, father)

Different levels of technological advancement in respondents' barns resulted in their different perceptions of the meaning of technology in agriculture. For some of them, technology led to the dehumanization of their relations with animals, as more of them populated the barns and were "not known by name." For others it was quite the opposite, and technological solutions were seen as an improvement in animal welfare and comfort. The cows were no longer tied to their stands during milking but only goaded to the milking platform at designated times.

When talking about investments in farm buildings and machinery, most of the respondents expressed a fear of excessive farm expansion. In almost every interview, the hypothetical situation of investing too much and having too many farm animals was seen as an example of the organization of farm work extending beyond the framework of the family farm. Mortgage and credit liabilities—as well as the increased production potential—were, according to the respondents, a trap that all farmers modernizing their farms could fall into. This was considered a notable threat to the desired model of farm and rural life.

> I have a friend who expanded their barn for the third time. This never ends. If you increase the number of animals then the area becomes too small. If you increase the area then the machines have to be changed as a bigger area requires a different kind of machines. This leads to a situation where, individually, you are not able to keep up with the work. (village of Główki Wielkie, son)

In one of the municipalities where the research was conducted, the respondents gave the example of a farmer who fell into the expansion trap. He took credits and mortgages and invested in various agricultural endeavors. He created a rather big farm but, at the same time, overburdened himself with so much work that his farming model was considered a negative example in the area.

> Around here, there is a man who tries everything. He breeds cows, sells corn, and also provides services. They can work there for 5 days straight, day and night . . . Bigger and bigger machines are being bought. This year, it looks like they may not be doing so great. They are not able to keep up with all that stuff. (village of Zastały, young farmer)

The progress, related to cultural changes, gradual gentrification, and better availability of material goods was, in the respondents' shared opinion, obviously parallel to the transformations stemming from agricultural modernization. The most important changes in that regard had to do with the increased availability of private motor vehicles. Institutions and amenities located in urban areas that had been relatively inaccessible for rural residents, could now be reached easily, as cars had become more common and affordable. Consequently, the rural residents who participated in the research did not express the notion that their lifestyle differed much from the lifestyle of people in urban areas.

Now, the countryside is not much different than the town. In terms of living conditions, dress style, food, it is not what it used to be. Today everybody has a car and you can get anywhere you want quickly. Today, even the attics are heated; there is Internet and three TV sets in almost every [rural] house. (village of Wola Krótkowska, group interview)

It is not difficult to note this factor as the cause of another phenomenon. It becomes obvious that a large portion of rural residents have, over time, fallen from the local, rural, and cultural field and now feel more connected to the urban world. Even the capital is not so distant anymore and it offers various types of employment and income that were never before possible. The city also provides incentives to have different lifestyles.

Another innovation frequently mentioned was the availability of communication tools and the Internet. For farmers, these were important tools for acquiring information, including news of techniques and technologies which could be applied in agriculture. New technologies can be viewed as factors that enhance the position of young people on the farm. It is usually the younger generation that is more familiar with these kinds of information tools, which invert the natural process for passing knowledge on to the next generation. While the older generation was still leading in having the hands-on experience and practice in working in the field, the identification and assimilation of new technological possibilities became the domain of the younger generation of farmers. Their competencies in that area could help them develop in the role as people who will eventually take over in leading and managing farms of the future.

—I took over the farm in 2015.
—What was it really like?
—It was gradual. Decisions about the machines were made by me… Obviously, I have better access and familiarity with the Internet. I know what was going on. My dad gets information from people. (village of Gadziny, young farmer)

In this context, the change in the mechanisms of recruitment for field work was also mentioned. In the past, one could speak of a negative selection causing the less bright males in rural families to stay on the farm. Today, the tendency is quite the opposite.

Back in the day, only dummies stayed on farms. Today, the one who stays on the farm must be smart to be able to manage all the stuff. (village of Gadziny, group interview)

I did not have to stay on the farm. As a child, I kind of liked the farm work and, with time, I got used to it. My grandfather became a farmer because he had to. My dad became a farmer because it was done from generation to generation. And I see a lot of potential in farming. I am very aware of my choice. It is not all about the money . . . I see the potential for my self-fulfillment . . . I do what I am capable of, and I like it. Furthermore, I work in very fine conditions. (village of Przeszkody Dworskie, the group interview)

The level of education of at least a good portion of the young respondents is rather high. They decided to return to farming of their own volition, after completing their college or university education, expecting to have better opportunities for both professional advancement and self-fulfillment. Young farmers in the study were more educated than their parents and their peers from the neighborhood who worked in urban areas. The educational gap did not harm their relations at home, and only solidified the position of the new head of the household after they took over the farm.

Education matters. The studies give you knowledge on lots of things. After I started attending university, I had some ideas about the barn/cowshed. I knew what would be better for the cattle and better for us. This was also because of school... (village of Gadziny, young farmer)

When asked about the working conditions, farm strategies, and the determinants for success on farms, the interviewed farmers were not convinced that much had truly changed in the countryside. One of the respondents summarized it in the following way:

If someone worked then, he was able to get by. Now, also, when somebody works, he will get by. You don't have to be a genius of economic science. If one wants to meet ends, he will meet them. The same as before...

I mean there are new technologies, modernity now... (village of Torby, young farmer)

The application of new technologies in plant cultivation and animal husbandry makes physical labor lighter and farms more efficient. Family life is altered as people now have access to technological solutions that bring changes to various areas of their everyday life. However, as indicated by the respondents, when it came to the most crucial conditions of family farm operation, technological solutions had a very limited influence. As was the case with previous generations, the success of the farm was strongly related to the diligence and assiduity of farm operators. In order to succeed, the farmer must get up early in the morning and, just like in the past, the rhythm of his life is determined by the rhythms of nature and seasonal duties relating to plant cultivation and animal breeding. Although the circumstances are different than before, farmers still need to adjust production to market prices and carefully calculate to make sure that there is a balance between the production inputs and the subsequent income from product sales. Farmers should maintain adequate profits to have enough money to support their families.

Interestingly, even the current system of agricultural supports did not appear to the respondents much different than the one organized for them during the socialist regime. This was also addressed by the young respondent, who was previously quoted. His comparison of the types of support was the following:

> Getting to the bottom of what was before and what is now in terms of aid, I do not see much difference. Dad was buying stuff thanks to the allotment. Now we have the applications for the EU aid. When dad wanted to buy something [not through the allotment], he searched and searched and then bought it. He paid three times as much, but he still bought what he needed. It is the same now, if you want to buy stuff for half the price you need to fill out the application. At the end of the day, it is all the same. (villege of Torby, young respondent)

Older farmers remarked that, during their youth, the state used to guarantee the purchase prices, which gave farmers some sense of stability. Although the younger respondents followed the price of milk with some apprehension, they seemed to be accepting of its lack of stability. It might have been helpful that, at the time of the interviews, the price of milk was at a level high enough to ensure a satisfactory income for milk farmers.

12.5.2. Agricultural Production Abandoned by Rural Residents

Civilizational changes and technological progress made a significant impact not just on the manner of farming but also on the social functioning of the countryside. Changes related to the demands of production profitability and, consequently, the methods of farm management, affected agricultural employment. Today, only a small minority of rural residents is engaged in agriculture and an even smaller percentage of them makes their living through agriculture. Agricultural activity is quite often a supplemental economic activity, and in such cases cannot be accurately considered commodity agriculture. This is mostly true for older farmers, whose children did not take over the farm. Such farmers live to see their last days on the farm and they do not invest in any serious agricultural production. A large number of rural residents work in towns and cities. Some have moved to urban and suburban areas, treating their countryside homes as weekend dachas. Reflections on the decline of the rural population in villages where the study was conducted could be noted in respondents' answers.

> The village is depopulating. In the winter there are hardly any people. There is this neighbor who moved here in his old age. He has two cows. Such vegetation. But he has a pension . . . The neighbor is not the one behind the fence but the one who really lives closest to you. And that may be a kilometer away. (village of Torby, group interview)

> There are not many farms. Most of them just try to survive as long as the European Union is paying. Without the subsidies, these farms would be completely liquidated. I saw our village mayor distributing the adjuration documents in the neighborhood. There were so many last names in his stack that I did not recognize half of them. The head of the village did not know them either. (village of Wola Krótkowska, group interview)

Such transformations occurring in rural areas can be linked to the fact that non--intensive and non-specialized farms, which used to be the norm for the Polish countryside, are no longer viable. According to the research participants, living off of agricultural work nowadays requires more labor input than before.

> In the past a person could hold a job and have a few cows at the same time. One could make some money and it would be enough. And now, we are workhorses. We have to work more. In the past, the one who did not want to work hard stayed on the small farm [which eventually phased out its production]. (village of Gadziny, group interview)

The current state is the result of the process of phasing out of agricultural production and changes in the directions of professional activities of rural residents. It is worth emphasizing, however, that the abandonment of agriculture by rural residents was not perceived by respondents as unnatural or wrong.

When identifying the main factor causing rural depopulation, respondents were more likely to name a lack of descendants willing to take over the farm from their parents than the decline of unprofitable farms. This problem was particularly well understood by research participants, as the majority of them had to recently face some moments of uncertainty regarding their own sons' life choices. The phasing out of agricultural production due to a lack of descendants was perceived as sad, but natural, even for farms that had been prosperous up to that point.

> [First case:] They did not have anybody to farm with. They did not have descendants. [Second case:] The son of our neighbors died very young at 32. [Third case:] It was a different farm. Bigger than ours but there were no descendants. (village of Zastały, mother)

Abandoning agricultural production was accepted by the respondents as a natural result of modernization and the increasing demands for production quality. It was also seen as beneficial for rural residents, in terms of their freedom of choice. They also noticed that, under current conditions, working the fields the in the way they had before could not guarantee viable levels of income. For that reason, they thought it was better if former farmers were able to find different occupations, more to their liking than agricultural production.

> The farmer must be an efficient producer to make his work profitable. Before, a farmer was living at a subsistence level. There was not much control over milk quality . . . Today, if a farmer is not industrious, he will end up at subsistence level, just like in the old days. (village of Gadziny, group interview)

> I think they are making good decisions. If someone has a small farm, it is hard to live off of it. It is better to stop farming and then find some suitable work that pays the bills. (village of Gadziny, young farmer)

In many cases, respondents were able to provide examples of misfortunes that farmers themselves could be blamed for. They named insufficient work engagement and lack of talent, as well as what Bourdieu would call "the feel of the game." They also mentioned alcohol addiction, which was not so uncommon in rural areas at that time.

> —Was it an issue of profitability of these farms?
> —The low profits from these farms was one problem, but the attitude of the farmers was also a problem. When someone takes good care of their farm, the income is better. They did not have much income because their farming was so-so. They resigned from farming because of a lack of profitability. (village of Młodziki, young farmer)

> There are two types of people in the countryside: those who like to abuse substances and those who work. (village of Wola Krótkowska, group interview)

As far as personal success was concerned, the participants did not have the notion that it was the result of some special talents (to recall the words of one respondent:

"one does not have to be a genius in economics"), or exceptionally precise strategies in farming. Success was thought to come from hard, systematic, and regular work, and the roadmap to success was available to any farmer who would choose to follow it.

12.5.3. The Path of Specialization

In the early 1990s, a so-called "mixed economy" was the most common type of farming. Today, such dispersed production would not be viable and, for that reason, there aren't many farmers able to support themselves and their families with this type of production. The history of the studied farms oriented to milk production is the history of disappearing farm diversity. In most cases, however, farm specialization was not something assumed or well-planned ahead of time. It could be described as a sequence of individual decisions taken in consecutive years over a farm history and based on regularly conducted calculations.

> There was some bigger money and that is why this farm developed. (village of Gadziny, young farmer)

To recreate the path of the studied farms that allowed them to achieve today's state of specialization and development, one should consider the fact that in the small, studied sample there were two types of farms. For most of them (7 to be precise), the respondents reported that the situation on their farm had been rather stable as long as they could remember and that included the early 1990s, which was generally difficult for farmers in Poland. On three other farms, it was noticeable that, until recently, they had been in a certain state of stagnation as the developmental investments made on these farms, securing their eligibility for research, were rather fresh. In one case, the period of farm stagnation was caused by a serious illness of the farm owner. These farms were visibly less affluent, and their cattle herds were the smallest, at 20–30 animals. The farm inhabitants had smaller cultural capital than the farming families whose farm development path had been more stable. This was illustrated by the rather low educational attainment level in this category of young respondents. None of them had agricultural education or any other type of college-level education.

For the farms with a more stable path of development, the history of specialization started in the 1990s and was characterized by investments. The interviews did not indicate that their farm development followed some far-reaching plan. For the most part, it was the effect of ad hoc calculations. Even the dairy cattle orientation was the result of improvised calculations. Farmers simply assumed that with the potential of their land and the machinery already in their possession, the investments in dairy production would be more cost-effective than other alternatives. The fact that neighboring farmers operating on sizeable farms had successfully raised dairy cattle seemed to confirm such assumptions.

> Considering the size of our farm, we can make a living from dairy production, but it would be hard to make ends meet from beef cattle. (village of Torby, young farmer)

The decisions, once made, favored the continuation of the specialization path. Phytosanitary regulations that required the separation of various types of animal production were an important component of that process. The farmers generally did not have additional buildings where they could engage in other types of animal production, which they continued to maintain. They did not have the financial means allowing them to quickly construct another farm building for other animals, and thus decided to phase out those types of animal production, as they would not bring adequate profits. As inferred from calculations, dairy production appeared more profitable than raising pigs or sheep.

> The barn that we still have on our farm used to be divided into two parts: one for cows and one for pigs. After the EU accession and introduction of new regulations we had to give up one type of animal production. We decided not to raise pigs anymore and dealt with cows only. There was no boom for pigs at the time. There were better prospects for milk production, so we switched... (village of Zastały, young farmer)

The investments mentioned by the respondents had very far-reaching consequences. However, at the time of their implementation, they seemed to be a natural choice, a matter of convenience, a solution to make the work easier. They could also be understood as a result of effective marketing.

> It was important to me to erect that building. I knew my parents had a lot of work and they were tired. I thought I was helping them and making things easier for myself . . . The neighbor also invested in a similar building. It can be said that he is doing better now. Things are easier... (village of Gadziny, young farmer)

> I still remember the cooling equipment. It cost 20 thousand old zloty. The salesman came several times and was very persuasive. In the entire village of Zastały only 2 people purchased it. It was a serious decision. (village of Zastały, mother)

The positive results that new machines and innovative solutions brought to the farm predetermined the progress of further specialization as the novelties truly made farm work easier:

> More convenient for cows, and more convenient for people. (village of Gadziny, mother)

They also increased farm profitability, justifying resignation from other forms of economic activity.

> At the time, in the 1990s, farming became profitable. We liquidated our store because the store could not generate the same profits as the farm. So, we switched to farming completely. (village of Zastały, mother)

It is well known that raising dairy cows can be labor-intensive. At a certain point in farm development, engaging in other types of production appeared to be an unnecessary dilution of power, energy, and financial means, which would not bring any income or other benefits.

The investments created a distance between farmers who modernized their farms and those who decided not to. Although it was not visible at first, the farmers who did not make such investments created developmental barriers for themselves.

> If we did not have that cooling equipment, we would probably have hit the wall like those farmers who ceased dairy production. We would think that this was the end and no further development could be possible. It was like a developmental barrier . . . For many years after we had our cooling equipment, farmers still delivered milk in those old-fashioned cans. We were more advanced. Then the buyers did not want to get milk by the road, and these farmers hit the wall. (village of Zastały, mother)

It should also be noted that once a sequence of decisions was taken, they contributed to a certain "path of dependency," leading their development in a certain direction. Credits, mortgages, and other financial liabilities incurred at the time of receiving subsidies, as well as large investments in modernization, limited farmers' options for changing their course in farm development. As mentioned earlier, once the dynamics of investments were started, it stimulated new investments that went even further. Farmers started to worry about the possibility of being sucked into a spiral of new obligations and liabilities which would make them expand their farms beyond their needs and abilities.

Despite the specialization, other ways of acquiring income had not been completely rejected by farmers at the time of the research. This should be seen as another illustration of a pragmatic and practical approach to the farming strategy. Firstly, all farms that were included in the study were involved in plant production, as a source of animal feed for their livestock. The respondents shared the opinion that such a solution was cost-effective and the quality of their own animal feed was better than the quality of feed that could be purchased from other farmers. Besides milk, the respondents also sold what could be considered agricultural "sidelines" from the main focus of their production, such as heifers, beef cattle, and surplus plant production. In some cases, farmers had made a conscious decision to keep certain elements of a "mixed economy." On more developed farms, income from auxiliary types of production did not make a significant impact in the overall farm budget. On two or three less developed farms, the operators intentionally preserved the principle of multifaceted production, as described in the 2014 research by Amanda Krzyworzeka. However, this could also be interpreted as one of the symptoms of the early stage of specialization.

Diversification could also be seen in the phenomenon of a certain portion of farm residents being employed outside of agriculture but, to some extent, still helping with farm work. It was fully accepted that some people who lived on the farm did not participate in agricultural work. In particular, young unmarried farmers declared that they were fine with their future wives having urban professions and not farming with them.

The model of internal organization of the farm did not change much over the years. Except for one farmer, all respondents did all of their farm work with the exclusive help of the family, without hiring permanent employees. Inside the farm, the division of labor and duties was based on traditional gender-related roles. The situation of farm residents not participating in farm work and working elsewhere was also readily accepted. On one of the weaker farms, there was even a case of a young farmer who had combined the function of being the head of the farm with a full-time job in a different sector. All of the above solutions resulted from the simple reproduction of well-established patterns. For all of the involved parties, such a system of organization was obvious and there was no need to question it.

12.5.4 Reproduction of Farming Patterns

The selection of farms for the study favored representatives of the younger generation as heads of farms. This had to reflect the true status of the farmer and the farm rather than just be a meaningless declaration connected with the need to fill in an application for EU aid. During the preparation of the research tools, it was assumed that the strategic ideas of the younger and older generations would be much different. However, this hypothesis was not confirmed. Despite the significant number of young farm operators having an agricultural education exceeding the educational level of their parents, they were more likely to rely on farming practices and methods observed on their family farm, rather than apply strategic solutions of external origin. They all stated that they had learned to farm from their parents and by following their examples.

> I learned farming from my father. Even if I do some things differently, I understand that difference and I know why that is. (village of Torby, young farmer)

The mechanism of passing on the farm to the younger generation is gradual, and favors the continuity of farming patterns. Although the head of the farm household has changed, the parents are typically still present on the farm and there is common consent about the directions of farming. Similar to earlier times in history, innovations are introduced in small steps, and not in contradiction to the opinions of older family members.

> [Division of responsibilities:] It is obvious how it was earlier: Dad [was the decider]. I mean, both parents. And now, more and more responsibilities are passed onto me . . . When it comes to investments, it is mostly me. I have some ideas... and I convince others about these ideas. They are always a bit afraid of them and would prefer that someone [else] had already tested them . . . There are no big changes. (village of Gadziny, young farmer)

Young farmers in the study often admitted that they did not spend much time thinking about various changes on the farm. The statements that the production conducted on the farm earlier was the most sensible and best adjusted to the existing conditions usually concluded this part of the conversation. This did not change the fact that young farmers were bringing some new solutions to farming. They also declared themselves to be greater followers of mechanization than their parents, and were more likely to apply novelties that appeared on the market.

On all but one of the studied farms, the young farmers confidently articulated their willingness to preserve such a model of farming that emphasized fulfilling the needs of the family, including having some margin for free, leisure time, rather than everyone just devoting their time solely to the maximization of profits. The examples of profit-driven farmers who transformed their farms into business enterprises were presented as negative. They were seen as threats to the family status of the farm and the interviewed farmers wanted to avoid this type of farm management.

This lack of conflict within farming families about the overall direction of farming had to do with the absence of a strategic perspective in farming, which was described earlier, and noted in older and younger farmers alike. The farmers participating in the research did not give much thought to the long-term trends in the economy. They did not notice and, thus were not able to name and describe, the changes taking place in agricultural policies. Although they took lots of variables into consideration when making their decisions, used various sources of expert knowledge, and observed world market trends for their production profile, they still based their planned activities on simple, short-term calculations. They were quite aware that refraining from investments would bring stagnation. Nevertheless, farm modernization was mostly seen as convenient, and a means for decreasing the workload. This all tied in well with the expressed idea that the goal of farming was not to maximize profits, but to maintain production to a degree which would provide adequate earnings for securing stability and dignity for the farming families. Farmers did not want to be forced to work too hard, beyond their capabilities.

> I am not able to say how my income would increase [after the barn expansion]. That's not it. It's not about the income. I could stay at the status quo: I have time for my family, I have time for everything. My thing is that I want to do as little physical work as possible. (village of Zastały, young farmer)

> [Optimal number of cows that the respondent would like to have is:] 40 heads is enough when you take the path of least resistance. For one person, 30 cows, and they would be known by name, and for a married couple, 60. Anything above 100 is an enterprise that is aiming at maximizing profits, where the numbers are in charge. (village of Torby, young farmer)

Such an attitude could possibly derive from a lack of workforce in rural areas. The respondents strongly indicated that hiring help in agriculture was quite difficult. Nevertheless, the model of the barn that was preferred by them, and described as satisfying the final stage of the developmental perspective for their farm, was the one where farm work could be done by the family without overburdening any of the family members.

12.6. Changes in the Rural Cultural Field

> Back in the day, when I was taking over the farm, there was this willingness to work. They were the happy times. You worked hard but then you went to the roadside and could talk to people. There were young people and there was joy. Now, there is less work but there is no youth and nobody to talk to. Sad times. The countryside has become destitute. (village of Przeszkody Dworskie, grandfather)

While the respondents described the changes related to farming with satisfaction, and saw them as a field of their economic successes, they were rather saddened, and even frightened, by the changes they observed in the rural community. In their nostalgic recollections, the older respondents presented their villages as communities of people who had frequents contact with each other, led similar lifestyles, and shared similar customs. The way farm work was conducted at the time required cooperation and mutual aid. Limited transportation possibilities made rural residents spend free time together in their villages, in self-organized forms.

> In the old days, it was normal that there was a dance event in some village nearby. It meant large gatherings of young people . . . One cassette player in the village, or even without it, people had fun. Just a melody and everybody was dancing . . . There were some comedy skits. Yes, my father acted in them, my mother-in-law as well . . . In the firehouse local people put on some performances, like actors . . . And everybody was from the villages . . . (village of Gadziny, group interview)

> In the past, there was a bench by every house. Anybody who walked by could come and sit down next to you... (village of Olinka, group interview)

Contrary to what can be observed today, the countryside used to be more populated, and the residents, in one way or another, were involved in agriculture. The rhythm of seasonal agricultural work set the rhythm of life and free time.

> People used to flock to our old house. Half of the village. They came and talked. This woman who had 6 kids also came, even though she was tired. She was half-asleep but she still came. When people finished farm work and tending to animals, they came. (village of Zastały, group interview)

Everything changed. The respondents observed depopulation of the countryside. A significant percentage of rural youth migrated to urban areas. The young people who did not move away worked in towns and cities. If new residents moved to the villages it was not for the purpose of working in agriculture. This was the case for villages in Eastern Mazovia in areas with a picturesque landscape, which was considered a good property asset and highly valued by people from Warsaw.

> Before, there were people who worked in towns and cities and the farm was an additional thing. Many of them left. Now, they sell their houses as summer houses. Now, people who live here commute to work, even to Warsaw. A family with four kids lives in the big house nearby and living here pays off. (village of Gadziny, group interview)

Once defined as a place where people worked in agriculture, rural areas have now become bedroom communities for people employed in urban areas, as well as places of residence for people who have finished their professional careers and retired.

> Question: Are people in the village divided into any groups?
> —Yes, farmers and tenants. Those who sleep in the village and those who spend time in the village. There is also a group of older people who do not farm. They are not professionally active; they just live through their final years here. They are quite old. The whole village is changing into a bedroom community with only one or two farmers left, who can be treated as a nuisance. This is because of the farm smells and the sound of the milking machines and so on. (village of Główki Wielkie, group interview)

The interviewed farmers felt that nowadays they definitely had less in common with other residents of their village than in the past. Some of the studied farms did not have any physically close neighbors. The ones who had neighbors reported that these neighbors were not always farmers. Non-farmer neighbors had very little in common with the research respondents; they did not share their experience, and had very few things to talk to farmers about. If the neighbors happened to be farmers, they usually had different production styles and different scales of operation. The respondents were generally more affluent than their newer neighbors, and suspected that there was some envy felt towards them. The respondents generally thought that their social contacts were satisfactory although they were meeting and mingling with very few people. This was very different from the past.

Cultural and lifestyle changes have limited the occasions for rural residents to meet, to be together. Work is no longer performed together. One exception is corn harvesting on one farm, where owners/operators do not have special equipment.

This particular work requires a larger number of people and its completion is followed by a social party. This happens only once a year.

> There is corn harvesting... Among the acquaintances... There is some mutual help, people have to drive there and do stuff together. At the end there is some dinner. (village of Gadziny, group interview)

In other farm interviews, such occasions were not mentioned. The respondents emphasized having good relations with neighbors. Nevertheless, it is rare today to ask neighbors for help. The time of the interviewed farmers is measured differently than the time of other rural residents. This is also a reason why it is so hard to meet.

> With the work, we pass each other and do not meet. In the past, everybody was milking cows, there was a time for milking. After milking there was free time. And now, he comes home from work at 5 pm and we start taking care of animals at 5 pm. We don't see the ones who commute to work. It is hard to meet each other. Each family is closed inside the walls of their home. (village of Zastały, group interview)

The respondents also noted the weakening of traditions which had integrated the rural community. They mentioned that farmers' wives associations existed in some villages, as well as volunteer fire departments. A significant number of respondents declared being volunteer firemen. This should be interpreted as an expression of their engagement in local, rural matters, rather than an argument for the continuous importance of volunteer fire departments in the studied villages. The statements of respondents indicated that the number of volunteer firemen had decreased noticeably in recent years and the local volunteer fire departments no longer played the cultural role to the extent they had before. Obviously, on weekends, young people did not come to the firehouse for entertainment, but looked for it in the city. The firemen were reported to be more likely to remove the trees that blocked the roads after heavy winds than they were to organize dances and social functions.

> The only parties that we have are meetings after firemen tournaments or firemen assemblies. But these are just sit-in functions. (village of Gadziny, group interview)

All of these factors contribute to what could be described as a lack of such spaces where the residents of one village or neighborhood could have a reason to meet and be together. Even the children of parents who work in the city rarely go to local schools. When neighbors go to church on Sunday, they do not meet on the way, as they drive there by car.

> After church, where people come by car, they stand for a few minutes talking. Mostly men. And then they drive home. (village of Wola Krótkowska, group inteview)

The mere fact of living in one area does not create a common social space. Farmers who live off of agricultural production increasingly consider themselves to be a minority which is not well-integrated into the rest of the village they inhabit. They notice that their lifestyle and ways of working, which used to be the norm, have now become anomalous in the eyes of other rural residents, who are not engaged in farming. For those residents, farm practices which were once typical for rural areas now appear problematic. They complain about them and even engage in protests. Furthermore, local authorities and legal regulations are usually not on farmers' side when these conflicts occur.

These conflicts relate to farm smells and forms of pollution connected with agricultural production, as well as consents and permits for agricultural investments. The use of manure is a particularly challenging issue, a flashpoint. The neighbors complain about the smell from the fields where the manure is spread and about the manure spillage on the roads, where the transport of manure and other material related to agricultural production takes place.

> The usual transport of manure becomes problematic. You need to be careful not to spill it and not to leave dirt behind you. If manure gets spilled on the road, somebody might be bothered by it. Looking back 2–3 years, the one who now complains used to keep a cow and dealt with manure every day and was not bothered by it . . . This is not a city, this is a village. If something gets dirty, it will be cleaned up. He should remember that, as in the past he sometimes spilled dirty stuff, too. Those were different times, and nobody made complaints. (village of Gadziny, group interview)

The remnants of manure that are found on the road are sometimes even reported to police. Sometimes the cattle might not be put out to pasture because the neighbors do not like the cattle walking in front of their windows. Similar protests are prompted by the noise of agricultural machines.

It was indicated several times (out of 10 studied farms!) that farmers have to face serious difficulties when applying for permits to build farm buildings. Their neighbors might be against these buildings, as they are seen as sources of bad smells and noise. In these disputes, farmers do not receive support from local governments.

The elected officials view farmers as only part of their constituencies. It should also be remembered that farmers, due to tax breaks, make a rather modest contribution to the budget of local communities and there is always some potential for conflict with other residents.

> Our municipality is the worst when it comes to the development of [farming]. They tell us we destroy the roads and stink up the area. They do not see any benefits of farming. I talked to the mayor. It was a friendly talk. I told him that I was starting things [investing]. He said no... he talked about the smell and how there was no need for more farming. (village of Zastały, young farmer)

> From the tax perspective, a farmer who makes investments is not interesting. Such a farmer is usually eligible for tax breaks and does not miss such an opportunity. For the mayor, a regular resident is more interesting. (village of Główki Wielkie, group interview)

In the eyes of the respondents, the new residents are the initiators of conflicts. Such inhabitants of rural areas treat the countryside as a space for recreation. For them, agricultural production, which disrupts their tranquility, is not something they are keen on. What is also important, they bring new customs to rural areas which might be strange or even offensive. It is the newcomers, who squirm at the sight of the barn and complain about the smell. It is the newcomers, who don't go to church and engage in serious gardening activities instead.

> **Farmer:** These people who moved here from the city cause the most problems. They have their tantrums. Sometimes they live here permanently but for the most part they only come here for summers.
> **Interviewer:** Do they have a different lifestyle?
> **Farmer:** On Sunday they cut trees and mow the grass. This is their [different] way to spend time. We try to limit our Sunday work to a bare minimum but they do the most serious rural labor such as cutting trees and chopping wood. They start on Sunday . . . Don't they have time to do it on weekdays? Don't they have any respect for Sunday? (village of Zastały, group interview)

Changes that have taken place on farms, including mechanization and new production manners, although serious, appear to be less important than the other changes in the countryside. Farmers can still say that, despite modernization and the numerous modifications introduced to their daily lives, they have, for the most part, been able to maintain their lifestyle. However, the villages, where they are based, have changed in terms of lifestyle. Social relations and customs have changed, as well, and people live differently than they did 30 years ago. In this context, despite numerous transformations, family farms are still perceived as a beacon of the traditional, rural way of life and traditional customs.

12.7. Continuity and Change on the Dairy Farms

According to Bourdieu, social life is always situated in a certain way. It is always oscillating around some social space that Bourdieu calls a field. There are historical rules that apply within the field. There is also some relative isolation from the external world, where others live. As can be seen in the collected material, the activities of the studied dairy farmers are taking place mainly within two "Bourdieuan fields," neither of which being the literal farm field of soil and machinery. One of them is the micro-space and micro-cosmos of a family farm. This is where the family members spend most of their time and where their economic activities take place and, at the same time, are tied to the family life. This is also a place for the reproduction of rural habitus, where older generations gained their social disposition and competencies and then passed them on to younger generations. The second field is the village understood as a local community, where farmers conduct these parts of their lives, which do not involve working in the farm field or the barn. This is the space where farmers traditionally fulfill their social life needs, gaining the group identity.

Farmers spend most of their time, and conduct most of their life activities, in these two fields. The material collected during the research has revealed that the respondents rarely engaged in activities occurring in different "Bourdieuan fields" and had not done so in the past. However, it was quite obvious that the fields were not independent from the environment surrounding them. Although farmers and rural residents did not consider themselves as having any weight in these changes within the historical process, they still admitted that their lives had changed significantly because of that process. The studied farmers, especially the generation that was concluding their vocational activities, did not have any doubts about the extent of the transformations that encompassed every dimension of their lives.

The consequences of these changes in each of the described fields were quite different. When considering just the field of the farm, one might notice that technological innovations and civilizational changes were fully absorbed by the traditional habitus. These changes made farm work much easier, cutting its arduousness and increasing its efficiency. Changes in the division of labor followed as well, and the possibilities for how to spend free time also increased, with access to cultural attractions and entertainment not being much less than that available to urbanites. What did not change, from the residents' perspective, were the rules and principles of family life connected to the farm. The respondents, within the space of their own farm, had an awareness of their participation in the progress of civilization, without particularly changing the cultural model they were raised in. This model was still attractive to them and could be reconciled with innovations, which appeared to support its functioning and make it even more convenient.

The traditional habitus of the "good farmer" also seemed to be adapting quite well to new economic conditions. During the research it was established that, even in the past, these farms had been thriving, and enjoyed good reputations and leading positions in their villages. The younger generation took the agricultural work customs from their parents, and these included being industrious and diligent. These characteristics turned out to be helpful and appropriate, even under the new conditions of farming. The farming economy continued to be based on a pragmatic, but hard to pin-point habitus, rather than following some external strategies. This approach to farming turned out to be an adequate tool for production of capital, one that could be equally appreciated in both the past and present, in a Polish economy based on new principles.

The respondents had awareness of, and a well-articulated need to reproduce, such a cultural model. It was within this cultural model that they identified the attractive lifestyle they would like to keep. The model oriented toward entrepreneurial profits was not appealing to them. The model of farming they were keen to continue was, in fact, a sustainability model and/or a model of responsible economy, in which various aspects of life were not dominated by a drive to make a profit. The respondents were willing to accept some hard duties of the good farmer but still able to find joy in family life and make good use of their free leisure time.

The picture is quite different when the consequences from changes in the socio-cultural field of the Polish countryside are analyzed. Here, what have constituted the rules of the game thus far are no longer functioning. Until now, just as the respondents indicated, rural customs were almost completely derived from the habitus of family farms, which, of course, are so prevalent in rural areas. At present, due to ongoing changes, the contemporary countryside is increasingly becoming a place where people of various habituses reside, and many of them are not interested in maintaining the previously existing rules.

Migrations, the easier commute between rural and urban areas, as well as changes in agricultural profitability, caused a shift in the rural residency structure.

Now, the majority of those who reside in rural areas are city-oriented in their customs and aspire to the urban habitus. The countryside, in its old form encompassing all the inhabitants in the community, as well as the space for fulfilling the inhabitants' socio-cultural needs, no longer generates much interest for new residents. Their needs are fulfilled elsewhere and in relation to the people of their choice, not necessarily the ones who happen to live in their neighborhood.

In the face of this increasing invasion of new habitus, the socio-cultural field of the Polish countryside is dying. Although the studied category of farmers, who have increasingly become a minority in rural areas, would still like to fulfill their needs, they are not finding partners for that. They seemed to be the only ones interested in the continuation of the social relations that existed before. Unlike other social categories, they did not have aspirations to be part of the urban space and non-rural customs. Under these conditions, they needed a new space for everyday social relations and ways to spend their free time, and finding such space could be quite difficult under new circumstances.

It should be remembered, however, that the countryside did not cease to exist as a social space. Although, for the new types of residents who are not farmers, the physical space of the countryside might not exist as such an important field as it has before for farmers' families, it is nonetheless the new residents who now play a dominant role here. In the old, traditional countryside that respondents referred to in the study, the norms and values were defined by the habitus of a good farmer. Such a farmer created or contributed to the type of capital that was appreciated in this space and the function of the field was to accommodate the farmer.

Today, farmers must fight for the basic recognition that "This is not a city; this is countryside. If something gets dirty, it can be cleaned later." This was reflected in the rather passionate statement of one respondent, who was angry at his neighbors, who were bothered by the sight of manure. The conflict between this farmer and his neighbors was not really about how they felt about the smell of cow manure, but about the ways to define what constituted the countryside, what was rural. The question was whether agriculture was a crucial element of this definition, or if it was just being barely tolerated under the existing circumstances. It is hard to expect that the wishes of the quoted farmer will be fulfilled: rural customs that he is a carrier of are not of any value in the eyes of his non-farm neighbors. Furthermore, even the financial capital in his possession, which was much greater than the financial means of his new neighbors, could not give him much influence over the local community, and did not play any critical role in the ongoing cultural game.

Returning to the question from the introduction of this text, regarding the further destinies of habitus carried on by traditional family farms, it can be said that, in the two fields that are the main spaces of the studied communities, these destinies might take quite different shapes. In the narrow field of the family farm, this habitus adapts very well to changes and it is, thus, reproduced. Its carriers consciously affirm it, wanting to live according to its principles.

In the socio-cultural field of the Polish countryside, the habitus of the traditional family farm falls onto barren ground, as this field is deteriorating in terms of its former principles. For farmers who considered the social space of the neighborhood to be an important addition to the farming space, this could be a serious problem. Their rhythm of life, marked by mornings and evenings tending to animals, makes it difficult for them to find a new pace of social life. In a way, they are tied to their farms. Making social contacts outside of the area of residence or the place of work might, in some limited way, become a substitute for a former life of local community. It is with a certain sadness, but also with an awareness of the necessity for changes, that farmers begin to look around for new forms of social life.

To some extent, we could say that farmers are continuing hysteretic behaviors. For example, they still try to invest in the traditional forms of the rural social life of yesteryear. Therefore, they are still present in the voluntary fire departments and farmer's wives associations even though the status of these organizations is in decline. A typical example of the hysteresis effect can be observed in the behaviors and conduct related to elections. In local elections, the farmers assume that the candidates from their village will adequately represent them and they are disappointed later if and when this does not occur. Not finding the tools for political articulation that could be adequate under the new conditions, these farmers helplessly reproduce the outdated patterns of representation based on a community that no longer exists.

It should also be added that these farmers search for new fields to fulfill their social needs. Some beginnings of such new spaces can be found in the new social customs of young farmers. Their social circle includes other well-educated farmers located in various parts of the region and the country, as well as agricultural advisers and experts in many fields that they know from university. They comprise a group of friends that cannot be found locally. They exchange information with each other or search for it online on Internet forums, the way their predecessors used to exchange information with their farming neighbors across property line fences. That being said, they are quite aware that these social spaces are not very satisfying and such friends are not easily and truly available every day. Nevertheless, they continue their involvement in these modern social spaces as they are more attractive than contacts with their new neighbors just beyond the fence, who lead entirely different lifestyles and with whom they do not have any bond or much of anything in common.

One of the possibilities that could be helpful in fulfilling the needs of more affluent farmers could be the establishment of producer groups or other organizational forms for the circle of farmers managed by dairy processors. Sometimes social dances, tourist excursions, and study visits are organized within such a framework. This is a fairly new phenomenon and, as can be inferred from the response of the study participants, their success is only partial. Farmers approach them with a certain distrust, seeing them as rather artificial endeavors of external entities and not as activities rooted in old customs. They consider these actions to be mercenary in nature, aimed at using farmers, and not contributing to the building of some credible forms of community.

When it comes to the rebuilding of some socio-cultural space which could potentially serve to unite enclaves of affluent family farmers currently dispersed throughout various villages, its future form remains an open question. The studied farmers raising dairy cattle are definitely experiencing this void. Young farmers, especially those in the phase of life that calls for intense social contacts, have this uncomfortable awareness that something is missing. It is possible that a new social space will be based on new technologies facilitating contacts despite large physical distances. It is hard to say what such a space will look like. It can be assumed that this void will somehow be filled, and its form will be essential to the continuity of customs connected with the operation of a family farm.

REFERENCES

Bourdieu, P. (2000). Making the economic habitus. *Ethnography*, 1(1), pp. 17–41.

Bourdieu, P. (2005). *Dystynkcja. Społeczna krytyka władzy sądzenia* [Distinction: A social critique of the judgment of taste]. Przekład P. Biłos. Scholar.

Bourdieu, P. (2007). *Szkic teorii praktyki, poprzedzony trzema studiami na temat etnologii Kabylów* [A sketch of the theory of practice, preceded by three studies on Kabyle ethnology]. Przekład W. Kroker. Wydawnictwo Marek Derewiecki.

Bourdieu, P. (2009). *Rozum praktyczny. O teorii działania* [Practical reason: About the theory of operation]. Przekład J. Stryjczyk. Wydawnictwo Uniwersytetu Jagiellońskiego.

Bukraba-Rylska, I. (2014). Polska wieś w UE: Esej o (brakującej) Wielkiej Narracji. *Wieś i Rolnictwo* [Village and agriculture] 2, pp. 65–81.

Crossley, N. (2003). From reproduction to transformation: Social movement fields and the radical habitus. *Theory, Culture & Society*, 6(20), pp. 43–68.

Fedyszak-Radziejowska, B. (2005). *Proces demarginalizacji polskiej wsi. Programy pomocowe, liderzy, elity i organizacje pozarządowe* [The process of demarginalization of the Polish countryside: Aid programs, leaders, elites and non-governmental organizations]. Instytut Spraw Publicznych.

Fedyszak-Radziejowska, B. (2008). Polska wieś w cztery lata po akcesji – wymiar demarginalizacji [Polish countryside four years after accession: The dimension of demarginalization]. In: J. Wilkin & I. Nurzyńska, *Polska wieś 2008: Raport o stanie wsi* [Polish countryside 2008: Report on the state of the countryside], pp. 59–76. Fundacja na Rzecz Rozwoju Polskiego Rolnictwa.

Gorlach, K. (2001). *Świat na progu domu* [The world at the doorstep]. Wydawnictwo Uniwersytetu Jagiellońskiego.

Gorlach, K. (2009). *W poszukiwaniu równowagi. Polskie rodzinne gospodarstwa rolne w Unii Europejskiej* [In search of balance: Polish family farms in the European Union]. Wydawnictwo Uniwersytetu Jagiellońskiego.

Halamska, M. (2015). Specyfika rolnictwa rodzinnego w Polsce. Ciężar przeszłości i obecne uwarunkowania [The specificity of family farming in Poland: The burden of the past and current conditions]. *Wieś i Rolnictwo* [Village and agriculture], 1(1), pp. 107–129.

Halamska, M. (2016). Zmiany struktury społecznej wiejskiej Polski [Changes in the social structure of rural Poland]. *Studia Socjologiczne* 1, pp. 37–66.

Hardy, C. (2008). Hysteresis. In: M. Grenfell, *Pierre Bourdieu: Key concepts*, pp. 131–148. Acumen.

Krzyworzeka, A. (2014). *Rolnicze strategie pracy i przetrwania. Studium z antropologii ekonomicznej* [Agricultural work and survival strategies: A study in economic anthropology]. Wydawnictwa Uniwersytetu Warszawskiego.

Maton, K. (2008). Habitus. In: M. Grenfell, *Pierre Bourdieu: Key concepts*, pp. 49–66. Acumen.

Ploeg, J.W. van der (2003). *The Virtual Farmer: Past, present and future of the Dutch peasantry*. Royal Van Gorcum.

Thomson, P. (2008). Field. In: M. Grenfel, *Pierre Bourdieu: Key concepts*, pp. 67–81. Acumen.

Think Locally, Act Globally: Polish farmers in the global era of sustainability and resilience, ed. by Krzysztof Gorlach and Zbigniew Drąg in collaboration with Anna Jastrzębiec-Witowska and David Ritter
Jagiellonian University Press, Kraków 2021, pp. 561–590
ISBN 978-83-233-4949-5
DOI: http://dxdoi.org/10.4467/K7195.199/20.20.12741

Chapter Thirteen: Food Safety and Food Security from the Perspective of the Functioning of Farms in Poland

Zbigniew Drąg https://orcid.org/0000-0002-9106-7758/

Piotr Nowak https://orcid.org/0000-0001-7991-5534/

Martyna Wierzba-Kubat

13.1. Introduction

Every social system has been known to have, and will continue to have, special institutions focused on fulfilling people's food needs. Since people started to develop sedentary lifestyles and engage in agriculture, the responsibility for providing sufficient quantities and quality of food has fallen on farmers. Thousands of years ago, people selected animals and used selected seeds in order to have control over where, how, and in what quantity their food was produced (see also Ploeg, 2016). The knowledge of animal husbandry and plant cultivation passed from generation to generation gave farmers relative autonomy in society because they were in charge of food security for the entire community. In early history, farmers as a social class played a significant role in the economic and political life of territorial communities and enjoyed a high social position related to the importance of their profession. However, with time and civilizational development, the inventions enabling farmers to cultivate the land more effectively, and thus allowed for the development of other areas of social life, without having to concentrate exclusively on the fulfillment of basic needs. As a result of these processes, the social position of farmers started to lose importance to such an extent that from free and respected people they became like slaves, bound to the land that was once "the feeding mother" and now appeared at times to be a "lifetime curse." Even though, in the 21st century, those who produce agricultural commodities (i.e., mostly raw materials for the food industry) are less often referred to as farmers and more commonly called agricultural producers, they still play an important role in the global food system and food security (see also Bello, 2011; Ziegler, 2013).

Food security was for many years determined by three variables: population size, farmers' hard work, and the "kindness of nature." For many centuries this last factor had the biggest impact on the quantity and quality of the food produced (see also Kowalczyk, 2009, 2014). However, humanity did whatever was possible to learn to acquire independence from nature. One of the most important innovations was the field irrigation system that, in many countries, has been functioning continuously for over three thousand years (see Koocheki & Ghorbani, 2005; see also Grochowska, 2014). It should be added that in contemporary production of agricultural crops irrigation systems that increase yields are quite common and widespread (see also Kirwan & Maye, 2013). More importantly, their usage guarantees the predicted efficiency of food production and minimizes economic risk. The development of civilization and the increased independence from the nature was closely tied with technical progress, which resulted in population growth. Inevitably, it became necessary to constantly increase food production. At the beginning of the 20th century, the human population on Earth was 1.2 billion people and just a hundred years later it was 7.5 billion people. It is expected that the world's population will reach 9 billion people in 2050. There are expert voices claiming that the Earth is becoming too small to feed the rapidly growing carnivore population (FAO et al., 2018). Even today, just in the United States, 9 billion farm animals are raised and slaughtered every year and such annual turnover is higher than Earth's entire human population. The increase in the pace of meat consumption is enormous: while the number of people has doubled since 1960 meat consumption has increased fivefold in the same period of time (Shapiro, 2018, p. 20). Despite the fact that meat production requires large quantities of land, water, energy, chemicals and other resources, its proliferation is possible due to scientific progress related to good knowledge of animal anatomy and physiology and its application in the system of industrial meat production. Thousands of farm animals are crowded into cages in order to lower production costs. They are given hormones and antibiotics. The food and water are automatically administered through special feeders and filling trays while the animals, introduced to various innovations, do not see daylight and receive heat and ventilation through a central system. Individuals working in such facilities are part of the industrial food production system. Unlike farmers in the past, these agricultural workers spend less and less time in the field, with very little exposure to nature. They concentrate instead on the efficient use of machines or even the monitoring and controlling of computer programs that operate those machines (see also Kołożyn-Krajewska & Sikora, 2010).

Thanks to the development of scientific knowledge and continuous improvements in the industrial food production system, the traditional farmer is replaced by machines and nature may be largely disregarded. Today, it seems that natural process related to food production can be fully controlled by people, and the newest revolution is called cellular agriculture and it involves production of food in laboratories. Just like many years ago, when people domesticated and raised animals in order to be able to eat them, today they are adapting and raising cells. In 2004, a research institute called New Harvest was established and its main goal was to start post-animal bio-economy (Shapiro, 2018, pp. 58–59). Its research was dedicated to finding ways to produce animal food products without using animals. The idea was to create animal tissues through cellular agriculture in vitro from cell cultures using a combination of biotechnology, tissue engineering, and molecular and synthetic biology to create animal food products such as meat, milk, skin, eggs, and gelatine. Raising meat without animals was pioneered by the Dutch scientist Mark Post (Shapiro, 2018, p. 79). In 2013, the participants of a press conference in London were able to taste a hamburger produced from stem cells originating from a cow, transferred to a lab and subjected to replication. The Dutch scientist estimated that the sample taken from one cow could allow for the production of 20 thousand tons of meat, which normally takes 40 thousand cows (Shapiro, 2018, p. 88). Compared to traditional meat production, raising "clean meat" in laboratories has 96% lower gas emissions, decreases the use of energy by 45%, and that of water by 96%. Most of all, it takes 99% less land than a traditional beef cattle farm (Tuomisto & Teixeira de Mattos, 2011). The serious barrier in the development of such technology has to do with high costs: the Mark Post team spent 330 thousand dollars to produce the first "clean" hamburger. However, if this type of food production technology would develop further in the future then farmers will no longer be necessary in agriculture and in society.

It should be added that cellular agriculture is not the only future direction for possible food production development. Nanotechnology is also seen as having promising potential. Thanks to this technology, food could have some "interactive" qualities, which could help with adjusting its sensory features—such as flavor, smell, consistency, and color—to customers' expectations. The next stage in the development of nano-foods production would involve its personalization, meaning an individual approach to foods, connecting the diet and food plan with genetic predisposition of particular consumers and production of "intelligent" food, which could on its own adjust to the consumers' food needs. It should be added that nanotechnology is already being used today in food packaging (see Idzikowska et al., 2012; Głód, Adamczak, & Bednarski, 2014).

The changes described above taking place in food production and possible directions for their development are met with criticism and protests in various segments of society (Wilkinson, 2015) and a significant number of authors have been reflecting on them. They identify a notable number of factors which, in the near future, could have impacts on food security and food safety. As pointed out by Julian Cribb (2010), there the issues related to climate change should be of interest but direct political actions taken by some governments and connected with trade protectionism are also important. Similarly, Lester R. Brown (2012) identified the source of the problems related to food security in government actions and in the increase in food demands in countries such as China, and the dilemmas related to producing bio-components for fuels. Frederick Kaufman (2012) somewhat differently emphasized that the source of many problems related to food shortages could be tied to its insufficient quality and treating food not as a necessary means but as one of many commodities included in market transactions. These issues were also analyzed by Raj Patel (2012) who stated that the fight between various actors of the world's food system brought about some negative phenomena such as food shortages in some areas of the world, and diseases related to overconsumption of unhealthy foods. The comprehensive analysis of the problems related to food shortages and improper distribution of foods was presented in the work of Martin Caparrós (2016). Other authors (see Keneally, 2011; Applebaum, 2017) have concentrated on political possibilities and actions leading to the social construction of hunger situations in various parts of the world. An overview of the contemporary reflection on food security and food safety suggests the increasing marginalization of the position of family farms in the food chain as well as their diminishing role in the contemporary food security system. This chapter is therefore devoted to these issues, with special attention given to the situation in Poland.

13.2. Food, Food Market, and Food Chains

One of the most difficult aspects of food security has to do with defining food. In the past, food meant physical substances created naturally without human intervention, which had nutritional value and did not harm human health. Nowadays, food consumed by humans is largely based on products made by men and not by nature. In the laws of the European Union the definition of food can be found in Regulation (EC) No. 178/2002 of the European Parliament, adopted by the Council of January 28, 2002. Article 2 reads:

> "[F]ood" (or "foodstuff") means any substance or product, whether processed, partially processed or unprocessed, intended to be, or reasonably expected to be, ingested by humans. "Food" includes drink, chewing gum and any substance, including water, intentionally incorporated into the food during its manufacture, preparation or treatment (EUR-lex, 2002, Document 32002R0178).

Foodstuff does not include animal feed, medicinal products, cosmetics, tobacco, tobacco products and drugs. The term "food" ("foodstuff") has a narrower meaning than the term "food product." Therefore not everything that gets consumed by humans, including medicines, tobacco products, and drugs can be considered food.

In the scale of the entire humanity, or concrete society, or even a smaller territorial community, foodstuffs—as indicated by their definition—are a commodity, which means that they may be sold or transferred from the producer to consumers with or without the help of a middleman. All parties participating in the production, distribution, and consumption of foods create a food market and are its actors. The functioning of various food markets, be they at the global, national, or local level, is often described through the system of food chains, where various actors operating within particular markets make the elements of the chain. Over the years, the markets, as well as food chains, have been undergoing constant changes. The characteristic phenomenon observed globally is the growing number of market actors, as well as the addition of new elements in food chains. Another accompanying process is the change of the order of elements in these food chains (see Chechelski, 2008, 2015; Szczepaniak & Firlej, 2015).

Today, the number of elements in food chains is quite diverse depending on the market scale. For example, in the European Union, about 12 million farms produce agricultural commodities to be processed by about 300 thousand food industry enterprises. Processors sell their products through 2.8 million distributors and other types of food services, including the catering industry and restaurants. Then, the food is delivered to 500 million consumers in the European Union (EU Agricultural Markets, 2015).

Scheme 13.1 presents the direction of changes that have been occurring in the food chain in the last century. What seems particularly important is the change in the position of the farmer. The farmer has been deprived of his previous dominant position and subordinated to the remaining food chain links. Today farmers produce mostly what is needed for agricultural products, which can later be sold in the stores, where such products end up. Farmer becomes a producer of agricultural raw products necessary for the food industry that needs to deliver products to market chains. Hence, the farmer becomes less involved in true food production.

Primary model	Traditional model in the 20th century	Contemporary model in the 21st century
Farmer	Farmer	Trade
↕	↕	↕
Consumer	Food industry	Consumer
	↕	↕
	Trade	Food industry
	↕	↕
	Consumer	Farmer

Figure 13.1. Food chain links

Source: Own work based on Chechelski, 2015, p. 57.

The current organization of the global food market as well as its contemporary situation is presented in Figure 13.2. It presents a rather large number of market actors with the dominant position held by the chemical industry, well-illustrated by the data on pesticides being used in the European Union. In 2016, almost 400 thousand tons of chemical products were sold in Europe and most of them were used in the agricultural sector (PAN Europe, 2019). The data on the situation in Poland are also quite meaningful and revealing. As of April 1, 2019, the number of plant protection product units including pesticides, fungicides and herbicides approved for sales by the Minister of Agriculture and Rural Development reached 2249 (*Rejestr środków ochrony roślin*, 2019), while in 1991 the same ministerial list contained only 441 units. According to studies conducted by the Ministry of Agriculture and Rural

Development plant protection products were used by 22% of Polish farmers (Raport Polska Wieś i Rolnictwo, 2018, p. 83). In making the decision to use plant protection products on their farms, 30% farmers generally took their harmfulness to people, bees, and the natural environment into consideration. Every fifth farmer used the existing support system when deciding on plant protection (Raport Polska Wieś i Rolnictwo, 2018, p. 84). It is worth mentioning that in 2017 qualified seed material was purchased by only 13% of farmers (Raport Polska Wieś i Rolnictwo, 2018, p. 79).

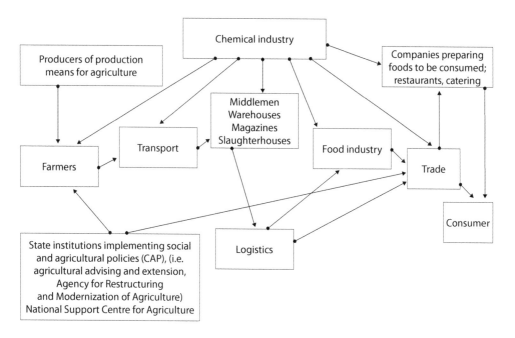

Figure 13.2. Organization of global food market

Source: Own work.

Changes in the food market organization had quite serious social and biological consequences. In the history of agriculture there were a number of important moments that should be mentioned in the context of food security. One such moment had to do with the introduction of new chemical products, the application of which was to revolutionize food production. They were supposed to be a long-awaited miracle solution freeing humanity from hunger and disease. One of these products was the pesticide DDT (dichlorodiphenyltrichloroethane), known in Poland under the name of Azotox, which had a special place in Polish agriculture. It was popular in the 1960s and early 1970s. Today, there is scientific evidence that this chemical is dangerous to human health through cumulative exposure and it may even be deadly. Currently, the situation with glyphosate, which is an active ingredient in some non-selective herbicides, appears to be similar to that of DDT in the past. Interestingly, the phase-out of DDT started at about the same time as glyphosate's introduction to the agricultural market by the Monsanto Corporation in 1974. Today, glyphosate is suspected to have carcinogenic effects. Another area where chemicals are applied in food production encompasses additives of the "E" group. They include preservatives, coloring, taste enhancers, emulsifiers, antioxidants, and food stabilizers. As indicated by a study conducted by Poland's Supreme Chamber of Control in 2019, the average consumer ingests about 2 kilograms of such substances every year. Some of these substances can cause allergies or even life-threatening anaphylactic shock, while others may be carcinogenic (Raport NIK, 2018, p. 5).

Another important issue in the history of agricultural development is the technology focused on genetically modified organisms (GMO) which started to be widely used in the mid-1990s. The use of GMOs causes a lot of controversies but their relatively short history of application makes it difficult to determine their safety regarding human health. It is possible that, just as with DDT, the negative consequences of this technology will only be known in the future. However, there is one negative outcome stemming from the introduction of GM technologies which can already be observed today: that is the process of the privatization of nature, with patented seeds and the selection of certain animal genomes, which excludes less affluent farmers (Parekh, 2004; Marzęda-Młynarska, 2014, pp. 182–187; Lucińska & Grajeta, 2015).

Advanced farming technologies exemplified by precision farming can be considered an important change in agriculture that impacts food security. The use of modern power generators controlled by special digital devices deprives farmers from having any influence on agricultural activities and procedures. The tractor administers the proper amount of plant nutrients and moves along the line mapped out by the GPS, while plant care is controlled by a special computer program. Such an advanced form of farming is not so much in need of a farmer but a computer scientist or an analyst to monitor the proper parameters. Possibly, this could be done via the Internet. In the near future, the entirety of food production and processing might become the domain of artificial intelligence, and farmers could be left with little or no influence on food production, so fundamental to human existence.

The described changes are proof of the continuous alienation of the farmer in the food production process. It should be stressed that the emergence of multiple movements objecting to such changes can be observed. These social movements are focused around initiatives such as urban agriculture, food cooperatives, La Via Campesina, Nyeleni forum, and other agri-ecological movements (see Ploeg, 2018).

The degree to which they might be able to influence future food production has not been determined. Could this be a turn towards neo-traditional agriculture, usually symbolized by the new social movement La Via Campesina, or perhaps an increasing appreciation of the "clean meat" produced in laboratories without farmers and promoted by Mark Post? A lot could depend on further elaboration of the food security concept or perhaps the sense that this term is equipped with.

13.3. Food Security and Food Safety in the 21st Century

It might be helpful to distinguish between the terms "food security" and "food safety." Food security deals with securing a sufficient volume of food products for a certain country or other territorial unit, e.g., the European Union, and ensuring that this food would not harm the people. Food safety is more about providing guarantees that the food consumed by people does not contain harmful substances such as traces of DDT dioxins, antibiotics, or dangerous food colorings and other additives, etc.

Food security is treated as an integral part of a social security system and it has a multidimensional character. It is often described by the following elements: global (encompassing the entire world), regional (continental; covering for example Europe or Africa or some other continent country-wide, national, local, family, and individual Food security as an element of the state or national security is often connected with political security, military security, ecological security, health security and cyberspace security. It can be stated that food security is often analyzed in its political, economic, cultural, and religious dimensions.

The most problematic aspects of food security relate to the problems of hunger and malnutrition. It is a real problem and acutely experienced even today. The newest data from the Food and Agriculture Organization (FAO) of the United Nations show that after a significant period of time when the number of malnourished people was decreasing, it is on the rise again, for the third year in a row. There were 806 million malnourished people in 2016 and in 2017 their number went up to 821 million (FAO et al., 2018, p. 2). The problem of food shortages globally affects about 1 out of 9 people and is present in every part of the world. Even in the 28 countries which constitute the European Union and which has reached the highest level of food security the percentage of malnourished people was estimated at 1.6% in the 2015 to 2017 time period (FAO, 2018, p. 9).

There is a productivity narrative regarding food shortages focusing on the thesis that "feeding the world" requires doubling food production by 2050 and making it one of the most important contemporary global issues. Questions about the character of food policy regarding what, where, and how it should be produced, and in what quantities, result in a multitude of wide-ranging answers addressing the matters of market privileges, politics, state authority, and social justice. To counterbalance the productivity theory, it is pointed out that rural areas, where most of the world's food is being produced, are not unlikely to be populated by people suffering from hunger-based malnutrition. This is particularly true for developing countries, where 75% of those who suffer hunger live in rural areas (IFAD, 2011, p. 16). Hunger currently gains recognition as a structural issue rather than a natural phenomenon. The problem lies in institutional decisions about who can receive food, and how much and at what price. Put simply, food is a commodity that is sold for profit. Regardless of the production decisions made by small farmers, the vast majority of world food trade is controlled by transnational corporations, whose goal is to generate profits for shareholders. Corporations usually prefer to sell relatively expensive and profit-making food products to affluent consumers and not basic food products made cheaply for the poor. Obviously, global food chains are quite complex, with markets and corporations not being the only institutions within them. It should be remembered that the governments of many countries subsidize agriculture, monitor market regulations and trade agreements, so their citizens can buy food at moderate prices (Shepherd, 2012).

The discussion on how much food should be produced without the knowledge on the population size is rather pointless. Therefore, one of the most important questions regarding food security is really a question of the maximal number of Earth inhabitants and the land resources necessary to feed them. In 1967, the Rockefeller Foundation estimated that global agricultural land available for cultivation totaled 14.3 million km^2 and pastures covered 25.8 million km^2. When this was converted to the standard land numbers it came to 108 million km^2. It was also estimated that the production of all ingredients for a well-balanced diet needed for a healthy American would require 2000 m^2 per person. It was calculated that the agricultural land on Earth can provide food for a population as large as 47 billion and in the case of food products necessary for survival, as many as 157 billion people could be fed (Zwoliński, 2006, p. 250). Considering the above calculations, and with the supporting data on the ongoing technological progress in food production over the last 50 years, one might conclude that land resources are not a barrier to providing food security for all of humanity today and in the future.

Along with the discussion on providing sufficient food quantity for an increasing human population, the approach to food security issues also went through an evolution process. One result of the evolution of the food security definition was that the dimensions and level of food security also changed (Table 13.1). Currently, the most frequently quoted definition of food security is the one proposed by Food and Agriculture Organization of the United Nations (FAO), which reads: "Food security exists when all people, at all times, have physical and economic access to sufficient, safe and nutritious food that meets their dietary needs and food preferences for an active and healthy life" (FAO, 1996).

Table 13.1. The evolution of the concept of food security

Year	Dimension	Level	Definition
1943	• physical	• national • global	The Conference in Hot Springs that established FAO 1943. The definition adopted at the time reads: "the goal of freedom from want of food, suitable and adequate for the health and strength of all peoples, can be achieved"
1974	• physical • economic	• national • global	World Food Conference FAO 1974. Food security was defined as "availability at all times of adequate world food supplies of basic foodstuffs to sustain a steady expansion of food consumption and to offset fluctuations in production and prices"
1983	• physical • economic	• national • global	Food security means "ensuring that all people at all times have both physical and economic access to the basic food that they need"
1996	• physical • economic • health-related	• individual • household • national • global	FAO World Food Summit 1996: "Food security, at the individual, household, national, regional and global levels [is achieved] when all people, at all times, have physical and economic access to sufficient, **safe and nutritious** food to meet their dietary needs and **food preferences** for an active and healthy life"
2009	• physical • economic • health-related • social	• individual • household • national • global	FAO 2009: "Food security [is] a situation that exists when all people, at all times, have physical, **social** and economic access to sufficient, safe and nutritious food that meets their dietary needs and food preferences for an active and healthy life"

Source: Marzęda-Młynarska, 2014, p. 102.

This concept has been discussed on four different levels: global, national, household, and individual, and refers to the four dimensions that need to coexist at the same time (Marzęda-Młynarska, 2014, pp. 102–105). These dimensions are:

a) physical dimension, encompassing the physical presence and availability of food; it is close to the concept of food sovereignty, which means having a sufficient amount of food being a priority with food produced in close proximity to people's homes being considered the most appropriate and most desired;

b) economic dimension, relating to the possibilities of acquiring the food, which means possessing the adequate financial resources; it is closely tied with individual incomes and food prices; c) health dimension, referring to the nutritional value of food, which should be stored, prepared and consumed in a certain manner; d) social dimension, connected with people's food preferences; it combines the necessity of having a sufficient volume of food with its quality and other characteristics being appropriate in terms of individual, cultural, and religious preferences.

In recent years, the problems of food safety have generated as much attention as food security issues. Food safety has gained importance in the agricultural policy of the European Union, especially after the 2007–2008 food crisis (Marzęda-Młynarska, 2014, p. 17). Food safety is usually viewed in a somewhat narrow sense, as a legal system comprised of various actors (Leśkiewicz, 2012) and sometimes as a kind of multifaceted discourse addressing issues such as hunger, malnutrition, and the establishment of global and local food chains, as well as various threats confronting the system of production and distribution of food (Mooney & Hunt, 2009).

13.4. Systems of Food Security and Food Safety

Food security and food safety are global issues and therefore they have been institutionalized. The world's most recognized institution dealing with food security issues is FAO, the Food and Agriculture Organization of the United Nations. Its work includes fighting hunger and poverty and increasing human welfare and well-being by redistribution of food and through rural development. It was established as an initiative by an anti-Nazi coalition of 44 countries that participated in the conference devoted to the problems of food and agriculture in the town of the FAO's earliest origins, Hot Springs, Virginia, in May 2019. The most important international and intergovernmental platform for all countries and institutions interested in achieving the goal of ensuring food and food security for all people is the Committee on World Food Security (CFS). It reports to the United Nations General Assembly through The Economic and Social Council and FAO Conferences. Applying a multifaceted, complex approach, the CFS elaborates upon and approves the directives on food policies and other issues related to food and food security as widely understood. These directives and recommendations are based on evidence found in academic reports elaborated by The High Level Panel of Experts on Food Security and Nutrition (HLPE) and/or works technically and logistically supported by the Food and Agriculture Organization (FAO), The International Fund for Agricultural Development (IFAD), The World Food Programme (WFP), and representatives of the CFS Advisory Group. The Committee on World Food Security (CFS) organizes an annual plenary session in October each year at FAO headquarters in Rome. It remains an open question whether the activities of the described institutions bring about the expected results on the global level. It would be very hard to provide an unambiguous answer whether the problem of world hunger would be wider without these organizations or if it would be the same regardless.

In Europe, one of the main institutions focused on the problem of food safety is the European Food Safety Authority (EFSA). It is one of the more visible EU institutions and its headquarters are in Parma, Italy. It sets standards for food products and prepares legal acts regarding food. In the European food safety system, an important role is played by Rapid Alert System for Food and Feed of the European Union (RASFF), the system for rapid dissemination of information on harmful food products, which is in operation in all countries of the European Union. This system allows for the exchange of information between the organs of food control in Europe, which are members of that system. The information on foods and feeds, as well as materials that have contact with same, and pose potential harm to people, animals, and the environment, is entered into the system. Then, actions are taken to identify and locate such products. In the European Union the RASFF system is based on the following legal regulations: Regulation (EC) No. 178/2002 of the European Parliament and of the European Council gathered on January 28, 2002 laying down the general principles and requirements of food law, establishing the European Food Safety Authority (EFSA), and laying down procedures in matters of food safety (EUR-lex, 2002), as well as Commission Regulation (EU) No. 16/2011 of January 10, 2011, laying down the implementation measures for the Rapid Alert System for Food and Feed (EUR-lex, 2011). The RASFF network of members consists of the European Food Safety Authority (EFSA), surveillance authority and at national level in member countries and in the EFTA countries (Iceland, Liechtenstein, Norway, Switzerland), and clearly identified contact points in the European Commission. The most frequent threats detected in food products in 2017 by the RASFF system were salmonella and a substance called Fipronil in eggs meant for human consumption. Fipronil was illegally used as an insecticide on chicken farms and traces of it were found in eggs in many European countries.

In Poland, the chief sanitary inspector monitors the situation of food safety in the entire country and provides information within the RASFF system, as well as cooperates with the chief veterinary officer to ensure food consumers' safety in Poland. Traceability is required by article 18 of ordinance 178/2002, in which traceability is highlighted as necessary to monitor the safety of food, feed, animal livestock, and substances being added to food at all stages of production, processing, and distribution. The ability to trace the path of a food product on the market back to the raw product it originated from is a necessary requirement to ensure a high level of human life and health protection as well as protection of consumer interests. Food that does not qualify for consumption due to contamination with foreign substances or because it's rotten, spoiled or in a state of decomposition should not be put on the market. In the case of primary agricultural production, health hazards might be due to residues of plant protection products, heavy metals, fertilizers, or microbiological contamination. Such raw products are recognized as harmful foods as they might negatively affect a consumer's health or not be suitable for human and animal consumption (see also *Rejestr środków ochrony roślin*, 2019).

Table 13.2. Examples of safety infringements of food products from 2008 to 2018

Year	Incidents of food adulteration	Country of production
2008	Melamine in infant milk formula	China
2009	Canned meat produced 26 year prior (?), still available for consumption	Poland
2010	Instant coffee with glass particles	Poland
2011	Fenugreek sprout seeds contaminated with Escherichia coli (initially it was reported as so-called "cucumber scandal")	Egypt (the country of origin of the seeds)
2012	Road salt sold as edible salt Fish marked with new expiration dates and returned to markets Microbiologically contaminated dried egg powder	Poland
2013	Traces of rodent poison in infant formula antibiotics on poultry farms larvae in chocolates	Poland
	Horse meat in beef products	unknown
	Alcohol for human consumption with some addition of methanol	Czech Republic
2014	Baby cereal containing unacceptable alkaloids	Germany
2017	Eggs contaminated with Fipronil	The Netherlands, Germany, Belgium
2018	Eggs contaminated with Fipronil	The Netherlands

Source: Own work based on: Omieciuch, 2016, p. 130; DoRzeczy, 2017; wolnosc24.pl, 2017; TVP Info, 2018.

The problem of food adulteration is the most important problem related to food safety that national and European institutions are currently dealing with. Table 13.2 presents examples of incidents that took place in recent years. The important conclusion that can be drawn from the analysis of cases of food adulteration is that chemicals used incorrectly or inappropriately are usually a source of danger.

In this context, it should be noted that in Poland the discussion regarding the necessity to reform the food safety system has been going on for many years. Most experts and politicians agree with the statements that there is a need for more uniform and more orderly control and supervision over food safety. In May 2017, the government adopted a bill for the establishment of a National Food Safety Inspectorate that had been by the drafted Ministry of Agriculture and Rural Development. Its main purpose was to merge various food safety inspection units into one institution. This new inspectorate was to combine the competencies of the Veterinary Inspectorate, the Trade Quality Inspectorate of Agricultural and Food Products, Main Inspectorate of Plant Health and Seed Inspection and some tasks of the State Sanitary Inspectorate and Trade Inspectorate. This new institution was to be supervised by the Ministry of Agriculture and Rural Development. The bill was sent to Polish Parliament but as of January 2019 it still had not been put to a vote. Currently, the legal basis for the functioning of the food safety system in Poland is the Act on Food Safety and Nutrition from August 25, 2006, with later amendments.

The current system of food safety in Poland is a typical example of the dispersed model. There is no leading institution and all the institutions within the system have a certain range of sole competencies. This leads to a situation where the competencies of some institutions overlap, while other functional areas in certain institutions in the food chains remain completely unsupervised. This system consists of five institutions:

a) the Veterinary Inspectorate, responsible for foodstuffs of animal origin at the production stage;

b) State Sanitary Inspectorate, controlling food safety of non-animal origin at the production site as well as all of the products on the shelves in stores and supermarkets;

c) the Main Inspectorate of Plant Health and Seed Inspection responsible for the quality of feed used in animal production;

d) Agricultural and Food Quality Inspection, monitoring the quality of food at the producer level;

e) Trade Inspectorate, responsible for food quality in retail settings (see Figure 13.3).

The food safety management system in Poland is based on the assumption that, according to the laws and regulations of the European Union, food production plants and food sales facilities are obligated to set up and implement procedures on food safety according to the HACCP method (Hazard Analysis and Critical Control Points). Some flexibility and simplification in the implementation of this system are possible depending on the size of the facility, nevertheless there are some non-negotiable requirements. They are: Good Hygiene Practice (GHT) and Good

Manufacturing Practice (GMT). In 2014, there were 187,031 entities which implemented the HACCP system and 309,593 facilities that applied the GHT and GMT principles in Poland (Obiedzińska, 2016).

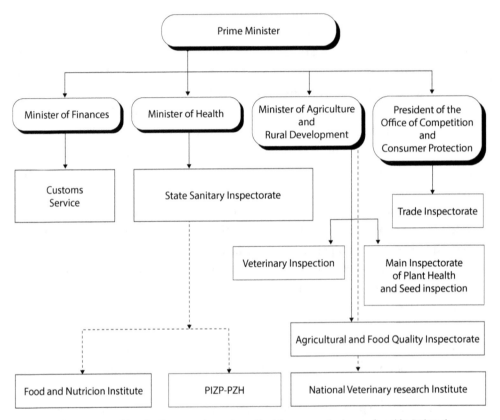

Figure 13.3. System of institutions and services that have control over food in Poland.

Source: Kowalczyk, 2016, p. 229. The acronym PIZP–PZH means National Institute of Public Health—State Institute of Hygiene.

While examining the literature on food security and safety it is not hard to notice that these issues are not analyzed from the perspective of the most crucial food chain link, namely farmers. Their role in achieving and maintaining both is treated rather marginally. To make up for this deficiency, this chapter, which already addressed the issue of food security attempts to provide at this point a view of food safety from the farm perspective and its operators in particular. Here, there are references to American authors (see Mooney & Hunt, 2009), who worked on food security analysis by discussing and comparing interpretive frames of various actors engaged in the processes of food production and distribution.

13.5. Food Safety at the Farm Level according to Polish Farmers

Considering that food security mostly deals with the quantity of food and food safety with its quality, it should be stated that each of these terms accentuates a different scale and different aspects of problems, which are analyzed in reference to different types of communities. Food security is mostly analyzed in the categories of the global community and within the context of hunger and malnutrition in the world's poorest countries. Food safety focuses on achieving and maintaining high quality standards for foods intended for consumption in the developed countries. However, even in these affluent societies it is still possible to find marginal groups of malnourished individuals, families, and even pockets of society. At the same time, thanks to advanced food production technologies, the quantity of food produced in such societies and countries not only eliminates food shortages altogether, but also allows for food waste. The above issues are also interesting in the context of Polish society, where more attention is given to food safety as it is assumed that the problem of food security in Poland is not significant and its importance is secondary (see also Mikuła, 2012).

Concentrating on food chain links, the problem of food quality can be analyzed in a variety of ways and in reference to many aspects. In this chapter, the attention is focused on this link, which—regardless of its character in the existing food chain—has essential meaning, even if it does not take the dominant position. This link is constituted by primary food producers or just farmers. Their decisions can be the first to determine the quality of the produced food. Not only do they make choices on what to grow and what animals to raise but also what chemical products to use and technologies to apply. Obviously, the decisions made by farmers, which might impact the quality of food products, are not just the result of individual preferences. An equally important role is played by the activities of institutions that control the quality standards of food produced by farmers. Referring to the 2017 study, it was analyzed to what extent internal and external mechanisms of control could have impact on farmers' production activity. The frequency of farmers' contacts with official institutions of food quality control was recognized as an indicator of the impact of external mechanisms. Farmers' opinions on incidents of forbidden production practices and their declarations connected with consumption of food they produced were seen as internal mechanisms.

In order to recognize the importance of external mechanisms of food quality control, farmers were asked about their contacts with the representatives of five basic institutions responsible for food safety in Poland: Veterinary Inspectorate, Main Inspectorate of Plant Health and Seed Inspectorate, State Sanitary Inspectorate, Agricultural and Food Quality Inspectorate, and Trade Inspectorate. The data on this matter are presented in Table 13.1.

Among the interviewed farmers, 46% reported that while conducting their agricultural activities they never had any contacts with the institutions mentioned above and close to 13% remembered only one such contact. Almost 42% of farmers met with the representatives of the above institutions at least twice. Farmers clearly had the most frequent contacts with employees of Veterinary Inspectorate: almost one third had at least two contacts with them and an additional 15% had one such contact. Much less frequent were contacts with the representatives of the remaining institutions. Only one fifth of farmers admitted that during their time in farming they had at least one contact with employees of the Main Inspectorate of Plant Health and Seed Inspection. Less than 15% had contacts with representatives of the State Sanitary Inspectorate. Contacts with other institutions listed were reported by just a small percent of farmers.

Table 13.1. Farmers' contacts with the representatives of institutions responsible for food safety in % (N = 3544)

Type of institution	Answer		
	Yes, many times	Yes, just once	No, never
Veterinary Inspectorate	32.4	14.7	52.9
Main Inspectorate of Plant Health and Seed Inspection	10.4	10.1	79.5
State Sanitary Inspectorate	7.2	7.3	85.5
Agricultural and Food Quality Inspectorate	2.1	3.1	94.8
Trade Inspectorate	1.1	1.7	97.2
Total	41.8	12.6	45.6

Source: Own research.

Frequency of farmers' contacts with the above institutions of control varied in relation to characteristics of the farms they operated. These characteristics included: the size of farmed land, type of production and type of farm (selected with regards to the size of the farmed land and sources of household income), as well as the educational level of the farm operators (Table 13.2).

The interviewed farmers who were raising animals had significantly more contacts with representatives of institutions responsible for food safety than farmers who cultivated crops exclusively. The respondents farming on land over 5 ha revealed that their contacts with employees of the control institutions were more frequent than that of respondents operating on farms smaller than 5 ha. Even greater differences could be seen between farmers with the highest (over 36 thousand PLN in 2016) or medium income (from 13 thousand PLN to 36 thousand PLN) and the lowest income to 12 thousand PLN. Furthermore, the study confirmed that respondents operating various types of farms had different contacts with such institutions. Respondents with farmer-type enterprises of over 5 ha and agricultural income only as well as those with the multi-business family enterprise farming the land of over 5 ha and drawing income from various sources were the most likely to have contacts with the institutions of control. Such contacts were rarer for modern farms operated by farmers/workers farming on up to 5 ha of land and having income from various sources and operators of peasant farms of less than 5 ha farm land with farm income only.

Table 13.2. Farmers' contacts with representatives of institutions responsible for food safety according to farm size, farm profitability, type of production, and farm type (%)

Variable	Farm characteristics	Answer		
		Yes, many times	Yes, only once	No, never
Farm size (N = 3544)*	up to 5 ha	31.6	13.2	55.2
	above 5 ha	59.1	11.7	29.2
Farm profitability in 2016 (N = 3434)*	up to 12,000 PLN	31.2	12.3	56.5
	13,000–36,000 PLN	46.0	14.5	39.6
	37,000 PLN and over	71.4	12.3	16.3
Type of production (N = 3543)*	livestock	56.0	14.7	29.3
	crop production only	22.8	9.9	67.3
Type of farm (N = 3511)*	peasant farm (up to 5 ha, income from farming)	37.3	14.7	48.0
	modern farm of farmer/worker (up to 5 ha, income from various sources)	31.3	13.3	55.4
	farming enterprise (over 5 ha, income from farming)	69.5	11.8	18.7
	multi-business family enterprise (over 5 ha, income from various sources)	54.0	11.9	34.1
Respondents' level of education (N = 3534)*	elementary	43.0	12.8	44.2
	vocational	43.3	13.1	43.6
	high school	42.0	12.7	45.3
	beyond high school	33.7	11.0	55.3

* $p < 0.01$.

Source: Own research.

The interpretation of the study results presented above appears to be rather clear-cut. The institutions responsible for food safety thoroughly monitor commodity farms with the largest farm area, diverse portfolio and the highest profitability, where agricultural production is the most important or the only form of professional activity of household members. Such monitoring is usually less detailed and partial in small, less profitable farms that focus on plant production and which are mostly peasant farms. The same applies to modern farms operated by farmer/workers. One issue should be given some attention here. Farmers who had the least frequent contacts with the institutions of control were usually the ones with the highest level of education. How to interpret such a situation? Perhaps this means having trust in one's own knowledge and own abilities to gather knowledge from various sources and/or the unwillingness of these farmers to ask for advice from the institutions mentioned above. It could also be possible that such farmers put emphasis on, and maintained, high standards for the quality of their products and, in return, received the trust of the regulatory institutions, which was reflected in their limited inspections and control activities.

Table 13.3. Respondents' contacts with representatives of Veterinary Inspectorate (%)

Variable	Farm characteristics	Answer		
		Yes, many times	Yes, only once	No, never
Farm size (N = 3544)*	up to 5 ha	23.9	14.8	61.3
	over 5 ha	46.7	14.6	38.7
Farm profitability in 2016 (N = 3434)*	up to 12,000 PLN	25.1	14.3	60.6
	13,000–36,000 PLN	34.3	15.2	50.5
	37,000 PLN and over	56.9	14.0	29.1
Type of production (N = 3543)*	livestock	48.8	19.9	31.3
	crop production only	10.5	7.7	81.8
Type of farm (N = 3511)*	peasant farm (up to 5 ha, income from farming)	27.0	14.9	58.1
	modern farm of farmer/worker (up to 5 ha, income from various sources)	23.9	14.8	61.3
	farming enterprise (over 5 ha, income from farming)	55.1	15.1	29.8
	multi-business family enterprise (over 5 ha, income from various sources)	42.7	14.5	41.8

* p < 0.01.

Source: Own research.

The data presented above show factors fostering farmers' contacts with the institutions of food safety control as confirmed when farmers' contacts with each of the above institutions are being analyzed. Tables 13.3 and 13.4 present data on contacts with the Veterinary Inspectorate and Main Inspectorate of Plant Health and Seed Inspection. These were the two institutions that the study participants had the most frequent contacts with. Farmers who had the most frequent contacts with the Veterinary Inspectorate were the ones who raised animals, had the highest income and the largest farms. The rarest contacts with the Veterinary Inspectorate were reported by respondents from the category of modern farmer/workers, who only cultivated crops on small farms and had rather modest income and profitability. Similar tendencies could be observed with respondents' contacts with representatives of the Main Inspectorate of Plant Health and Seed Inspection. The only exception was in the production type's lack of impact on the frequency of contacts. Farmers who only engaged in crop production were just as likely to have contacts with this institution as animal producers.

Table 13.4. Contacts with the representatives of Main Inspectorate of Plant Health and Seed Protection (%)

Variable	Farm characteristics	Answer		
		Yes, many times	Yes, only once	No, never
Farm size (N = 3544)*	up to 5 ha	5.6	7.2	87.2
	over 5 ha	18.4	15.0	66.6
Farm profitability in 2016 (N = 3434)*	up to 12,000 PLN	4.8	6.3	88.9
	13,000–36,000 PLN	11.1	13.2	75.7
	37,000 PLN and over	27.3	17.7	55.0
Type of production (N = 3543)**	livestock	10.7	11.2	78.1
	crop production only	10.0	8.6	81.4
Type of farm (N = 3511)*	peasant farm (up to 5 ha, income from farming)	6.7	12.0	81.3
	modern farm of farmer/worker (up to 5 ha, income from various sources)	5.5	7.0	87.5
	farming enterprise (over 5 ha, income from farming)	26.9	16.3	56.8
	multi-business family enterprise (over 5 ha, income from various sources)	14.5	14.4	71.1

* p < 0.01, ** p < 0.05.

Source: Own research.

The above data results show that only Veterinary Inspectorate and, to some extent, the Main Inspectorate of Plant Health and Seed Inspection are the institutions that could have an impact on decisions made by farmers and leading to a certain type of farm such as a farmer-type enterprise raising animals, and their acquiring high incomes, as well as on the system of food quality control. The remaining institutions had such minimal contacts with the farmers participating in the study that they seemed to be facade institutions mostly dealing with making regulations rather than enforcing them, monitoring farmers and cooperating with them to ensure food quality and its adequate oversight. Therefore, it should be stated that the exogenous mechanisms of food quality control at the level of the individual farm in Poland are not sufficiently effective.

Knowing the limited effectiveness of the above described exogenous mechanisms of food quality control on farms, it seems essential to give some attention to endogenous mechanisms, which are generally connected with the use of forbidden practices in agricultural production.

Table 13.5. Respondents' opinions on the statement: "Farmers in Poland often use forbidden practices in food production in order to acquire higher revenues" (scale from 1—"I strongly agree" to 5—"I strongly disagree"; responses "I strongly agree" and "I somewhat agree" were combined in one "I agree" column, while responses "I strongly disagree" and "I somewhat disagree" were placed in one "I disagree" column) in %

Variable	Farm characteristics	Answer		
		Yes, many times	Yes, only once	No, never
Total (N = 3544)	×	20.9	39.4	39.7
Farm size (N = 3544)*	up to 5 ha	21.1	37.1	41.8
	over 5 ha	20.6	43.1	36.3
Farm profitability in 2016 (N = 3434)	up to 12 000 PLN	21.1	38.8	40.1
	13,000–36,000 PLN	21.1	40.3	38.6
	37,000 PLN and over	19.5	44.2	36.3
Type of production (N = 3543)*	livestock	19.9	41.5	38.6
	crop production only	22.4	36.4	41.2
Type of farm (N = 3511)*	peasant farm (up to 5 ha, income from farming)	21.3	41.4	37.3
	modern farm of farmer/worker (up to 5 ha, income from various sources)	21.0	37.1	41.9
	farming enterprise (over 5 ha, income from farming)	16.4	45.9	37.7
	multi-business family enterprise (over 5 ha, income from various sources)	22.4	42.0	35.6
Respondents' level of education (N = 3534)*	elementary	14.8	46.0	39.2
	vocational	21.0	38.8	40.2
	high school	23.4	37.2	39.4
	beyond high school	20.8	40.0	39.2

* p < 0.01.

Source: Own research.

First, the respondents were asked to express their opinions on Polish farmers using forbidden practices in food production to acquire higher revenues (Table 13.5). About one fifth of respondents approved of such practices, while two fifths disapproved. Another two fifths of respondents were not able to provide a clear-cut opinion. The distribution of opinions in various groups of farmers was unfocused, however those most disapproving of the above statement were farmers and peasants who had the highest income, as well as those with the lowest educational level.

While the issue analyzed above mostly related to the internal motivation of food producers to use forbidden practices in order to maximize their revenues, the next question takes up the problem of farmers being forced to engage in such practices through market mechanisms. The respondents were asked to respond to the view that in order to stay afloat on the market and maintain some revenues, farmers sometimes have to engage in forbidden practices. The data results revealed in the answers to this question are not much different than the results for the previous question (Table 13.6). Slightly more than half of farmers rejected such an opinion and, at the same time, held the view that market forces do not necessitate the use of forbidden practices by farmers. The opinions of farmers using forbidden practices in order to keep us with market conditions was expressed by 15% of the study participants, who recognized that the market situation had an impact on the usage of forbidden practices by food producers. About one third of respondents did not have a clear-cut opinion. The groups who rejected the use of forbidden practices the most generally included farm owners or operators who had the highest agricultural incomes.

When discussing farmers' internal motivation and market requirements as determining the use of forbidden practices in food production, it should be added that only a small group of farmers (about 15–20%) entertained the idea of intentional lowering the quality of food produced by Polish farmers. The group that rejected such a possibility was larger (40–50%), and by expressing such an opinion they confirmed their confidence in the high quality of Polish food. It should be noted, however, that about one third of respondents had no opinion on this matter. Finally, it needs to be stated that farmers with the highest revenues from agricultural production, with the largest production volumes were the least likely to notice the problem of forbidden practices.

An important indicator of whether farmers producing food products trust their quality is their consumption of what they produce and their reason for such consumption. Therefore, the study participants were asked for their opinion on two statements related to potential reasons for consumption of food they produce. The first statement addressed the issue of consuming the food from one's own production to save money on household food expenses and the second referred to the conviction that the food products from one's own production were of better quality than the products available at stores. The obtained results were quite explicit. Farmers and farm operators who consumed their own products to save money on food purchases made up 77% of respondents (Table 13.7). At the same time, over 83% of respondents reported that they often consumed the food products from their own production as they were better than those sold in stores (Table 13.8).

Table 13.6. Respondents' opinion on the statement "In order to stay on the market and have some revenue it is sometimes necessary to use forbidden practices in food production" (scale from 1—"I strongly agree" to 5—"I strongly disagree"; responses "I strongly agree" and "I somewhat agree" were combined in one "I agree" column, while responses "I strongly disagree" and "I somewhat disagree" were placed in one "I disagree" column) in %

Variable	Farm characteristics	Answer		
		Yes, many times	Yes, only once	No, never
Total (N = 3544)	×	15.4	52.6	32.0
Farm size (N = 3544)*	up to 5 ha	15.3	51.4	33.3
	over 5 ha	15.6	54.6	29.8
Farm profitability in 2016 (N = 3434)*	up to 12,000 PLN	14.9	51.8	33.3
	13,000–36,000 PLN	17.4	53.4	29.2
	37,000 PLN and over	15.2	57.6	27.2
Type of production (N = 3543)*	livestock	14.2	54.8	31.0
	crop production only	17.1	49.6	33.3
Type of farm (N = 3511)*	peasant farm (up to 5 ha, income from farming)	20.0	54.7	25.3
	modern farm of farmer/worker (up to 5 ha, income from various sources)	15.3	51.3	33.4
	farming enterprise (over 5 ha, income from farming)	13.5	58.0	28.5
	multi-business family enterprise (over 5 ha, income from various sources)	16.7	52.9	30.4
Respondents' level of education: (N = 3534)*	elementary	15.3	50.5	34.2
	vocational	14.9	52.2	32.9
	high school	17.6	52.4	30.0
	beyond high school	12.0	57.0	31.0

* $p < 0.01$.

Source: Own research.

Table 13.7. Respondents' opinions on the statements: "To save money on food, we often eat what we produce on our farm (scale from 1—"I strongly agree" to 5—"I strongly disagree"; responses "I strongly agree" and "I somewhat agree" were combined into one "I agree" column, while responses "I strongly disagree" and "I somewhat disagree" were placed in one "I disagree" column) in %

Variable	Farm characteristics	Answer		
		Yes, many times	Yes, only once	No, never
Total (N = 3544)	×	77.0	10.8	12.2
Farm size (N = 3544)	up to 5 ha	76.3	12.7	11.0
	over 5 ha	78.2	11.0	10.8
Farm profitability in 2016 (N = 3434)*	up to 12,000 PLN	86.6	11.4	2.0
	13,000–36,000 PLN	81.0	9.3	9.7
	37,000 PLN and over	74.7	10.4	14.9
Type of production (N = 3543)*	livestock	83.9	5.6	10.5
	crop production only	67.7	17.8	14.5
Type of farm (N = 3511)	peasant farm (up to 5 ha, income from farming)	83.2	7.8	9.0
	modern farm of farmer/worker (up to 5 ha, income from various sources)	75.9	10.8	13.3
	farming enterprise (over 5 ha, income from farming)	77.2	11.1	11.7
	multi-business family enterprise (over 5 ha, income from various sources)	78.4	11.1	10.5
Respondents' level of education (N = 3534)*	elementary	79.2	7.5	13.2
	vocational	78.6	10.1	11.3
	high school	77.7	11.3	11.0
	beyond high school	67.5	15.8	16.7

* p < 0.01.

Source: Own research.

Based on the data in the last two tables, it can be inferred that although the two selected factors strongly motivate farmers to consume the food from their own production, not all farmers who eat the food products of their own production do so for the economic reasons such as poverty or rationalization of the costs of supporting the family. There are also farmers who are convinced that their own products truly have the highest quality. It should be highlighted that the economic reasons were mostly indicated by respondents who operated the smallest farms and had the lowest revenues. The respondents with larger farms, reporting the highest income, as well as livestock producers and farmers operating agricultural or multi-business enterprises which had their greatest share of business activity in food production, reported consuming the products from their own production due to its better quality. This should be considered as a positive indicator of the product quality from Polish food producers.

Table 13.8. Farmers' opinions on the statement: "We often eat what we produce on our farm because our products are of better quality than the ones sold in stores" (scale from 1—"I strongly agree" to 5—"I strongly disagree"; responses "I strongly agree" and "I somewhat agree" were combined in one "I agree" column, while responses "I strongly disagree" and "I somewhat disagree" were placed in one "I disagree" column) in %

Variable	Farm characteristics	Answer		
		Yes, many times	Yes, only once	No, never
Farm size (N = 3544)*	up to 5 ha	80.3	8.1	11.6
	over 5 ha	88.7	5.2	6.1
Farm profitability in 2016 (N = 3434)*	up to 12 000 PLN	80.8	8.9	10.3
	13,000–36,000 PLN	87.0	4.3	8.7
	37,000 PLN and over	88.7	4.0	7.3
Type of production (N = 3543)*	livestock	89.4	3.2	7.4
	crop production	75.3	12.1	12.6
Type of farm (N = 3511)*	peasant farm (up to 5 ha, income from farming)	86.8	1.3	11.9
	modern farm of farmer/worker (up to 5 ha, income from various sources)	80.0	8.3	11.7
	farming enterprise (over 5 ha, income from farming)	89.0	2.9	8.1
	multi-business family enterprise (over 5 ha, income from various sources)	88.7	6.1	5.2
Respondents' level of education (N = 3534)*	elementary	83.1	5.5	11.4
	vocational	84.2	7.5	8.3
	high school	84.6	6.1	9.3
	beyond high school	78.0	9.8	12.2

* $p < 0.01$.

Source: Own research.

13.6. Conclusions

How to characterize the issue of food safety at the level of the farm, in lieu of what was discussed above? The consciousness of farmers usually has an impact on their decisions regarding what to produce. Not without the meaning is the general confidence about the high quality of currently produced food. There is also a fear or some uncertainty among the majority of farmers about the fierce competition on the food production market and the possibility that some producers may use forbidden practices. As indicated by the study results, half of participating farmers thought that using forbidden practices was not a necessity but fewer respondents were certain that Polish farmers did not engage in such practices. Perhaps a "phantom" of unfair competition has its source in the rather ineffective, barely visible functioning of the institutions responsible for food safety. To be clear, these institutions do have an impact on the part of the most crucial groups of agricultural producers (farmers and peasants who have the most profitable farms with the largest area) by having some contacts with them. The institutions do not in any sense reach the majority of farmers. Perhaps, in order to maintain the high standards of food produced in Poland, the effectiveness of these institutions should be improved in the first place. The institutions should be activated in the field through intensification of their contacts with direct food producers.

People usually want, and try to have, sources of food relatively close to their residence, and would like to ensure unlimited access to them. In the pyramid of people's needs, food belongs in the category of the most fundamental needs and is therefore in the center of interests of local communities and their inhabitants, as well as executives of large multinational corporations. The problem lies in the contemporary food chains that are so complex, containing many various links, that their effective monitoring is hard to execute. This, in turn, can lead to the lowering of the food safety level. Can, within this context, the fact that Polish farmers willingly and regularly consume products of their own production be recognized as a sufficient guarantee of the high-quality food produced in Poland? Yes, to some extent. It should also be stressed that the quality of food products that reach consumers is increasingly less consistent than the quality of products that leave the farm or agricultural holding. Therefore, it seems accurate that each society is safest when food sovereignty is achieved and its own food can be produced using the local raw materials and local workforce in the food industry, thus benefitting local communities. This way new jobs are created, the costs of transport and waste are minimized, and stronger guarantees of maintaining high quality food production are given to consumers. Such an economy, fostering small, healthy communities, makes these countries self-sufficient and rather independent in terms of food production and food availability. In other words: the shorter the food chain, the greater the food safety. This statement is very much in line with the conclusions presented by Mooney and Hunt (2009, pp. 492–494), who identified the local character of food production and local control of this process as the most important conditions for food security and food safety.

REFERENCES

Applebaum, A. (2017). *The Red Famine: Stalin's War on Ukraine*. Penguin Random House.

Bello, W. (2011). *Wojny żywnościowe* [Food wars]. Przekład P.M. Bartolik. Instytut Wydawniczy Książka i Prasa.

Brown, L.R. (2012). *Full Planet, Empty Plates: The new geopolitics of food scarcity*. W.W. Norton.

Caparrós, M. (2016). *Głód* [Hunger]. Przekład M. Szafrańska-Brandt. Wydawnictwo Literackie.

Chechelski, P. (2008). *Wpływ procesów globalizacji na polski przemysł spożywczy* [The impact of glo-balization processes on the Polish food industry]. IERiGŻ-PIB ("Studia i Monografie," 145).

Chechelski, P. (2015). Ewolucja łańcucha żywnościowego [Evolution of the food chain]. In: I. Szcze-paniak, K. Firlej (eds.), *Przemysł spożywczy. Makrootoczenie, inwestycje, ekspansja zagraniczna* [The food industry: Macro-environment, investments, foreign expansion], pp. 45–63. UEK, IERiGŻ-PIB.

Cribb, J. (2010). *The Coming Famine: The global food crisis and what we can do to avoid it*. University of California Press.

DoRzeczy (2017). *Skażone jaja dotarły do Polski. Toksyczna substancja w 40 tys. sztuk* [Contami-nated eggs have reached Poland: A toxic substance in 40 thousand pieces]. https://dorzeczy.pl/kraj/38348/Skazone-jaja-dotarly-do-Polski-Toksyczna-substancja-w-40-tys-sztuk.html (access: April 10, 2019).

EU Agricultural Markets (2015). *You Are Part of the Food Chain: Key facts and figures on the food supply chain in the European Union*. Briefs No. 4, June, 2015. https://ec.europa.eu/agriculture/sites/agriculture/files/markets-and-prices/market-briefs/pdf/04_en.pdf (access: March 28, 2019).

EUR-lex (2002). Document 32002R0178, Regulation (EC) No. 178/2002 of the European Parliament and of the Council of 28 January 2002 laying down the general principles and requirements of food law, establishing the European Food Safety Authority and laying down procedures in mat-ters of food safety, OJ L 31, February 1, 2002, pp. 1–24. https://eur-lex.europa.eu/eli/reg/2002/178/oj (access: April 10, 2019).

EUR-lex (2011). Document 32011R0016, Commission Regulation (EU) No. 16/2011 of 10 January 2011 laying down implementing measures for the rapid alert system for food and feed. Text with EEA relevance, OJ L 6, January 11, 2011, pp. 7–10. https://eur-lex.europa.eu/eli/reg/2011/16(1)/oj (access: April 10, 2019).

FAO [Food and Agriculture Organization of the United Nations] (1996). *Declaration on World Food Security and World Food Summit Plan of Action*. World Food Summit, Rome, November 13–17, 1996. http://www.fao.org/3/w3613e/w3613e00.htm (access: April 10, 2019).

FAO (2018). *Publication Regional Overview of Food Security and Nutrition in Europe and Central Asia 2018*. Budapest. 110 pp. Licence: CC BY-NC-SA 3.0 IGO.

FAO, IFAD, UNICEF, WFP, & WHO (2018). *The State of Food Security and Nutrition in the World 2018: Building climate resilience for food security and nutrition*. Rome, FAO.

Głód, D., Adamczak, M., & Bednarski, W. (2014). Wybrane aspekty zastosowania nanotechnologii w produkcji żywności [Selected aspects of nanotechnology applications in food production]. *Żyw-ność Nauka Technologia Jakość* 21(5), pp. 36–52.

Grochowska, R. (2014). Specyfika koncepcji bezpieczeństwa żywnościowego jako „problemu bez rozwiązania" [The specificity of the concept of food security as a "problem without a solution"]. *Zagadnienia Ekonomiki Rolnej* [Agricultural economics] 3, pp. 95–106.

Idzikowska, M., Janczura, M., Lepionka, T., Madej, M., Mościcka, E., Pyzik, J., Siwek, P., Szubierajska, W., Skrajnowska, D., & Tokarz, A. (2012). *Nanotechnologia w produkcji żywności – kierunki rozwoju, zagrożenia i regulacje prawne* [Nanotechnology in food production: Directions of development, threats and legal regulations]. *Biul. Wydz. Farm. WUM* 4, pp. 26–31. http://biuletyn-farmacji.wum.edu.pl/

IFAD (2011). *Rural Poverty Report 2011*. Rome, International Fund for Agricultural Development.

Kaufman, F. (2012). *Bet the Farm: How food stopped being food*. Joseph Wiley & Sons.

Keneally, T. (2011). *Three Famines, Starvation, and Politics*. Public Affairs.

Kirwan, J. & Maye, D. (2013). Food security framings within the UK and the integration of local food systems. *Journal of Rural Studies* 29, pp. 91–100.

Kołożyn-Krajewska, D., & Sikora, T. (2010). *Zarządzanie bezpieczeństwem żywności. Teoria i praktyka* [Food safety management: Theory and practice]. C.H. Beck.

Koocheki, A. & Ghorbani, R. (2005). Traditional agriculture in Iran and development challenges for organic agriculture. *International Journal of Biodiversity Science and Management* 11, pp. 1–7.

Kowalczyk, S. (2009). *Bezpieczeństwo żywności w erze globalizacji*. Szkoła Główna Handlowa w Warszawie.

Kowalczyk, S. (2014). *Prawo czystej żywności. Od Kodeksu Hammurabiego do Codex Alimentarius* [The law of clean food: From the Hammurabi Code to the Codex Alimentarius]. Oficyna Wydawnicza Szkoła Główna Handlowa w Warszawie.

Kowalczyk, S. (2016). *Bezpieczeństwo i jakość żywności* [Food safety and quality]. WN PWN.

Leśkiewicz, K. (2012). Bezpieczeństwo żywnościowe i bezpieczeństwo żywności – aspekty prawne. *Przegląd Prawa Rolnego* 1(10), pp. 179–198.

Lucińska, M. & Grajeta, H. (2015). Wpływ modyfikacji genetycznych na jakość i bezpieczeństwo żywności [Impact of genetic modifications on food quality and safety]. *Problemy Higieny i Epidemiologii* 96(4), pp. 705–712.

Marzęda-Młynarska, K. (2014). *Globalne zarządzanie bezpieczeństwem żywnościowym na przełomie XX i XXI wieku* [Global food security management at the turn of the 20th and 21st centuries]. Wydawnictwo UMCS, p. 641.

Mikuła, A. (2012). *Bezpieczeństwo żywnościowe Polski, Rocznik Ekonomii Rolnictwa i Rozwoju Obszarów Wiejskich* [Yearbook of Agricultural Economics and Rural Development] 99(4), pp. 38–48.

Mooney, P. & Hunt, S. (2009). Food security: The elaboration of contested claims to a consensus frame. *Rural Sociology* 74(4), pp. 469–497.

Obiedzińska, A. (2016). *Seminarium: Bezpieczeństwo żywnościowe i bezpieczeństwo żywności – wielowymiarowe podejście* [Seminar: Food security and food safety: A multidimensional approach]. https://www.ierigz.waw.pl/aktualnosci/seminaria-i-konferencje/1454682190 (access: April 10, 2019).

Omieciuch, J. (2016). Jakość i bezpieczeństwo żywności w Polsce [Food quality and safety in Poland]. *Społeczeństwo i Ekonomia / Society and Economics* 2(6), pp. 123–134.

PAN Europe (2019). *Pesticide Use in Europe*. https://www.pan-europe.info/issues/pesticide-use-europe (access: March 28, 2019).

Parekh, S.R. (ed.) (2004). *The GMO Handbook: Genetically modified animals, microbes, and plants in biotechnology.* Springer Science + Business Media.

Patel, R. (2012). *Stuffed or Starved: The hidden battle for the world food system.* Melville House.

Ploeg, J.D. van der (2016). *Family Farming in Europe and Central Asia: History, characteristics, threats and potentials.* Working paper No. 153. http://www.fao.org/3/a-i6536e.pdf (access: April 1, 2019).

Ploeg, J.D. van der (2018). From de-to repeasantization: The modernization of agriculture revisited. *Journal of Rural Studies* 61, pp. 236–243.

Raport NIK (2018). *Nadzór nad stosowaniem dodatków do żywności.* Nr ewid. 173/2018/P/18/082/LLO.

Raport Polska Wieś i Rolnictwo (2018). Ministerstwo Rolnictwa i Rozwoju Wsi. https://www.gov.pl/documents/912055/913531/Wyniki_badania_PWiR_2018.pdf/60865ce2-cab8-e775-4511-00cc23c5caf9 (access: April 10, 2019).

Rejestr środków ochrony roślin (2019). [The register of authorized plant protection products]. https://www.gov.pl/web/rolnictwo/rejestr-rodkow-ochrony-roslin (access: April 1, 2019).

Shapiro, P. (2018). *Czyste mięso. Jak hodowla mięsa bez zwierząt zrewolucjonizuje twój obiad i cały świat* [Clean meat: How breeding meat without animals will revolutionize your dinner and the world]. Przekład N. Mętrak-Ruda. Wydawnictwo Marginesy.

Shepherd, B. (2012). Thinking critically about food security. *Security Dialogue* 43(3), pp. 195–212.

Szczepaniak, I. & Firlej, K. (eds.) (2015). *Przemysł spożywczy. Makrootoczenie, inwestycje, ekspansja zagraniczna* [The food industry: Macro-environment, investments, foreign expansion]. UEK, IERiGŻ-PIB.

TVP Info (2018). *Skażone jaja trafiły na niemiecki rynek* [Contaminated eggs found their way to the German market]. https://www.tvp.info/37626576/skazone-jaja-trafily-na-niemiecki-rynek (access: April 10, 2019).

Tuomisto, H.L. & Teixeira de Mattos, M.J. (2011). Environmental impacts of cultured meat production. *Environmental Science & Technology.* http://2014falleng114.pbworks.com/w/file/fetch/88945115/cultured_meat_lca_es_t_published.pdf (access: March 28, 2019).

Wilkinson, J. (2015). Food security and the global agrifood system: Ethical issues in historical and sociological perspective. *Global Food Security* 7, pp. 9–14.

wolnosc24.pl (2017). *Holandia i Belgia obrzucają się jajami. Afera jajowa rozlewa się na Europę, a Bieńkowska milczy* [The Netherlands and Belgium are pelting each other: The egg scandal spills over Europe, and Bieńkowska is silent]. https://wolnosc24.pl/2017/08/11/holandia-i-belgia-obrzuca-ja-sie-jajami-afera-jajowa-rozlewa-sie-na-europe-a-bienkowska-milczy/ (access: April 10, 2019).

Ziegler, J. (2013). *Geopolityka głodu* [The geopolitics of hunger]. Przekład E. Cylwik. Instytut Wydawniczy Książka i Prasa.

Zwoliński, A. (2006). *Jedzenie w relacjach społecznych* [Eating in social relationships]. WN Papieskiej Akademii Teologicznej.

Think Locally, Act Globally: Polish farmers in the global era of sustainability and resilience, ed. by Krzysztof Gorlach and Zbigniew Drąg in collaboration with Anna Jastrzębiec-Witowska and David Ritter
Jagiellonian University Press, Kraków 2021, pp. 591–640
ISBN 978-83-233-4949-5
DOI: http://dxdoi.org/10.4467/K7195.199/20.20.12742

Chapter Fourteen: The Impact of Mechanization and Digitalization of Agriculture on the Farm Development in Poland

Adam Dąbrowski (iD) https://orcid.org/0000-0003-3620-2184/

Maria Kotkiewicz (iD) https://orcid.org/0000-0001-5500-4728/

Piotr Nowak (iD) https://orcid.org/0000-0001-7991-5534/

14.1. Introduction

The notion of ongoing social change involving the progressing digitalization of social life is a starting point for the reflection on the use of new technologies, including information and communication technologies, as well as mechanization and digitalization on farms and their impact on agricultural development. Digitalization of agriculture can be perceived as the next process of radical change of a qualitative and quantitative character in agriculture that comes after agricultural mechanization, which is understood as the equipping of farms with engine-run machines. Digitalization is understood not only as the transition from the analogue to digital format of recording information, but also as changes of consciousness among the users of technology and equipping farms with technologies that use microprocessors and the Internet. This is another stage of agricultural development, where new opportunities for solving various problems related to production and distribution of agricultural products on both a local and global scale may be seen.

At the root of the assumption about the growing importance of the digitalization of social life lies a thesis about the increasing tendency to use new technologies in various dimensions of social life. Promulgation of the Internet was a decisive phenomenon for this. In 1991, the Internet became available for individual users and this started the Internet revolution that encompassed the entire world. New applications, social networks and portals, programs, Internet forums and other forms of communications involving the Internet emerged one by one and became commonly used tools of social mobilization. Another important moment deepening the described changes was the introduction of mobile technologies. In 2014, the "mobile revolution" became a fact as the number of mobile Internet users exceeded the number of those who used stationary computers connected to the net (Pieriegud, 2016, p. 17). As is well-known, the mobile technologies facilitate access to almost unlimited information resources and allow for fast communication free of physical barriers. Mobile technologies have had a huge effect on everyday life, including the emergence of the so-called "Internet of things" and wearable devices.

The digital and mobile revolution has also spread over the domain of the economy. As indicated by Warsaw-based economist Jana Pieriegud (2016, p. 11) digitalization of the economy and society means we are going through one of the most dynamic changes of our times. It also carries the uncertainty and various kinds of threats related to social consequences of the automation of production processes or generally understood safety issues (Pieriegud, 2016, p. 11). The described process becomes one of the important causes of progressing innovative changes in the majority of economic sectors. Among the key factors driving the development of the digital economy are: the automation combined with the use of robots, development and diversification model of the distribution of products and services, increasing the role of knowledge as the main product determining the competitiveness of enterprises, and the development of communication technologies facilitating non-synchronous communication independent of any physical barriers. According to Pieriegud (2016, p. 12), the digital transformation of various spheres of economic activities is the answer to the challenges related to the rapid development of digital technologies.

Digital progress makes its way into agriculture as well. This is a widely spread phenomenon and it goes way beyond the studies conducted so far. However, the phenomenon of digitalization may serve as a theoretical frame encompassing the theme areas directly stemming from the main premises of the project "Think Locally, Act Globally." The scope of the research on new technologies in agriculture for the project conducted at the Institute of Sociology of the Jagiellonian University was narrowed down to several issues. Farmers operating the farms and actors who implemented innovations in agriculture, as well as agricultural advisors were chosen as research subjects. Three research questions were presented within the defined research problem. The most important question was the question about farmers' attitudes towards new technologies, including the ones that were applied in farm management. The second question dealt with changes caused by the use of new technologies in functioning of farms and future strategies examined in the local and global context. The third question addressed the relation between new technologies in agriculture and sustainable development. This last research question was: what is the impact of the new technologies used on farms on sustainable development?

The process of acquiring knowledge and dissemination of innovation is another important matter connected with the possibilities of neo-endogenous development of agriculture based on local and global resources, which has its driving force in technological changes in the manners of communication and transfer of information (Nowak, 2012, p. 31). In the academic literature numerous definitions for the term innovation can be found. However, for the purpose of this chapter a rather broad definition has been chosen, in which innovative processes as the effect of cooperation and interaction between people, organizations and their social environment. Innovation is therefore the set of activities and procedures that can be implemented in the field of technology, in the processes and logistics of management, in organization of production or services (Chyłek, 2012, p. 42). The definition featured above shows that innovation is a complex phenomenon. First of all, it is worth noting where farmers acquire the knowledge, which they can later transform into innovative practices on their farms. The summary of the 2012 study conducted by TNS OBOP and ordered by Martin & Jacob on a group of one thousand farmers in Poland revealed ". . . growing credibility of the Internet as a source of information on the subject of farm operating." At the same time, it should be stated that farmers still declared that agricultural programmers on TV had the first place in terms of providing the needed information on operating of the farm. The second source of such information were the opinions of their acquaintances. The reason for changes in farmers' purchasing behavior was recognized as connected with the need to confront the opinions on the product with a larger number of users of this product than it is possible in the local environment (Uchaniuk, 2012). The results obtained in the discussed study were confirmed by the responses of the project participants. They indicated the significant role of traditional and digital media as well as the role of acquaintances in acquiring information on new technologies in agriculture. They also mentioned the significance of state institutions, nongovernment organizations and commercial companies in spreading information on new technologies. The

14.2. Through Farmers' Eyes

14.2.1. Technological Elite and Their Farms

The farmers selected for the research interviews were in some way an elite among Polish farmers and could possibly map out the path in the processes of mechanization and digitalization of agriculture in Poland. The analysis of the conducted interviews pointed to several common features of those who could be found among the 5% of those applying mechanization and computer technologies on their farms and therefore were diagnosed in quantitative studies (see part two of this publication). All interviewed farmers came from farming families, which meant that the farms they operated were passed to them by their fathers or relatives. It might be valuable to quote one of the respondents, whose experience was somewhat different:

> I was born on the farm. First, I lived with my parents at my grandmother's and my uncle's farm, so I have farming in my blood. Later I kind of separated myself from farming. My high school and my studies were related to technical science. Then I did some postgraduate agricultural studies. So, there was some agriculture in my education but the real adventure with farming started after I got married (R5).

Other respondents mostly inherited the farm or a part thereof from their parents:

> So, I have worked on a farm since childhood but I have personally operated the farm since 2007. From the day of my father's death, I started to operate the farm on my own (R2).

Respondents' background had a positive impact on their experience of working on the farm. All of them noted that they had started working on the farm at a really young age, initially helping in simple agricultural tasks. With age they gained experience, skills, and abilities as they engaged in more advanced works and decision-making processes related to farming.

> We can say that in the countryside children practically help parents from the youngest age. I started operating the farm professionally when I was 17 after they signed the farm over to me. I was modernizing and I was going in that direction. I did not look for happiness elsewhere but wanted to modernize and improve what I had to make living from, to get by here in Poland. I did not want to move abroad to feed myself and wanted to stay in Poland (R3).

One of the respondents said that during his agricultural studies he was so engaged in farm work that it had to change the way he studied.

> During my first semester I was a full-time day student but such studying took a lot of my time, so I switched to weekend classes after the first semester. Now, it is my fifth semester (R6).

This statement reflected significant involvement in the work on family farms and orientation toward practical knowledge as well as human capital that respondents received from their families and family farms.

Another interesting characteristic shared by respondents had to do with their self-evaluation regarding the use of computers. Despite the fact that the respondents had different computer skills and competencies they usually recognized them as low. Even the respondents who used databases available on the Internet and were able to create their own databases with the use of calculation spreadsheets to increase the efficiency of their farm production, downplayed their computer skills. One of the respondents noticed the following:

> I am a master of Excel ... First of all, Excel, I would perish without Excel. Here, I have all the data basis related to animals. As far as the Internet is concerned, I see it as a database and source of knowledge, I also use it (R4). In another situation the same respondent said: I would say that I am a bit behind with this. Things like searching for information, checking something, improving my skills, watching something on YouTube are still within my capabilities ... The smartphone I would gladly sink in the river (R4).

Another respondent, who used modern agricultural machinery on a daily basis, prepared his applications for direct subsidies, and used the Internet to expand his agricultural knowledge as well as exchange opinions on managing the farm, said bluntly that the computer was "black magic" to him (R1). A similar opinion was expressed by a different respondent, who also used modern technologies on his farm but was quick to state that he "did not have much love for the computer" (R7). Respondents could be seen as people convinced of their low computer skills, which they worked to improve, and because of that could effectively operate their farms. It should be stressed that the respondents were oriented toward continuous developments which they defined as an open process and not the acquired state of things.

As was shown, the orientation of respondents towards development was a unifying characteristic. This related to the individual dimension of personal development as well as to farm development. Therefore, in the respondents' comments there were statements alluding to the modernization processes taking place on the farm as well as far-reaching future investment plans. One of the respondents referred to the investments made on account of his newly established farm. As he financed these investments with his own money, farm development proved to be a tremendous challenge due to the high costs of land and machinery purchases. He also had to buy seedlings and wait a few years for them to bear fruit, which he could later sell:

> First, the field needs to be prepared. It takes about a year for field preparation, then there is planting. Three years pass before any significant fruit can be seen (R5).

The quoted farmer recently had some large expenses from starting the farm but at the same time he already had plans for future investments:

> An orchard tractor is a costly investment. I am postponing it for a more distant, unknown future point in time. For now, whatever machinery my family has, should be enough (R5).

Another respondent reported large farm investments but indicated they were made with the help of external funding:

> In the last 10 years everything went upside down. The barn was erected in 2009 and then the machines were to be purchased one after another because there were such needs. A side car for harvesting, reaping machines, macerator, two tractors and a larger spreader were bought. Then the telescope charger—everything seemed useful and needed and the EU funds were used. I do not know if my parents omitted any fund that was available. The entire Rural Development Fund and our regional Local Action Group called the Garner of Upper Silesia provided the fund. Diversification of production and other programs were also used. The machinery is slowly but surely being replaced. Obviously, we must be looking into new ones (R6).

This farmer was not going to stop the investment activities right there and had rather interesting future plans:

> Generally, I am quite interested in detector devices that monitor the movements of cows, the number of their chews, their rut activities or their problems with metabolism. The problem is that such technology is currently very expensive. Without the aid funds it is difficult to introduce such technologies. They would add a gigantic cost on top of other costs, which are already massive. After several years of constant investments, it is very

much needed to have one year of breathing easy, having some distance and later starting to invest from the beginning (R6).

The quoted comments of respondents reflected their ability for strategic thinking.

Interestingly, all the respondents had agricultural education at least at the secondary level but some were more educated than others. Among the research participants were individuals with vocational school, people who acquired agricultural education at the high school level, as well as those who completed university studies:

I graduated from the Agricultural Academy in Krakow, The Department of Forestry. Additionally, I finished the animal husbandry program and got my engineering certificate in animal husbandry. Before that, besides my technical high school with a focus on forestry I also finished technical high school with a specialization in beekeeping and agricultural technical high school (R4).

Regardless of the educational level and actual computer skills, all respondents had a great need for searching and acquiring knowledge through the Internet. All these distinguished characteristics were the components of the innovative type. This was confirmed by the respondents' answers, which reflected their attitudes towards new technologies.

14.2.2. Farmers' Attitudes towards the New Technologies

Individual actions are the necessary elements of social change. Ultimately it is the farmer's attitude towards absorbing, adopting, and using new technologies that determines whether the neo-endogenous model of rural development will be implemented and whether the farm will use the new technologies and will be able to respond to global trends while using local resources.

According to a classic approach from Stefan Nowak the attitude of certain individuals towards certain objects is the entirety of relatively permanent dispositions toward evaluating this subject and having an emotional reaction to it. These emotional and evaluative dispositions are accompanied by relatively stable and permanent convictions about the nature and attributes of this object and relatively permanent dispositions for certain behavior towards the object (Nowak, 1973, p. 23). According to the author of the structural definition of the attitude, this abstract theoretical construct consists of three elements: (1) a cognitive component (relatively permanent convictions about the nature and the attributes of the object), (2) an affective component (dispositions to evaluate the object and emotionally react to it) and (3) a behavioral component (relatively permanent disposition for certain behavior towards the object). Various configurations of the distinguished elements make quite diverse attitudes towards something or somebody. Therefore, it would be justified to talk about the emotional attitude that is not grounded in the cognitive basis and not leading to any actions. A different attitude with strong cognitive background on certain issues and weak emotional attitude with no action could also be observed. In the attitudes of the interviewed farmers certain clues could be found related to "whether," "how," and "why" farmers did not use new technologies on their farms.

The cognitive component of the attitude towards the new technologies could be seen as the state of knowledge on the subject. The analysis of answers given by respondents suggested that in the investigated group of farmers there were three types of the cognitive components of the attitude. The responses of one of the interviewed farmers indicated that rather impressive knowledge on new technologies and his ability to use them and apply on the farm. Additionally, that respondent suggested that he was not only buying some computer devices but also ordered the manufacture to do some custom-made alterations to his specifications. He was also trying to implement optimal solutions for agricultural technologies that he applied on his farm. The use of new technologies on this farm should be viewed quite broadly, from the use of new agro-environmental solutions, through the use of modern machinery of agricultural production and the use of various computer programs that serve to inform, facilitate animal husbandry, and sell agricultural products.

> This means that the machines were needed so I could develop production. In the qualitative sense and in terms of improving production quality... Practically all the needed studies were conducted on my farm: botanical studies, ornithological studies, and other types (R4).

The respondent representing such an attitude also described having a very good agricultural education, which he constantly improved. This type of cognitive attitude could be easily recognized as a knowledge-oriented attitude. The second type of cognitive attitude could be seen as profit-oriented. The difference between these two attitudes was defined by the quality and extent of acquired knowledge. With the first type of cognitive attitude, the knowledge was the main resource on which the farmer capitalized through constant development of the farm, investing in widely understood new technologies which would bring economic profit to the farmer. In the second attitude type the knowledge was acquired only when there were external premises which imposed the new equipment purchase on the farmer, or there was a desire to sell agricultural products:

> Sometimes I check agricultural magazines, mostly TopAgrar, but I use the Internet more often because I look for what I actually want at the moment. These are the main sources of information (R2).

The third type of attitude could be viewed as the backwards attitude. Farmers presenting such an attitude did not look for knowledge and were convinced of their infallibility stemming from experience and long-term work in agriculture. They did not have knowledge of new technologies and used them in a very limited way, characteristic for the era of modernization and not the era of digitalization:

> Computer is like black magic to me . . . The human factor is better . . . the best machine won't do it and won't shorten the pre-harvest interval. The best machine won't identify and throw out the wheat or corn that was on the sludge and got contaminated. But pigs eat that stuff and kids put it in their mouths, too. What do the computer and mechanization have to do with it? If there were computers detecting heavy metals and other things, I would be in favor of that (R1).

The second attitude component that Nowak (1973) distinguished was the emotional approach to a given issue or phenomenon with the tendency to evaluate. Interestingly, the collected material did not allow for two extreme types of attitude: positive or negative towards technologies. Regardless of the state of the knowledge (i.e., the cognitive element of the attitude), the respondents were able to name positive and negative sides of the use of new technologies in agriculture with better or worse proficiency. It seemed that the respondents had various reasons to provide ambiguous evaluations. What was characteristic of the attitude described as knowledge-oriented was the formulation of a critical evaluation towards the application of new technologies in agriculture. A rational analysis of collected knowledge was made and the decision on using the technologies was seen as having more benefits than negative effects. This type of attitude could be viewed as ambivalently rational:

> The only negative effect that I can talk about is greater soil compaction because of heavier machinery; a larger tractor weighing more, so the total impact on the soil is greater (R4).

The research participants, who only occasionally had the need to acquire new knowledge in order to increase the profit or found themselves in situation where they were forced to do so, were more likely to refer to inconveniences in using the technologies and machines that were in their possession than to notice their impact on the functioning of the farm. Such an attitude could be described as ambivalently functional:

> The functions are entirely different stories and one needs to be prepared in how to use them. The preparation can also vary. The salesman should provide instructions in how certain devices should be used but it does not always happen. Perhaps not all of the functions are used but I think that the person can learn that on his or her own and it should not be a big problem (R2).

It should also be noted that every respondent who at the level of their own knowledge mostly used their own experience was also able to identify the positive sides of the use of technologies on farm.

The last component of the attitude dealt with the behavior related to the application of the new technologies. There were no respondents who did not use technologies to operate their farms but the extent to which they did varied greatly. This was strongly related to the adopted research sample, for which the farmers applying new technologies on their farms were selected. The scale of this phenomenon was an arbitrary issue and therefore, for this publication, three types of actions connected with technologies could be distinguished in order to simplify things. First, there were people who implemented new technologies on their farm to a great extent and at the same time tried to modify these technologies to serve the farm needs. This sort of action could be described as active use of technologies. Such farmers were characterized as resourceful, creative in implementing the selected useful technologies in software and hardware versions. In the active use of computer programming, it was mostly MS Excel that farmers declared a familiarity with:

> ... the easiest solution to create such a basis for your own use is through the Excel program. You do what you need, you do not have ready-to-use tables but you create your own table according to needs (R6).

The works connected with machinery, equipment, and infrastructure dealt with various aspects of improvements, refinements, and even rebuilding the whole systems servicing the crops and adjusting them to the conditions of particular farms. It might be useful here to excerpt some respondent-researcher dialogue.

> **Respondent:** The whole plantation was established with our own money.
> **Researcher:** Is this the system that you built?

Respondent: Yes. By myself. I used the already tried and tested solutions but I built it myself.

Researcher: This is a very interesting issue for me. Where did you get the knowledge to use this cable or this extension cord?

Respondent: Luckily for me, I am not the first one in Europe to have a plantation of blueberries. There are others who have been planting blueberries for 10, 15 years and I have learned from them (R5).

Interestingly, none of the respondents recommended the adaptation of software or hardware on the farm he operated, unless it dealt with purchase of new equipment, which was reflected in the dialogue quoted below:

Respondent: I think that everybody knows that common sense matters, right? It is not a problem to come up with money but I want what I want. Most of the machines, I would say, 90% are custom made for me.

Researcher: What does it mean?

Respondent: Whatever such machine should have, according to my preference. All tractors can be custom-assembled from the ground up (R7).

The participating farmers proved to be quite resourceful in terms of solving problems on their farms and only by using their own power, skills, and ideas. These were people with technical and manufacturing knowledge, which allowed them to perform various tasks on the farm and various upgrades to improve the machines. The extent of these competencies related to mechanics, and assembly of non-digital machines allowed for many significant projects to be undertaken with the machines according to farmers' own ideas. Farmers understood the mechanisms of machine operation and they were experts in that regard. Digitization was in some way a cause of certain stress for farmers, as the introduction of computers to the machines left them unable to repair the machines as they could before the machines' computerization. They were no longer the self-reliant experts in that discipline and therefore their role was limited while the importance of external services increased.

What characterized other respondents who used technologies on their farms was the passive use. For these farmers, implementation of production technologies as well as their choice of machines, equipment and computer programming was based on default parameters. The extent of technological implementation could be moderately high or moderately low but overall, they were people who did not shun the overall technological progress taking place in agricultural labor. As one of the respondents said:

... whether it is about one machine or another, it could be the one that was most needed, or the new machine because the old one was too worn out and had to be replaced or we did not have the machine that had to be used and the rental service was too expensive in our area (R2).

For analytical purposes, a third category of respondents should be distinguished. They were the ones who did not use computer information technology on their farms. However, this group of respondents was not considered in the qualitative research described here.

14.2.3. Changes on the Farm due to Implementation of New Technologies

The respondents' attitudes toward new technologies and development would justify the statements that they were innovators who played a significant role in the process of social change, on both the micro and macro levels. The research material presented so far by the interviewed farmers indicated that farms they operated changed over the years in terms of infrastructure and the use of machines or computers. The question of whether the investment decisions made by them brought some expected and/or unexpected changes on the farm and in the farmer's life might be an interesting one to explore. Keeping the perspective of the innovation theory, the analyzed problem can be viewed from four different angles of innovation; 1) **product innovation** involving all sorts of changes perfecting the product already produced by a particular enterprise, 2) **innovations of production and distribution methods**, 3) **marketing innovation** meaning introduction of new marketing methods including significant changes in designing the product and packaging, product promotion, and price strategy, 4) **organizational innovation** defined as introduction of all new organizational methods in a company's business practices, organization of workplace and/or external relations as well.

The types of innovations enlisted above were reflected in respondents' practices but only one respondent was able to apply all four types of innovations. He was a cattle farmer who made the decision to start an ecological farm with the use of agroforestry techniques and started a separate new company that specialized in distribution and marketing of the livestock cattle. Additionally, the respondent, his brother, and his father operated three formally separate farms as one. Although there was no formal cooperation it contributed to the increased efficiency and profitability of the cattle farm. The dialogue below illustrates a rather interesting way to organize work on a farm.

> **Researcher:** Your business deals with sales, trade, and marketing, right?
> **Respondent:** Yes, but it also deals with distribution and disassembling in some ways. This is a cattle farm, so farm produces live cattle and the rest is done by the company . . . I see how this can bring some benefits, concrete benefits, and I am not just talking about economic profits but they are also our goal. For example, it is good for us if we have some additional lumber production without taking a lot of space (R4).

Another respondent implemented product innovation changing the direction of production and he also applied technological innovation by purchasing new machines and new computer programs.

> There was canola and cereals but now it is mostly changing. I was pondering the whole thing for a while as cereals were always in first place. Now it is more about corn for silage production. There are also green pastures on arable land and meadows. It is important to provide enough food for animals as our main direction of agricultural production is milk cattle. This is what we do now (R2).

Another respondent specializing in cattle farming (R6) pointed out some interesting components of organizational and product innovations (R6). He was an example of the farmer who, in the globalization era, could use innovation to achieve a competitive advantage on local and beyond local markets. The so-called "new economy," which meant an economy based on knowledge is currently recognized as

> . . . superior economic structure powered by the stream of innovations of computer information and telecommunication technologies, which affect all areas and branches of economy accelerating the efficiency and the pace of economic development (Kozera, 2013, p. 35).

The "new economy" is a certain response to the clash of global trends and pressures stemming from international market competition with the functioning of local economic systems and the increasing role of communication technologies in all types of enterprise. Within the context of global changes of the "new economy" that result in the increasing importance of the 2nd and 3rd sectors of the economy, the changes taking place in rural areas could be seen as leading to the *New Rural Economy*, which is the type of economy "in which there are numerous small and medium enterprises besides agriculture" (Halamska, 2012, p. 15).

The activities of the farmer quoted below well illustrate the process described above. The respondent stated:

> . . . the situation on the Polish and global market allows the farmer who bought bigger agricultural machines with the help of UE funds to work more effectively and in the time, he gains this way he performs services on neighboring farms that do not have such modern equipment or sometimes a farmer may purchase the machines with the purpose of providing services to others (R6).

As indicated by the respondent, the technologies used on farms brought about two type of markets for services connected with agriculture. First, there was the emergence of businesses oriented to providing services for technologies used on farms. These businesses included mechanics, electronics specialists, electricians, and other types of specialists responsible for the proper and smooth functioning of technologies, as well as stores with machine parts and technological services as well as online agricultural services including the possibility to purchase raw material for production, participation in online auctions, and the exchange and trading of machines and equipment, etc. There were also advisers and experts whose role was to improve the processes taking place on the farm by applying new solutions in agriculture, and IT specialists creating useful computer applications and securing data entered by farmers. Second, the market of services performed by farmers and based on the already existing technology which farmers had invested in also came into being. In this type of service farmers who performed them charged other farmers money. They services were the following: collection of crops such as cereals, canola and hay, spraying and precision fertilization, renting machines, sowing cereals (often with GPS), harvest collection with delivering to the farm facility or piling harvested crops on the field, transport services of agricultural products. The described situation of increasing application of new technologies created the alternative activities dealing with supporting and servicing of these technologies. These activities and services added some possibilities for diversification of revenues for those farmers who already had technology. This on one hand could be seen as a favorable change and on the other as a process deepening diversification within the discussed professional category.

The use of computers in securing competitive advantage and in implementing innovative solutions is a crucial issue for every enterprise, including the agricultural one. Referring to the classic theory of Alvin Toffler (1980), who analyzed the development of civilization using the metaphor of waves that "wash away" previously existing basic mechanisms of human development, it might appear Polish farmers are not keeping up with civilizational changes. Toffler distinguished three waves in the development of humanity: agrarian, industrial, and post-industrial connected with the Information Age. However, the dynamic changes of the 20th century compelled researchers to consider adding a fourth wave of development to the concept. In the fourth wave, knowledge would be recognized as the main factor of development (Sadler, 1988). The fourth wave society is seen as having the ability to use acquired knowledge and transfer it into added value. To make this process effective, highly specialized human resources would be needed with intellectual capital being the most sought after type of capital (Drucker, 1994). In this situation, the role of the dominant agent of change is assigned to people who, regardless of macroeconomic conditions, can get access to information, acquire knowledge and apply it in practice. This means that having a competitive advantage in today's world is not possible without a certain level of expertise, acquiring, processing, and applying knowledge. The process of arriving to the stage of development described as the society of knowledge is a logical consequence of experiencing the Information Age wave with computerization as its domain. Therefore, it is not possible to "skip" any phase of development and that realization, in the context of the study results described in this chapter, might cause some concerns.

Some of the study participants were quite attuned to technical innovations and, to various extents, they were utilizing modern machinery and computer programs on the farm. One of the respondents emphasized what he was able to do by himself:

> On my own. With the use of already tested and tried solutions I built my own system for watering bushes (R5).

What was common among all respondents was the fact that farmers made autonomous decisions on investments and this meant increasing risk:

> But the risk is mine. I search as long as necessary and I do things as long as necessary, until all the options are out. There is no way to do things without any risk. The risk is a given. If things work out, the success is mine and if they don't, that's just too bad (R1).

It was characteristic of respondents that they treated the investment risk as intertwined with their job.

> There are upsides and downsides. If you want to invest, you have to take a risk and take a loan. But this is still better and before people had to work much harder. Farmers were so overworked (R3).

Respondents pointed to similar consequences of implementing technological innovations on their farms. First of all, the study participants stressed that introduction of technologies resulted in making work on the farm easier. Some respondents were particularly vocal about cutting on economic expenses:

> Lower costs. Lower cost, less time and one person can do all the work with one machine, one tractor (R1).

Other respondents emphasized the benefits beyond the economic ones and mentioned having more free time:

> So, there is some free time, to put it colloquially, to make life easier (R2).

The analysis of statements related to "reclaimed" time, which respondents refer to as "free time" suggested significant diversification of changes that occurred in the lives of farmers due to technologies. One of the respondents said that the time he regained thanks to greater automation of farm work was now spent on other matters also related to the farm.

> I do not have free time but I have time to do other things. There is no such thing as free time in agriculture, you simply need to do something else. I mean that such time can be spent on other farm matters (R5).

Another respondent presented this issue in the following manner:

> I would say that there is more time indeed. The question is whether this time is moved to other areas of farm activities. It seems to me that when it comes to work it has no bearing on the time with the family. There is more time because of some decisions and not so much due to the change in how farming is conducted. The fact is that I have more time now and this is because of the machines. At the moment, I can say that, for a while, I used that time to develop my business. Now, I would put more emphasis on family time and having time for my passions such as agro-forestry and agro-ecology. This means having passion at work and working with passion (R4).

The way farmers used their "regained" free time depended on their particular aspirations and current financial situation. Age was another important issue when it came to managing free time. It was highlighted by another respondent and illustrated by the dialogue below:

> **Respondent:** One wants to rest a little especially when one is older. Young people see things differently and they chase time. In the past, we worked 14, 16 hours a day and it was the standard.
> **Researcher:** And now?
> **Respondent:** No, we can't. Now 10 hours and we are done. I will tell you this: there are physical therapy sessions, doctors' appointments and hospital visits that one needs to show up for. So now, a lot of time is spent on one's health (R1).

Another respondent reported that while technologies allowed for faster and more effective work, resulting in having more free time, there was also a downside to technologies "such as TV and computer eat up your time" (R7). Ultimately, despite all the facilitations offered by technologies, free time is still a scarce resource. "A lot of things are done over the phone but there is still a shortage of time" (R7).

The ambiguous aspects of changes occurring on farms because of applied technologies could be seen in the cooperation between farmers and/or in changing of social bonds. Analysis of the respondents' answers indicated that farmers talked to others about agricultural matters but the number of people they kept close contacts with was shrinking. There was one respondent who had a lot of sentiment for the old times before the introduction of the general use of agricultural machines.

> . . . it won't be the way things used to be, because nowadays everybody keeps to himself. People lie that they do not sow, pretend they do nothing. There is a lot of lying going on. The things aren't the way they used to be. People do not visit each other much (R3).

According to many respondents, digitalization of agriculture negatively impacted social bonds and solidarity within the farming profession and within their local community. The act of borrowing agricultural equipment from someone else was no longer a common practice and had become quite rare:

> . . . now nobody loans stuff to others. We used to loan our manure spreader to other farmers. I would still loan it. The combine harvester I would not loan because one needs to know how to work it. Combine harvester, absolutely not (R3).

Another respondent admitted that the number of close friends in the long-term view was shrinking. This was reflected in the dialogue with the researcher:

> **Respondent:** The circle of friends has shrunk. One or two friends and that's it.
> **Researcher:** Were things more social in the past?

Respondent: Back in the day, half of the village would come when there was one TV set in a village (R1).

At the same time farmers reported tightening bonds within the narrow group of their friends, relatives and family, which was possible with having more free time:

> I try to devote more time to the family . . . I have friends and neighbors who also have such machines and I can count on their help. In return, they can expect that I will help them when needed, so there is cooperation. Perhaps different now than before, but it is still going on (R2).

It should also be noted that nowadays with the Internet and social media farmers are being able to expand the network of contacts. Farms seem to be part of the trend of increased reliance on the Internet and counting on the power of weak bonds as shown in the concept of Granovetter (1973). Polish author Olcoń-Kubicka notices that in contemporary times

> . . . weak bonds are also important for cooperation and not just strong bonds as the weak bonds create various possibilities for actions, enrich the existing resources, and help with providing information (Olcoń-Kubicka, 2009, p. 81).

The emergence of the division of competencies is an important aspect of changes occurring in Polish agriculture, due to the impact of new technologies. In the academic literature, the process of changes concerning the abilities to use new technologies is also connected with the phenomenon of "digital exclusion," defined as the gap between those who have access to computers and the Internet (Dijk, 2006, p. 178). Exclusion is a rather complex issue and does not mean simple access to technology. In fact, three degrees of exclusion can be distinguished.

Digital exclusion of the first degree involves the classic understanding of this type of exclusion, meaning both the access to a computer and the Internet as well as the quality of that access. This obviously can influence the abilities to use these technologies. Concerning access to broadband Internet there are no significant differences in Poland between urban and rural areas. In 2017, an estimated 81.5% of large city residents, 77.2% of small-town dwellers, and 74.1% of those residing in rural areas had good quality Internet access. It should be therefore assumed that the current state of Internet infrastructure in rural areas does not exclude farmers digitally (Wagner, 2017, p. 120).

The second degree of digital exclusion involves differences between Internet users who have the abilities to use available Internet technologies without problems and those who do not have such abilities (Hargittai, 2002). The competencies analyzed with this type of exclusion in mind are quite wide and include reading and writing, knowledge of terminology and understanding of computer programming languages, abilities to find information and awareness about the consequences of online activities. The study described in this chapter did not include the collection of quantitative data that would allow one to draw conclusions about the computer competencies of the studied population. However, the conducted qualitative research indicated that the problem of second-degree digital exclusion might occur and could affect the quality of using the new technologies. The interviewed farmers did not have proficiency in computer programming languages even if they used computers and programs on their farms:

> I had some Internet forums and I also used a website called Farmer.pl or something like that, if I remember well. That would be it, and sometimes I use "Uncle Google," I can tell you that (RP4).

This also resonated with the ability to find information on the Internet. Farmers who used the Internet to search for agricultural knowledge used rather basic and sometimes even accidental sources of information. As one of the respondents reported:

> I use YouTube, Onet.pl, and OLX for machines or something like that, Allegro.pl sometimes. The basic websites. That's it (RP3).

From the point of view of the described phenomenon, the digital exclusion of the third degree, described as the effect of having or not having the pertinent computer and Internet[1] skills, might be the most crucial here. This form of exclusion is challenging to measure and even the of digital exclusion of the third degree had a rather weak theoretical grounding, which indicated a lack of empirical research in that regard, as well. It might be assumed that effectiveness is closely correlated with competencies and skills enabling people to work on the computer and use the Internet and that is how this dimension of digital exclusion can be evaluated. There is also a mental dimension, dealing with the use of information technologies in various aspects of everyday life, an openness to innovative solutions and readiness,

---

[1] Jan van Dijk presents digital exclusion in a similar manner, describing four successive kinds of access to digital technology: motivational, physical, skills, and usage (Dijk, 2006, p. 179).

not only for the use of adopted schemes but also at personalization and expanding of available options to accommodate the needs of individuals. This dimension is particularly important for rural areas because, as indicated by Polish sociologist Andrzej Kaleta, their residents had

> ... [a] higher level of cultural conservatism and resistance against the innovation, which has always been noted in sociological studies and recognized as one of the immanent characteristics of rural areas resulting in lower readiness to adapt modern technological solutions and fewer possibilities for work, studies and other types of life activities, which could appear in rural areas through popularization of ICT[2] (Kaleta, 2016, p. 39).

Regardless of the degree of digital exclusion of particular respondents, the successive changes on the farm stemming from introduction of technical solutions could be noted. This was reflected in respondents' opinions related to increased access to sources of knowledge on matters such as: farming and crops, means of production, world market prices, and the information on technological novelties. This had a bearing on respondents' decisions on farm matters:

> Ok, there are some social media and they are fine. Because of them agriculture appears on social media. You can observe agricultural profiles on Instagram and perhaps you can see something interesting. Perhaps Germans have something new, new technology that you can see as interesting and you already like the profile of some machine producer there. Maybe they invented something new. It might also be Facebook, where you like pages related to agriculture again and again. I do not check Interia.pl, I do not check Onet.pl to know what is going on in the world. Ok, I listen to the radio to have the awareness what is happening in the world. It is not that I only care about the farm and nothing else. To find out what is happening in Poland, whether there is some new scandal or if somebody stole from someone or not, I have a radio. I also check TopAgrar for the novelties in the agricultural domain that I am interested in. Periodicals that I read are mostly agricultural periodicals so . . . (R6).

The use of modern technologies in acquiring knowledge was even better described in the two dialogues below:

---

[2]   ICT—acronym for Information and Communication Technologies.

Researcher: Do you read TopAgrar in the print version or online?

Respondent: I use the print version for the time being but I am slowly and carefully considering the online version. Sometimes, when I need to, I check the TopAgrar website and I usually find what I need, so…

Researcher: What other Internet websites do you visit? As I said previously, farmers' Internet use would be of interest to me. Do you memorize the addresses of the websites that you use, or do you keep virtual bookmarks and just click those? What websites do you visit to get the information you mentioned?

Respondent: You know, I had some Internet forums and I also used a website called Farmer.pl or something like that, if I remember well. That would be it and sometimes I use "Uncle Google," I can tell you that (R2).

Respondent: Today, without the Internet you are stuck.

Researcher: Exactly. And can you tell me what do you use? What functions?

Respondent: What do I use from the Internet? Auctions are good and helpful and I can buy some used things or I can ask for advice. I can also find information on plant diseases and pests . . . (R5).

The changes described above are taking place as a result of modernization and technological progress on farms and they show the role of technology in respondents' lives and on their farms. Technology not only impacts the production methods, increasing the profitability of the farm and improving work organization but it also brings about better development possibilities and quality of production. At the same time, there is a loosening of bonds between people, which used to be particularly important in small local communities and in the farmer's profession. This phenomenon is presented in academic literature but with no clear interpretation. A certain sentiment for the "old times" was expressed by some respondents when they referred to the times "before" technology. One study participant emphasized that:

There were more people to help and they were more willing. The times have changed a bit. It used to be that someone would come to your farm to help with farm work and get some potatoes, cereal or whatever in return. In the past there were more small farms and fewer machines. We had the machines and we helped the others. For example, we had a combine harvester that was about 40 years old but it could still do the work and we helped the neighbors. Then the neighbor would help us with harvesting or with hay collecting. There used to be more cooperation in the village. Now, things are more private as I can see. Although, my colleagues and I still help each other. If there is some action, we call on each other and help each other, but this is not the same as it used to be. In the past, there were moments with lots of people around the table. At breakfast people would come to help and cooperation was greater, I can say. Now, it seems to me that we act more individually. Recently a veterinarian was making fun of me for buying the tool for drying off cows. He said that it was not helping integration in the village. Back in the day, farmers would come, help and then they would have some vodka. There was more comradeship and now everybody is alone. This has changed as well. I think the village is different now in that regard than it used to be . . . (R2).

This opinion was seconded by another respondent, who stated that:

. . . we would go to the fields, talk to people, we would meet people in the field. Today you won't meet people there. Practically one person comes, does the sowing and he is gone. All of them are landowners and practically there is no contact with these people. That's how things are now. I used to go to the fields with another person, now I go alone. Why would two people need to go? Back in the day people farmed more land; they had potatoes, beets. There was a lot of work done by hand, such as weeding, and now it is different. There was more contact with people. That contact was different, everybody went to the field and today there are only 4 farmers left practically. Back in the day, everybody in the village lived off of farming, so everybody went to the field. People had small chunks of land but everybody went to the fields and everybody had some stuff growing . . . (R1).

Another respondent made reference to ludic aspects of farming:

. . . yeah, there was some drinking after they came to help. It was fun. If they came to help, it should not be for nothing. During the day there was sour curd milk, and in the late afternoon a gallon of wine on the table . . . (R3).

Another respondent pointed out the weakening of social bonds rather than the decreasing the number of contacts. He said that in the past one could acquire knowledge through such contacts:

Respondent: . . . from neighbors . . . information was passed from person to person orally. Somebody went somewhere and saw something and then the whole village knew it.
Researcher: Does it still function that way now? Do you often meet with other farmers to discuss your problems?
Respondent: Now, I just hear gossip, rather than other things. Nowadays, there is no discussion of problems (R5).

The same respondent reported that he would probably be able to borrow agricultural equipment from someone else if he ever had such a need, for example, if his own equipment broke,

> ...it would be possible. I think it would be possible. I think, if you smile nicely, you will always be able to get stuff loaned to you (R5).

14.2.4. Changes on the Local Market

The role of new technologies in operating the farm is a key to competing on the global agricultural market. What is the relationship between the technology and increased competition? The answer to such a question is not obvious. The relations are well explained by concepts that note various benefits of using computers, the Internet, and intelligent machines in everyday life (see Levinson, 1997; Castells, 2006). The concept of neo-endogenous development defined as "a lasting and partnership type of cooperation of all subjects that contribute to the making of a certain social space and use three types of knowledge: scientific, managerial, and local" provides an even more interesting explanation of this relationship (Nowak, 2012, p. 48, after: Adamski & Gorlach, 2007, p. 28). The interaction between local and global forces is the key component of development in the neo-endogenous perspective. The driving force of the development in rural areas can therefore be seen in globalization and in technological changes related to communication and dissemination of information, which means a knowledge-based economy (Galdeano-Gómez, Aznar-Sánchez, & Pérez-Mesa, 2010, p. 62). The referenced authors point to the progressive development of new technologies as an important factor in neo-endogenous development. It seems that this category should be expanded to include modern and intelligent digital technologies in general. Such a wider understanding of the role of new technologies in neo-endogenous development not only stresses the importance of communication abilities that are at the core of networks but also in the functions of the machines. Thanks to the use of new technologies, farmers get on the one hand, a chance to change the model of farming and their own lifestyle and, on the other, might be able to build a network of relations around these technologies. In this sense, what is individual and local can be defined with the process of technological development in mind and it is understood that society undergoes changes because of two forces: global (technological progress, digitalization, forcing competition through global markets) and local (i.e. local culture, family, and local resources).

In their responses the study participants emphasized changes caused by introduction of technology to the farms; not just changes occurring on their own farms, but also changes affecting the entire agricultural market. The majority of the interviewed farmers mentioned technologies improving the quality of agricultural products as a positive consequence of the introduction of technology to the farm:

> Yes, if everything is so hygienic—that's ok. Maybe not antiseptic, but all cleaned up and secured. Well, it's not that you can walk around the cow barn with coffee or camera in your hand but everything is kept clean. The milk is probably safer now, health-wise. You have an idea whether the cow is healthy or not and the product that leaves the farm is surely better (R6).

Another respondent, who was also a milk producer, stated that the better quality of agricultural products was possible because of machines:

> . . . milk, definitely better because the system of leveraging milk and raising cows has changed. Back in the day the cows had dirt on them and they were kept in one place. Today, they can lay down on clean beddings. It can be said that the quality of milk is definitely much better. I can see in the test results of somatic cells that the cows' udders are healthier. No doubt that the way the barn is built has a bearing on the health of the animals. And this is reflected in the quality of the milk that is being sold (R2).

The respondent who grew blueberries also reported the positive influence of technology on his crops, stating that without the technologies that he applied, the blueberries would be "smaller, less luscious and with the use of technologies they are better" (R5).

There was also a respondent who had a different opinion about the influence of technologies on the quality of offered products:

> Quality-wise? I don't know. Perhaps the crops are more fertilized now. They were better before. Maybe there were fewer products but they were healthier for people and the environment. Now, there is a lot of everything but if you ate less, the effect would be just the same. So, now there are these open-top cargo trailers but also more herbicides. Everything is over-fertilized and chemical because everybody wants to make money and does not care about the quality. This is all modified, and different (R3).

Another respondent had a more nuanced and more neutral opinion, stating that the technology had no bearing on the quality of his production:

. . . quality maybe was not affected because we always did think on time, there was no slowing down or accelerating because of machines, nothing got worse (R1).

Explaining further he added:

It is not about computers, it is about mentality. There is some threat. As I say, a generation would need to pass for people working in agriculture to be engineers at least. Then things will be good in agriculture . . . The human factor is better . . . the best machine won't do it and won't shorten the pre-harvest interval. The best machine won't identify and throw out the wheat or corn that was on the sludge and got contaminated. But pigs eat that stuff and kids put in their mouths, too. What do the computer and mechanization have to do with it? If there were computers detecting heavy metals and other things, I would be in favor of that. Then I would be happy but the mentality would have to change. Or the computers would need to be changed to find what I was talking about. Then the seizure of entire property . . . (R1).

This was a very valuable insight as it highlighted the problem of thinking about the machines as direct agents of change in an extremely determinist perspective. Quality of production is tied to competitiveness, which creates another important aspect of change in the agricultural market. In reference to this issue one of the respondents presented the opinion suggesting links between competitiveness of the farm and an increase in production quality:

As far as factors determining my [market] competitiveness I would say it is the lasting high quality of production and the reputation that I have worked for over the last 5 years (R4).

Among the respondents, answers positively assessing the improved position of the farm on the competitive agricultural market prevailed. Farmers saw this as a consequence of technological progress. Within this context one of the respondents used the term "healthy rivalry" rather than competition:

I think rivalry is what brought us to development. The building of a barn can be an example. We became less competitive, right? I think this is about healthy rivalry . . . In the village, it might not be so, but in the region, I think this is the case as we are the only ones in the region to sell milk to the cooperative, right? There is also a thing where small farms disappear and larger ones emerge. In our municipality there are several farms like ours and we sort of followed their pattern of doing things. One is looking at the other.

If someone is doing well, the other wants to be even better and that is a healthy rivalry. This is how it should be, that one is stimulating the other in terms of development (R2).

Respondents also indicated that technical development of a farm allowed for increasing the volume of production:

Technology? Technology has a rather not-so-great influence on the matter of competitiveness and it is more about higher yields. I mean technology is impacting the yields, so theoretically technology has that impact on competitiveness (R5).

Another respondent said that with increased production he could count on more preferential offers from processors concerning distribution of his products:

It is easier I would say because they now weigh in our presence on the market. When we had 30 cows, there was less milk and milk processing plants were not interested to come down for the milk. And now with the crisis of our milk processing plant I ask in other milk processing plants whether they would be able to come and there was great interest from these other milk processing plants. They would gladly come to get our raw product. This was due to the fact that there was now a concrete volume of product and it was economically sound to come and get it. I think that this improves the possibilities for selling products (R2).

The effect of scale, which the respondent could use to gain preferential treatment in distribution of his product, was, according to other study participants, causing increased divisions among farmers. There were notable differences between farmers who had large farms and those who had smaller holdings. The former were able to buy more land and had a continuous increase of revenues. The owners and operators of small farms could not keep up with the competition and were forced to change their production profile or had to abandon farming altogether. This, according to the same respondent, translated to the specialization of farms:

Yes, it can be said that the technology had an impact on all of this, as I can see farmers who change their qualifications and production profiles. One farmer switched from goat production to work in the service sector, for example . . . Now, there is greater specialization, I would say. Specialization for sure, and now there is less of such dispersion and diversity than what we saw in the past. It used to be that one did a little bit of this and a little bit of that. This has changed a lot (R2).

Another respondent also pointed to the sharp differences between farms:

> . . . these farms that raise animals are eating up the farms that have little land and these farms with a sufficient amount of land are doing fine (R6).

In this statement, another aspect connected with changes on the agricultural market demands our attention. The respondent mentioned that farmers were leaving raising animals in favor of growing crops. He explained it in the following way:

> Yes, generally the technology of growing crops is much simpler than raising animals. Let's not pretend that farming land, growing crops is much harder than doing so in addition to raising animal. This is partially due to technology, as back in the day you could do nothing without manure. At the time, artificial fertilizers were not so readily available. Things were not as precise as they are now. There were not so many products for plant protection and not so much spraying. Technological progress changed all that. Everybody can now have storage or a silo and this is fine and simple. So, now, raising animals got lost somewhere. Looking at the machines and other things, there are only a few bigger barns. There are no farmers with let's say 10, 15 cows. Perhaps in Wielowieś there is one such farm (R6).

The reflection on changes affecting the farms and related to the application of new technologies suggests that these processes are ongoing and have significant impact on sustainable agricultural development. Respondents referred to qualitative and quantitative development of production and the application of global solutions as leading to further development of individual farms, which could increase their competitive abilities on local and global markets. Describing the mechanism of competition and specializations, the study participants perceived it as a rather positive process, causing a continuous progress of the agricultural sector while ensuring high quality production. Thanks to the use of computer programs and the greater ability for acquiring knowledge online, farmers increase the effectiveness of their work. They might do so through limiting the use of artificial fertilizer, fewer sprayings, or reasonably fast but more environmentally friendly growth of animal weight. The great example of sustainable agricultural production of continuous growth while ensuring ecological security could be seen in the respondent who completely converted his farm to an ecological enterprise. Such change in a farm system facilitated the maintenance of good agricultural practices and reaching economical goals at the same time.

14.2.5. Importance of Agricultural Advising Institutions and the Institutions of the European Union in the Process of Digitalization of Agriculture in Poland

Public agricultural advising turned out to be an important theme within the context of knowledge transfer in regards to new technologies implemented on farms. Such institutional landscape as a whole has influence on the state of farmers' knowledge. Getting the knowledge as objective as possible and adjusting it to the potential and needs of a particular farm is highly important from a farmers' point of view. Therefore, the knowledge most desired by farmers should be the knowledge passed on by state institutions through a system of agricultural advising. The respondents indicated lack of trust in commercial advisers:

> . . . we have a nice word for them, we call them "pushers." They are not advisers. First of all, they want to sell (R6).

For that reason, farmers need access to reliable, sound, and objective knowledge, which can only be passed to them through institutions that do not have financial interests in selling particular products but share the interest of optimal production and better farm efficiency with the farmer. As noted by Polish author Eugeniusz K. Chyłek, the system of agricultural advising has its various problems. Among the eight important factors determining the effectiveness of Farm Advisory System—FAS[3] in Poland are human capital and legal system. Also important are finances and availability of solutions based on science and research. Conditions for the FAS's functioning in Poland are not easy, with numerous barriers that make its advisory mission hard to fulfill. Human resources of agricultural advisory units to some extent need better professional training. Access to technical infrastructure and to the tools of modern communication is not sufficient and below the expectations of advisers (Chyłek, 2012, p. 77). Farmers can read the current situation of advisory institutions and this makes them lose some trust towards them. This is rather unfortunate as these institutions are still important for a knowledge-based development process. One of the respondents noted that:

---

[3] Among the most important factors shaping the agricultural advisory system, the author enumerates: human capital, technical infrastructure, finances, policy on family, consistency of science policy with agricultural policy, availability of solutions based on science and research, availability of information, new communication technologies, and legal system framework (Chyłek, 2012, p. 77).

... the advising in the agricultural advisory centers is not functioning the way it used to, right? The centers are well prepared for writing applications and filling out forms. This is probably going in the direction of what is more needed now. The agricultural advisory centers used to function quite differently. Their goal was to advise people on how to farm the land. Now private companies do that, not the advisory centers (R2).

Another respondent reported that the public advisory system had its role reduced significantly and now mostly dealt with filling out applications for farmers:

Agricultural advisory centers are only good when it comes to filling out the application form, right? It is usually the advisers from agricultural advisory center who fill out the application forms as they are familiar with that stuff. It is very hard for a farmer to write the application himself from scratch. If you are unfamiliar with legal regulations, you will need to complete many instructional courses to write one application (R6).

Respondents were asked directly about the importance of the European Union in the process of technical development of farms. Among the study participants there was no unanimous opinion on the role that the European Union played in the modernization of Polish rural areas. While they helped farmers to acquire various types of subsidies and funds for farm investments and activities, one of the respondents noted that:

... subsidies are in place now but the machines became so expensive after Poland's EU accession that they have eaten up all the subsidies (R7).

Another respondent was blunter when discussing the increase of machine prices related to the proliferation of co-financing mechanisms for farm purchases:

I will say this: before [Poland joined] the union machines were cheaper and a tractor cost 1000 PLN per horsepower. With Poland in the EU, a tractor costs 1000 Euro [per horsepower] (R1).

A different group of respondents addressed the difficulties of obtaining the EU funds for development and operation of a farm. The respondent pointed out that:

... the European Union perhaps could offer better opportunities if there were more possibilities in the D area[4] (R5).

---

[4] The Agency for Restructuring and Modernization of Agriculture conducts application enrollments for subsidies pertaining to the "Modernization of Farms" in four areas: area A (development of piglet production), area B (development of dairy production), area C (development of beef cattle production), area D (purchase of agricultural machines).

A similar tone was presented by another respondent, who was overall satisfied with his work in agriculture. He said:

> It is alright. I do what I like. It is somewhat difficult in today's world with all these requirements that one needs to meet and the requirements of the European Union in particular (R2).

Among the study participants there were also individuals who expressed full satisfaction from the opportunities they had thanks to the EU funds, which they used for the development of their farms:

> We are talking about the machines, so it needs to be said that the machines were bought with the subsidy money from the EU programs (R4).

The next respondent in the same line of thought said that:

> . . . for everything that was needed the EU funds were used. I do not know if any fund was skipped by my parents. The entire Program of Rural Development [was utilized] and there was also something regional (R6).

The opinions of the respondents indicated that the impact of the European Union on the technical development of farms were somewhat ambivalent. There were some objections towards the European Union in regard to strict requirements that made it difficult to get the funds, as well as toward the propelling of the demand mechanism stimulated by the large availability of subsidies, which resulted in the price increase of agricultural machinery.

14.3. Through the Eyes of the Advisers

14.3.1. About the Advisory System in the Context of Technology: Addition to farmers' opinions

As was described above, the public advisory in regards to the use of modern technology did not meet farmers' expectations. Various problems related to this very issue were also reported by representatives of the public advisory system. Their responses indicated quite clearly that the public agricultural advisory was not an institution implementing and promoting innovative solutions in the application of modern technologies and computer programs in agriculture. The advisers were not innovators in that regard:

> Public advisory system does not include advisory services on helping farmers to choose the right computer programs or the machines available on the market, through comparison of market offers (R8).

When it comes to the purchase of machines, the advisers fill out the forms for subsidizing the purchase and have to follow the rules regarding the technical parameters that the machines must meet for any particular farm:

> … the following things are being considered: the farm area to calculate the engine power of the tractor, for which a farmer can get financial aid (R8).

The services related to computer programs are not provided either. Farm advisers use applications to fill out the applications for financial aid or subsidies:

> Additionally, at the website of the Center of Agricultural Advisory one can find applications for nitrogen fertilization plans and calculations of maximal nitrogen doses as well as preparation for herd turnovers based on data from Agency for Restructuring and Modernization of Agriculture (R8).

At the Centre of Agricultural Advisory, a document entitled "Computer Programs for Agriculture" and updated in June 2018 can be found. It contains description of 26 computer programs available for farmers for free or through a commercial offer. The programs are divided by scopes: farm management, farm management in plant production, management of animal farm, management of agricultural land, fertilization plans, balancing of nutrients and optimizing of food plans, agricultural subsidies application service, supporting veterinary services and practices.

The study participants working in the public farm advisory sector did not have knowledge of new technologies or possibilities to obtain such knowledge at work as part of career and credential development. The ones who did have the knowledge reported lacking opportunities to test it in practice:

> The system of agricultural advising in Poland does not finance instructional courses of this kind and does not purchase the newest solutions to be tested in order to later have the advisers utilize such knowledge for innovative farm development (R8).

Due to the fact that the technological development of agriculture is so dynamic, the public advisory is not able to keep up with acquiring the knowledge and its systematization, especially when it comes to the newest discoveries. Therefore, the public advisers are not able to give farmers the knowledge for facilitating decision-making to implement digital technologies on their farms.

When asked to compare public advisory services with commercial ones, representatives of the advisory sector had similar observations as the farmers:

> . . . public advisory does not provide service dealing with the implementation of new technologies and their optimal adjustment for the abilities and needs of the farm. It does not provide objective assessment to that effect and does not fulfil the role of a mediator between farmers (equipped with knowledge and experience) and commercial advisers, who present their offers to farmers (R8).

Commercial advisers try to sell products from their line while aiming to maximize their profits. When the farmer comes to the public adviser to get help with the application form for financial aid related to the purchase then the public adviser can only verify the offer made by the commercial adviser if the proposed purchase does not meet formal requirements for financial aid (e.g., engine power too great for the agricultural area that the farmer is cultivating).

It is a common practice of representatives of public advisory services to rely on the knowledge and abilities to organize presentations conducted by advisers from commercial circles. The representatives of various companies producing technologies for agriculture are sometimes invited to workshops organized by public advisory services to present their offer. As a result, farmers are educated by them during meetings organized as part of public advisory.

It is worth noting that the use of computer programs on farms is not within the scope of specific advisory (neither commercial nor public). Commercial advisers offer the computer programming as long as it is a part of the larger technology sold by their companies (e.g., computer programming is needed for operating the machine).

14.3.2. Suppliers of Machines and Programming: Global or local market?

Respondents were asked to characterize the market of machines and agricultural software in Poland. The question regarding the market of agricultural machines did not pose any serious problems for farmers but knowledge of computer programs was not very common. One of the respondents stated the following:

> The market in Poland is open to all brands but some farmers have certain preferences as they are used to some brands or know about them from other producers. So, there are only a few producers and brands that matter. For the most part, they are not producers from Poland. There are some pieces of equipment such as lawn mowers or sprayers where farmers might be in favor of Polish products (R9).

Respondents stressed that the machinery market was dominated by foreign companies. Farmers' habits and mentality played a significant role in the shaping of the machinery market, which another respondent noted as well:

> The market of agricultural machines, equipment, and software in Poland is completely dependent on the global situation of producers. There are very few companies offering domestic products. In farmers' opinions, domestic machines and equipment are of rather low quality. Putting aside the objective potential of domestic machines and equipment, there is a noticeable preference among farmers to buy them from foreign companies. This is strongly influenced by farmers' mentality, for whom profitability of the farm is just as important as its status in the eyes of others. The status is thought to be seen through the lens of machines and equipment in the farmers' possession that are produced by well-known and well-regarded foreign producers (R8).

Interestingly, farmers thought of themselves as rational people making sound decisions and using available information and knowledge (also obtained online) when making a choice on a machine purchase. The advisers seemed to have different opinions in that regard. The study participants who were part of the advisory sector reported that machines chosen by farmers for the most part were not purchased from Polish producers. The lack of popularity of Polish products was tied to their perceived poor quality, limited effectiveness, and durability problems. One of the respondents stated that:

Machinery and farm equipment produced in Poland are considered to be of inferior quality. Even if their prices are attractive these products do not generate much interest as farmers value reliability and the certainty of continuous, uninterrupted work more than saving some money at the time of purchase. The low price is not encouraging. It is quite the opposite. If farmers have to pay more for a product, they will be more likely to consider such a product reliable and sure (R9).

According to the respondents above, the cultural factors had quite a significant impact on the choice of machines and the foreign brands (marques) would be preferred. The respondents' answers suggested quite clearly that the market of agricultural machines in Poland was highly globalized and Polish producers applying the strategy of price competitiveness were not winning many customers this way. The specific character of work in agriculture combined with the availability of subsidies for the purchases of agricultural machines influenced farmers' preference for foreign brands. Additionally, the availability and quality of service offered with the machine was also an important aspect in choosing a machine brand. One of the interviewed farmers expressed it in the following way:

I want to buy a machine and have no problems with it. I mean, I oil the machine, I service it, and I use it. I do not have to repair it (R6).

Another respondent complained about the insufficiency of the service networks of Polish producers:

So, we are saying that Polish machines as so good and back in the day farmers only bought Polish machines. And I say our taxes and the service matter. We went to Ursus (Polish company known for production of tractors) and I would tell the guy there what I needed. And he asks me if I am from Silesia. I say yes, you can probably recognize from the way I talk. And then he tells me in all honesty that they don't have a proper (service) representative there. He can sell me the tractor, no problem, but if something happens, I would have to wait about two weeks for service (R1).

These words were confirmed by the responses of advisers participating in the study. They judged the product assets combining information on availability and service locations of particular companies producing agricultural machines and farmers' opinions of these machines.

The lower prices of Polish machines do not attract buyers and this is to some extent very specific to the agricultural sector. Farmers cannot take a break from fieldwork during the time of harvest. For that reason, they have to have reliable machines and equipment as delays in fieldwork (like not collecting crops before the bad storm) can cost them lots of money. Changing weather forces farmers to be vigilant and well-prepared, and that includes having reliable machines and equipment (R9).

One of the respondents noted an interesting problem:

... not using all of the abilities and functions of the machines that farmers have in their possession is quite common. Farmers' falling into routine and maintaining old habits can be seen as the reason why farmers do not fully use the potential of their machines (R8).

Therefore, in such cases, it can be said that farmers pay for machines that apply solutions which are not tuned to their expectations.

Although respondents had rather extensive knowledge related to the machine market, the questions about the producers of agricultural computer software proved problematic. One of the study participants reported that "programs that the advisers work with and fill in application forms with are Polish" (R9). Another respondent noted that the sales range of many Polish agricultural institutions includes various programs which are available to farmers. This is not, however, the only programming available on Polish market.

Both Polish and foreign producers offer computer programs easily available on the market that farmers could use. However, it is the Excel program that farmers conduct the most work in (R8).

Excel is obviously not a typical program designed for farmers, so it is hard to consider it a product targeting farmers. It is quite interesting that the non-agricultural program of a transnational producer is the one that farmers use the most.

Describing the market of computer software and its producers, the respondent used the catalog of the Center for Agricultural Advisory listing the programs designated for farmers. Among them were various producers such as:

- Department of Electronic Computing Technology in Olsztyn (Zakład Elektronicznej Techniki Obliczeniowej w Olsztynie) (PL),
- AgroPower LLC (UK),
- The Institute of Soil Science and Plant Cultivation (IUNG Instytut Uprawy Nawożenia i Gleboznawstwa w Puławach) (PL),

- CeeS Programs (PL),
- Polish Agricultural Institute (PL),
- Meteoryt Software (PL),
- Agroboss (PL),
- Land-Data Eurosoft LLC (DE), Branch in Poland,
- IT Service Leszek Mroczko, PhD (PL),
- Michał Cupiał, PhD. Eng., Faculty of Agricultural Engineering and Information Technologies, Agricultural University in Krakow (PL),
- A-Lima-Bis LLC (PL),
- BayWa Agro Polska (DE),
- BitComp Polska (FI).

Among the computer programs designed for farm operations, the product range from public research institutes and their employees deserves special attention. The programs and applications prepared by them are available in Polish which, for a large group of farmers, is a must if they are to use them. These programs are also free of charge. However, the respondents were not able to say which programs or applications were popular among farmers and which were not. The above tally of computer program writers/producers shows that among the programs recommended in the brochure of the Centre for Agricultural Advisory Polish companies were prevailing. It does not mean, however, that these producers were the ones that farmers chose the most. In their responses, farmers were more likely to name programs of foreign producers.

14.3.3. Use of New Technologies by Polish Farmers on their Farms

The interviewed advisers had rather positive opinions on the technological progress that was taking place in rural Poland:

> The times when farmers went to Western Europe to see modern agricultural solutions are gone. Currently, it is the farmers from Western Europe, who can come to us to see modern solutions (R9).

This was confirmed by another respondent who added that:

Polish agriculture is quite technically advanced when it comes to using the machines. There are lots of examples of quite modern solutions that can be seen inside agricultural buildings (R8).

In the opinion of respondents, Polish producers did not adequately participate in the agricultural progress, which made Polish agriculture dependent on foreign technologies.

Analyzing the issue of computer programs used by farmers in their work, certain trends could be seen. One of the respondents reported that:

Small farmers do not use the computer programs beyond the necessary filling out of applications for financial aid and subsidies. The situation is quite different with large farms, where more advanced tasks such as leading the herd are necessary. It is even more different with the use of programs calculating fertilization of soil—here, small farms usually rely on the knowledge and experience of the farm operator or/and the neighbors. In the case of big farms, in addition to the knowledge of the owner or operator, there are also suppliers of the means of production who visit the farm several times both during and outside the growing season and, based on conducted calculations, make a joint decision on further procedures for the farm (R9).

It should be stressed that farmers' ever-expanding application of new technologies of farm management, combined with their Internet use for acquiring agricultural knowledge provides—according to agricultural advisors—notable evidence that farmers actively implemented modern technological solutions. One of the respondents perceived that to be tied with the farmers' age and educational level.

Farmers often use the Internet as their source of knowledge. They use specialized agricultural portals and also have their own groups or YouTube channels that they observe. Their use of the Internet is to a large extent connected to their education and age. Older people are more likely to build upon their experience and it is harder for them to accept and implement technological innovations (R9).

The prestige that public agricultural advisers have among farmers, especially those of older generations, appears to be quite serious. Older farmers trust the public advisers more and thus, the adviser could be a source of innovations for them.

14.4. Final Remarks

The farmers who participated in the study were considered to have innovative personalities, which influenced their decisions related to the digitalization of farms. They also had adequate knowledge and experience enabling them to make rational decisions connected with farm development. Having sufficient capital, they could focus on day-to-day functioning of the farm allotting certain funds to investments and long-term plans. The cases described earlier in this chapter presented a significant role of the previous generation who managed the farms in the past and before passing them to the respondents. In each of these cases the respondents were the ones who made decisions about investments, only consulting with their fathers or grandfathers, who operated the same farm in the past. The older generation did not interfere with the decisions of young and "inexperienced" farmers after they took over the farm and started making investments. Intergenerational cooperation in that regard seemed quite important for the farmers who inherited the property as they learned the farm work and agricultural issues from previous generations. Had the older generation been against investments and changes on the farm that was passed to the descendants then the mechanization progress and digitalization would have been quite limited. It could be said that the progress, at least within the mental dimension, has its intergenerational sources. The open attitude of the older generation could have a vital influence on shaping the decisions in the next generations. The generation currently in charge of the farm could not only count on the knowledge of older generation and its experience in farm work but also on the support of older generation in these types of activities.

What should be given more attention is the level of development of the so--called information society and knowledge society or, to use the metaphor of Alvin Toffler, the process of the third and fourth wave among Polish farmers. As indicated by Polish economist Krystyna Krzyżanowska, in order to increase the quality of production and to foster competition, agricultural producers take up innovative activities (Krzyżanowska 2016, p. 52). It pertains to product innovation and innovation at the stage of the production process, as well as marketing innovation and organizational innovation (OECD, 2005). Due to the rather low use of computers in the operation of farms, a future where competitive advantage is acquired on the basis of applied knowledge might look somewhat daunting. Digital exclusion of the second and third degree should be seen as a troubling state of affairs, which can limit the abilities of Polish agriculture to compete on the global market. The examples presented in this chapter show that there are various possibilities for reaching out and providing support to farmers who currently do not use new digital technologies. Good examples of innovative farmers could become an inspiration to others. Such innovators who could find value in acquiring knowledge and implementing it should be recognized as advisors. A public advisory system that does not have resources to maintain the excellence of instruction courses and classes on new technologies in agriculture could certainly benefit from the experience of the people who already started following the path of digitalization. It seems important for such instructional offers to be directed not only to the largest farms but also to the smaller ones. As presented in the earlier examples, the introduction of new technologies to the farms might bring many benefits, not just economic ones, but also social. New technologies can contribute to an increase in safety, improvement of work quality, and increase the amount of free time. The type of "farmer to farmer" help was certainly not a new solution. Respondents regularly indicated remembering the times when they talked to neighbors with whom they had worked through various problems and issues. Currently the model of work on farm is undergoing significant changes. Social bonds are loosening up and it causes the progressive individualization of work. Introduction of farmer-advisers could be a good response to continuous formalization of social contacts in the professional category.

Another problem related to investments in new technologies is the issue of agricultural market stability and a certain predictability of agricultural development on the national level. Investments in new technologies require significant financial means, and that goes together with the risk of credit obligations. Farmers who will be investing in technological development have to have certain conditions allowing them to rationally calculate the risk and in the case of investment failure they need to have access to various financial aid instruments offered by the state. Greater market stabilization, the mapping out of general directions of agricultural development in Poland, and the instruments of passing the risk from farmers to the state appear to be some stumbling blocks on the way to the digitalization of Polish agriculture.

The system of co-financing of agricultural modernization seems to be another structural barrier in the development process of digitalization in agriculture. There is an opinion among farmers that domestic funds as well as EU funds are meant for farmers who have already accumulated certain financial capital. Therefore, these funds do not reach smaller or less affluent farmers that are not able to co-finance machines, even with the external support. Furthermore, the study participants expressed the opinions that increased prices of agricultural machines could be seen as a negative effect of the EU financing schemes. Consequently, the current functioning of the system of financial aid for farmers might marginalize farmers who cannot afford using the financial aid offered by the state and by the EU, as they do not have the required initial capital.

The diverse attitudes of farmers towards digitalization of farms show that advisory process aimed at technological development of farms should allude to various motivations. This is illustrated by significant emotional ambivalence. For some respondents the arguments to be made for introduction of new technologies have an economic character. For others, functionality is crucial. There are also social arguments that may be interesting to others. A good system of instruction classes should be implemented with taking into account various levels of digital exclusion among farmers. Consequently, a diverse offer should reach and be attractive to these farmers who do not have much experience and are lacking knowledge in technology as well as the ones who already have such knowledge but want to develop it further.

The weight of the described problem is highlighted by the theory of neo-endogenous development which emphasizes the ability to implement new solutions. It also values adaptation and the use of knowledge and new technologies as crucial to much desired competitive rivalry on the global agricultural market. Such competition is possible through local resources and it is also connected with the way farmers think. They are not just oriented toward the individual maximization of economic profits but also toward networking possibilities and using the progressive specialization within the farm. Therefore, attitudes of farmers towards the use of technology in agriculture are so important. It should be noted as a positive phenomenon that regardless of the knowledge, motivation, or presented attitude towards technology, all of the respondents used the technologies, with some of them being more advanced than others. The use of technologies was diverse as much as diverse was the ground for an effective approach to technology and the quality of knowledge on agricultural matters. This indicated that education and the mentality of farmers still could be a significant obstacle in neo-endogenous development. One of the respondents offered the following remarks:

> It is not about computers, it is about mentality. There is some threat. As I said, a generation would need to pass for people working in agriculture to be [agricultural] engineers at least. Then things will be good in agriculture. Then everything will be good in agriculture. As long as there will be people with just elementary school, high school and vocational education, agriculture will not be doing very well (R1).

Grudziński (2006) noticed that the most common factors responsible for the inadequate use of computer programming in agricultural activities were the following: (1) fears that the costs of equipment and software would not balance out, thus outweighing any benefits from using the equipment, (2) problems related to the servicing of computers and the programs, (3) low credibility of concepts offered by programs of simplified models which ensure their universal character, (4) problems with adjusting the programs to specific features of a given farm.

Digital exclusion is clearly another interesting issue to consider in the analysis of digital progress in agriculture. The available data show that digital exclusion of the first and second degree can be a significant barrier to the development of Polish agriculture. These two types of exclusion can negatively impact the abilities of economic, social, and ecological optimizing in operating of a farm. The fact that the most popular form of innovation among the described cases was technological innovation indicates that Poland is in the early stages of implementing a digital revolution in agriculture. For farmers, the security of hardware is more important than software or the ability to use both creatively. This suggests that the need for investments in machines and technologies is still significant. However, in order to accelerate the digitalization process, it is necessary to develop the competencies related to the creative use of digitalization and work out new solutions, adjusting to local or individual conditions. The negative opinions on Polish companies producing agricultural machines expressed by Polish farmers should be mentioned here. According to agricultural advisers, there were objective and subjective arguments confirming that Polish producers of agricultural machines did not enjoy a good reputation among farmers. To change that, the transfer of knowledge and technology is much needed and it should not only involve farms but also producers of agricultural machines. Currently, this market in Poland is global, dominated by foreign concerns. Changing that would require improving the quality of the products as well as more openness from farmers, which could only be possible through positive experience stemming from work. This should not be limited to investments in technology of production but also in the network of suppliers and servicing which, from the farmer's point of view, are seen as important factors influencing the decision to purchase a certain machine.

Empirical material analyzed here allows us to state that among the interviewed farmers, the first signs of the introduction of the program Economy 4.0 could be observed with processes of production controlled by special programs placed on net servers. Precision farming, which is already present on some farms (R6 and R7), proves that (Samborski, 2018). Integrated activities of various programs supporting animal and plant production based on original systems of monitoring confirm that the technological revolution is taking place in Polish agriculture as well.

REFERENCES

Adamski, T. & Gorlach, K. (2007). Koncepcja rozwoju neo-endogennego, czyli renesans znaczenia wiedzy lokalnej [The concept of neo-endogenous development, i.e. the renaissance of the importance of local knowledge]. In: K. Gorlach, M. Niezgoda, & Z. Seręga (eds.), *Socjologia jako służba społeczna. Pamięci Władysława Kwaśniewicza* [Sociology as a social service: In memory of Władysław Kwaśniewicz], pp. 137–150. Wydawnictwo Uniwersytetu Jagiellońskiego.

Bringer, J.D., Johnston, L.H., & Brackenridge, C.H. (2004). Maximizing transparency in a doctoral thesisl: The complexities of writing about the use of QSR*NVIVO within a grounded theory study. *Qualitative Research* 4(2), pp. 247–265.

Bringer, J.D., Johnston, L.H., & Brackenridge, C.H. (2006). Using computer-assisted qualitative data analysis software to develop a grounded theory project. *Field Methods* 18(3), pp. 245–266.

Castells, M. (2006). *The Theory of the Network Society*. MPG Books.

Chyłek, E.K. (2012). *Uwarunkowania innowacyjnego rozwoju sektora rolno-żywnościowego i obszarów wiejskich w ramach polityki rolnej* [Conditions for the innovative development of the agri-food sector and rural areas within agricultural policy]. Agencja Reklamowo-Wydawnicza Arkadiusz Grzegorczyk.

Dijk, J. van (2006). *The Network Society: Social aspects of new media (2ⁿᵈ ed.)*. Sage Publications.

Drucker, P. (1994). *Post-Capitalist Society*. HarperCollins Publishers.

Galdeano-Gómez, E., Aznar-Sánchez, J.A., & Pérez-Mesa, J.C. (2010). The complexity of theories on rural development in Europe: An analysis of the paradigmatic case of Almería (South-east Spain). *Sociologia Ruralis* 51(1), pp. 54–78.

Granovetter, M. (1973). The strength of weak ties. *American Journal of Sociology* 78, pp. 1360–1380.

Grudziński, J. (2006). Technologie informacyjne w systemach doradczych w zarządzaniu gospodarstwem rolnym. *Inżynieria Rolnicza* 58(80), pp. 207–213.

Halamska, M. (2012). Nowa gospodarka wiejska: Socjologiczna konceptualizacja i próba analizy zjawiska w Polsce [New rural economy: Sociological conceptualization and an attempt to analyze the phenomenon in Poland]. In: A. Rosner, *Rozwój wsi i rolnictwa w Polsce. Aspekty przestrzenne i regionalne* [Development of rural areas and agriculture in Poland: Spatial and regional aspects], pp. 13–38. Instytut Rozwoju Wsi i Rolnictwa PAN.

Hargittai, E. (2002). Second-level digital divide: Differences in people's online skills. *First Monday* 7(4).

Kaleta, A. (2016). *Społeczeństwo informacyjne na obszarach wiejskich*. Wydawnictwo UMK.

Konecki, K.T. (2000). *Studia z metodologii badań jakościowych*. WN PWN.

Kozera, M. (2013). Rozwój polskiego rolnictwa w realiach gospodarki opartej na wiedzy [Development of Polish agriculture in the realities of the knowledge-based economy]. *Roczniki Ekonomiki Rolnictwa i Rozwoju Obszarów Wiejskich* [Annals of agricultural economics and rural development] 100, pp. 35–43.

Krzyżanowska, K. (2016). Innowacyjność w grupach i organizacjach producentów branż rolniczych [Innovation in groups and organizations of producers of agricultural industries]. *Studia Komitetu Przestrzennego Zagospodarowania Kraju* [Studies of the national spatial development committee] 173, pp. 51–59.

Levinson, P. (1997). *The Soft Edge: A natural history and future of the information revolution*. Routledge.

Niedbalski, J. (2013). Komputerowe wspomaganie analizy danych jakościowych (CAQDAS) w projektowaniu i prowadzeniu badań [Computer-aided qualitative data analysis (CAQDAS) in designing and conducting research]. *Nauka i Szkolnictwo Wyższe* 41, pp. 185–202.

Nowak, S. (1973). Pojęcie postawy w teoriach i stosowanych badaniach społecznych. In: S. Nowak (ed.), *Teorie postaw* [Attitude theories], pp. 17–88. Państwowe Wydawnictwa Naukowe.

Nowak, P. (2012). *Rozwój obszarów wiejskich w Polsce po integracji z Unią Europejską w opinii lokalnych elity* [Development of rural areas in Poland after integration with the European Union in the opinion of local elites]. Wydawnictwo Uniwersytetu Jagiellońskiego.

OECD (2005). *Oslo Manual: Proposed guidelines for collecting and interpreting data*. Oslo.

Olcoń-Kubicka, M. (2009). *Indywidualizacja a nowe formy wspólnotowości* [Individualization and new forms of community]. Wydawnictwo Naukowe Scholar.

Pieriegud, J. (2016). Cyfryzacja gospodarki i społeczeństwa – wymiar globalny, europejski i krajowy [Digitization of the economy and society: Global, European and national dimension]. In: J. Gajewski, W. Paprocki, & J. Pieriegud (eds.), *Cyfryzacja gospodarki i społeczeństwa. Szanse i wyzwania dla sektorów infrastrukturalnych* [Digitization of the economy and society: Opportunities and challenges for infrastructure sectors]. Publikacja Europejskiego Kongresu Finansowego.

Sadler, P. (1988). *Managerial Leadership in the Post-Industrial Society*. Gower Publishing Ltd.

Samborski, S. (ed.) (2018). *Rolnictwo precyzyjne*. WN PWN.

Toffler, A. (1980). *The Third Wave*. Bantam Books.

Uchaniuk, M. (2012). *Internet zmienia życie rolników* [The Internet is changing farmers' lives]. http://www.farmer.pl/fakty/polska/internet-zmienia-zycie-rolnikow,33493.html (access: April 22, 2019).

Wagner, M. (2017). *Społeczeństwo informacyjne w Polsce. Wyniki badań statystycznych z lat 2013–2017* [Information society in Poland: Results of statistical research from 2013–2017]. GUS.

Wiltshier, F. (2011). Researching With NVivo. *Forum: Qualitative Social Research* 12(1), art. 23. http://www.qualitativeresearch.net/index.php/fqs/article/view/1628/3156 (access: December 10, 2019).

Addendum: Descriptions of respondents

Respondent 1 (Farmer)

He is a Silesian farmer, who manages the close breeding cycle of 150 pigs. Additionally, he grows wheat and canola selling the surplus. He grows beans and potatoes as well. His farm encompasses 37 ha of the land that he owns. The farm was passed from father to son 34 years ago. The respondent has an agricultural high school education. He has been working on the family farm since childhood and after he took over the farm he never stopped investing in its development. He likes to say that he "built a farm all over again." Among the most important investments there are: modern pig facility, with an appropriate feeding system, a tractor, pneumatic planter, and sprayers. The profile of his farm has changed over the years from vegetables and wheat to animal breeding and wheat. The farmer uses both traditional and modern sources of information. The former include: exhibitions, magazines, and opinions of close acquaintances and the latter relate to Internet sources such as portals on agricultural trade. The farmer does not use computer programs to operate the farm but uses them for accounting and bookkeeping purposes as well as for filling out the applications for subsidies. He claims that he approaches the computer as if it was a form of "black magic."

Respondent 2 (Farmer)

He is also a Silesian farmer who has worked on the family farm since childhood and inherited that farm in 2007, after his father's death. He graduated from technical high school in Silesian Nakło and also completed studies in agricultural engineering at the Agricultural University (formerly Agricultural Academy) in Kraków. In addition to obligatory agricultural courses, he also finished courses on sprayings and inseminations. This farmer is a member of the Agricultural Chamber, an organization which, in his opinion, helps associated farmers to effectively manage their farms. He currently has 55 ha of land in his possession, including forest. He rents out about 25 ha of farm land. The profile of the farm currently leans toward milk cattle. This is the effect of what the owners consider a necessary narrowing of the formerly diverse farm profile. In the last five generations the farm combined growing potatoes, beets, with breeding cows and pigs but now it is being steered towards raising dairy cattle. Thus, the farmer currently grows canola, cereals, and corn for silage. His farm also includes meadows and other agricultural areas. The farmer's herd consists of 120 cows, of which 64 are dairy cows and the rest are calves. His biggest investment in recent years has been the construction of a modern barn. He also recently bought some agricultural machines such as a feeding trailer, a disc mower, a round baler, a bale wrapping machine, a tractor loader, and a new tractor. This farmer was using computer programs on the computer and on his smartphone, as well as on-line applications (Wirtual Zootechnik—Virtual Animal Husbandry Technician) to manage his herd. He is, however, planning to switch to the SOL program offered by the Polish Federation of Cattle Breeders and Milk Producers because it is connected with the Symlek system, which generates the lineage application documents for every newborn calf. Additionally, bookkeeping, accounting, and payment applications are used on this farm, as well as Internet which serves as a source of knowledge on managing the farms, machinery, etc. During the interview the farmer was also talking about buying a special robot for milking cows with built-in programming for herd management, which would facilitate individual treatment of cows in the herd. The purchase was to be made possibly later that same year.

Respondent 3 (Farmer)

He is a Silesian farmer who has been operating the farm for 17 years. Although he makes autonomous decisions on farm matters, he still lives with his parents who signed the farm over to him. He worked on his parents' farm since childhood. This work, as well as graduating from a technical high school that focused on car mechanics, helps him to manage the farm and be efficient. This farmer is not socially active within his community and does not belong to any agricultural organization. The farm encompasses 55 ha of agricultural land, of which 20 ha is rented out. The land is quite dispersed and fragmented with plots of 3–4 ha. The cereals, such as wheat, barley, and corn, are grown on the farm, as well as alfalfa and grass. In recent years, the profile of the farm has not changed as far as crops are concerned. However, the farmer mentioned that his father grew vegetables, canola, and beans, which, in his view, were no longer profitable today. Besides crops, there was some animal production of dairy cattle and pigs and part of the arable area was devoted to growing plants for animal feed. The farm is highly diversified and can react relatively quickly to changes in markets, adjusting its profile even annually. The farmer had considered infrastructural farm investments such as barn modernization, expansion and remodeling of farm buildings such as sheds, storage warehouses, and silos as the most important. He also stressed the importance of recently purchased machines such as a tractor loader with an on-board computer panel and a combine harvester. Admitting that he refused GPS installation when recently buying the combine harvester he also said that he did not use additional program functions and perceived them as unnecessary and expensive gadgets. The farmer also owns new sprayers and a planter which, compared to other equipment, are more electronic than mechanical.

The respondent uses computer programs in their basic forms, mostly for acquiring knowledge. His brother is an IT specialist and he is involved in all the tasks related to setting up programs of the machines and updating them. The farmer uses a computer for online shopping but only for transactions that do not exceed 500 PLN as he is aware of the dangers related to Internet transactions.

Respondent 4 (Farmer)

This respondent is a farmer from Lesser Poland, where he has operated the farm since 2004 and, for the last 10 years, he has been doing so independently. He is a graduate of a technical high school with a specialization in forestry as well as a technical school focusing on beekeeping (evening and weekend classes) and an agricultural technical school (extramural system of studying). He graduated from Agricultural Academy in Kraków (now Agricultural University), where he studied in the Department of Forestry. Additionally, he finished engineering studies in zootechnics. He belongs to several organizations, including the Lesser Poland Association of Polish Ecology (Małopolskie Stowarzyszenie Polska Ekologia), the Polish AgroForestry Association (Ogólnopolskie Stowarzyszenie Agroleśnictwa), the Polish Horse Breeders Association (Związek Hodowców Koni), and the Galician Association of Cattle Breeders and Producers (Galicyjski Związek Hodowców i Producentów Bydła). The respondent purchased the farm in 2003 and has operated it since. The farm includes permanent green areas, meadows, and pastures for cattle of the Limousine breed. The entire livestock production is ecological. The farm of 274 ha, of which 72 are allotted to forest and scrubland, is divided into three smaller parcels. The respondent is the owner of 96 ha of the farm. Additionally, he operates a company that specializes in marketing and distribution. The other parts of the farmland belong to the respondent's brother and father, but the decisions affecting the entire farm are made jointly, with active participation by the respondent. Respondent is proficient in using computer programs, but he mostly uses the Microsoft Excel program, which is not a typical program designed for farmers. He is able to create his own Excel spreadsheets and calculations and he archives the data from operating the farm. He uses the Internet regularly to acquire knowledge and databases. He also uses mobile applications. The respondent is highly competent in using the computer, but he does not need to use commercial computer programs to operate the farm. The farmer also uses modern agricultural machines such as loader wagon, automatic water trough, self-loading bale wrapper, spreader, and a machine for manure removal. Quite significant innovations implemented by this farmer go beyond modern technologies. The agro-forestry system of production is applied on the farm, which means that the land is used simultaneously and in parallel fashion for agriculture and forestry. The forestry system not only determines the production of biomass and lumber mass but could also mean orchard production. On this farm, trees are planted in a loose truss—8 by 10 and 8 by 12 meters. Old fruit tree varieties are maintained with the assumption that the orchard production will be developed in the future and the taller, old trees will provide shade for cows grazing in the field. Some grass will be cut and processed for silage.

Respondent 5 (Farmer)

This farmer manages an orchard of about 3 ha in Lesser Poland and he comes from an agricultural family. He finished technical high school and technical university studies as well as agricultural studies. For the last seven years he has been a co-owner (with his wife) of the farm they bought together. Currently the farm specializes in production of American blueberries and honey and the owner belongs to the beekeepers' association in his area. Nothing was cultivated on the farm before the blueberries and the first few years after buying the farm were a time of investments and preparation for the first plantings. Importantly, the main investment so far has been the irrigation system that the respondent built independently following the good practices of a farm located in Oleśnica that he was familiar with. The irrigation system provides a sufficient amount of water for the entire plantation, even though it is not yet computerized and requires manual regulation of the irrigation level. The Internet plays an important role in application of new technologies on this farm, both as a source of general agricultural knowledge and specific knowledge on blueberry planting, as well as systems needed to work such a farm. The entirety of needed knowledge the farmer acquires from the Internet and then checks other sources to ensure objectivity. Additionally, he uses Internet for shopping (auctions), making payments and to exchange opinions on agricultural matters. Besides the already mentioned irrigation equipment, the farmer owns a mulching mower and frequently analyzes the soil at the regional agricultural station to ensure his proper use of fertilizers. This farmer also uses computer programs quite often, mostly Excel for accounting purposes, fertilizer calculations, and archiving agricultural procedures and treatments. Additionally, he fills in the subsidy application form on the computer.

Respondent 6 (Farmer)

This farmer comes from an agricultural family and operates a farm in Silesia jointly with his father. He has been involved in farm work since childhood and graduated from an agricultural technical high school. He currently studies agronomy; it is his final year in an agricultural engineering program. The total area of his farm is 100 ha and the soil is classified as 4th and 5th category. The farm crops include potatoes (covering around 5–8 ha), canola, and cereals (mostly for animal feed). About 25 ha of the farm land is allotted to wheat, which is later sold. The farm is able to fulfill the entire demand for animal feed for the herd of 60 milk cows and their offspring. The farm is divided into two plots: one owned by the farmer and one owned by the farmer's father. Decisions related to the particular plots are made autonomously but the decisions affecting the farm as a whole are made jointly by the father and son. The farm is equipped with a modern milking parlor, tractors, loader, spreader, mixer wagon, mowers, collecting trailers, and a receiving hopper. The farmer is also interested in getting some special sensors for the barn to monitor the movements of cows, number of grazes, procreation activities, as well as metabolic problems. The farmer currently uses the Farmnet 365 program to manage plant production, but in the future, he would like to use SOL, the program of the Polish Federation of Cattle Breeders and Milk Producers, to manage his herd. The farmer is considering buying an active boxing module for Farmnet 365, so he can monitor the machine movement in the field and collect data. Additionally, together with his cousin who is an IT specialist, the respondent is planning to design a special farm management program of his own. First, he would use the program on his own farm and then, in a long-term perspective, he could license it and sell it to others. In addition to typically agricultural computer programs, the respondent also uses Excel for archiving information and to prepare quantitative reports on work conducted and on the state of his farm.

Respondent 7 (Farmer)
This farmer operates a farm in Lesser Poland. Around 1990 he received the first parcel of farm land of 5 ha from his father. Five years later he bought another 50 ha. Currently he owns a bit over 80 ha and rents an additional 40 ha. Over the years the profile of his farm has changed from monoculture to the current state where canola makes up 40% of crops, wheat comprises another 40% and soy and corn constitute 10% each. The farm has a rather experimental character and the farmer is still searching for new solutions. Respondent is a delegate to the Agricultural Chamber in Lesser Poland region. He has over 40 years of agricultural work experience and graduated from a well-known agricultural high school in Czernichów. He owns a modern tractor, sprayer, and a spreader. He emphasizes that he personally chooses the model of the machine as well as all the accessories that the tractor is equipped with: "90% of the machines are custom-made for me." The farmer uses computer programs, including an e-mail program to send documents. He also uses the Internet for making payments online, filling out online applications and to check the weather

with a special meteorological application. All of his machines have built-in computer systems (GPS, parallel running, search function for information on proper use of sprayer and spreader). The respondent admits that his use of computers involves "hardly any love." Respondent invests a lot in his land (not just agricultural land) and therefore uses the geo-portals for identification of land plots and their purchase.

Respondent 8 (Adviser)

Respondent is an employee of one of the agricultural advisory centers. He does not specialize in technological or computer advising. This particular one, Center for Agricultural Advisory, has a few computer applications available and farmers and field advisers can use them with no charge. The Centre also has an online publication which reviews computer programming for farmers. Even with such offers relating to software, the respondent considers the public agricultural advisory in Poland as mainly connected with agricultural matters.

Respondent 9 (Adviser)

The respondent is an employee of the Regional Agricultural Advisory Centre. In his daily work, he deals indirectly with technological advising but only in regards to proper filling out of applications for financing of agricultural machines. It is within the scope of the adviser's work to follow certain guidelines in selecting the parameters of the machine that the farmer can purchase within certain co-financing mechanisms. In that sense, agricultural advisory and extension help with the modernizing of Polish agriculture but this is not the type of advising that would allow the farmer to choose the machine that would be most appropriate for farm features and conditions. The public advisory of today means selecting a certain category of machine which would meet some complex criteria. According to this respondent, the public agricultural advisory in Poland is not suited to deal with issues of mechanization and digitalization of agriculture as there are insufficient funds allowing the advisers to be trained with the newest equipment in how to apply innovative solutions and new programming for farmers.

Respondent 10 (Adviser)

Respondent is an employee of one of the field offices of the Agricultural Advisory Center. He describes a rather ambivalent approach to public advising on technology. He claims that Agricultural Advisory Centers could in the future engage in this kind of advising but large investments would be needed. Then advisers could conduct ongoing and up-to-date tests on technologies introduced to agriculture. In his opinion, the level of qualifications of advisers on new technologies is rather high.

Think Locally, Act Globally: Polish farmers in the global era of sustainability and resilience, ed. by Krzysztof Gorlach and Zbigniew Drąg in collaboration with Anna Jastrzębiec-Witowska and David Ritter
Jagiellonian University Press, Kraków 2021, pp. 641–677
ISBN 978-83-233-4949-5
DOI: http://dxdoi.org/10.4467/K7195.199/20.20.12743

Chapter Fifteen: Polish Farmers' Protest Activities and Mobilization Process: Past and present

Grzegorz Foryś https://orcid.org/0000-0002-9411-2681/

15.1. Introduction

Farmers' contesting activities and peasant contesting actions that took place in the past have been the subject of interest to many researchers. A significant number of them have concentrated on the role of collective contestant actions in peasant revolts. In their studies these researchers made frequent references to social movement theory. They attempted to find the causes of social upheavals of this kind but were not able to come up with a deeper reflection on their regularity and their patterns. This came from the lack of belief that peasants can become an autonomous social class able to influence modernization processes or even their own social position. This way of thinking about peasants was not only expressed by Karl Marx, who described this social category as "a sack of potatoes" (Marks, 1979, p. 457). Similar views, albeit less radical, were expressed by Theodor Shanin (1971), who argued that the main cause weakening the mobilization potential of peasants is the situation of domination and exploitation that they are subjected to. Another author, Eric Wolf (1973), pointed to the large internal diversification of peasants and their lack of experience in organizing contesting activities as weakening the potential necessary for collective action. It should be remembered that the opinion of Wolf did not stop him from calling the 20th century the time of "peasant wars."

Modernization was the process which increased peasants' mobilization abilities and at the same time took peasants out of their own isolated world. It brought some progress in education of this social group, increased the social consciousness of its members, and resulted in emergence of peasant political organizations as well as created the leaders of this circle. Peasants followed the path that Barrington Moore described in these words: "the process of modernization begins with peasant revolutions that fail. It culminates during the twentieth century with peasant revolutions that succeed" (Moore, 1966, p. 453). As a result of the modernization processes, peasants became empowered to take up collective action, not only aiming to change their position or protect their own interests, but also to influence and shape the modernization processes.

The emancipation processes of Polish peasants had a similar trajectory. Within the last century, the contesting activity of this social strata (meaning peasants and then farmers) could be sorted into three stages, which were characterized by the changing socio-economic and political conditions, in which the peasants were functioning. They were: the interbellum period (1918–1939) the time of real socialism (1945–1989) and the times after the regime change (1989 onward). In all of these time intervals, an emancipation of peasants was taking place as well as their professionalization and transformation into farmers and later into agricultural entrepreneurs. What also changed were the conditions and mobilization abilities of this social category.

The contesting activities of Polish peasants and farmers in the above-mentioned historical time periods could be characterized by several prominent features. First, they were cyclical, which would seem obvious, as it might be impossible to be permanently on standby for contesting activities. Second, the cyclical character of these activities was determined by economic factors. Third, besides the factors of an economic nature, national and political issues played a crucial role in the mobilization of peasant protest activities. These issues clearly marked their presence in the period between the wars, as well as in the time of People's Republic of Poland and, most recently, in the last three decades. Fourth, initially, in the interwar period, such actions were mostly organized by political organizations. With time, particularly during the People's Republic of Poland and over the last 30 years, these organizations were replaced by peasants' trade unions. Fifth, the repertoire of protest activities went through an evolution process, starting with rallies, mass meetings, marches, and strikes as dominant forms in the interwar period. Next were strikes and petitions during the time of real socialism and then after 1989 the role of road blockages, occupation of public institutions, and destruction of agricultural commodities increased noticeably. Six, the demands of contesting activities conducted by peasants and then farmers were always of an economic nature and this confirmed the earlier thesis on the economic grounds for the peasants' protests. Peasants' and farmers' protests rarely had political goals (Foryś, 2019). Seven, looking into the protest activities during the three periods of time mentioned above, it could be argued that, over time, peasants' protests became more successful in achieving their goals.

The following chapter mostly focuses on specific factors spurring farmers and agricultural entrepreneurs to take up protest actions in contemporary Poland. It also deals with the model of articulating interests that would be appropriate for the description of contesting activities. This seems to be the most important research aspect for farmer protests and for the relation between farmers and the state. It appears that in Poland, the professionalization, routinization, and institutionalization of farmers protests have been going on for the last 30 years. As a result, in many of their characteristics, farmers' protests have become similar to actions taken by farmers in Western Europe. This applies to, among other things: the protest organizers, applied forms of protests, and demands. As a consequence, the meaning of the protests diminishes and this is the price paid for a more permanent relationship of cooperation between the farmers and the state. Open contesting becomes an example of an unusual, extraordinary activity that is supplemental to the activities conducted by farmers in the institutionalized space of producer groups, farmers' trade unions, and interests groups.

Before the reader becomes familiar with the analyses of farmers' contemporary contesting activities, two other issues should be addressed, as they are linked to the main motif of the ongoing reflection of farmers' protests. Firstly, several remarks need to be made in reference to the theory of social movements to provide an explanatory role for the analyses of collective contesting activities. Secondly, the most important characteristics of peasants' and farmers' protest should also be addressed in the two periods of time described above, namely the interwar years and the time of state socialism. This will allow for a better understanding of a certain continuity (or changeability) in peasants' and farmers' contesting actions, including the contemporary ones. Finally, a later/the last part of this chapter will be devoted to an analysis of farmers' protests in the last 30 years, as well as farmers' contemporary mobilization potential.

15.2. Analyses of Peasants' and Farmers' Contesting Activities in the Theoretical Frame

The already mentioned activities of mass peasant revolts taking place in the middle of the 20th century, especially in many countries of the Third World, became an area of research interest. These resulted in the flourishing of various theories explaining peasants' contesting activities (see, among others, Stinchcombe, 1961; Moore, 1966; Wolf, 1973; Migdal, 1974; Paige, 1975; Scott, 1976; Jenkins, 1983). These theories can be arranged into two groups, with the first group encompassing structural theories (Arthur L. Stinchcombe, Jeffrey M. Paige) and the other concentrating on historical theories (Barrington Moore, Eric R. Wolf, Joel S. Migdal, James C. Scott).

The dominant motif of structural theories deals with class relations and especially the exploitations that the peasant strata are subjected to. Jeffrey M. Paige (1975) attempted to explain the causes and the character of agrarian conflict with such variables as: source of income (land ownership, capital and remuneration) and the scheme of class structure in the agricultural sector (character of the division between land owners and land users/operators). In each case he analyzed he determined that peasants' contesting mobilization derived from existing social inequalities and conflicting interests among land cultivators and land owners who, as non-cultivators, constituted a higher agrarian class. However, this might not in every case have resulted in peasant revolution. If small farms were operated by people who were renting the land, then the class of land owners aimed at maximizing profits deriving from their land ownership and peasants were fighting with the owners to increase their share in the income. This could happen, of course, if peasants had sufficient mobilization abilities. When they did not, they might still have had the chance for a successful fight if they had used help from outside sources, such as political parties. This could have led to agrarian revolts. The probability of conflict was much smaller on large farms. Their owners were usually able to maximize their profits not just from financial means that came from renting the land to others, but also from the capital already in their possession, as well as investments in farm modernization, which were followed by the intensification of agricultural production. Therefore, they could take a more liberal course towards those who rented from them and also conduct concessional economic activities. A reform commodity movement oriented toward the expectations of dependent people would be more likely in such cases than revolutionary activities. Other mobilization conditions could be seen on larger farms that are in the possession of individual land estate owners or owned by corporations that hire farm labor. Their relationship was one of a worker and an employer but could also be seen as the relation of subordinate and dominant. The participants in potential conflict usually concentrated on salary conditions and workers' rights and the protest often took the forms of strikes or some other action typical for the contemporary work environment and reform labor movement. Far reaching mobilization might emerge in the fourth type of conflict. It could apply to land owners who hired land cultivators and renting was the only source of income for the dominant class. In such cases, revolting peasants would most likely try to take over the state power in a revolutionary action (agrarian revolution) and demand agricultural reform.

A somewhat different approach to the political mobilization of peasants was expressed by Arthur Stinchcombe (1961). He considered land ownership and lifestyle as crucial characteristics determining the level of farmers' political mobilization. These characteristics determined class division within agrarian relations between higher and lower class. These two features were, in his view, supplemented by two more, useful in analyzing peasants' mobilization potential. They were: owners' and operators' technical culture of farming and their level of political organization and engagement. As a consequence, four basic characteristics of rural social structure could be distinguished: diversity based on privileges (owners *vs.* subordinates); wealth as reflected in the lifestyle; the technical culture of farming and its diverse presence on farms (traditional *vs.* modern); political activity and peasant organiz-ing (apathy *vs.* activity). On this basis, Stinchcombe proposed distinguishing five types of farms and evaluating their propensity for political mobilization. Among them were: 1) large farm (manorial system), 2) tenant family farms (family—size tenancy), 3) family smallholding), 4) plantation agriculture, 5) ranches. He noted, however, that only in the case of family tenant farms was the mobilization level high. It resulted from the combination of the high level of exploitation of peasants who operated the land that belonged to land owners and peasants' political sensitivity. Also, not insignificant was the fact that this category of peasants had nothing to lose, which could explain its greater propensity to engage in political movements. Such a scenario could usually result in a nationalist and/or populist kind of movement. At the opposite end of the mobilization potential there were owners of small family farms, who as the only ones had the status of land owners. They had the most to lose and, according to Stinchcombe, they were rather reluctant to engage in collective political actions.

The concepts of the two authors above stressed different factors as decisive in the mobilization of peasants. Although Paige emphasized external factors of mobilization and Stinchcombe did not, both these theories had a lot in common. Both concepts identified the line of division within agrarian relations that existed between peasants and dominant social categories. The conviction about the proclivity of tenants of small farms for contesting activities was also present in both theories.

Historical theory might serve as a supplement to structural theories. From this perspective, the key role triggering the collective actions in peasants could be found in an external factor (socio-cultural), which determined the quality of life and lifestyle of this social category. At the foreground were modernization processes, which transformed the existing socio-political order and, at the same time, changed the conditions of peasants' existence and, thus, brought about the disintegration of traditional communities and a loss of support within them. Peasants might need to leave them as they are lacking the sense of safety and might want to abreact.

According to J. Craig Jenkins, historical theory functions in two forms: market theory as represented by i.e. Eric Hobsbawm, Joel S. Migdal and Eric R. Wolf, and exploitation theory elaborated by Barrington Moore and James C. Scott (Jenkins, 1983). According to the "market" variant, the origins of discontent in the peasant category lied in breaking the old economic order and the introduction of market relations. This resulted not only in the commercialization of economic and social life but also in losing the sense of safety, culminating in frustration and increased unrest among peasants. Joel S. Migdal called this process a transition from "inward orientation" to "outward orientation" (Migdal, 1974, p. 87). In the "exploitation" variant, the reasons for peasants' contesting activities were thought to derive from their economic exploitation. The increase of fiscal services towards the state or land owners could lead to a diminished sense of safety and the increasing unrest among the members of the peasant class. It could be said that such an approach to origins of contesting was very much in tune with the premises of the "market" version of the historical theory. Taking cultural factors into consideration was a certain novelty in the exploitation theory, as presented by Moore (1966) and James C. Scott (1990). The first of them proposed consideration of a "conservative and radical solidarity" of rural communities. Conservative solidarity could emerge when small farm owners, tenants, or hired workers were dominated by rich peasants or land owners who controlled the resources and organizational abilities of rural communities, thus also limiting the mobilization abilities of other social groups. Radical solidarity could occur in a situation where peasants had in their possession the resources needed for building an organization which allowed them to resist the state or landowners. As Scott noticed the dominance of other social categories over peasants, he used the term "discourse" to describe their dissent. Contrary to open confrontation, which occurred rather rarely, this discourse accompanied peasants every day. It took the form of a hidden transcript, through which peasants' relation to social classes dominant in society was expressed. This unjust (according to peasants) social order based on submission to others, as well as subordination and acceptance of existing conditions, created a public transcript of the discourse (Scott, 1990). The content of the hidden discourse enabled peasants to assess the reality and evaluate their mobilization potential. Nevertheless, having sufficient cultural and socio-organizational autonomy from elites was a necessary component for peasants' mobilization.

Comparing structural and historical theories, it might be concluded that they had complementary visions of mobilization factors and the abilities of peasants. The former mostly focused on internal factors of mobilization, such as peasants' class position and internal diversity, and the latter on external factors, including modernization, which caused disintegration of traditional institutions and social relations.

The classic concepts of mobilization of peasant masses have very limited explanatory potential for contemporary mobilization of peasants and farmers in developed countries. This stems from several facts. Most of all, farmers in highly developed countries are nowadays quite often the owners of the land, which they can use in the way they like as market game participants. Moreover, agricultural production is specialized and conducted to a large degree on sizeable farms and in agricultural enterprises. Farmers in developed countries today are of the professional category that organized farmers' trade unions defending the interest of that group or certain sector of the group. Farmers might organize around the production of a certain commodity (e.g., milk, beef, pork, poultry, apples, cabbage, etc.) Although contesting farmers might refer to peasant ethos in their action, in practice the elements of this ethos are not that well rooted in consciousness and actions. Therefore, expanding the theoretical base for analyzing the contesting activities of contemporary farmers with more up-to-date theories of social movements is sorely needed.

Without going into great detail, theoretical aspects of social movements can be divided into three theoretical concept groups: collective actions, resource mobilizations, and new social movements. The first one is the theory of collective actions, the second is the theory of resource mobilization, and finally there is a theory of new social movements. Such division is very simplified and it does not consider more detailed theoretical proposals that would synthesize the output of the concepts mentioned above. Thus, the division does not include interpretative theories of social movements, cognitive theories, or neo-institutional theories. However, this approach is broad enough that it helps with the building of an analytical scheme useful in explaining the collective mobilization of contemporary farmers. This is supported by Patrick H. Mooney, who expanded the application of this scheme to other rural residents, not just farmers (Mooney, 2000, pp. 47–52). He stressed that the development of communication technologies help with overcoming the isolation of rural residents and, at the same time, increased their mobilization abilities. Furthermore, Mooney identified the following issues: a) social categories of rural residents were less diverse than in urban areas and this made their mobilization easier; b) there were links between urban movements (especially ecological movements) and the ones forming in rural areas; c) rural residents were just as much interested in access to collective consumption as urban dwellers; d) rural residents also had the need to defend their identity and this could be the basis for mobilization; e) finally, and most importantly for the following work, agriculture was a crucial resource for the rural population, and due to its certain role in the economy and national policy, farmers had political power which they could use through collective mobilization. The most essential argument, which should be addressed in using the theory of social movements in the analysis of collective contesting actions of farmers, was that social mobilization remained at the center of interest of these theories.

With so many different theories of social movements in the analysis of farmers' contesting activities, each of them has some useful explanatory aspects. The theory of collective behavior is quite strong in explaining the causes of social movement. It can be said that the theory of collective behavior gives some clues about the reasons for political mobilization. This is done in its initial form within the psychology of the community with the emphasized role of emotional and psychological state in the emergence of collective actions. In the structural version of the theory of collective behavior proposed by Neil Smelser (1962), these issues could be seen as networking. The differences in these two approaches can be summarized is such way that the psychology of collective community identifies mobilization factors on the individual level, emphasizing the role of the psychological and emotional state of individuals in mobilization processes, while the structural version places these factors on a macrostructural level. In the structural concept, these factors stem from strains in the social system, which result in psychological tensions among a certain number of individuals, oftentimes bringing a collective reaction to the situation. The theory of collective behavior in both presented versions treat the contesting action and the social movements that use them as non-routine activities that disrupt the social order and, in some way, can be seen as a socially undesirable deviation.

The theory of resource mobilization might put a different light on social movements and contesting actions. According to this theory, such actions reflect rational choices, those which are normal and acceptable in a social sense. The development of this theory was connected with the social practice of numerous mass protests in Europe and America in the 1960s and the 1970s, which generated interest among researchers. Its role in explaining contesting activities dealt with the collective mobilization process and the internal, organizational dimension of the functioning of social movements and their relation to the environment. In other words, the question of how the social movements functioned was crucial, and not so much attention was given to the reasons for the movements' emergence. In this approach, collective actions, including protest, could be seen as a conventional form of political activity. An important aspect of the analysis of social movements is the concept of political opportunity structure (Tarrow, 1998, p. 7), which places emphasis on the political contexts of protest activities within. On one hand, social mobilization is perceived as political activity, and on the other, the obvious target of contesting activities is the state.

Finally, there are new social movement theories, which bring two significant elements into the analysis of mobilization and organizational processes. The first of them is connected with the role that collective behaviors play in the shaping of collective identity. Theories of new social movements see a crucial role for this process in the building of community and solidarity, uniting people in common action and allowing them to achieve common goals. The second element is the concept of learning through collective actions at the root of these theories. The role of past experience in organizing protest actions is emphasized and certain forms of protests are well explained. This seems to be the case with peasants in the past and now farmers, who act as a social and professional category, and these qualitative aspects of their political mobilization should be taken into consideration and the new social movement theories should attempt to explain these aspects.

As can be seen, the social movement theories significantly expand the possibility of analysis of protest mobilization among peasants and farmers. They contribute to explaining the causes of protest behaviors, their course, and various aspects of protesters' relation with the contextual environment, as well as organizational processes inside the protesting groups. Therefore, they should not be treated as an alternative for each other but complementary, and helpful with understanding the various aspects of collective contesting actions.

15.3. Contesting Activities of Peasants and Farmers in Poland

Protest activities of Polish peasants go back to the Middle Ages, but at the time they were just local. The first notable protest activity of a peasant character took place in 1846 in the area of Polish Galicia, which, at the time, was part of the Austro--Habsburg Empire, as a result of the partition of Poland. This peasant uprising is often referred to as the Galician Slaughter. It made history due to its intensity, rather extensive range, and atrocities. The main reasons for this upheaval included exploitation and serfdom that had become unbearable for the peasants. The uprising was triggered by the hunger crisis caused by the poor harvest and floods of 1844–1845. In the later period of time, political organization and political parties, in particular, played an important role in awaking the consciousness on economic and social matters and a sense of self-worth among peasants. Therefore, in the period of time between World War I and World War II (1918–1939) the contesting activities of Polish peasants were concentrated on two main goals. The first goal was the striving for satisfactory results in agrarian reform, which would guarantee ownership of cultivated land in of an area adequate to the protesters' needs, in terms of giving them the ability to support their families through farming. The second goal was the recognition of peasants as a valuable part of the Polish nation and as citizens with full social and civil rights.

In the period of time between the world wars, peasants' protests could be described as modest. In the 1920s contesting activities were territorially dispersed, and involved the following issues: agrarian reform, the process of farm merging, divisions of communal land (belonging to an entire settlement unit such as a village or a hamlet), organizing affordable and long-term credits and loans, as well as regulating the contracts of agricultural workers. Maintaining and strengthening Poland's independence, as well as improving the quality of education in rural areas, were also among the demands of peasants.

An important change in peasant movements in Poland occurred in the beginning of the 1930s. It was connected with the intensified activities of protesters caused by the acute perception of the Great Crisis (1929–1933) which, in Polish rural areas, continued into the mid-1930s. Peasants in southern Poland were particularly active at the time and this could be linked to subsistence farming in the area with lots of small, dispersed farms. Farmers' demands at the time included: cancellation of debts in rural areas, some relief with credit repayments, and higher purchase prices for agricultural products. Protesting activities of agricultural workers that took the form of strikes or boycotts on the trade of plant and animal products are also worth attention. Peasant marches involving several thousand people often clashed with police and there were even some casualties. The protest organized in Łapanów that took place on June 5, 1932 could serve as an example here. It involved the participation of 10 thousand people in peasant festivities that had been banned by authorities. The protesters were engaged by the police with deadly violence resulting in 5 peasant fatalities and 10 seriously injured. There were also 200 people with minor injuries on both sides of the conflict. Peasants engaged in similar activities in Lesko County, where 5 thousand of them protested unpaid forced labor in road construction. In that confrontation with police 7 farmers were killed and several hundred were arrested.

The Great Peasant Strike of 1937 had a particular significance among protest activities of Polish farmers in the interwar period. The strike was of a general character, public and universal, with millions of people protesting. It covered a significant area of Poland where the revolting peasants lived. The demands were of an economic nature, with emphasis on agrarian reform that peasants wanted to be complete. Political demands also played a role and included: democratic parliamentary elections, return to the March 1921[1] Constitution, and the cessation of power by the Piłsudski camp.[2] The state pacified the protesters; 44 peasants were killed, 5 thousand were arrested, and 617 were sentenced by court orders (Wierzbieniec, 2008).

In the time of state socialism, after World War II the key issues determining peasants' protest activities included: 1) Structural fragmentation of farms, which was partially a result of unfinished agrarian reform from the time between the wars and unfavorable for peasants' reform that the Communist authorities introduced after the war. This reform made the farm fragmentation even worse; 2) Introduction of state socialism's main economic principles in agriculture and rural areas. It led, on one hand, to numerous attempts of authorities to push for collectivization of agriculture and, on the other, it involved development of state-run agriculture.

In such contexts the collective actions of peasants took place and with increasing modernization after World War II peasants became farmers. The campaign of Communist authorities to promote collectivization had the hidden goal of disintegrating the peasant class. The Communists emphasized differences between rich and poor peasants, fostering hostility and lack of mutual trust. It can be stated that, symbolically, the history of peasants' resistance in Poland started in post-war Poland as a reaction to the decision of the Communist Information Bureau of the Communist and Workers' Parties (Cominform), made during the Bucharest meeting of June 28, 1948, which started the process of establishing production cooperatives in the Communist Bloc, including Poland (Gorlach, 2001, p. 114). These collectivization efforts of the state were only partially successful. Finally, the state took ownership

---

[1] The first modern constitution of the Polish state as restored in 1918, was in effect from March 17, 1921 until April 24, 1935. It proclaimed Poland to be a democratic republic with a parliamentary and cabinet system of governance.

[2] The political camp focused around Józef Piłsudski was often called *sanacja*. It thrived from 1928 through 1939 and its goal was to establish/maintain a moderately authoritarian form of government with some abuse of democratic civil rights. This materialized in the constitution from April 1935, which eliminated the separation of powers (legislative, executive and judiciary) giving full authority to the president.

of 30% of the farms in Poland and almost 25% of the utilized agricultural land. In the countries of the Communist Bloc with individual farm ownership becoming a small minority, Poland had an exceptional farm structure.

The first peasants' post-war protests were conducted in the atmosphere of collectivization and nationalization. The collective actions were motivated by the state policy. For example, in 1953 there were over 160 cases of peasants' protests. Their frequency dropped after 1956 when the state backed away from the plans of making land ownership communal. Then, the contesting activities were triggered by obligatory deliveries of agricultural products, enforced by the state until 1972. Peasants' and farmers' protests in the period of time between World War II and the emergence of "Solidarność" in 1980 could be described as passive resistance. In practice this meant not following the state orders for establishing cooperatives, deliberate reduction of agricultural production, and concealment of its real volume (Gorlach, 2001, p. 119).

The change in methods which farmers utilized to fight for their interests could already be observed in the late 1970s. The nature of this change related to contesting activities as being conducted through political organizations emerging at the time. Organizations such as Committee of Peasants' Self-Defense (Komitet Samoobrony Chłopskiej—1978), Temporary Committee of Farmers Trade Union (Komitet Związku Zawodowego Rolników—1978), Centre for Peasant Thought (Ośrodek Myśli Ludowej—1979) should be mentioned here. However, the breakthrough moment for institutionalized resistance of peasants and farmers came with the workers' strikes of 1980, which provided some opportunities to fight for farmers' interests. As a result, new farmers' organizations emerged in order to oppose the government policy course. Among them were: Peasant Trade Union of the Dobrzyńska Territory and Kuyavia (1980), later renamed Independent Self-Governing Trade Union "Peasant Solidarity" and, most of all, Independent Self-Governing Trade Union of Individual Farmers "Solidarność" (1981). In terms of membership, these organizations were rather multi-faceted and non-uniform and so were their programs. Their goals focused on: providing permanent and stable development of the peasant economy, recognition of agriculture as an important and durable component of the Polish economy, making farmers equal with other social groups in terms of social benefits and services, preparing conditions for the emergence of peasant governance (Halamska, 1988, p. 150). Social demands of a general nature should also be added here, such as freedom to establish and operate political parties, free democratic elections, and freedom to nominate candidates in elections. A quite important change in the peasant/farmer movement occurred in the 1980s and resulted in diminishing participation among traditional peasants and increased interest from farmers, including the owners of large farms. This was reflected in the protests organized by the "Solidarność" camp after 1989.

The dynamics of peasants' protests in post-1989 democratic Poland were mostly determined by economic factors. Free market principles began to influence the process of agricultural products and at the same time the conditions of home budgets of family farm owners. In the 1990s, two waves of farmers' protests can be identified. The first took place from 1989 to 1993 and the second started in 1997 and lasted through 2001. The time between the two waves of protests was rather peaceful, as Poland was governed by the coalition of the Democratic Left Alliance and the Polish Party of Peasants. The policy conducted by this coalition attenuated the negative consequences of the reforms that were experienced by farmers in the first years of the system transformation. In both protests' waves certain protest cycles should be acknowledged. In the first wave, there was a Solidarity cycle and a post-Solidarity cycle and these cycles were named after the protests' organizers. In the same wave there was also a cereal cycle and pork and sugar cycles named after the main issues of discord. The names were not coincidental, as in the first wave of protests there was a division between those who initiated them and whether they had Solidarity connection. In the second wave of protests such divisions were not so crucial and the interests of other producer groups became more of a priority (see Gorlach, 2001, p. 154; Foryś, 2008, p. 177).

Both protest waves were dominated by similar demands for change in economic policy regarding agriculture (i.e., agricultural support, establishment of agricultural institution, protection of agricultural market) and expected material compensation (i.e., minimal prices, tariffs on agricultural imports, low percentage credits). In that period of time, throughout the waves of protests, a repertoire of protest actions was established. Initially, they took the form of strikes and strike emergencies, then they escalated to demonstrations and the occupation of buildings. The radicalism of some of these actions increased and there were some cases of violence and even skirmishes with the law (such as the destruction of agricultural commodities and road blocks). From 1997 to 2001, the repertoire of farmers' protest actions became solid and it might even be used presently with road blockages, demonstrations, and destruction of property.

A turning point for farmers and their protest activities came in 2004, when Poland became a member of the European Union. On one hand, farmers gained the status of being the biggest beneficiary of this membership, while on the other they joined protest activities, which Sidney Tarrow called "euro-protests" (Imig & Tarrow, 2001, p. 32). In regards to farmers' protests taking place in Poland over the last fifteen years, some remarks should be made (Foryś, 2019, pp. 618–622). First, the causes of the protest activities always had an economic character and were tied to the worsening financial situations of farms. Second, the protests were related to particular sectors of agriculture and were conducted by groups of agricultural producers. Third, the political demands were rather rare, and if they appeared they were mostly related to expected legal changes that could potentially hurt some categories of producers (e.g., in 2017 poultry and fur producers protested against the planned change in legal regulation on ritual slaughter). Fourth, the Europe-wide protests emerged, aiming at the institutions of the European Union, mostly the European Commission. Polish farmers participated in them alone or with farmers from other countries (i.e. in 2008 there were protests for milk and cereal producers, in 2012 protests against milk quota liquidation, in 2014 protests against insufficient compensation of farmers' losses caused by the embargo on Russia, in 2016 dissent about the problems related to sales of fruit and vegetables).

In the longer perspective, changes in protest activities taken up by farmers could be noted. Change in the repertoire of protests forms also occurred, with a diminishing number of rallies, marches, and strikes and an increase in more radical actions such as road blockages, occupation of public space, and destruction of agricultural commodities. As a result, the effectiveness of protests in terms of goal achievement increased over the consecutive decades. In the interwar period, the protests were conducted by political organizations which, with the passing years, conceded their leading role to trade unions and producer associations. The more permanent characteristics were: the cyclical character of farmers' protests, economic grounds for protests, and the character of farmers' demands, with the dominant issue being financial compensation.

15.4. Contexts of Farmers' Potential for Protests

Observing contemporary contesting activities of farmers and agricultural producers in Poland, their sector divisions are quite noticeable. Protesting activities are taken up by these groups of agricultural producers who, at the time, are having economic difficulties such as low buying prices, overproduction, and unfavorable legal solutions. They can be seen as interest groups that are organized under the umbrella of certain commodity producer associations and defend their own interests. To a great extent, this is the result of internal diversification in the socio--professional category of farmers and agricultural producers. The interests of farmers are only shared to a certain extent. What benefits one producer group might not necessarily be good for another group, or will have neutral results, at best. There are also territorial differences between farms and this also causes diversification of farmers' interests. To overcome these problems some attempts are made to establish one organization which could unify the mobilization efforts of these groups. The process of establishing such an organization, with formal membership, under the name AgroUnia, started in 2018. The question arises whether or not this organization will be able to become the main representative of farmers' interests and if the level of dissatisfaction among farmers and agricultural producers is sufficient to awaken their mobilization potential and initiate an active and effective fight for their own interests?

While analyzing some data collected in the nationwide research conducted on the representative sample of 3551 farm owners, attempts were made to evaluate the mobilization potential and level of dissatisfaction among Polish farmers at the end of the second decade of the 21st century. General statistics related to respondents' approach to Poland's membership in the European Union were analyzed. Also considered were their opinions on functioning of the political system in Poland as well as evaluations of whether the Polish state ensured social justice and whether respondents thought that Poland was a country of equal opportunities. The independent variables were the following: sex, age, education, and size of the farm owned. These data are significant as they relate to three important contexts of farmers' functioning (i.e., issues of subsidies, implementation of certain economic and agricultural policies, ways of organizing Polish society to ensure equal opportunities for people). Farmers' evaluation of these contexts might serve as an indicator of their dissatisfaction and to later decide on the mobilization potential of this socio-professional category. Obviously, these contexts should not be treated as direct reasons for an eventual frustration outburst, but could be seen more like a permanent base on which direct reasons for disappointment might be activated. In other words, the analysis of respondents' answers can serve as a diagnosis of what makes favorable conditions for farmers' contestation. The answers to some questions were compared to those obtained in the research conducted in earlier years. They included evaluations of benefits related to Poland's membership in the EU, approach to organizations defending farmers' interests and farmers' methods of fighting for their own interests. The data are presented in Table 15.1.

Among the respondents, the most numerous was the group that considered Poland's EU membership as beneficial to all farmers and these respondents agreed with the statement that the EU accession gave all farmers the access to financial means. Second largest was the category of respondents who pointed out that Poland's EU membership only benefited the owners of large and modernized farms.

The answers suggesting that operators of small farms benefited by receiving some money "at hand" (i.e., direct cash payment) or that all farmers lost as a result of the EU membership oscillated around the level of 6.9%. The remaining research participants had no opinion on this issue. As can be seen, farmers had a rather realistic evaluation of the consequences of Poland's membership in the EU for their social category. In reality, they became the largest beneficiaries of Poland's participation in the Common Agricultural Policy. It should be added that there was no statistical correlation between the answers and the sex/gender of respondents. Both men and women (with women comprising 36.6% of all respondents) presented similar opinions on this matter in each category of answers.

Table 15.1. Who, in respondents' opinion, benefited from Poland's accession to the European Union (in %)

Year	All farmers	Large farms	Small farms	Everybody lost	Hard to say	Total
2017	57.7	32.4	3.1	2.8	3.9	100.0
2007	37.3	57.2	2.9	2.5	–	100.0
Respondents' age in 2017						
18–34	62.6	25.7	0.8	4.2	6.6	100.0
35–54	56.6	33.3	3.4	3.4	3.4	100.0
55 and up	58.2	33.1	3.3	1.5	3.8	100.0
Respondents' education in 2017						
Elementary	56.0	32.5	4.5	1.8	5.1	100.0
Vocational	56.6	33.4	3.4	3.1	3.4	100.0
High school	58.8	30.7	2.8	3.1	4.5	100.0
Above high school	61.2	32.5	1.2	2.4	2.6	100.0
Farm size in 2017						
1–5 ha	56.1	35.9	2.9	2.1	3.0	100.0
6–10 ha	58.2	31.3	3.0	2.3	5.2	100.0
Over 10 ha	62.9	21.3	3.9	6.2	5.7	100.0
Respondents' sex/gender in 2017						
Female	57.5	34.3	2.7	1.6	3.8	100.0
Male	57.8	31.3	3.3	3.6	4.0	100.0

Source: 2017: own research, 2007: Gorlach, 2009, p. 133.

Age of respondents played a more significant role here. Younger people aged 18–34 were more likely to state that Poland's membership in the European Union was beneficial to all farmers than respondents aged 35–54. Interestingly, the category of the youngest respondents was the one most inclined to think that Poland's EU membership brought some losses to all farmers. In the other age categories these values were 3.4% and 1.5%, respectively. It is hard to find the link between the education of respondents and their evaluation of the consequences for farmers from Poland's EU membership. The only trend that could be discerned here is that with higher education acceptance of the statement that "all farmers benefited from this membership" was increasing slightly, but in a rather regular manner.

Reactions to pre-formulated statements on benefits experienced by farmers as a result of Poland's joining the EU were clearly diversified, reflecting the size of the farms owned by respondents. As farm size grew, so did acceptance of the view that all farmers benefitted from Poland's EU membership. Distribution of acceptance of the statement that only owners of the largest farms benefited from Poland's EU accession was also interesting, as three farm size categories presented a tendency directly opposite to the one described above: the smaller the farm size, the higher the level of acceptance of the statement on Poland's EU membership being beneficial to all farmers. What is clearly seen here is an indication that the owners of the largest farms did not consider themselves as significant beneficiaries of Poland's EU membership to the extent that owners of smaller farms see them to be. This could confirm the tendency observed in the results of the 2007 study conducted by Krzysztof Gorlach (2009, p. 134). At the time, Gorlach stated that such a situation was caused by a certain socio-psychological mechanism, which prevented beneficiaries from admitting to having made some gains as a result of the situation they became part of. This mechanism is still quite common among Polish farmers in the second decade of the 21st century.

In comparison with the statement that small farmers had made gains after Poland's joining the EU there was no significant difference between the answers in all categories and the share of positive declarations oscillated between 3.0% and 4.0%. Among owners of large farms there was noticeable support for the statement that all farmers lost because of Poland's EU accession (6.2%) with remaining categories agreeing with this opinion only at the level of 2.0%.

Based on the above data, one can state that farmers' evaluation of Poland's EU membership is becoming slightly but consequently diversified and this could be linked to the size of the owned/operated farm. The evaluation of such diversification is not unambiguous. It was quite obvious that the scale of problems that owners of small and large farms had to face was not uniform and the benefits incurred by owners of small and large farms were not the same. It could be cautiously asserted, however, that despite significant profits related to agricultural subsidies, the owners of large farms were less satisfied than the owners of the smallest farms. This could come from their being strongly entangled in market mechanisms and exposed to market forces, which led to the thesis that the market could be an important factor in generating dissatisfaction in this category of farmers.

Referencing the general data from 2007 reflecting the opinions of respondents regarding benefits that Polish farmers receive because of Poland's EU membership, it might be stated that the important flow of opinion between the 2007 and 2018 research editions applied to only two categories. This meant that the percentage of people who noticed benefits stemming from EU membership increased significantly, from 37.3% to 57.7%. The percentage of respondents who only noted gains on the side of the owners of large farms declined from 57.2% to 32.4%. To explain that, it could be stated that after fifteen years of Poland's presence in the EU, the positive effects of the EU programs targeting rural areas, including the owners of small farms, have a more salient impact. Their effects, such as development of agri-tourism and development of multifunctional farms are noticeable. It might be worth emphasizing that the percentage of respondents who thought that EU membership brought losses to all farmers became marginal over time. Another issue is how the respondents evaluated the functioning of the political system in Poland (Table 15.2).

Table 15.2. Respondents' opinions on the statement that the political system in Poland is working well (in %) in 2017

"The political system in Poland works the way it should"	I strongly agree	I rather agree	I neither agree nor disagree	I rather disagree	I strongly disagree	Hard to say	Total
	3.4	14.2	27.0	29.7	18.8	6.8	100.0
Respondents' age in 2017							
18–34	1.1	16.3	27.9	25.5	23.9	5.3	100.0
35–54	3.3	13.7	26.7	30.8	19.3	6.2	100.0
55 and up	4.5	14.6	27.3	29.3	16.2	8.0	100.0
Respondents' education in 2017							
Elementary	2.9	17.9	25.5	30.3	13.2	10.2	100.0
Vocational	4.5	13.1	27.9	29.6	19.0	5.9	100.0
High school	3.1	12.8	26.7	31.9	18.5	7.0	100.0
Above high school	1.2	18.2	26.4	24.4	25.2	4.8	100.0
Farm size in 2017							
1–5 ha	3.8	15.3	27.8	29.2	16.9	6.8	100.0
6–10 ha	3.0	13.2	25.1	31.5	20.4	6.8	100.0
Above 10 ha	2.4	11.6	26.4	29.5	23.8	6.4	100.0
Respondents' sex/gender in 2017							
Female	2.6	13.0	29.3	29.8	16.3	9.0	100.0
Male	3.9	14.9	25.7	29.7	20.2	5.5	100.0

Source: Own research in 2017.

In the responses to the statement that the political system in Poland works the way it should, negative answers prevailed. The cumulative percentage of "I rather disagree" and "I strongly disagree" answers was 48.4%, while the combined percentage of answers "I rather agree" and "I strongly agree" was 17.6%. The percentage of people with a rather ambivalent response was rather high (27.0%), but relatively few respondents had no opinion. Considering variables such as sex, age, education, and farm size, certain conclusions might be drawn. First, sex had no bearing on the opinions of whether the political system works the way it should. Second, the diversity of opinions in various age groups was not highly pronounced. Among those who "strongly agreed" with the above statement the category of the oldest respondents (over 55 years of age) was the most prevailing. The support for this statement was much weaker in younger age groups (aged 35–54) and (aged 18–34). The opposite situation could be noted with the "I strongly disagree" answers, where the viewpoint was the highest among the younger respondents and decreased in the two older age groups. In the remaining categories of answers such diversity was not discernible. It can be stated that the evaluation of the proper functioning of the political system in Poland improved slightly with the age of respondents. Third, education was also a variable that differentiated the respondents in terms of evaluation of the statement quoted above. The level of disapproval of the functioning of the political system in Poland increased with the educational level of respondents. This was confirmed by two tendencies; the first being the decreasing support for the statement on the proper functioning of the political system in Poland among those with a higher level of education, and the second being a lack of acceptance for this statement (answers: "I strongly disagree" and "I rather disagree") among those with a higher level of education. The group of respondents with the highest level of education is somewhat off this trend. This could be interpreted as a manifestation of a more thorough knowledge among this group regarding the functioning of the political system and thus, more nuanced opinions on the subject. Four, there was also a noticeable tendency of growing disapproval for the functioning of the political system correlated with the increasing size of farms operated by respondents. In this case, the distribution of answers in all three categories of farms reflected similar tendencies as with education level. The acceptance of the discussed statement decreased and disapproval got stronger with larger farm areas. Such a distribution of answers might be the consequence of stronger "politicization" of the owners of larger farms. What comes to mind here is the significant dependence of larger farm operators on political decisions regarding state policy, and agricultural policy in particular. This relates to the already mentioned strong ties of the large farm operators to the economic market, where they conduct their activities.

To sum up the analysis of responses on the functioning of the political system, it can be stated that the respondents had a rather negative opinion of the system. It is hard to explain why this might be so. There were no additional research questions which would allow for obtaining information on the reasons why such evaluations were formed. Two possible explanations might be carefully considered here. The first, and less probable, one could indicate farmers noticing limited civic liberties and threats to freedom of speech and linking these problems to the state policy of the moment. This could bring about the viewpoint of the political system as unable to defend such values. The more probable second suggestion could indicate evaluation of political system containing a significant economic component. Worsening of economic situation goes along with a negative evaluation of the political system, which, according to farmers' mind-set should correct these negative tendencies but does not do so. Such an explanation would be very consistent with the acceptance of protectionist policy of the state that is quite strong among farmers.

The state can play an important role in redistribution of burdens and benefits between various segments of society. This often influences the sense of social justice or lack thereof. Thus, the question about respondents' approach to the statement "our society is organized with regard for justice" and farmers' evaluation of the issue of equal opportunities in Polish society seems also important (Table 15.3). First, the approach to the statement on justice in Polish society will be analyzed.

Table 15.3. Respondents' opinions on the statement that Polish society is organized with regard for justice (in %) in 2017

"Polish society is organized with regard for justice"	I strongly agree	I rather agree	I neither agree nor disagree	I rather disagree	I strongly disagree	Hard to say	Total
	1.6	9.8	22.9	37.3	21.8	6.6	100.0
Respondents' age in 2017							
18–34	1.3	14.7	24.7	32.5	21.0	5.8	100.0
35–54	1.2	9.9	22.3	37.6	22.5	6.5	100.0
55 and up	2.5	8.1	23.4	38.4	20.8	6.8	100.0
Respondents' education in 2017							
Elementary	3.1	12.0	25.1	35.8	15.3	8.8	100.0
Vocational	1.7	9.3	21.0	40.2	22.5	5.3	100.0
High School	0.9	9.3	22.8	35.8	23.1	8.1	100.0
Above High School	1.7	10.3	27.3	33.5	23.2	4.1	100.0
Farm size in 2017							
1–5 ha	2.1	10.0	23.2	37.9	20.7	6.1	100.0
6–10 ha	0.9	9.1	20.9	38.2	23.7	7.4	100.0
Over 10 ha	0.7	10.1	24.1	34.5	23.4	7.3	100.0
Respondents' sex/gender in 2017							
Female	1.0	9.6	23.1	38.4	20.1	7.7	100.0
Male	2.0	9.9	22.8	36.7	22.7	5.9	100.0

Source: Own research in 2017.

The data in Table 15.3 showed that the respondents unambiguously evaluated Polish society as unfairly organized. The aggregated distributions of "I rather disagree," "I strongly disagree" answers, negating the statement of Polish society being organized with regard for justice, came to 58.9%. The opposite views ("I rather agree," "I strongly agree") were expressed by only 10.4% of respondents. The percentage of people who neither agreed nor disagreed with the above statement was rather high. Similar to what was observed in the previous question, sex was not a variable that differentiated the answers. Age, however, was such a variable that diversified the opinions of respondents. It could be noticed that the people belonging to the younger age category were more likely to accept the statement that Polish society was organized with regard for justice. This was confirmed by the tendency that the older the respondent, the more likely they were to negate the statement. Interpretation of these results could not be easy, especially that the tendency related to the evaluation of the functioning of the political system was the opposite of the one presented immediately above. The young people were evaluating the political system more harshly than the justice issues. It might be assumed that in the mind of a significant number of respondents, especially the younger ones, there was no connection between the political system and society being organized according to the principles of justice. The sense of justice could be seen as a matter of individual human activity and perception, and not the results of state policy or politicians' activities.

The tendencies observed in the answers to the question on justice within Polish society were not clear-cut among the respondents with different educational levels. There was a weakly marked polarization of opinions. The highest support for the above statement could be seen among people with the lowest educational level, and this category was also the least likely to negate this statement. A similar tendency, but considerably weaker, was visible among people with post-secondary high-school education. Parsing the answers by the size of farms owned by the respondents, the highest percentage of support for the statement that "society is organized with regard for justice" could be seen in the group of farmers utilizing the smallest farms and then among the owners/operators of medium-size farms. The lowest percentage of support was in the last category of farm owners (over 10 ha). The level of negation of the statement on social justice presented the opposite trend with the lowest level of disapproval among the owners/operators of the smallest farms and the highest in the category of operators of large farms.

The next question could also be seen as providing some insight into an evaluation of the most important aspects of the functioning of farm owners. The question below addressing equal opportunities could also provide contexts that could form the basis for building permanent dissatisfaction in this socio-professional category.

For that reason, the question presented below deals with a deeper evaluation of how true the statement is that currently in Poland "everybody has more less the same opportunities to be happy and rich" (Table 15.4).

Table 15.4. Respondents' opinions on the statement that "today in Poland everybody has more or less the same opportunities to be rich and happy" (in %) in 2017

"Everybody in Poland has equal opportunities"	I strongly agree	I rather agree	I neither agree nor disagree	I rather disagree	I strongly disagree	Hard to say	Total
	3.3	12.0	17.3	31.6	30.9	4.8	100.0
Respondents' age in 2017							
18–34	3.9	12.4	22.4	30.5	26.8	3.9	100.0
35–54	2.9	12.2	17.5	31.8	31.6	4.0	100.0
55 and up	3.9	11.5	15.3	31.7	31.1	6.6	100.0
Respondents' education in 2017							
Elementary	3.9	8.9	17.7	31.5	29.7	8.3	100.0
Vocational	3.1	13.1	17.6	30.4	31.8	4.0	100.0
High school	3.9	10.3	17.1	33.0	31.5	4.2	100.0
Above high school	2.4	16.0	16.3	32.3	27.6	5.5	100.0
Farm size in 2017							
1–5 ha	3.8	12.2	18.3	31.6	29.2	4.9	100.0
6–10 ha	1.4	9.6	17.0	32.1	35.0	4.8	100.0
Over 10 ha	4.1	13.8	13.8	31,2	32.5	4.7	100.0
Respondents' sex/gender in 2017							
Female	3.6	10.6	18.7	33.3	29.3	4.5	100.0
Male	3.2	12.8	16.5	30.7	31.8	5.1	100.0

Source: Own research in 2017.

The distribution of answers to that question showed explicitly that the majority of respondents did not agree. The aggregate percentage of the answers "I rather disagree" and "I strongly disagree" was 62.5%. Approval for the statement was expressed by 15.3% of respondents. As many as 17.3 % of respondents were ambivalent about this issue and 4.8% had no opinion. When compared to the statements previously discussed, this particular statement on equal opportunities generated the most negative attitude of respondents.

What was the distribution of answers in regards to variables such as: sex/gender, age, education, and farm size? Similarly to previously discussed statements, sex/gender did not differentiate the respondents' answers. In both sex/gender categories the distribution of answers nearly mirrored the general tendency. Age had a rather minimal bearing on the respondents' answers and negative responses prevailed in all age categories. In the youngest age category (18–34 years) those who "rather agreed" or "strongly agreed" with the above statement on equal opportunities comprised a slightly higher percentage than those who approved this statement in other age categories. The youngest respondents were also the least likely to differ with the statement on equal opportunities, while the respondents in the oldest age category were inclined to "rather disagree" or "strongly disagree."

Statistically significant relations in the distribution of answers could be seen with education. The proposed statement was most rejected by people with the lowest level of education. At the same time, such respondents were the least likely to agree with it. Respondents whose education went beyond high school were the most likely to agree with the above statement and the least critical of it. In this category of most educated respondents, the undecided comprised a smaller percentage than for other statements. The opinions on the issue of social justice from respondents of various age groups could be explained in many different ways. It seemed to be confirming the well-known tendency of people with low educational levels to display a fatalist approach toward reality and the feeling that very little depended on them in a world that was, essentially, unjust. The links between farm size and the answers of respondents were hard to discern. There was more acceptance among operators of the largest farms (i.e., those over 10 ha) for the statement on equal opportunities than from other categories of farm owners/operators. It should be noted, however, that in the same group the aggregated percentage of respondents who rejected this statement was also the highest (63.8%). The distribution of responses among the owners of the smallest farms (1–5 ha) seemed hard to analyze with both approval and disapproval for the presented statement being only slightly lower than in the category of large farms owners/operators. There were no patterns in the responses worth further consideration. Among all the data variables considered in the research, educational level was the most differentiating.

Based on analysis of responses to the four discussed statements relating to the benefits coming from Poland's membership in the EU, proper functioning of the political system, social justice, and equal opportunities, several conclusions could be drawn. First, the situation in Poland regarding certain aspects of the political, social, and economic systems received a rather negative evaluation, which could be seen as pessimistic contesting. Second, variables that differentiated the views of respondents were mostly related to education and farm size and rarely correlated with age. Sex had no significant meaning for the choice of answers. Third, for issues containing an economic component, farm size was the most differentiating variable, while age and education had a less significant effect. Fourth, for the differentiation on political and social issues, variables such as age and education were quite meaningful. It was noticeable that the younger generation was more critical towards the functioning of the political system than to the organization of social order. Fifth, based on the above analysis, it could be stated that the level of dissatisfaction among Polish farmers, as measured by their attitudes towards the benefits of Poland's EU membership and their disapproval for the functioning of the political system, was relatively high. The decisive variable was farm size and, to a lesser extent, age and educational level.

15.5. Organizational Aspects of Farmers' Protesting Activities

In the previous subchapter, the level of farmers' deprivation was evaluated on the basis of farmers' attitudes toward the political and social system. This system provided somewhat general contexts for how farmers and their farm functioned, and were crucial for stimulating the sense of dissatisfaction. It might be worthwhile to determine farmers' potential to change what farmers evaluate so negatively and what methods of action they might consider as most appropriate for achieving these goals as seen in Table 15.5.

Analyzing past contesting activities conducted by farmers, it could be seen that their mobilization was primarily based on the organizational factor. This was the case in the 1990s, when trade unions, such as Independent Self-Governing Trade Union of Individual Farmers, "Solidarity" (NSZZ RI "Solidarność"), National Union of Farmers' Circles and Agricultural Organizations (Krajowy Związek Rolników, Kółek i Organizacji Rolniczych), and Self-Defense (Samoobrona) dominated in the organizing role. At the beginning of the 1990s these organizations competed with each other, which meant that they organized separate and autonomous protests. In the second half of the 1990s, they changed their action strategies and started to cooperate in organizing farmers' contestation. With protests' animators becoming quite prominent, a new protest quality emerged. This meant that farmers' trade unions and producer organizations started to play an increasing role in farmers' mobilization. This could serve as evidence of the growing diversification of farmers' interests that narrowly specialized interest groups had fought for. Therefore, it should be stated that achievement of farmers' goals was increasingly connected with the level of their membership in these types of organizations. Research results focused on this issue might be worth a closer look (Table 15.5).

Table 15.5. Respondents' membership in political organizations, trade unions, and producer groups (in %)

Year	Present membership in trade unions, producer groups and cooperatives		Present membership in political parties and trade unions		Past membership in political parties and trade unions		Total
	Yes	No	Yes	No	Yes	No	
2017	3.3	96.7	2.2	97.8	5.2	94.8	100.0
2007	5.3	94.7	–	–	–	–	100.0
Respondents' age in 2017							
18–34	5.3	94.7	1.8	98.2	0.8	99.2	100.0
35–54	3.4	96.6	2.1	97.9	2.9	97.1	100.0
55 and over	2.7	97.3	2.5	97.5	10.4	89.6	100.0
Respondents' education in 2017							
Elementary	1.8	98.2	0.0	100.0	5.3	94.7	100.0
Vocational	2.5	97.5	1.3	98.7	4.4	95.6	100.0
High school	4.3	95.7	2.7	97.3	4.6	95.4	100.0
Above high school	5.3	94.7	6.9	93.1	9.1	90.9	100.0
Farm size in 2017							
1–5 ha	1.3	98.7	2.2	97.8	5.3	94.7	100.0
6–10 ha	3.3	96.7	1.0	99.0	5.0	95.0	100.0
over 10 ha	10.9	89.1	3.7	96.3	5.1	94.9	100.0
Respondents' sex/gender in 2017							
Female	2.2	97.8	1.2	98.8	3.1	96.9	100.0
Male	4.0	96.0	2.8	97.2	6.3	93.7	100.0

Source: 2017: own research; 2007: Gorlach, 2009, p. 210.

The question on membership in organizations representing farmers' interests addressed membership in producer associations, producer groups, or cooperatives, as well as other types of membership in trade unions and political parties. As presented in the research results, only 3.3% of respondents declared membership in the first type of organizations such as trade unions, producer groups and cooperatives and 2.2% had membership in trade unions and political parties. The question on past membership in political parties or trade unions was met with a positive response from 5.5% of respondents. The obvious conclusion to be made is that there was a clearly decreasing tendency of farmers' membership in trade unions and political parties. It could be cautiously assumed that such a drop in membership indicated farmers' weakening trust in such organizations as truly representing farmers' interests. This could be consistent with earlier statements about the growing role of branch organizations and producer groups in organizing protest activities. Furthermore, it could be useful to compare the presented level of farmers' membership in trade unions and political parties with the general tendency seen in Polish society. According to the research of *European Values Study* conducted at the end of 2017 and the beginning of 2018, membership in political parties in Poland was declared by 1.4% respondents, while 4.9% of respondents admitted to being members of trade unions (CBOS, 2017, 2018). After averaging out these values to compare them with our own research presented above, the received value is 3.15%. With this in mind, the 2.2% level of farmers' membership in these types of organizations confirmed that farmers not only did not engage in the development of civil society but also did not see these types of organizations as defending their interests.

Analyzing membership in the first type of organizations (i.e., producer associations, producer groups and cooperatives) according to the variables discussed in the previous subchapter, such as sex, age, educational level, and farm size, several statements can be made. First, men were more likely than women to be members of such organizations. Second, younger people, under the age of 35, were more likely than older ones, aged 35–55 or above, to be members of such organizations. Third, the membership in such organizations increased with educational level. Fourth, there is a strong connection between membership in such organizations and farm size, as owners/operators of large farms are more likely to belong to such organizations. As indicated by the data in Table 15.5, 10.9% of the owners of farms larger than 10 ha were members of such groups and organizations. In the remaining categories of farms, the membership in trade unions, producer groups, and cooperatives was much lower: 3.3% among the respondents operating farms of 5–10 ha and 1.3% of those operating farms of 1–5 ha. This last tendency should be viewed as a rather obvious consequence of the functioning of large farms on the market. Their connection with the market was quite strong, as they needed to deal with the competition and had to cooperate within the framework of producer groups or cooperatives. The role of age and education in determining membership in such organizations, although smaller, could be explained by the larger cultural and human capital of younger and more educated people. Whether a similar tendency might be observed within the membership of political parties and trade unions could be worth further exploration.

As was presented before, membership in political parties and trade unions was rather low among the interviewed farmers and this was also noticeable when compared with general, nationwide data. Differentiation of membership based on sex could also be observed. Men were much more likely than women to be engaged in these types of activities, which replicated the tendency observed with associations and producer groups. There was a rather weak growth tendency for participation in parties and trade unions among higher age respondents but this still remained at a very low level. This was an opposite tendency to what was observed with producer associations and groups. It could be argued that young people preferred more a professional organization connected with grassroots activities rather than political parties. The analyzed activities within parties and trade unions had a somewhat stronger connection with the educational level. With a higher level of education, there was a visible and statistically significant increase of engagement with political and trade organizations. Nobody with just an elementary school education declared such activity, while among people with education beyond high school, almost 7.0% did. In the middle educational categories (vocational, high school), the percentage holding membership in parties and unions was much lower. It could be stated that the statistics confirmed the general trend that people with a higher educational level were more likely to be engaged politically and socially. The last independent variable, namely farm size, did not have as significant an influence on the distribution of answers as education did. A tendency among the owners/operators of the largest and the smallest farms to get involved in political parties and trade unions could be noted but it was still very weak and not pronounced. This could be explained from the economic point of view, as the operators of the smallest farms not being able to actively participate in producer group and associations. They could be more inclined to become active within political parties and trade unions, which they might recognize as greater potential defenders of their interests. At the same time, the owners of larger farms were strongly entangled in the economic market and quite dependent on it. Therefore, they more frequently remained under pressure to defend their interests than owners of smaller farms. Large farm owners/operators were more likely to look for allies among various organizations and public entities, including political parties and trade unions.

To summarize this part of the reflection, it should be stated that the participation of Polish farmers in various organizations, especially trade unions and political parties, was at a relatively low level when compared to Polish society as a whole. Additionally, this diminishing tendency was accelerating. It should be emphasized that membership in agricultural organizations such as producer groups and associations remained an important aspect of farmers' activities. Membership in them was mostly declared by owners of large farms, which seemed quite logical, as farm area determined the possibilities for market activities. These agricultural organizations have a market-based, professional character and organizers of agricultural contesting could look for support in them for protest actions.

An interesting observation could be made by juxtaposing the data from 2007 with that of 2018 regarding membership in producer organizations. Such a comparison was only possible in this category. Two conclusions can be made on this basis: 1) the percentage of membership in these groups declined from 5.3% to 3.3%, 2) the decline only applied to men. In 2007 the participation level was 8.5% (Gorlach, 2009, p. 210) and in 2018 it dropped to 4.0%. The membership of women increased slightly from 2.0% to 2.2%. This state of affairs could be linked to a diminishing tendency in the number of producer groups in Poland. The data in the agricultural reports indicated a growth trend until 2015 and then their number started to drop. Their numbers in the selected years were as follows: 2005—34; 2007—154; 2011—751; 2013—1200 (Chlebicka, 2013, p. 33); 2014—1349;[3] 2018—962.[4] However, the factors specified by Andrzej Pilichowski (2018, p. 118) could seem more important here, and among them were: treating membership in producer groups as an occasion for one-time financial support, and a considerable lack of mutual trust among the group participants. This last factor could be the aftermath of the negative experience related to jointly conducted activities at the time of state socialism, as well as today.

Distribution of answers to two other questions related to farmers' defense of their interests is presented below. The first question deals with evaluating the role of farmers' organizational habitat in defending their interests and the second question addresses various forms of fighting for farmers' interests (Table 15.6).

In the first question, the skepticism of farmers regarding the abilities and engagement of organizations to defend farmers' interests was rather obvious. About every tenth respondent agreed that such organizations existed, while 76% doubted their existence and 13% had no opinion. Interestingly, men were more likely than women to give positive answers to this question. There was no correlation between the age of respondents and the recognition of organizations able to defend farmers' interests. In the youngest category of respondents (18–34 years), the perception of the existence of organizations defending farmers' interests was at a level below the overall result. In other age categories, the percentage of positive answers to that question was close to 11%, which was close to the general level of positive answers in the overall studied population. What clearly differentiated the respondents was educational level. There was a growth tendency in identifying organizations defending farmers' interests with respondents' higher educational level. As seen in Table 15.6, the distribution of answers were as follows: respondents with a primary education—5.7% positive answers, respondents with vocational school education—10.3% positive answers, high school graduates—12.4% positive answers, and people with education above high school level—15.3%. Such distribution of answers should be seen as the effect of the knowledge and consciousness regarding the functioning of agricultural organizations and their context of activity. Such awareness tended to grow along with educational level.

---

3   http://biblio.modr.mazowsze.pl/Biblioteka/Rozwoj/WSPARCIE _GPR.pdf.
4   https://www.raportrolny.pl/news/item/3313-naukowcy-wspieraj%C4%85-rolnictwo.

Table 15.6. Opinions on the existence of organizations defending farmers' interests and the most effective forms of defense of farmers' interests (in %)

Year	Are there any organizations defending farmers' interests (all respondents)			The most effective forms of fighting (all respondents) in 2017					
	Yes	No	Hard to tell	Road blockages, demonstrations, occupations	Discussions and pressure on politicians	Organizing associations and cooperatives	Does not make sense to do anything	Hard to tell	Total
2017	11.0	76.0	13.0	9.6	15.5	35.7	31.1	8.1	100.0
1994 *	13.6	86.4	–	–	–	–	–	–	–
1999 *	41.1	58.9	–	16.0	16.9	35.2	38.7	–	100.0
2007 *	17.5	82.5	–	7.2	13.1	48.0	29.7	–	100.0
Respondents' age in 2017									
18–34	9.2	76.6	14.2	10.8	13.7	42.9	27.9	4.7	100.0
35–54	11.0	76.4	12.6	9.3	15.7	36.0	31.7	7.3	100.0
55 and up	11.5	75.2	13.3	9.7	15.8	32.8	31.2	10.5	100.0
Respondents' education in 2017									
Elementary	5.7	82.5	11.8	10.2	14.4	26.4	35.0	14.0	100.0
Vocational	10.3	76.5	13.2	10.2	15.3	32.5	34.4	7.6	100.0
High school	12.4	74.2	13.4	8.8	15.9	40.7	27.5	7.2	100.0
Beyond high school	15.3	71.8	12.9	8.4	16.7	44.9	24.1	6.0	100.0
Farm size in 2017									
1–5 ha	8.5	78.5	13.0	8.5	15.3	35.7	31.7	8.9	100.0
6–10 ha	11.8	75.6	12.6	10.6	16.5	33.3	31.6	7.9	100.0
over 10 ha	19.0	67.6	13.3	12.0	15.0	38.5	28.8	5.7	100.0
Respondents' sex/gender in 2017									
Female	7.2	79.1	13.6	6.5	17.3	37.0	29.9	9.3	100.0
Male	13.2	74.2	12.6	11.3	14.4	35.0	31.8	7.4	100.0

Source: 2017: own research; 2007: Gorlach, 2009, p. 128.

An even stronger role in the distribution of answers was held by the area of the operated farm. The larger the farm, the better the recognition of organizations engaged in fighting for farmers' goals. The tendency to answer the questions in such a manner derived from farmers' engagement in these kinds of organizations. These respondents who had larger farms due to their specialized production, market connections, and vulnerability to fluctuations in the market of agricultural products were more likely to be interested in following the activities of such organizations. They had the knowledge of such organizations as representing the interests of particular producer groups. Some farmers who operated large farms were members of such organizations.

When comparing the data just analyzed above with that collected in 2007 (Gorlach, 2009, p. 128) it could be stated that the number of respondents noticing the existence of organizations able to defend farmers' interests was declining. Going back into the older studies, the time frame between the consecutive editions was much longer as the earlier study was conducted in 1994. At the time, such organizations had been recognized at the same level as now but by the end of the 1990s, farmers' perception of them was at its highest. There could be three possible explanations. First, the end of the 1990s was the time of the second wave of farmers' protests, with the leading roles played by "Solidarity" ("Solidarność") and "Self-Defense" ("Samoobrona"). The style of actions by these organizations, and the radicalism of "Self-Defense" in particular, left an impression, not only in the consciousness of farmers, but in the collective mind of the entire society. This experience of the late 1990s, when farmers protested most frequently and in the most radical manner, was the reason why farm owners easily pointed to these organizations as defending their interests. Second, in the first half of 1990s, farmers also protested rather frequently but they were represented by several organizations that originated from different political circles and various groups of farmers (KZRKiOR—as a post-Communist organization; NSZZ "Solidarność"—as an organization trying to encompass the interests of all farmers; "Samoobrona"—representation of large farm owners). For that reason, it might be stated that, to some extent, farmers could have been disoriented and uncertain as to whether anyone represented their interests. Additionally, at the time of the first wave of farmers' protests, they had not yet achieved the expected goals. Third, while the end of the 1990s could be perceived as a time of some consolidation in the "market" of organizations representing farmers' interests, then the second decade of the 21st century was a time of pluralism and proliferation for such organizations. The role of farmers' trade unions of a general character was weakening and the number of producer organizations and associations had increased. The multitude of such organizations did not go hand in hand with their effectiveness in fighting for farmers' interests, especially that these interests were more specific and narrowly defined. Therefore, it could be concluded that such organizations were no longer perceived as representing the interests of farm owners.

Another issue worth a closer inspection is the distribution of answers regarding the forms of farmers' self-defense that could bring them the most desired results (Table 15.6). This question is indeed about the effectiveness of farmers' activities. As was mentioned above, in the 1990s a certain repertoire of protest forms was established. It was the result of experience gained during the two waves of farmers' protests which took place from 1989 to the beginning of the 21st century (Foryś, 2008). Initially, as clearly stated before, the actions similar to the ones of the workers' movement were being organized and they included letters of protests and strikes. With time, and as a result of gaining new experiences and observing the protests of farmers in other European countries, Polish farmers developed a relatively stable and consistent set of protest forms and methods. It included blockages of roads and public places as well as demonstrations and destruction of agricultural commodities. Analysis of the answers to these questions allows us to evaluate how farmers themselves perceived the effectiveness of the various protest forms.

As can be inferred from the collected data, in addition to recognizing direct forms of protests, a significant percentage of respondents thought that the most effective form of fighting for farmers' interests was establishing producer associations and cooperatives in order to become independent from middlemen. The second most popular answers were the ones rejecting doing anything, so as to express fatalism and hopelessness, with farmers being known to say, "They will do with us whatever they want." Respondents were supportive of putting pressure on parliamentarians or regional authorities. The least popular protest form was direct actions such as road blocks, demonstrations and occupations of public buildings.

The presented results were quite surprising. It turned out that although farmers were frequently involved in the discussed methods of direct action, they did not think of them as being the best choices in terms of the most effective form of fight. Based on these answers it might be assumed that exposure to market mechanisms determined their action strategy to a significant extent. For that reason, they were more inclined to appreciate the grassroots work connected with membership in producer groups, which could help them with prospering on the market, rather than involvement in direct actions, such as road blocks and demonstrations. As seen here, grassroots work was treated as an every-day activity allowing for the true fight for farmers' own interests. Other forms of protest were reserved for other occasions. Generally, it was well-known in Poland that, thanks to protest actions such as road blockages and demonstrations, farmers were able to achieve a lot during the previous years. However, the evaluation of these actions in the research did not confirm that the respondents still saw them as effective.

The next question worth answering is how the variables that were analyzed so far differentiated the answers of the respondents. With the last question, sex/gender was a rather important variable. It could be stated that the men were more radical than women and more often identified road blockages as an effective form of fighting. Women, more than men, were in favor of applying pressure on politicians and farmers organizing themselves in producer groups and cooperatives. Age was also a variable that differentiated respondents' answers on organizing in producer groups and cooperatives. The youngest respondents (ages 18–34) had the most favorable outlook towards producer groups and cooperatives. The respondents in the middle age category (34–54) were less likely to point to these organizations and respondents over the age of 55 were even less interested in them. It should also be added that the youngest group of respondents was also the most supportive of the most radical forms of protests. Nevertheless, the margin between this age category and the other ones was very small.

The differentiation of answers was particularly noticeable in the education category. The dominant tendency was that radical actions were the most popular among respondents with elementary and vocational education. The support for these forms of protest was lower in the categories of high school graduates and above. At the same time, the preference for organizing farmers into producer associations and cooperatives grew, which could be connected with the tendency that with educational level also grew the size of the owned/operated farm. Larger farms were more likely to have a *stricte* market character and for their owners/operators the most rational forms of fighting for farmers' interests made the most sense. With a higher level of education, the sense of hopelessness about the futility of taking any actions also declined. It could also be noted that farmers with better education were less inclined to engage in radical actions. They also were more likely to feel the need to do something, and thought that the best way to do so was through integration within producer groups. Not without significance was the fact that among the best educated respondents the percentage of those who had no opinion on the matter was declining.

Considering that the owners of the largest farms were the ones who in previous years had the highest participation in farmers' protests, it should be expected that the owners of the largest farms would be the biggest supporters of the radical forms of actions in fighting for their interests. This was confirmed by the collected data. Support for these types of actions (road blocks, demonstrations, and occupations of public buildings) grew with the farm size owned or operated by respondents.

Following the same tendency, the support for farmers' organizing in producer associations and cooperatives also grew with farm size, which seemed understandable, as this type of cooperation was the everyday reality for many owners/operators of large farms, and the trusted form of activity. However, it should also be mentioned that even with the observed tendency, there was no statistically significant relationship between the answer to this question and the size of the farm operated/owned by respondents.

What could be seen, in reference to data from 1999 and 2007 (as data from these years could be compared) was a rather important change in evaluating the effectiveness of various forms of protest actions. Radical forms of fighting were the most referred to as effective at the end of 1990s, which confirmed the general contestation tendency in this entire period of time as marked by intensive fighting by farmers for their interests. These interests, as well as the forms of protests, had played quite an important role in the history of farmers' contestation. Furthermore, two other tendencies were also noticeable. The first one indicated the relatively high level of farmers' resignation expressed in 1999 by the opinion that "it does not make sense to do anything as they will do whatever they want, anyway." It was related to the actions of the state, which at the time were perceived by farmers as far removed from their expectations. This meant that in the beginning of the 1990s, when farmers fought for solutions to some concrete financial problem (e.g., credit, subsidies, minimum purchase prices) the state mostly operated in the institutional area (establishing institutions important to the functioning of agriculture and the agricultural market). Then, in the second half of the 1990s, when farmers demanded organizational and institutional changes, such as the establishment of sugar holding, the state focused on financial demands (Foryś, 2008). The second tendency was seen in 2007, and it meant approval for the establishment of producer organizations and associations as a form of fighting for farmers interests. This tendency was the result of Poland joining the European Union in 2004 and the introduction of numerous financial incentives for such forms of organizing. As seen in Table 15.6, such activities are not so popular at the moment, which could be explained by the already discussed arguments, such as: lack of trust in cooperation, which was the effect of past experience; lack of leaders; and poor understanding of the idea of cooperation among people, who were instrumentally interested in one-time advantage in the form of a financial premium for starting such a cooperation. As a result of such experience, the level of positive evaluation for such a method as an effective form of fighting for farmers' interests is currently quite similar to that of twenty years ago.

15.6. Summary: Contexts of farmers' contesting and the system of interests' representation

The reflection in this chapter has not yet presented a full picture of farmers' contesting abilities. A large number of factors has not been considered even though they could have influence on farmers' contesting potential. These factors include: cooperation network between organizations, organizational financial resources, protesters' goals and programs, or types of social support. However, this chapter is not aiming to analyze these types of resources, and mostly focuses on the possible causes of dissatisfaction in this socio-professional category, rooted in the wider contexts of their existence and as reflected in their consciousness. It should be added that the cause of the increasing social dissatisfaction originated in the system itself. The data analysis conducted in this article allowed for the identification of these categories of farmers who comprised the most probable dissatisfied group. Furthermore, such analysis allowed for the reconstruction of the model of relations between disappointed farmers and the state and its institutions.

Remembering what was stated about farmers' protest activities in the past, and particularly before 1989, and the conclusions already formulated on the basis of the data analyzed in this article, several statements can be made. On one hand, they would describe the mobilization foundation for farmers' actions and, on the other, they would answer the question about the possible relationship between the state and the protesting farmers.

First, looking at farmers' mobilization contexts it should be said that the factors meaningful for future protest mobilizations stem from the negative perception of the functioning of the political, social, and economic systems. The data in several tables presented respondents' rather negative reception of these aspects of life. The one factor alleviating this situation was Poland's membership in the EU and profits for farmers resulting from this membership. All categories of respondents noticed the positive aspects of this membership. This was confirmed by the verifiable fact that farmers were the socio-professional category which gained the most from Poland's EU membership. This is an important finding, especially since farmers had been the most vocal opponents of joining the European Union immediately before the accession was to be considered. It should be remembered that farmers organized protests to express their anti-EU sentiments. Second, with the observed diversification of answers to the first four answers and the selected independent variables, the key variable was the size of the farm owned/operated by the respondents. This is particularly important for the evaluation of the consequences of Poland's EU membership, as well as economic issues. It can be assumed that market connection and exposure to its influences caused farm size to be the main variable in farmers deciding on engagement in contesting activities. The larger the farm, the more likely it was that the owners would engage in protests. Third, young people formulated more realistic, more nuanced, and more positive opinions of the social system, rather than of the political system. Respondents with a lower level of education had rather extreme and more negative opinions on both of these systems. Four, the weak engagement of farmers in civil society was quite noticeable. Farmers were less engaged than other socio-professional category in membership in political parties and trade associations and this tendency was deepening. It could be assumed that among contesting activities were such forms of participation, which were treated as an effective way to fight for their interests at least by some farmers' circles.

In evaluating farmers' view of potential support, they could receive from outside, as well as the methods of action they thought were most effective, some conclusions on possible forms of mobilization can be made. First, farmers did not see in their surroundings the entities that could represent their interests and fight for their implementation. Such perception indicated that farmers were most likely alone in their fight but not helpless. This factor could be seen as integrating farmers and as an antidote for gradual differentiation of interests of particular groups of agricultural producers. Second, being both a surprise and a reflection of the change in farmers' collective thinking that occurred over the last three decades, organizing in producer groups and cooperatives became the key form of fighting for farmers' interests. Protest activities had kept some moderate support, especially among farmers with lower educational level who operated large farms. Farmers noticed changes in the agricultural surroundings and various forms of fighting for their interests and preferred to connect them with an institutional platform, adjusting them to requirements of the economic market. Establishment of producer groups was the *de facto* attempt at organizing into a group of interests that had greater effectiveness in influencing the market, protecting farmers from its negative aspects, as well as putting pressure on politicians and the deciders of agricultural policy. Third, in light of what was just stated, it should be expected that farmers' protests would still be a part of the picture of farmers' fight for their interests, but they would play the role of last resort only when more formalized methods of fighting had become ineffective.

Considering the content of this chapter it could be stated that in the 21st century and from the moment of Poland's entry into the European Union in 2004, relations between the state and farmers related to the directions of agricultural policy and securing farmers' interests went through the processes of institutionalization and professionalization. Continued modernization of Polish agriculture and marketing of production relations strengthened the position of agricultural producers and, at the same time, weakened farmers' mobilization potential, as well as re-formulated the dominant patterns of relations between farmers as the interest group and the Polish state. While analyzing the period of time of state socialism one could speak of corporate socialism combining the flexible economy with the monopoly which represented the interests of economic organizations, then after 1989 there was a gradual transition towards the model of organized cooperation. Farmers' protests played an important role in this process, empowering farmers in their relations with the state. At the time, these types of actions were rather spontaneous, unruly, and of a contesting nature. In the first two decades of the 21st century, institutionalization and petrification continued to shape relations between the farmers and the state. This relationship took the final form of organized cooperation. This could also be confirmed by farmers' collective state of consciousness on the issues discussed in this chapter.

REFERENCES

CBOS (2017). Działalność związków zawodowych w Polsce [Activities of trade unions in Poland]. *Komunikat z badań* [Research report] 87.

CBOS (2018). Aktywność Polaków w organizacjach obywatelskich [Activity of Poles in civic organizations]. *Komunikat z badań* [Research report] 29.

Chlebicka, A. (2013). Diagnoza rozwoju i perspektywa wsparcia grup producenckich w Polsce [Diagnosis of development and the perspective of supporting producer groups in Poland]. *Studia Ekonomiczne i Regionalne* [Economic and regional studies] 4, pp. 31–36.

Foryś, G. (2008). *Dynamika sporu: Protesty rolników w III Rzeczpospolitej* [The dynamics of the dispute: Farmers' protests in the Third Republic of Poland]. Wydawnictwo Naukowe Scholar.

Foryś, G. (2019). Od ruchów chłopskich do protestów rolniczych [From peasant movements to agricultural protests]. In: M. Halamska, M. Stanny, & J. Wilkin (eds.), *Sto lat rozwoju polskiej wsi: Ciągłość i zmiana* [One hundred years of development of the Polish countryside: Continuity and change], pp. 597–625. IRWiR PAN, WN Scholar.

Gorlach, K. (2001). *Świat na progu domu* [The world at the doorstep]. Wydawnictwo Uniwersytetu Jagiellońskiego.

Gorlach, K. (2009). *W poszukiwaniu równowagi: Polskie rodzinne gospodarstwa rolne w Unii Europejskiej* [In search of balance: Polish family farms in the European Union]. Wydawnictwo Uniwersytetu Jagiellońskiego.

Halamska, M. (1988). Peasant movements in Poland, 1980–1981: State socialist economy and the mobilization of individual farmers. In: B. Misztal & L. Kriesberg (eds.), *Research in Social Movements, Conflicts and Change*, Vol. 10, pp. 147–160. JAI Press Inc.

Imig, D. & Tarrow, S. (2001). Mapping the europeanization of contention: Evidence from a quantitative data analysis. In: D. Imig & S. Tarrow (eds.), *Contentious Europeans: Protest and politics in an emerging polity*, pp. 27–49. Rowman & Littlefield.

Jenkins, J.C. (1983). Why do peasants rebel? Structural and historical theories of modern peasant rebellions. *American Journal of Sociology* 88, pp. 487–544.

Marks, K. (1979). *Człowiek i socjalizm: Pisma wybrane* [Man and socialism: Selected writings]. Przekład J. Ładyka. PWN.

Migdal, J.S. (1974). *Peasants, Politics and Revolution: Toward political and social change in the third world*. Princeton University Press.

Mooney, P.H. (2000). Specifying the 'rural' in social movement theory. *Polish Sociological Review* 129, pp. 35–55.

Moore, B. (1966). *The Social Origins of Dictatorships and Democracy: Lord and peasant in the making of the modern world*. Beacon Press.

Paige, J. (1975). *Agrarian Revolution*. Free Press.

Pilichowski, A. (2018). Grupy producenckie w rolnictwie: Refleksja socjologiczna [Producer groups and producer organisations: Sociological reflection]. *Wieś i Rolnictwo* [Village and agriculture] 1(178), pp. 97–122.

Scott, J.C. (1976). *The Moral Economy of the Peasant: Rebellion and subsistence in southeast Asia*. Yale University Press.

Scott, J.C. (1990). *Domination and the Arts of Resistance: Hidden transcripts*. Yale University Press.

Shanin, T. (1971) Peasantry as a political factor. In: T. Shanin (ed.), *Peasants and Peasant Societies*, pp. 238–263. Penguin Books, Harmondsworth.

Smelser, N.J. (1962). *Theory of Collective Behavior*. The Free Press.

Stinchcombe, A.L. (1961). Agricultural enterprise and rural class relations. *American Journal of Sociology* 67(2), pp. 165–176.

Tarrow, S. (1998). *Power in Movement: Social movements and contentious politics (2nd ed.)*. Cambridge University Press.

Wierzbieniec, W. (ed.) (2008). *Wielki strajk chłopski w 1937 roku: Uwarunkowania i konsekwencje* [The great peasant strike in 1937: Conditions and consequences]. Wydawnictwo Uniwersytetu Rzeszowskiego.

Wolf, E.R. (1973). *Peasant Wars of the Twentieth Century*. Harper and Row.

Internet sources

http://biblio.modr.mazowsze.pl/Biblioteka/Rozwoj/WSPARCIE _GPR.pdf (access: March 15, 2019).
https://www.raportrolny.pl/news/item/3313-naukowcy-wspieraj%C4%85-rolnictwo (access: March 16, 2019).

Think Locally, Act Globally: Polish farmers in the global era of sustainability and resilience, ed. by Krzysztof Gorlach and Zbigniew Drąg in collaboration with Anna Jastrzębiec-Witowska and David Ritter
Jagiellonian University Press, Kraków 2021, pp. 679–689
ISBN 978-83-233-4949-5
DOI: http://dxdoi.org/10.4467/K7195.199/20.20.12744

Conclusion: Some Final Remarks from the First Editor

Krzysztof Gorlach ⒾⒹ https://orcid.org/0000-0003-1578-7400/

The reflection of this final part of the monograph should not be seen as its summary or set of consistent conclusions. Neither is it an attempt to draw a portrait of Polish farmers in 2017 at the time, when the empirical survey was conducted, which served as groundwork for the entire project. This final part contains the set of remarks related to the surveyed population, as well as covers other issues connected foremost with methodology of conducted studies.

The monograph does not have a homogenous character. It is not the monograph of a single author, prepared by the head of the project. Quite to the contrary, it is an attempt to write in several different voices and present a contemporary, albeit still incomplete, depiction of Polish farmers. Such a character for the monograph was announced in the first chapter, which is a thorough presentation of a theoretically analytical concept that was the result of a group debate. Quite wide frames for the analysis of farmers' situation in Poland, presented in this concept, concentrated on global development matters and ideas of sustainability and resilience. These very issues are also widely discussed in the international subject literature and mark their presence in numerous studies conducted in various places of the world.

The processes and phenomena described in this monograph are quite complex and multidimensional. This goes well with having a diversity of approaches due to the individual and joint work of members of the research team, whether directly employed in the project or having loose ties to it and collaborating in it in a more flexible way. Thanks to their engagement, the analyses of selected topics, resulting in a rather discerning picture of the world of Polish farmers, elaborated from their own vantage points, can be found in this monograph. As this complex picture was created with the use of various research methods it could be metaphorically compared to a painting created simultaneously by various artists using various techniques of applying paints and having different methods for depicting the lights and shadows of sketched objects, as well as different color preferences.

Due to the characteristics of the research matter presented above, as well as the specifics of the study, the structure of this volume also has a diverse character. The first part contains reflections on the analytical, theoretical, and methodological characteristics of the matter. The second part presents quantitative analyses of Polish farmers based on the 2017 study, which include general characteristics of the studied population, as well as the diversification of farms according to farm area or sex/gender of the main farm user (owner). In this part of the book, there are numerous references to earlier works and publications, elaborated during the three study editions of the research program devoted to Polish farmers in the years 1994–2007. The third part of the monograph contains the analyses of characteristics of surveyed populations of farmers, residing in particular units (counties) with the use of advanced methods of statistical analysis. Finally, there is the last, fourth part with four detail studies, focused on selected issues, such as: 1) lifestyle, 2) role and importance of information technology in operating the farm, 3) problems of food safety and security, as well as 4) the political mobilization of farmers. The work of these parts of the monograph was conducted on the basis of selected results of an elementary survey of the population of Polish farmers in 2017, as well as carefully selected cases, which sometimes went beyond the scope of the 2017 study.

The main message of this monograph is expressed in the idea to "think locally, act globally," which is a play on words and a reversal of the well-known statement in the literature on globalization, for one to "think globally, act locally." This line and its various contexts are described in detail in the first chapter. What seems even more important from the point of view of the authors of this chapter are the attempts to identify among farmers these elements of the mindset that could serve as the ground for decisions contributing to the implementation of sustainability and resilience.

To make these measures effective one should start with the identification of the core of social reality behind the motif of the reversed approach to globalization. The traditional theme of "thinking globally, acting locally" presents a vision of the world that is widely shared, where humans think in a more or less similar way and, as a consequence, adapt shared convictions, values, ideas or principles to certain local conditions in such a way as to give some shape and sense to their actions. Reversing the above slogan according to the title of this volume and transforming it into the slogan of "think locally, act globally" causes the inversion of the entire scheme. Humans think differently but they share their own, specific convictions, evolving in their own familial, local, regional or national habitats of values, ideas, or principles and, while using them, reach for various means and resources, which are located outside of their habitat/environment in the global net. This is the globalization scheme from below or the bottom-up globalization type also known as agential globalization, which can be juxtaposed with the "down from above" type of globalization, or the concept of the "flat world."

Such an approach to the globalized world is close to the understanding of the sustainability concept presented in this volume. Its essence lies in the pluralism of action references and pluralism of relations, in which the acting agents engage. This not only means that actions taken should follow the principles of economic rationality but also that they need to consider the wider scheme of references such as conservation of national and cultural resources and elimination of some social problems, or at least alleviating them. The essence of actions taken by agents means entering the net of various social relations (see, e.g., Castells, 1996 [2000]) mostly outside of the local context, where the tissue of support is created. Diversity of this tissue fosters firmer rooting in social reality equally for the actions, as well as the social agents or actors who take them up. In this sense, it encourages their resistance to various crisis situations. It can be said that functioning within the framework of sustainable development principles, in and of itself, might partially be about building the resistance to the upcoming crisis. Furthermore, actions that have to be limited through entering various relations can translate into more judicious use of natural resources, preventing the elimination of various cultural resources, even if they don't bring economic benefits. Policies against excluding or marginalizing other humans and treating them not as subjects or sources of interests realization, but as partners, are also important. It might be stated that sustainability leads directly to resilience.

Various cases described in the monograph show a complicated and quite ambiguous path as to how ideas of "sustainability" and "global actions" appear in the consciousness and actions of Polish farmers. The predominant conviction among them is that mostly through affiliation to farm work and sufficient effort can one achieve success in the contemporary market. It should be emphasized that other types of thinking also appear here, with the particular role of the eco-farmer type that is quite popular today appearing within the context of the sustainable agriculture concept. In the schemes of thinking in both categories of surveyed farmers (big and small) the coexistence of modern thinking with peasant thinking can be noted. It turns out that farmers declare a generally low level of trust toward other people. They do not differ much from the general representation of Polish society, which could be a significant factor limiting implementation of the idea to "act globally" as well as that of "sustainability." However, this is not accompanied by any general reluctance to cooperate with other people. Cooperation is treated as one of the characteristics, and a mechanism, of sustainable development. In this sense, farmers who are operating larger farms appear to be more likely to implement the sustainable concept for conducting economic activity. Farmers operating smaller farms identify more with the strategy of farm management that might be described as "local harmony" and maintaining an orientation toward local production and, therefore, are close to the ideas of sustainable development. Furthermore, both categories of surveyed farmers do not practically differ in their declaration related to the statement that they use their own food products, because they are better than the ones available in stores.

Could women be more inclined to "global actions" and implementation of the sustainability idea? At first sight they seem to have better general education and civilization competences than men, and therefore are more predisposed to search for off-farm employment. However, it is the men who are more obliged to secure the financial functioning of the farm and they are the ones who must seek ways to do so and perhaps foster the idea of sustainable development. In recent years the rural areas observed a significant increase of interest in matters related to immediate social surrounding and this was particularly the case with female farmers. Their social activity increased threefold in recent years, almost reaching the same level of engagement as men. Female farmers were active in solving local problems mostly through fulfilling the functions in local associations, local social committees, and parish organizations. They were also more active than men in acquiring information on things happening in their village and municipalities from diverse sources, bother traditional and modern. Although, they preferred traditional forms of social contacts, they were also more likely than men to search for local news on the Internet and take up some action to benefit their local communities.

Farmers taking positions within the positively privileged class differently evaluated Poland's membership in the European Union. They were more frequently using various opportunities and financial instruments available within the framework of the Common Agricultural Policy. They expressed the conviction—more often than other farmers—that Poland's entering the European Union benefited all farmers and all farms. Finally, these farmers were decidedly more inclined to invest in the operated farm, which was the example of "global actions" and cooperation.

Wider local and regional characteristics, related to particular counties, also determined the functioning of farms as well as the families operating them. The use of the advanced method of multilevel analysis is aimed at presenting the multitude of existing conditions, as well as diversity of strategies, adopted by farmers to manage their farms as well as the lives of their families. Together, they draw a diverse, multihued picture of Polish agriculture at the time of globalization.

Certain tensions between the attempts to implement the ideal of "sustainability" and the reality of the contemporary market economy could be seen in the case of surveyed dairy farmers in central Poland. The model of farming that young farmers there would want to continue is a consciously articulated model of a sustainable economy, in which various aspects of life are not subjected to the pursuit of profit but also include possibilities for enjoying family life and free time. This case shows tensions related to taking up "global action" in the social realm. This presents social customs produced by young farmers whose natural habitat does not consist of their closest neighbors but well-educated farmer-friends located in various parts of the province and the country, agricultural advisers or the experts of various disciplines and professions that the respondents knew from their university studies.

In surveyed farmers' collective consciousness, which could have a strong influence on making production decisions, the general conviction regarding the high quality of currently produced foods as one of the aspects of sustainability constitutes a rather important aspect of sustainability concept. Nevertheless, this is reflected in the fear shared by most farmers that some producers might be applying forbidden practices as they are facing sharp competition on the market of agricultural products. As indicated by the respondents, keeping high quality standards for food produced in Poland would first require increasing the efficiency of the institutions that control production quality. This could be done through their greater activation in the field or through intensifying their contacts with direct food producers. In other words, implementation of this aspect of sustainability means that farmers should take "global actions" such as cooperation with "global" institutions.

It should be stressed that "global actions" in the contemporary world also apply to cyberspace. The sample population of farmers presented in this monograph is said to have a generally "innovative" personality, as well as appropriate experience and knowledge. This applied for the most part to the representatives of the young generation but not all of them and only those who received support from their parents and grandparents. The role of other farmers who are in the digitalization process of farm management could play a significant role as a special type of adviser was also noted in this context. To create such a network of cooperation and to facilitate decision-making for young farm users, financial instruments offered by the state, including those from the European Union, are necessary.

To finalize the remarks on the idea of "sustainability," it should be added that the concept of cooperation has been growing particularly important in the consciousness of farmers in the recent time. Protest actions are no longer the key forms of fighting for farmers' interests, although they still received moderate support, especially among farmers with a lower level of education who also operated large farms. Farmers organizing in producer groups and cooperatives have better opportunities to express their interests.

Certain elements of summaries and conclusions presented above, which come from selected chapters of the monograph, create a patchwork picture of the contemporary population of Polish farmers. It only presents selected characteristics and the statements formulated to support the overall depiction represent quite a diverse range and, as a consequence, hold various explanatory powers in regards to particular matters and research problems. The monograph contains analyses made on the basis of the studied representative sample for the general population of farm owners and farm users in Poland elaborated on the basis of particular case studies, deliberately selected by the authors of particular chapters. These analyses also used the results of the studies conducted in the previous years (1994, 1999, and 2007) and presented in earlier publications, which were summarized in the third chapter of this monograph, as well as the results of other studies referred to by authors of particular chapters. One issue is particularly important in this context: all of these analyses apply to only one society, which is Polish society and, to be more precise, they apply to only one social category, namely Polish farmers.

This matter appears particularly important, as in the contemporary literature on the subject there are opinions that studies limited to only one case or one nation-state society are prime examples of disregard for the most crucial feature of the contemporary world, namely globalization processes and the emergence of the new global—or, perhaps we should say globalized—society. Such a viewpoint can be found in the works of German sociologists Ulrich Beck and Edgar Grande (2007), a short description of which is presented below.

In their work devoted to the quite difficult formation of cosmopolitan Europe, the authors devote a quite significant part of their attention to the issue of appropriate research of these processes. They present the thesis that to understand the functioning of societies in Europe requires abandoning the perspective of what they call "methodological nationalism." In their opinion, this concept leads to limiting research studies to the level of a particular country or state and, sanctioning such a strategy as obvious, which is not adequate to the type of reality that the authors describe as European cosmopolitan society. In their view, such reality requires a new approach to studies, in which neither national case studies, nor even international comparative analyses are sufficient. First of all—as indicated by Beck and Grande—efforts should be made to sketch a horizontal Europeanization. The authors also emphasize the need to develop a so-called post-social theory of social Europeanization.

In order to at least initiate the implementation of this ambitious program it might be necessary to start with changing the definition of the society. According to the quoted authors, understanding of the term "society" is still strongly connected with the nation-state, which forces the methodological nationalism presented above. Some attempts to abandon the focus on national societies in sociology (social sciences) include the proposals of comparative studies and focusing on the convergence processes of particular societies, as well as analyses of common social elements and analysis of historical similarities of particular European societies.

It should be stressed that Beck and Grande did not examine this problem within the framework of globalization analysis, but in a very specific reference system, that of the process of Europeanization. This process has two dimensions: vertical and horizontal. In the first dimension, the vertical, attention is given to relations between particular societies and European institutions. The process of Europeanization includes adjusting various national norms and policies to directives and regulations adopted by European institutions, but also establishing various supranational institutions under assorted pressures from national societies. The horizontal dimension also encompasses two processes. On one hand, this could be seen as the creation of "post-social" spaces of Europe, where changing relations with external surroundings occur and, on the other, as producing new relations between particular European societies. In other words, these are the processes of networking and mixing of these societies and, most of all, their particular elements, such as: economy, educational systems, family, etc. This horizontal dimension seems to be more appreciated by the authors, as these processes create a new type of reality, which can be described as transnational reality.

The most characteristic manifestations of this new reality, according to Beck and Grande, are: language, identity, education, and economy. The language aspect of the new reality involves a multitude of coexisting languages, as well as the privileged position of some of them, especially English. The authors suggest that cosmopolitan Europe should be built on the basis of what they see as a well-designed multilingual model. The identity aspect, enlisted as the second manifestation of the new reality of Europe, could not simply mean the replacement of already existing national identities with one European identity. The new identity should develop but it should be treated as a continuous construction, a permanent creation open to diverse values and the world. However, such openness should not be equated with too much latitude but with a set of values based on the foundation of individual freedom and include procedures that would preclude forcing upon others any principles that could harm them. The third aspect for developing a cosmopolitan Europe relates to educational content freed from the narrow focus of concentration on national histories. Contrary to that trend, in the new reality, Europe should appreciate education referring to its common ancient, Christian, and early modern foundations, its political history of revolution and what the consequences are of the creation of that value system based on human and civil rights, democracy, peace, and tolerance. It should also present the political sources of nationalism, militarism, totalitarian violence, terror, and war as anti-values. Europeanization in education can also be observed in the process of educational mobility, as illustrated by the example of the Erasmus program or international research projects financed by the European Science Foundation. The fourth aspect of Europeanization, involving the economy, focuses on migration related changes in the labor market and the trans-nationalization of business enterprises and corporations. An additional aspect of Europeanization means viewing society through the prism of social processes in their dynamics, as well as acquiring the perspective, in which the concept of cosmopolitan Europe might be treated as a certain manifestation of the critique of its own history and taking the opposite stand towards various negative phenomena, so frequent in the history of the old Continent.

The process of Europeanization understood in such a way requires appropriate analytical and theoretical tools. Beck and Grande propose a theoretical cosmopolitanism and a theoretical perspective of experimental dialogue concentrated on global changeability, interdependence, and intercommunication. For the analytical perspective, this could mean the rejection of treating national society as the essential and ultimate research area and conducting analyses that would consider various approaches, which together could ensure a multilateral view of the given studied reality. Such study should be focused on analyses of changes and include descriptions and explanations of mutual relations and influences between particular elements of reality.

How in this context, then, should one treat the study on Polish farmers in this monograph? Without a doubt, this study is only limited to one national society and therefore, according to the concept of Beck and Grande—could be recognized as an example of methodological nationalism. Nevertheless, it also has some characteristics, which potentially—to use the wording of Beck and Grande—give it a bit of a cosmopolitan project profile. First of all, the primary frame related to various aspects of the situation of Polish farmers is the perspective described as agential globalization. The message presented above, stemming from analyses in this very volume, apply to various aspects of this problem, additionally connected with the idea of sustainable development and resilience. In that sense, while this is still an analysis of the situation of Polish farmers, it is conducted in the context of cosmopolitan global processes.

Analysis of opinions, attitudes, and actions of Polish farmers in the context of global processes comprises only one dimension of the problem that Beck and Grande address. However, this is a quite imperfect way of ensuring a transnational approach to the character of studied issues. What could be seen as the second dimension of translational analysis is ensuring the application of the theoretical perspective of experimental dialogue, as mentioned by Beck and Grande, which could be a rather difficult task. To follow such a suggestion in this very research project, it was decided to add the works of two external theorizing evaluators to the monograph. These authors, Jan Douwe van der Ploeg and Patrick Mooney, who are outstanding representatives of sociology in international sociological circles, prepared a certain context for the dialogue, referring to analyses presented in particular chapters of this monograph based on their own research experience. Including their writings in this monograph ensures, to at least a certain extent, the transnational character of scientific discourse.

These efforts might be continued in a more advanced manner with the use of existing research methods. Their goal is to overcome the limitations of simple international comparative studies which, according to Beck and Grande, do not truly result in abandoning methodological nationalism. Two such methods should be presented here. They are: qualitative comparative analysis (QCA), mostly formulated by Charles Ragin, and extensive case method (ECM) by Michael Burawoy. Their short presentation below is not a detailed elaboration or evaluation but only a recommendation as to which direction the further analyses of materials, presented in this monograph, should go in order to meet the expectations of Beck and Grande. Full use of the research potential of both methods would require including additional data and materials or even entire works and elaborations on farmers' situation in other countries. In that sense, this would go much beyond the goals of the project, the selected results of which are presented in this monograph.

The crucial message that emerges from the general proposal of Beck and Grande, to abandon methodological nationalism, appears to proceed in two directions. On one hand, research studies should be the ground for formulating conclusions much broader than the scope of nation-state societies and, on the other, various research perspectives should be applied at the same time in the study, according to the ideas of empirical and dialogical methodology. It appears that both of the above proposals have potential to meet these expectations. Burawoy's concept is an attempt to combine the consideration of the concrete case with a reference to the more general theoretical reflection. This means going beyond one particular case, such as the nation-state society, and assuming the approach in which a confrontation (or, to put it differently, dialogue) occurs between empirical analysis and theoretical reflection. A different emphasis can be found in the concept of Charles Ragin, combining quantitative and qualitative approaches and applying them to a particular case, which might be developed into a simultaneous analysis of many cases from the perspective of configuration of factors characterizing them.

The presentation of the research methods should start with the concept of Ragin as it has the earliest origins of the three, fully taking its primary form in the second half of the 1980s (see Ragin, 1987). The subject of the author's consideration mostly involves matters of comparative studies, whose specificity indicates that of sociology. According to this concept, comparisons serve foremost for analyses of diversity and the complex character of causality of social processes. The essence of the methodological approach to analysis combines synthetic approaches based on case-oriented (qualitative), as well as variable-oriented (quantitative) perspectives. In the entire process, there are constant references between ideas and evidence, or to put it slightly differently—between general terms or statements and empirical data. The Ragin concept continuously goes through the process of refinement and further defining. As a result, it has been developed into a comparative method with very precise principles and concentrated on the analysis of various factors bringing about various phenomena and social processes. It goes far beyond the framework of particular nation-state societies, focused foremost—as suggested by Beck and Grande—on their similarities and differences (see Rihoux & Ragin, 2008).

Burawoy's concept is also undergoing various modifications. Its initial version (see Burawoy, 1998) combined participant observation and ethnographic description of particular elements manifested locally in everyday life with consideration given to its supralocal and historical context. According to the author, this is an example of reflexive science, which is the opposite of so-called positive science, well exemplified by survey research. Reflexive science has its limitations and they are caused by so-called "power effects" such as domination, silencing, and objectification, as well as normalization. Positive science is limited by so-called "context effects," and Burawoy names the following of them: interview, respondent, field, and situational effects. The essence of reflexive science can be seen as the intersubjectivity of scientist and subject of study, but for positive science the main principle is the separation between scientists and the subjects they examine. Other characteristics of conducting science according to the principles of reflexive science involve attention given to the intervention, process, structuration, as well as theory reconstruction, while positive science emphasizes the importance of reliability, replicability, and representativeness. It turns out that the extended case method does not have to be applied to the analysis of only one case. In the collection of essays published twenty years later (see Burawoy, 2009) this author analyzes four various historical processes. They are: changes in post-colonial Zambia, the functioning of advanced capitalism in the United States, the functioning of state socialism in Hungary in the 1980s, and then the collapse of the Soviet Union and the destinies of the working class in the collapsing Soviet economy of the 1990s. Interestingly, Burawoy applies the same theoretical and methodological concept, based on the principles of the extended case method, to all cases.

The three strategies presented above, namely, the external theorizing evaluators, Ragin's strategy, and Burawoy's strategy, provide a chance to move beyond methodological nationalism. They also provide an opportunity for more than one person to make statements on certain subjects, presenting the products of the simultaneous application of qualitative and quantitative approaches, as well as to emphasize the importance of continuous confrontations of general, theoretical reflection with the results presented in the descriptions of particular cases. One of these strategies was partially utilized in this project and the two remaining strategies may be considered for future use. The starting point for such endeavors might be the data availability for public use, prepared as one of the results of the project presented in this monograph.

REFERENCES

Beck, U. & Grande, E. (2007). *The Cosmopolitan Europe*. Polity Press.

Burawoy, M. (1998). The extended case method. *Sociological Theory* 16(1), pp. 4–32.

Burawoy, M. (2009). *The Extended Case Method: Four countries, four decades, four great transformations, and one theoretical tradition*. University of California Press.

Castells, M. (1996) [2000]. *The Rise of the Network Society*. Blackwell Publishers Ltd.

Ragin, C.C. (1987) [2014]. *The Comparative Method: Moving beyond qualitative and quantitative strategies*. University of California Press.

Rihout, B. & Ragin, C.C. (eds.) (2008). *Configurational Comparative Methods: Qualitative comparative analysis (QCA) and related techniques*. SAGE.

Think Locally, Act Globally: Polish farmers in the global era of sustainability and resilience, ed. by Krzysztof Gorlach and Zbigniew Drąg in collaboration with Anna Jastrzębiec-Witowska and David Ritter
Jagiellonian University Press, Kraków 2021, pp. 691–703
ISBN 978-83-233-4949-5
DOI: http://dxdoi.org/10.4467/K7195.199/20.20.12745

Afterword: Renewing a Sociology of Agriculture

Patrick H. Mooney https://orcid.org/0000-0003-4716-4590/

Based on interviews with over 3500 Polish farmers, selected with regionally representative sampling, with nearly 500 variables available for the analysis, this is the single most sweeping sociological documentation of a modern nation's agriculture with which I am familiar. The preceding reflects the Jagiellonian University's rural sociological scholarship in an ever-changing Poland: reaching back to the days of state socialism, through "Solidarność," the so-called "shock therapy," EU integration and the reaction to that integration. The team assembled for the project reflects several cohorts of scholarly witness to Poland's transformation. If my comments here too heavily reflect a tendency toward the view from North America, rather than Europe, or the planet as a whole, I beg forgiveness. I will make an effort to think and write both globally and locally.

This impressive and wide-ranging sociological portrait of the current state of Polish agriculture is a portrait well-grounded in historical perspective. Yet it is surely not definitive. I expect and hope that this work will generate both provocative investigations of: 1) more detailed analyses of these findings, perhaps with more qualitative analyses, as suggested by Mielczarek's Chapter 12 and Dąbrowski, Kotkiewicz, and Nowak's Chapter 14; and 2) comparative sociological analyses of other nations' agricultural systems, especially in the EU. In that sense, though it is an extensive work to be sure, this exercise in "bringing farmers back in," as argued in the introductory chapter, may represent a resurgent sociology of agriculture; a new beginning, rather than the final word.

Definitions matter. Much effort has been devoted over the years to defining family farming. Academics, in their need to operationalize terms at the outset of their study, usually find a way to accommodate a definition that is more or less useful to their particular interests. I would suggest, however, that this is not simply an academic exercise. The term offers a consensus framing that seems to resonate with a wide range of societies and cultures. The current identification of The Decade of Family Farming (2019–2028) by the United Nations reflects the potency of this consensus. As such, quite a diverse set of actors have an interest in "owning" the concept, that is, of being identified with family farming. Thus, we can observe the perhaps ironic but not unusual contestation of a consensus frame (Mooney & Hunt, 2009). Agribusiness firms continuously advocate the next mechanical, chemical, or biological technology that will help the family farm, even as that next step on the technological treadmill will surely lead to the demise of many family farms. Correspondingly, policy makers often invoke the defense of family farming even as the very policies they advocate will likely drive many such families out of farming. Though differentiating by scale (gross income) of family farms, the United States Department of Agriculture (USDA), for example, offers a definition that finds 98% of U.S. farms to be family farms (USDA-ERS, 2019). Lacking much discriminatory power, such a definition is nearly meaningless (except to note that the remaining 2% account for 12% of agricultural production). However, finding 98% to be family farms is politically useful, ensuring the public that, despite all the moaning and groaning, the family farm is not really under threat.

The United Nations' declaration of the Decade of Family Farming conveys a global interest in the importance of family farming. A definition was in order. The English language web site (FAO, 2020) offers first, a functional definition (in the sense of what family farms "do"):

> Family farming offers a unique opportunity to ensure food security, improve livelihoods, better manage natural resources, protect the environment and achieve sustainable development, particularly in rural areas.

The site continues with the following definition of what family farming "is":

> Family farming includes all family-based agricultural activities. It is an integral part of rural development. Family farming is agricultural, forestry, fisheries, pastoral and aquaculture production managed and operated by a family and is predominantly reliant on family labour, including both women's and men's.

Who could possibly oppose such an institution?

Akin to the USDA position noted above, another recent publication (Graeub et al., 2016, p. 1) examined and compared multiple definitions of family farming but came to rest on a definition that also found 98% of world's agriculture to be family farms, though producing only approximately 53% of the world's food. Not surprisingly, Graeub et al. found, as did Mooney and Tanaka (2015) that different definitions lead to different conclusions about the nature and condition of the family farm. Graeub et al. compare their definition with UN FAOs (2014) State of Food and Agriculture (SOFA) report which found 90% of farms to be family farms but producing at least 80% of the world's food production (Graeub et al., 2016, p. 2). As with USDA, this definition of family farming found a need to differentiate within the category. While USDA (2019) makes 3 distinctions on the basis of scale (reduced to gross sales: small, midsize, large), Graeub et al. (2016) follow a more nuanced distinction: Group A is "well-integrated and well-endowed;" Group B "has significant assets and favorable conditions but lack critical elements (like sufficient credit or effective collective action;" Group C tend to be outside the market and lack "social safety nets."

Another strategy has been to reduce the definition to one of scale, that is, to consider the "agriculture of the middle" (Stevenson et al., 2014). While this approach has offered some useful insights, policy proposals, and opportunities for sociological contributions, it is a definition grounded in spatial or monetary gradations rather than social relationships, hence a bit less satisfying from a sociological perspective. Nevertheless, the approach does highlight the practical and ethical significance related to several problems brought about by the "disappearing middle."

But, honestly, can we *expect* to find some single, universally agreed upon definition? The concept of *family* itself is hardly conducive to standardization. Likewise, *farming* is always evolving, adapting and so varied across the planet's diverse environments and agri-cultures as to present never-ending obstacles to one definition. Indeed, 25 years from now, technological "advances" may mean that a good deal of food production may not look anything like what we think about today as farming. Indeed, some of it may literally be "factory farming," cloning biomass into food. Thus, even as most seem to agree on the beneficial importance of family farming, the definition continues to resist consensus. I doubt that we can come to some consensus between the interests of sociologists, politicians, economists, publics, etc. on a single useful definition. Pushing this line of thought, perhaps we *should not* seek some single definition.

In the end, of course, despite the best efforts of capital and agricultural engineering, no two farms are identical anyway. That is a good thing. Diversity is fundamental to sustainability. In the spirit of Scott's *Seeing like a State* (1998) or Busch's

Standards (2011), we might consider that a consensus definition may be prelude to standardization, regulation and rationalization; processes that are not likely to be friendly to family farmers or the environment. Perhaps our attempts to define the family farm, should consider these definitions as a multiplicity of ideal types constructed with respect to the particular interests of the investigator. Taking that a step further, perhaps the present work gives us a new opportunity to develop an understanding of "family farm" that is constructed inductively, rather than deductively. The thorough-going nature of this study reveals the complexity and diversity of family farms. This extensive work may give us an opportunity to work toward an understanding (perhaps definition is too strong a term) that is built from the ground up, complex though that may be. Weber's position regarding the matter of ideal types and definition might come in handy here:

> Such an historical concept, however, since it refers in its content to a phenomenon significant for its unique individuality, cannot be defined according to the formula, genus proximum, differentia specifica, but it must be gradually put together out of the individual parts which are taken from historical reality to make it up. Thus, the final and definitive concept cannot stand at the beginning of the investigation, but must come at the end (Weber, 1996, p. 47).

The sociology of agriculture: Renewed. The introductory chapter to this volume notes an interest in bringing agriculture back from the margins of rural sociology. I will consider how this work is especially important for a rejuvenation of a *sociology of agriculture*. Though most good sociology must be attentive to substantive work in multiple disciplines, a sociology of agriculture should not lose sight of what sets sociology apart. I would argue that, whatever else sociology may be, it seems that most can agree that it is grounded in social relationships or interactions between persons or groups of persons. In that sense, it is always relational. After that, of course, it gets rather complicated. I will focus on that basic claim.

First, how did we get here? In the U.S. and somewhat more broadly, English-language rural sociology, the discipline had only begun to re-discover agriculture in the 1970s. What had been done in the post-World War II era on agriculture from a sociological point of view seems to have been little more than marketing research for agribusiness, masked as the diffusion of innovations. In the U.S, the rediscovery of agriculture in rural sociology came with a critical eye that reflected new understandings of peasantries, especially peasant mobilizations, such as the work by Paige and Wolf, as well as movements in civil society of the more developed economies, subsequent to the 1960s protest cycle. Regarding the latter, writers such as Wendell

Berry and Jim Hightower offered critiques of agriculture and, often, its unholy alliance with the land grant college complex that reflected concerns associated with large scale "industrial farming" and the subsequent deterioration of the quality of farm work, the environment, and food itself. The framing of smaller, family farms as more environmentally and consumer friendly, shared an affinity with a rural population turnaround and with the interests of a baby boom generation struggling to get access to enough land to farm, given the ever-increasing scale of production, rising land prices as well as interest rates. All of this came to fruition in the disastrous farm crisis of the early 1980s, when the emergent academic work by scholars in the U.S. and Britain, such as, Heffernan, Buttel, Newby, Friedland and others began to get more and more attention. Of course, the situation in Europe was somewhat different and in Poland it was quite different, although Galeski's *Basic Concepts of Rural Sociology* (1976) offered some of the flavor of a (less than welcome) Polish rural sociology at that time. Subsequently, Gorlach was among the leading figures of the era to offer a sense of what the sociology of agriculture looked like in Poland. Some of his valuable analyses of that time are presented in the preceding work.

Guided by attention to Marxian-inspired interest in the dialectical tension between the social relations of production and the forces of production, there were hopes, especially along the lines of commodity analysis that sociologists could build an understanding of agri-food systems (see, e.g., Buttel & Newby, 1980; Friedland, 1984). However, like a Tower of Babel, that project got lost in the specification of each "box": research and development; class structure and inequalities; genetic engineering; farm labor; the role of the state; vertical integration; mobilizations; environmental issues; identities; family life; etc.; or in each commodity—the structural organization of each commodity being so complex that it could lose the capacity to compare; apples could not be compared with oranges, as they say. Substantive concerns with agriculture began to be addressed from a multiplicity of theoretical perspectives and methodologies. The sociology of agriculture began to lose a coherent and systemic focus. In Mills' terms, we might say the grander theory moved toward an abstracted empiricism. Meanwhile, the "cultural turn" in sociology facilitated a further shift to focus on relations of consumption (food) rather than relations of production (agriculture). Though we often talked about the agri-food system, food and other concerns gained prominence as agriculture receded. This is the condition which Chapter 1 above identifies as having marginalized not only farmers but agriculture itself, calling for rural sociology's return to a concern with agriculture.

At the outset, that "new" sociology of agriculture was largely a political economy of agriculture, with social relations largely restricted to the sphere of economic relations and environmental impacts (see Buttel, Larson, & Gillespie, 1990). However, that initial focus on the *social relations of production* as an entry point was, in retrospect, quite useful to a general sociological participation in a conversation long dominated by economists at both the academic and the policy levels. Those analyses often externalized many of the variables of greatest concern to sociologists, such as family relations, the gendered nature of agricultural work and land ownership, the fabric of rural community, and the environmental and health costs of economic "efficiencies." A sociology of agriculture needs to recognize that agriculture is embedded in a much larger sphere of society and culture. Lyson's (2004) notion of *civic agriculture* may have made some headway along these lines, seemingly directly and indirectly influencing a considerable body of work as well as collective action in North America and Europe. Much of the preceding analysis addresses broader implications of agriculture's embeddedness within communities and regions, as well as families' embeddedness in farming with its' particular attention to age, educational, and gender distinctions.

Attention to the "social relations of production" can open the door to multiple sociological approaches, some of which have been relatively neglected in the sociology of agriculture. I would like to explore further some other sociological approaches in relationship to the preceding chapters. I appreciate the authors' following Abbott's advice to flip or reverse common expressions in order to discover novel insights, in this case, the reversal of "thinking globally and acting locally" into a model that focuses on "thinking locally and acting globally." This meme works well as a collective action frame. However, I might suggest that, for some sociologists, such a distinction between thinking and acting constitutes a misunderstanding of social action or perhaps as with Marx, an aspect of alienated existence. In Mead's "theory of the act," the thinking which we might associate with impulses and perceptions, are dialectically intertwined with the action, which we might associate with manipulation and the consummation of the act. For Mead, "an act should be seen as an organic whole, something which unfolds in a number of phases; these can be separated through analysis but cannot be understood except by reference to the whole act" (Silva, 2007, p. 65). These elements are only separable in our analysis, in practice they constitute a unified whole. As Mead put it:

> A perception has in it, therefore, all the elements of an act—the stimulation, the response represented by the attitude, and the ultimate experience which follows upon the reaction, represented by the imagery arising out of past reactions (Mead, 1938, p. 3).

However, this flipping of "thinking global / acting local" to "thinking local / acting global" can reveal the sociologically false distinction and points to the dialectical quality of thought and action as well as the role of space and place in that action.

Thus, we can say that local action has global implications and global action has local implications. Though it is possible to imagine actors who only think locally and act locally (e.g., Marx's misrepresentation of a provincial peasantry) or actors who only think globally and act globally (e.g., heads of the largest transnational corporations), such action seems more indicative of habitus on both counts. The reflexive actor brings the spatial dialectic between local action and global action to consciousness, similar to C. Wright Mill's famous insistence that the sociological imagination facilitates the transformation of personal troubles into public issues.

The findings in the preceding work, especially Drąg's Chapter 10, suggest that region functions as a mediation between local and global. His analysis demonstrates that farmers operate with "region" mediating their thought and action (perhaps consciously and strategically; perhaps less consciously, at the level of habitus). Drąg shows this with respect to regions within Poland. However, much of the above analyses point to the EU as another significant intermediating region between Poland and the global economy. Fligstein and McAdam (2012) develop the notion of a nested spatial hierarchy of "strategic action fields" that might be useful for further pursuit of this matter. That work, a revision of field theory, leads me to a consideration of the above text's welcome consideration of Pierre Bourdieu.

In the foregoing chapters, the authors have intermittently utilized the French sociologist, Pierre Bourdieu. Mielczarek's Chapter 12 takes on this task most explicitly. I found this endeavor to be a productive and provocative pursuit. I would like to take this opportunity to explore Bourdieu a bit further. Rural sociologists have often examined the capital metaphor, but primarily focused on social capital and seldom in association with field and habitus, as Bourdieu used the concepts. For whatever reason, agricultural sociology also, at least in the English language presentations, has a rather glaring absence of incorporating Bourdieu. Social capital is, practically speaking, ineffective in absence from other forms of capital (economic, cultural, political, etc.) with which it may be exchanged as needed. Further, social capital is also only meaningfully deployed in some field. Those fields themselves contain variable opportunities for thinking/acting which are, in turn, dependent on the level and forms of capital most effective in a specific field at an opportune moment. The notion of habitus suggests that some of what takes place on farms, in agricultural communities, etc. is not so much "thought" as enacted without much reflection. It is a "feel for the game" derived from repetitive practice. Some characteristics attributed to region in the above analysis may reflect habitus.

The present analysis raises serious concerns about the nature and "amount" of social capital in rural Poland. The data presented suggest that rural Poland represents a sort of "social capital desert." Most strikingly this appears in membership in organizations, such as political parties or farmer organizations, implying an absence of influence in economic and political fields beyond the value of market production per se. Gorlach, Drąg, and Foryś in Chapter 6 document low levels of trust among Poles. Foryś in Chapter 15 recognizes some of the difficulties for developing organizational networks created by the lack of trust. Even weak ties cannot offer much strength in the absence of trust. Much has been written about the determination of state socialist society to hamper intermediary organizations and eviscerate civil society in a jealous protection of Party dominance. My own experience, though as very much an outsider, is that rural Poland's civil society is not quite as bleak as the present portrait suggests. Nevertheless, the absence of tertiary organizations, the increased level of commuting to workplaces, the low level of party membership, the absence of farm organizations, much less cooperatives, are indicative. The seeming decline of the Church, as well, suggests an ever greater vacuum that deserves further research consideration.

Several chapters recognize that the fields in which various capitals might be deployed have clearly been shaken repeatedly. The events of 1980s, of course, were an unsettling of multiple fields of world historical significance, with Poland playing such an important role in the demise of state socialism in Eastern and Central Europe. In the broader literature, Polish agriculture often appears as a relatively neglected participant in that revolution, with attention focused on the rather dramatic confrontations at factory gates or urban streets of protest and confrontation. However, those strikes were surely enabled by the relative independence of Polish farmers from the control of the food system held by state authorities in most of the region. The Balcerowicz Plan, sometimes known as "shock therapy" of the 90s, and then the integration into the EU have continuously "unsettled the field" (see Fligstein & McAdam, 2012) creating both threats and opportunities for Polish farmers. However, those opportunities would seem to be diminished by the above noted low levels of social capital, trust, and "indigenous organizational" networks that might serve as mobilizing structures. I remain convinced that an important fork in the road for Polish agriculture is the development of marketing and input service cooperatives that constitute such an important aspect of most developed agricultural economies. The low level of social capital in civil society can be said to constitute both an obstacle as well as an opportunity to fill that vacuum and become a significant player representing the interests of Polish farmers. At the moment, based on the above text, the obstacles seem largely ideological. Perhaps we might read this as an example of Mielczarek's positing of a scenario characterized by a hysteresis in the disruption between an altered field and a no longer fully functional habitus.

It strikes me that the notion of habitus might also alert us to caution with respect to using the terms "profit" or "profitability" in the context of "family farms." To the extent that the family's labor and capital are unified, as in most understandings of family farming, it is difficult to sort out the extent to which financial rewards are due to capital or to labor, short of imposing categories and assumptions from other fields. The labor might be altogether different labor, in terms of quality and meaning, when capital is held by the family and when it is held by an external agent. One strategy might be to assume the average rate of return to capital, or what that return might be if held by someone outside the farm but that denies the *de facto* unity of the habitus. I would hypothesize that a good many "family farmers" simply do not engage in a calculation of distinct returns to labor and to capital investments, unless coerced by external entities, such as creditors, tax forms or, perhaps, social scientists. We may need to remember that often, the conceptual schemes are ours, the consciousness is theirs.

I surely concur with the hesitation of the authors' to buy into the "death of classes" argument. I maintain that class represents a particularly significant form of social relationships in agricultural production and that many other concerns are not unrelated to the dynamics of the class structure. Before I close, I would like to draw attention to two particularly interesting aspects of the present analysis: 1) the commodification of farm labor; and 2) the increased willingness to use financial instruments such as credit and the use of EU subsidies to expand the scale of production.

One finding of interest involves the apparent reframing of the meaning of hired labor in rural communities. This analysis examined the traditional Polish attitude of resistance toward family members' employment on another farm (beyond the informal exchange of labor). This is discussed here as a matter of peasant honor. However, the trend seems to be that this matter of distinction is losing strength and gaining acceptability.

This analysis has given a great deal of attention to the importance of technology in the restructuring of Polish agriculture. They have offered us detailed examination of the technological impacts from increased mechanization and even digitalization on scale. Increased dependency on credit and the use of EU subsidies seem to provide the primary means of financing the acquisition of new technologies. Interestingly, this analysis found respondents making decisions to use EU subsidies to invest either in the expansion of the farm or in the education of their children (perhaps *en route* to exiting agriculture).

It must also be noted that, historically, such subsidies, while appearing to be to the benefit of individual farmers, has also very much benefited the agribusiness manufacturers of such technologies. Two aspects of this seem significant. First, the finding that these EU payments seem to be finding their way back to western Europe in the documentation of Polish farmers distaste for the quality of Polish products. Second, Polish farmers' claims that EU subsidies are increasing the cost of both equipment as well as land. This raises question as to whether the subsidy is of Polish farmers as a whole or agribusiness manufacturers. Further investigation into the function of these technologies for sustaining or subverting family farms is warranted. Which technologies might enhance sustainable family farm production and which might subvert it?

There is much in the preceding analysis to suggest that Poland is moving toward an increasingly bifurcated agriculture. This should not be surprising as it seems to be something of a characteristic of the development of capitalism in global agriculture. I don't mean to suggest that this bifurcation is simply an increasingly bimodal distribution of farms with respect to some variable measuring scale. These can begin to form distinct fields of agricultural production and food consumption. I am suggesting that this bifurcation often tends to reflect two different agri-cultures, characterized by different technologies, organization of labor, rationalities, even possibly different markets. That these may not even be competitive with one another but complementary is reflected in the expression of consumption as "food from somewhere" and "food from nowhere." When this occurs in conjunction with increasing societal inequality driving divergent consumer markets, we may also talk about "food for someone" and "food for no one, or at least no one in particular." This seems to be another important issue raised by this study to be pursued in additional research.

Breaching trajectories. Much contemporary sociology of agriculture, especially that which tends to see itself as "critical" or "progressive" has proceeded as if two particular trajectories in the global agri-food system were nearly inevitable though mostly unacceptable. These trajectories, often related, have been neo-liberalism and ever more complex, tightly coupled supply chains for a global food regime. However, at the risk of dating this essay, I note that I am writing this afterword in the midst of two, possibly quite significant breaches in those trajectories. First, and perhaps more temporary, I am writing during the lockdown due to the coronavirus. This pandemic has revealed multiple fragilities of the U.S. agri-food system. I suspect some of these weaknesses are becoming apparent elsewhere. To be sure, this shock should not take all by surprise. Charles Perrow (2007) argued as much about the vulnerability of the food system years ago.

The pandemic here has made it clear that the willingness to externalize the costs of concentration and ever larger scales have revealed *de facto* vulnerabilities, especially in the meat, milk and fresh fruit/vegetable sectors. Hendrickson (2020) writes that the pandemic illuminates the lack of resilience to the systemic shock, and more importantly, "exposes the faults of concentrated power in food systems in coping with other slower-moving ecological disasters." I would suggest further that the situation may also render transparent the structural power of laborers in these fields. Their inability to work may reveal that power disrupt the supply chain both back to the farm and forward to the grocery. That power is often masked, at least in the U.S., by the marginalized ethnic and legal status of such workers. I do not know how this is playing out in Europe, but in the U.S. this may have considerable implications with respect to altering the trajectories of scaling up, particularly at the processing level, with implications for the farm level. Again, Hendrickson (2020) treats this as an opportunity to push for a transition to a more decentralized, diverse and secure food system.

The other breach has to do with the emergence of a "new" right wing in several nations. For some time, most critics have assumed the primary conservative impediment to progressive causes was neo-liberalism. The most significant aspect of this, perhaps ironically, is that these conservative movements have tended toward disrupting the neoliberal trajectory that were often assumed to be inevitable. That this is a phenomena in multiple nations, manifest in, for example, Trump's isolationism, Brexit, and in the present case, Poland's Law and Justice Party (PiS) tendency toward nationalism and Euro-skepticism. Whether this is a brief interlude or indicative of a substantial backpedaling is an important question.

Conclusion. The preceding text adopts an approach to defining the subject of family farming that reflects prior work by van der Ploeg, Gasson and Errington, and myself. My focus long ago (Mooney, 1988) examined the social relations of production, particularly those relationships in which capital was able to appropriate value and control of family farm enterprises. I have argued here and in Mooney and Tanaka (2015) that a focus on the social relations of production, though surely understood more broadly, can provide considerable room for a renewed sociology of agriculture. These other definitions complement that extension. Van der Ploeg's *Foreword* to this volume (Ploeg, 2021) identifies 10 variables related to family farming. Most of these refer to various aspects of social relations of production, though *expanding outwardly* toward relations with place (nature/landscape; domus) and history (a flow through time). Gasson and Errington's definition also relies heavily on social relationships, though *extending inwardly* toward relationships within the family. Most significantly, this directs us to greater attention to the role of gender within farm families, especially as this intersects with inheritance patterns.

It is not often that rural sociologists have the opportunity to conduct a survey of a nation's agriculture of this scope. Informed by these initial understandings of family farming, but not constrained by them, they have offered us a detailed and extensive examination of the social relations (family-based or otherwise) underpinning Polish agriculture with great attention to spatial and historical contexts. The authors of the above work have accomplished a significant contribution. They have contributed to building a case for reconstructing a sociology of agriculture, especially if this can be a beginning to: pursue some issues in more detail; pull some of the pieces together into a more holistic view; and facilitate agricultural sociologists in other nations to seek comparative analyses.

REFERENCES

Busch, L. (2011). *Standards: Recipes for reality.* MIT Press.

Buttel, F.H. & Newby, H. (1980). *The Rural Sociology of the Advanced Societies: Critical perspectives.* Allenheld, Osmun.

Buttel, F.H., Larson, O.F., & Gillespie, Jr., G.W. (1990). *The Sociology of Agriculture.* Greenwood Press.

FAO [Food and Agriculture Organization of the United Nations] (2020). *Introducing the UN Decade of Family Farming.* http://www.fao.org/family-farming-decade/home/en/ (access: June 27, 2020).

Fligstein, N. & McAdam, D. (2012). *A Theory of Fields.* Oxford University Press.

Friedland, W. (1984). Commodity systems analysis: An approach to the sociology of agriculture. In: H.K. Schwarzweller (ed.), *Research in Rural Sociology and Development: A research annual,* pp. 221–235. JAI Press.

Graeub, B.E., Chappell, M.J., Wittman, H., Ledermann, S., Kerr, R.B., & Gemmill-Herren, B. (2016). The state of family farms in the world. *World Development* 87, pp. 1–15.

Hendrickson, M.K. (2020). Covid lays bare the brittleness of a concentrated and consolidated food system. *Agriculture & Human Values* 37(3), pp. 579–580.

Lyson, T.A. (2004). *Civic Agriculture: Reconnecting farm, food, and community.* Tufts University Press.

Mead, G.H. (1938). Stages in the act: Preliminary statement: Essay 1. In: C.W. Morris, with J.M. Brewster, A.M. Dunham, & D. Miller, *The Philosophy of the Act,* pp. 3–25. University of Chicago Press.

Mooney, P.H. (1988). *My Own Boss? Class, rationality, and the family farm.* Westview Press.

Mooney, P.H. & Hunt, S.A. (2009). Food security: The elaboration of contested claims to a consensus frame. *Rural Sociology* 74(4), pp. 469–497.

Mooney, P.H. & Tanaka, K. (2015). The family farms in the United States: Social relations, scale and region. *Wieś i Rolnictwo* [Village and agriculture] 1(166), pp. 45–57.

Perrow, C. (2007). *The Next Catastrophe: Reducing our vulnerabilities to natural, industrial and terrorist disasters.* Princeton University Press.

Ploeg, J. van der (2021). *Family Farming: A foreword.* In: K. Gorlach, Z. Drąg (eds.) in collaboration with A. Jastrzębiec-Witowska & D, Ritter, *Think Locally, Act Globally: Polish farmers in the global era of sustainability and resilience,* pp. 9–21. Jagiellonian University Press.

Scott, J.C. (1998). *Seeing like a State: How certain schemes to improve the human condition have failed.* Yale University.

Silva, F.C. da (2007). *G.H. Mead: A critical introduction.* Polity Press.

Stevenson, G.W., Clancy, K., Kirschenmann, F., & Ruhf, K. (2014). Agriculture of the middle. In: P.B. Thompson & D.M. Kaplan (eds.), *Encyclopedia of Food and Agricultural Ethics.* Springer. http://agofthemiddle.org/wp-content/uploads/2015/05/AOTM-Encyclopedia-of-Food.pdf (access: June 2020).

USDA-ERS [United States Department of Agriculture, Economic Research Service] (2019). America's diverse family farms: 2019 edition. *Economic Information Bulletin* 214 (December). https://www.ers.usda.gov/webdocs/publications/95547/eib-214.pdf?v=2180.3 (access: June 2020).

Weber, M. (1996). *The Protestant Ethic and the Spirit of Cultural Capitalism.* Roxbury Publishing Company.

Biograms

About the editors

Krzysztof Gorlach (first editor and contributor)—received an MA in sociology in 1978 from Jagiellonian University in Kraków, Poland and was awarded his PhD in sociology, also from Jagiellonian, in 1986. In 1995, he successfully completed his postdoctoral examination, becoming a professor in the Department of Philosophy at Jagiellonian. His academic interests revolve around rural sociology generally and, specifically, the sociology of agriculture, social structure, and social movements. He currently works as full professor at the Institute of Sociology of the Jagiellonian University. He heads the research project "Think Locally, Act Globally: Polish famers in the global world of sustainability and development."

Zbigniew Drąg (second editor and contributor)—received an MA in economics in 1983 from the Academy of Economics in Cracow (now Cracow University of Economics) in 1983. Three years later he also received an MA in sociology from Jagiellonian University in Kraków. He also earned his PhD in sociology from Jagiellonian in 1998. His research interests include: macrosociology, theory of elites, issues of democracy and political propaganda, as well as the methodology of social research. Zbigniew is employed as an assistant professor for the research study "Think Locally, Act Globally: Polish famers in the global world of sustainability and development."

Anna Jastrzębiec-Witowska (first collaborating editor and contributor)—received her PhD in humanities/sociology from Jagiellonian University in Kraków, Poland in 2010. She received an MA in European studies in 1999 from the University of Exeter in England and an MA in sociology in 1998, also from Jagiellonian. In her academic work, she has specialized in rural sociology and sustainable food systems, with an emphasis on the civic aspects of alternative agriculture and the building of sustainable communities and food economies. Her professional experience includes working with international NGOs and local government in Dąbrowa Tarnowska in Małopolska Province, Poland.

David Ritter (second collaborating editor)—is a 1998 graduate of Antioch College in Ohio, where he received his BA degree in Philosophy, History and Religious Studies. He is also a certified teacher of English as a second language with 17 years of experience working with diverse groups of students. In the past, David was a policy analyst at the non-profit organization Public Citizen in Washington, DC, where he worked in the Critical Mass Energy & Environment Program, specializing in nuclear power and waste issues. He shares his time between Poland and United States, often engaging in English language academic projects in a variety of subjects to provide help with proofreading and editing.

About the external theorizing evaluators

Jan Douwe van der Ploeg (first external theorizing evaluator)—received an M.Sc. degree in Agrarian Sociology and Development Economics from Wageningen University in the Netherlands in 1976. He earned his PhD in Social Sciences from Leiden University, also in the Netherlands, in 1985. The scope of his interests encompasses rural development, sustainable agriculture, rural transition and change, and peasant agriculture. He currently works as adjunct professor of rural sociology at China Agricultural University/College of Humanities and Development Studies (Beijing) and as professor emeritus of rural sociology (in the Rural Sociology Group) at Wageningen University in the Netherlands.

Patrick H. Mooney (second external theorizing evaluator)—is a Professor of Sociology at the University of Kentucky (Lexington) in the United States, where he has worked since 1985. Though a native of rural Illinois, he received his MA in Sociology from the University of Northern Iowa in 1977 and his PhD from the University of Wisconsin-Madison in 1985. He has held Visiting Faculty positions at Jagiellonian University in Krakow, Poland; the University of Calabria in Cosenza, Italy; and Kyoto University in Kyoto, Japan. His research has included work in rural sociology, the sociology of agriculture, rural stratification, social movements, agricultural cooperatives, and food security issues. He has recently initiated research on the relationship between religion and environmentalism with particular interest in the role that religions may play in the possibilities for a more sustainable agriculture and planet.

About the contributors

Adam Dąbrowski—obtained an MA in sociology at Jagiellonian University in 2010. In the same year he also received a BA in European studies, at the same university. He is currently employed in the Institute of Sociology of the Jagiellonian University as a research assistant for the project "Care farms in the development of rural areas and in the face of demographic challenges," financed by the National Center of Research and Development, and conducted under the supervision of Professor Piotr Nowak in the Institute of Sociology of the Jagiellonian University. Adam is greatly interested in the sociology of rural areas, local development, new technologies, and civil society.

Grzegorz Foryś—is a sociologist and professor at Pedagogical University of Cracow (Institute of Political Sciences). In 1999, he was awarded an MA in Sociology, with first class honors, from Jagiellonian University in Kraków. He also earned his PhD in sociology from Jagiellonian in 2006, and received tenure at Pedagogical University in 2017. His academic interests center on social changes, social movements, and rural sociology, as well as, specifically, farmers' movements in Poland. He has recently been investigating the role of different social movements in promoting multifunctional rural development in Europe and, particularly in Poland, investigating them from the perspective of sociology and political sciences. He was recently elected dean of the Human and Social Sciences School at the Pedagogical University of Cracow.

Marta A. Klekotko—received an MA in sociology in 2004 from University of Silesia in Katowice, Poland. In 2009, she earned her PhD in sociology at Jagiellonian University in Kraków. Marta is currently employed as an assistant professor at the Institute of Sociology of the Jagiellonian University in Kraków. Her academic interests encompass community studies and urban studies, as well as development studies, with a special focus on the cultural aspects of development. She is a vice-president of Research Committee 03 (Community Studies) of the International Sociological Association.

Maria Kotkiewicz—received an MA degree in political science, with a specialization in local government administration and social work, from Pedagogical University of Cracow in 2011. In 2013, she completed postgraduate studies in mediation and negotiation at Bogdan Jański Academy in Cracow, and in 2015 obtained a BA degree in public sector management at Jagiellonian University in Kraków. She is currently a PhD candidate in political science at Pedagogical University of Cracow. Her academic interests include matters of nongovernmental organizations, social activism, civil society and participation therein. From 2018 to 2019, Maria was one of the fellows involved in the research project "Think Locally, Act Globally: Polish famers in the global world of sustainability and development."

Daria Łucka—studied philosophy and sociology at Jagiellonian University in Kraków. In 1998, she received an MA in sociology and an MA in philosophy, both from Jagiellonian. In 2003, she earned her PhD for the thesis entitled *Between Liberalism and Communitarianism. Civil Society in Poland, 1989–2000*. She is currently an assistant professor affiliated with the Institute of Sociology of the Jagiellonian University. Her research interests include: the history of ideas, theoretical sociology, political sociology, sociology of law, civil society, citizenship, nationalism, and transnationalism.

Adam Mielczarek—obtained both his MA and PhD in sociology, in 1988 and 2009, respectively, from Warsaw University, Poland. From 2013 to 2019 Adam was employed as a scientific assistant professor at the Institute of Sociology of the Jagiellonian University in Kraków, Poland, and was involved in several academic projects, including, most recently, "Think Locally, Act Globally: Polish famers in the global world of sustainability and development." While not affiliated with any academic center at the moment, he specializes in social movements and, in particular, the "Solidarność" (Solidarity) movement of the 1980s.

Piotr Nowak—received an MA in sociology in 1992 from Jagiellonian University, followed by a PhD in physical culture in 2001 from The University School of Physical Education in Cracow. In 2013, he obtained a social sciences post-doctoral degree in the discipline of sociology from the Philosophy Department at Jagiellonian University. His academic interests revolve around matters of rural development, with a particular focus on the changes and processes taking place in Polish agriculture after the country's integration with the European Union. In his research studies, he has conducted analyses on the condition of the cooperative movement in rural Poland, agricultural extension, rural tourism, farm strategies, food security, the functioning of local action groups, as well as social and care farming. Piotr is currently engaged in studies on aging in rural areas. He is employed in the Institute of Sociology of the Jagiellonian University as an associate professor and is also serving as deputy director for the research programs in the institute.

Martyna Wierzba-Kubat—received an MA in sociology from Jagiellonian University in Kraków, Poland. Her academic interests focus on rural development and, more specifically, on the role of information and computer technologies in agriculture. She is currently a PhD candidate at the Institute of Sociology of the Jagiellonian University in Kraków. From 2016 to 2018 Martyna was one of the fellows involved in the research project "Think Locally, Act Globally: Polish famers in the global world of sustainability and development."

Editor
Zofia Sajdek

Final proofreading
Katarzyna Borzęcka
Alicja Dziura
Irmina Gajewicz

Typesetter
Hanna Wiechecka

Jagiellonian University Press
Editorial Offices: Michałowskiego 9/2, 31-126 Kraków
Phone: +48 12 663 23 80, Fax: +48 12 663 23 83

CPSIA information can be obtained
at www.ICGtesting.com
Printed in the USA
JSHW051658100523
41478JS00008BA/19